Wood-Fired Ceramics
100 Contemporary Artists

Amedeo Salamoni

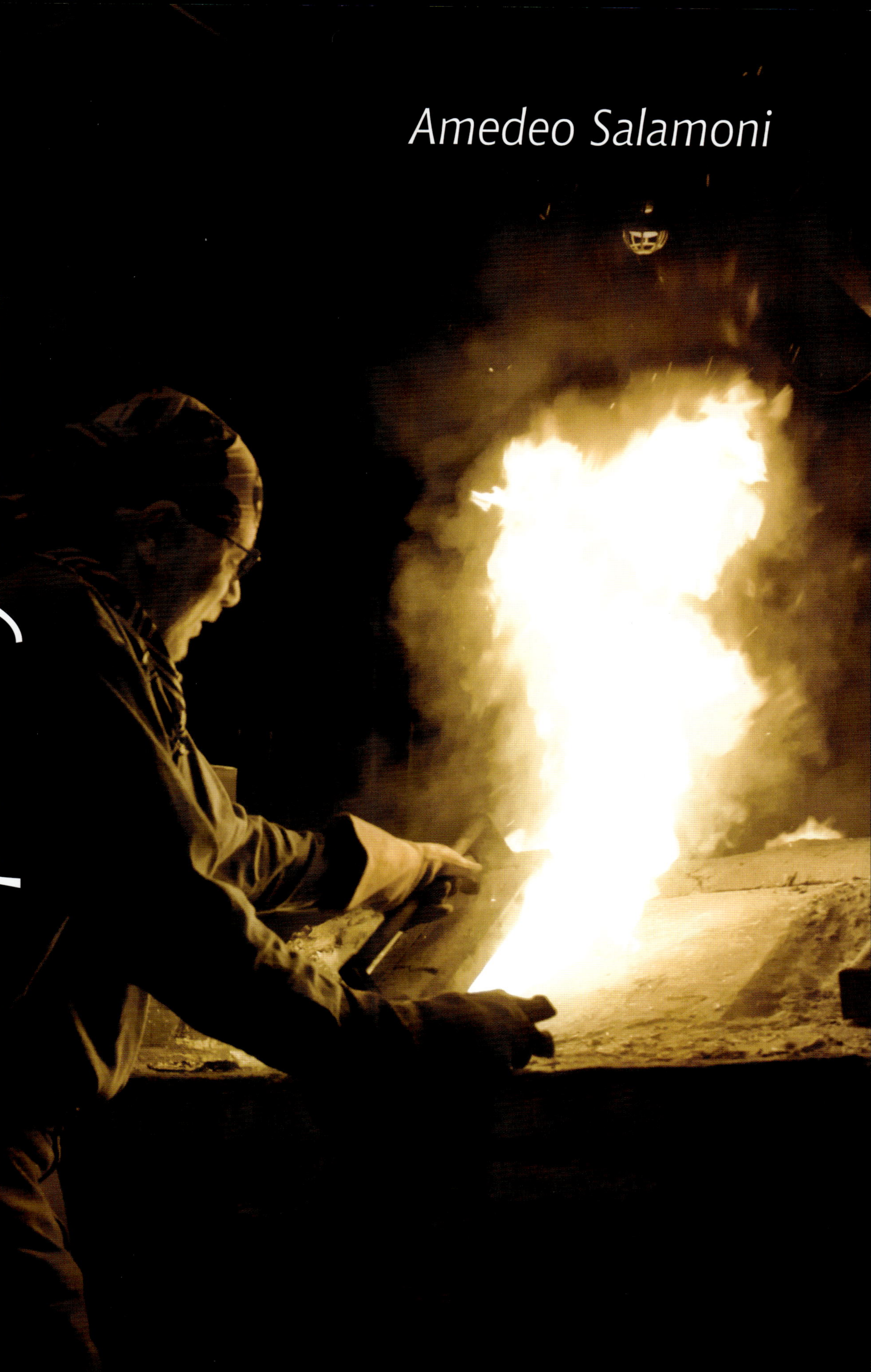

Schiffer Publishing Ltd®
4880 Lower Valley Road • Atglen, PA 19310

Other Schiffer Books on Related Subjects:
100 Artists of the West Coast. Douglas Bullis. ISBN: 978-0-7643-1931-0. $39.95
Ceramics Today. Jeffrey B. Snyder, editor. ISBN: 978-0-7643-3465-8. $50.00
The Golden Age of Karatsu Stoneware. François Villemin. ISBN: 978-0-7643-4405-3. $69.99
Mata Ortiz Pottery Today. Guy Berger. ISBN: 978-0-7643-3470-2. $49.99
Mexican Folk Art: From Oaxacan Artist Families. Arden Rothstein and Anya Rothstein. ISBN: 978-0-7643-2673-8. $39.95
Mosaic Art Today. Jeffrey B. Snyder. ISBN: 978-0-7643-4001-7. $50.00

Library of Congress Control Number: 2013945584

Designed by John P. Cheek
Cover design by Bruce Waters
Type set in GeosansLight/Cooper Lt BT

ISBN: 978-0-7643-4533-3
Printed in China

Published by Schiffer Publishing, Ltd.
4880 Lower Valley Road
Atglen, PA 19310
Phone: (610) 593-1777; Fax: (610) 593-2002
E-mail: Info@schifferbooks.com

Dedicated to the memory of Antoinette Salamoni, my mother, for encouraging me to stay true to what I believe in, and to David Jones, my teacher and friend, for introducing me to clay!

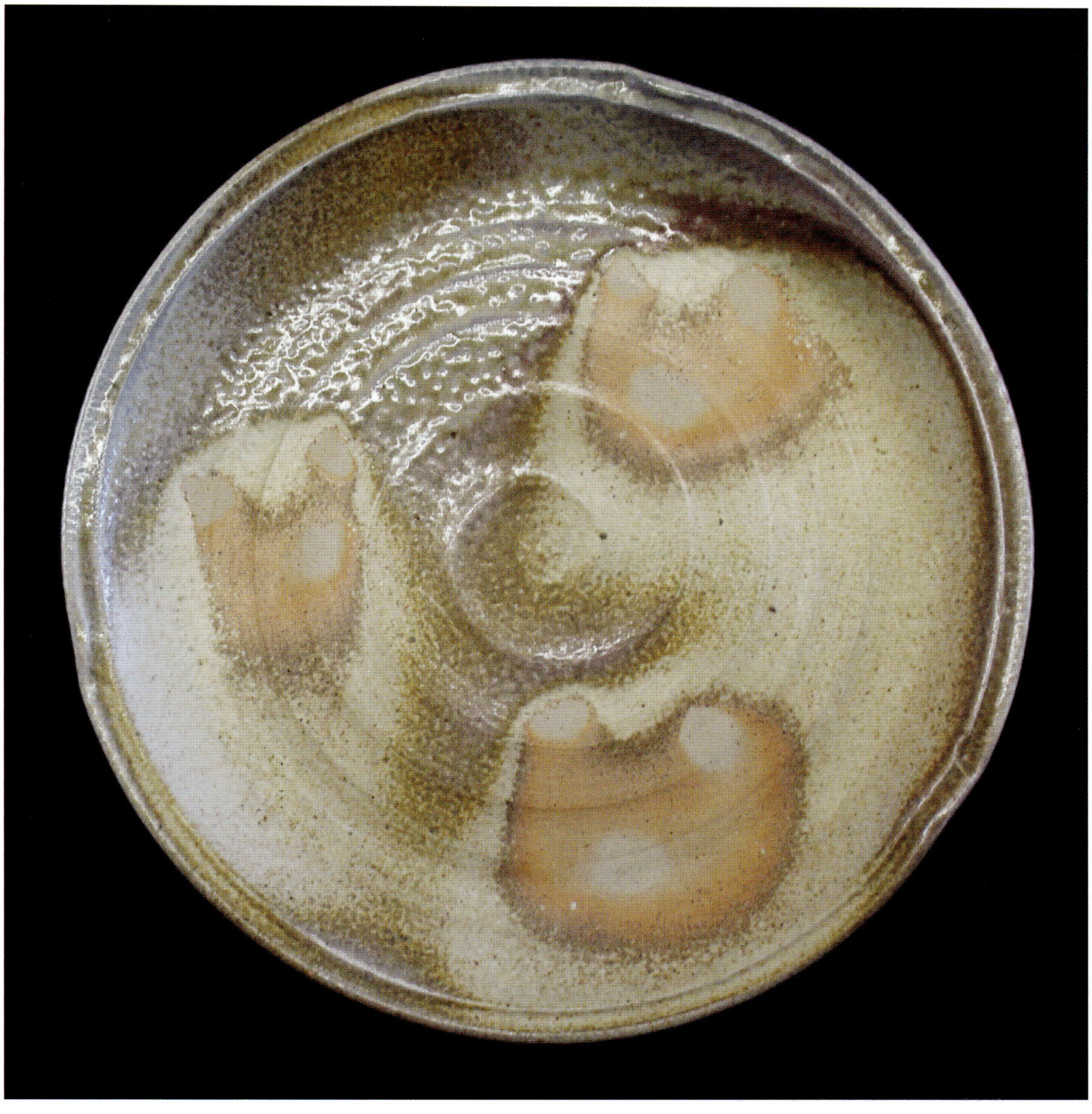

Amedeo Salamoni, *Plate with Comet Trail Pattern*, 2003. Wood-fired stoneware, wheel thrown, 14"d.
Photo credit: Amedeo Salamoni.

Epigraph

It is the person working with the clay that matters to me; the connection each of us makes to it. It's awesome, this clay; as if it were the "stuff" of imagination.[1]

—*Paulus Berensohn*

1. Paulus Berensohn. *Finding One's Way With Clay*. (New York, New York: Simon and Schuster, 1972.)

Contents

Foreword

Perhaps the best way to visualize the proliferation of wood-firing as we know it in the early twenty-first century is to imagine a single organism—a banyan tree—with its complex intertwining network of trunk, branches, and roots. Each time someone participates in a firing, bonds may develop: a node forms on the tree's branch. Wood-fired pieces are lived with, books change hands, other kilns and museums are visited, and acquaintances become friends: the node develops into a branch. Piece by piece and firing by firing, commitments develop, helping shape personal styles of work: the branch's identity evolves. Subsequent cycles of work and firings quicken one's attentiveness, relating tangibles with intangibles—clay, fire, cause and effect, aspirations: the branch, like a botanical stalactite, roots, enlivening the entire structure.

The true history of contemporary wood-firing is a series of uncollected stories relating shared experiences among thousands of participants through a borderless community shaped only by the Internet. If two people can be said to have been most influential in furthering interest in the genre, they would be Gerry Williams, the editor/founder of *The Studio Potter* magazine and Janet Mansfield. Williams' article, "The Phoenix Woodburning Kiln," appeared in *TSP* (Vol. 7, No. 2), in 1979, and was followed by an issue of the magazine the following year devoted to wood-firing. Williams also sponsored the first conference in the U.S. on wood-firing, held at the Japan Society in New York in 1983. In addition, he published an extensive list of all the known wood kilns in the U.S. in the early 1980s.

Janet Mansfield, in addition to sponsoring large-scale participatory wood-firing events at her sheep-station in Gulgong, Australia, she became, as founder/editor of *Ceramics Art and Perception* and *Ceramics Technical* magazines (ceramicart.com.au), the first person to feature articles and images of wood-fired ceramics in an international publication with wide circulation. In 1990, Coll Minogue and Robert Sanderson published *Wood-Fired Ceramics, Contemporary Practices*, in addition to issue #1 of *The Log Book* (thelogbook.net), creating an opportunity for wood-firers around the world to read and write about the field. Conferences in Europe, Australia, and the U.S. have brought together thousands of participants for camaraderie, panel discussions, exhibitions, and informal discussions, furthering our curiosity and willingness to discover solutions to common problems, as well as widening aesthetic sensibilities, if occasionally testing their resiliency. Wood-firers seemingly have no interest in hoarding or laying claim through intellectual property rights to what they have learned. "Generosity" might be the best word to characterize the overall approach to this way of firing.

When Rob Barnard, two students, and I built an anagama at Juniata College in Huntingdon, Pennsylvania, in 1978, it was due to pure curiosity—the fuel for learning's vehicle. Looking over the pieces from our first firing was confrontational, to say the least; the kiln, fired with green pine slabs, reached a soft c/8 with no appreciable soaking time. "All that work, for this?" we asked ourselves when we unloaded, but Rob had cautioned us about inflating our expectations: "Put some flowers in that one. Move it around; it belongs someplace. Give it a chance. Don't think of it being on the cover of *Ceramics Monthly*." Humbling words for humbling pots. Learning to see was going to be as important as the type of wood we used, or how we fired.

After returning from New Zealand in 1983, where I fired a large anagama, my students and I rebuilt the prototype, tripling its size. We were all involved in every aspect of the process, from clay-mixing, wood-stacking, kiln-loading, shelf-cleaning, and tidying up after each unloading. I had been unprepared for the *esprit de corps* radiating from this unique teaching tool, and our non-competitive relay over a period of four days. Time to stoke at 3 AM might mean a lull in a story about a student's grandfather, or the remembrance might continue while I handed her the wood. For several years we planted a pine tree after each firing, in a little clearing behind the kiln. Students brought their friends, who often turned up, stoked by the stoking, in the next semester's introductory class. A lad in southeastern Pennsylvania will someday learn his parents met at one of our firings.

We connected to our own terrain, to guys named Lester, Donnie, Dan, and Karl, who ran sawmills, drove log trucks, and wondered why a college they had driven past but never visited wanted their slab-wood. Curious enough to bring their families to firings, they took pots home with them, pleased to have been instrumental in the process. We shared firings with students and teachers from half a dozen other colleges and universities, several of whom built kilns of their own.

Subsequent firings always produced some work that needed talking about. We set the pieces out in the order in which they'd been stacked, passing them around, learning their language, trying to listen with our eyes and hands. Watching steam rise from tea in wood-fired cups made the ones in our parents' china cupboards look prissy. We confused one another in productive ways, sometimes talking up pieces whose goodness eluded their makers. The sense that, "We did this," rather than "I made this," contrasted with our response to work from other kilns. These pots and sculptures didn't give themselves up easily; they put our brakes on, the way some poetry baits us rather than going down in one gulp—pots that told us about what they'd come through; what we'd put them through. How could we not identify intimately with these insistent, understated offerings?

What we needed to know about the process seemed to lie just beyond our grasp, and with each unloading came an intriguing ratio of disgust, tentative accomplishment, and persistent inquisitiveness. We were fellow learners in this theater where cooperation, naïve curiosity, and invitational uncertainty shared the spotlight. We couldn't wait to do it again. Unfailing politeness was the inevitable response when the work was shown.

Elsewhere, on every continent, wherever trees grow, similar histories unfold, as wood-firing gains momentum, becoming a means of aesthetic expression in addition to functioning as a teaching tool. The metaphorical banyan tree continues to enlarge, strengthen, and shape the commitment of its practitioners and supporters, who form the purposeful root-stock of any enterprise done caringly, and with zeal.

—Jack Troy
Huntingdon, PA
January 3, 2013

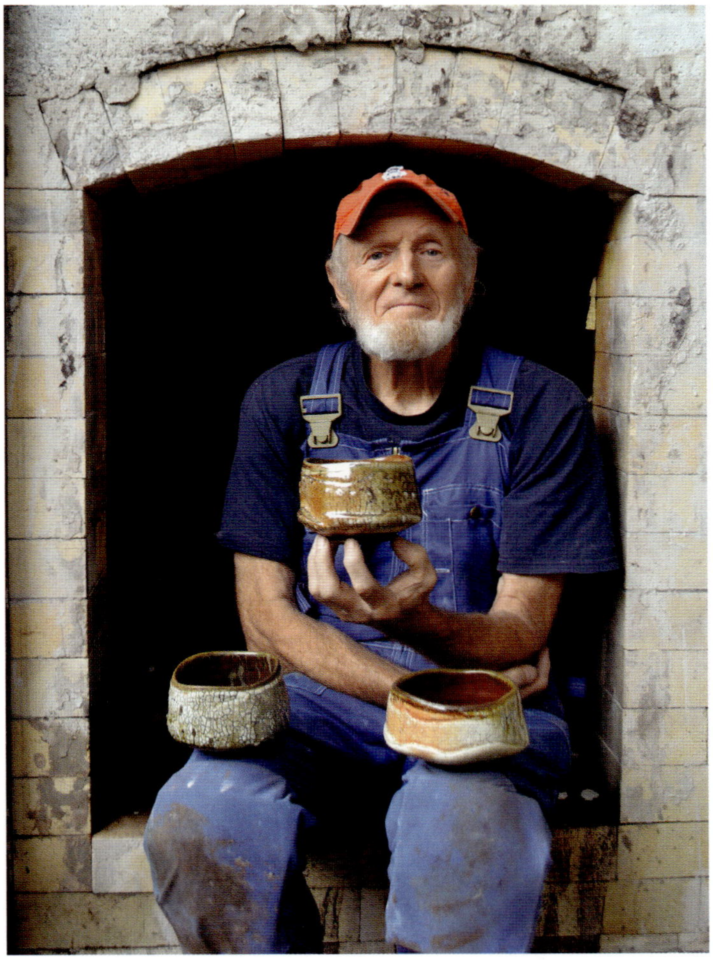

Jack Troy in front of Train Kiln, Watershed Center for the Ceramic Arts, Newcastle, Maine. 2012. *Photo credit: Sharon Williams.*

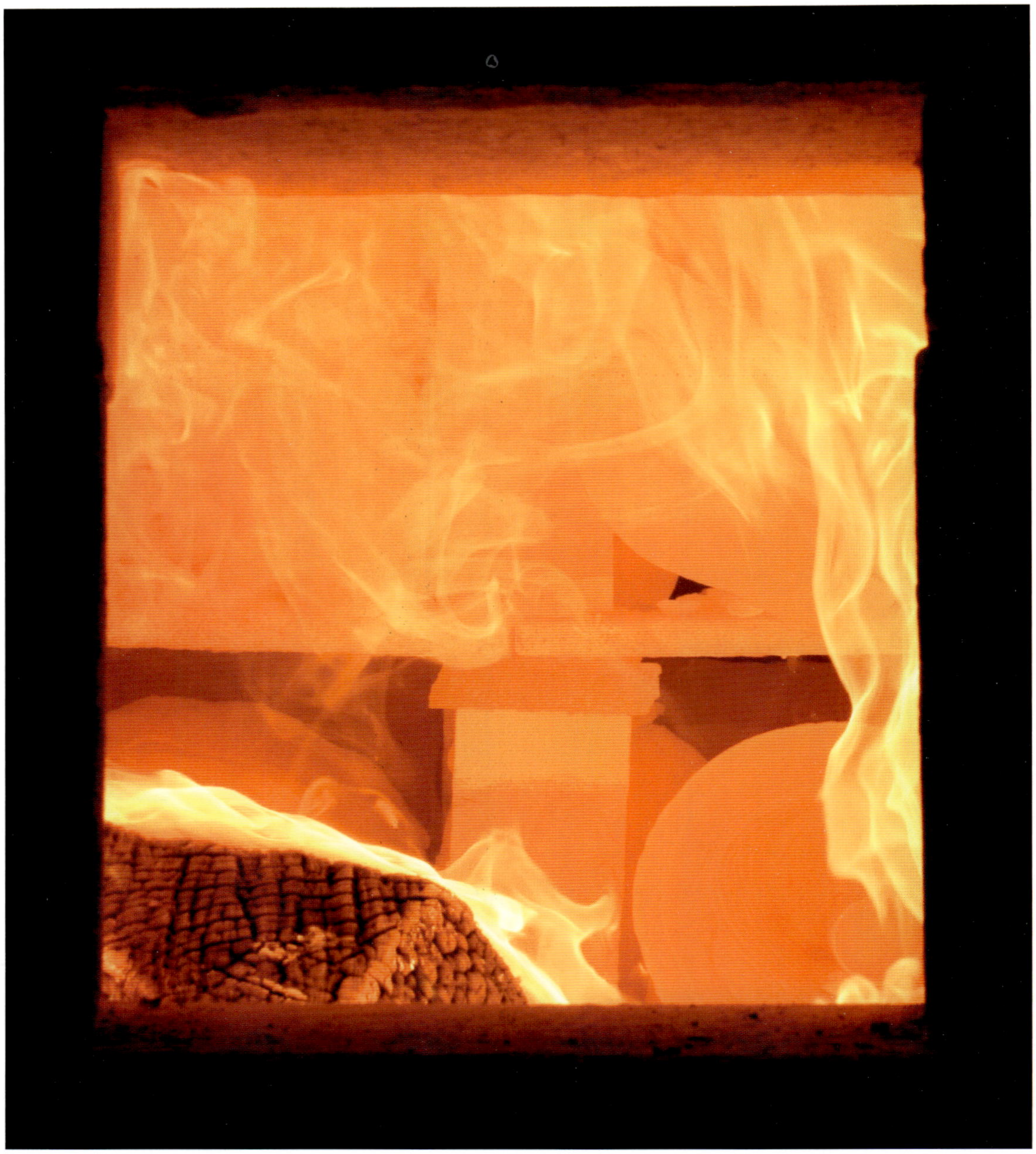

Amedeo Salamoni, *view into anagama firebox*. 2001. *Photo credit: Amedeo Salamoni.*

Acknowledgments

I am indebted to all the artists who contributed both photographs and information for this project, as well as the galleries, craft centers, colleges, universities, and individuals for forwarding my call for artists.

I would be remiss without thanking the countless participants and students who have been involved in my firings—I have learned from each of them. I am especially grateful for the George School community and their support of my work and of the arts—I consider myself lucky to live and teach where I do. I am grateful to my many friends and colleagues for looking over the manuscript with me along the way.

I would like to thank the following individuals who have helped and inspired me in many different ways: Richard Buncamper and David Jones, my first teachers in clay; Jack Troy, for adding his insight to this book; Kevin Lehman, my firing partner, for helping me stay focused during a difficult transition in my life; John Ground for his support in re-building our kiln at his location.

Most importantly, this book could not have been written without the encouragement, constant support, and assistance of my wife, Diane, who knows all too well the life of a potter's spouse—the countless days and nights of firing, cutting wood, making pots, and putting up with me!

Introduction

This project has been one of the most arduous and satisfying endeavors I have ever undertaken. It is not my intention to provide a comprehensive overview of the wood-firing process, but instead my goal in organizing this book is to present the work of many ceramic artists who choose to wood-fire their art. Rather than focusing on methodology, I have decided to let the work speak for itself. The artists represented in this book share a personal vision and design choice that has been affected by their various environments and influences in ceramics.

There are many strong traditional influences in wood-fired ceramics. These artists have adapted, transformed, and sometimes strayed from these traditional methods of firing and kiln designs, finding ways to make something old new again.

There is a mixture of work represented in this book from educators, potters, and sculptors, both well known in their fields as well as fledgling artists. Many have built and fire their own kilns. Others fire their work in kilns at art centers, colleges, universities, and in those of other artists. It amazes me how many people are firing their artwork in wood-fired kilns compared to when I began wood-firing.

Over the years I have formed many bonds with other potters and ceramic artists who share my passion for wood-firing. In meeting some of the artists whose work is represented here, I found them to share this tremendous sense of commitment towards their art. There was a great sense of community and willingness to share their wood-firing journeys and how they make their art.

This sense of community and shared vision is where the similarities end; this is a diverse group of artists who apply many different approaches and techniques in their work. Take, for instance, Willi Singelton, who makes functional pottery and fires in a multi-chambered Noborigama kiln. At first glance, many of his pots don't appear to be wood-fired at all, but with closer inspection one can see how the fire has left its mark, gently altering the surface, leaving the subtle beauty of the flame.

In contrast, artists Kristin Müller and Dan Murphy make wares based on function, but fire them in very different ways. Their pots are a result of an ongoing development; their forms reflect a gestural approach to clay. The finished works are fired without glaze. The colors, textures, and complexities are a result of the interaction of wood, fire, and clay.

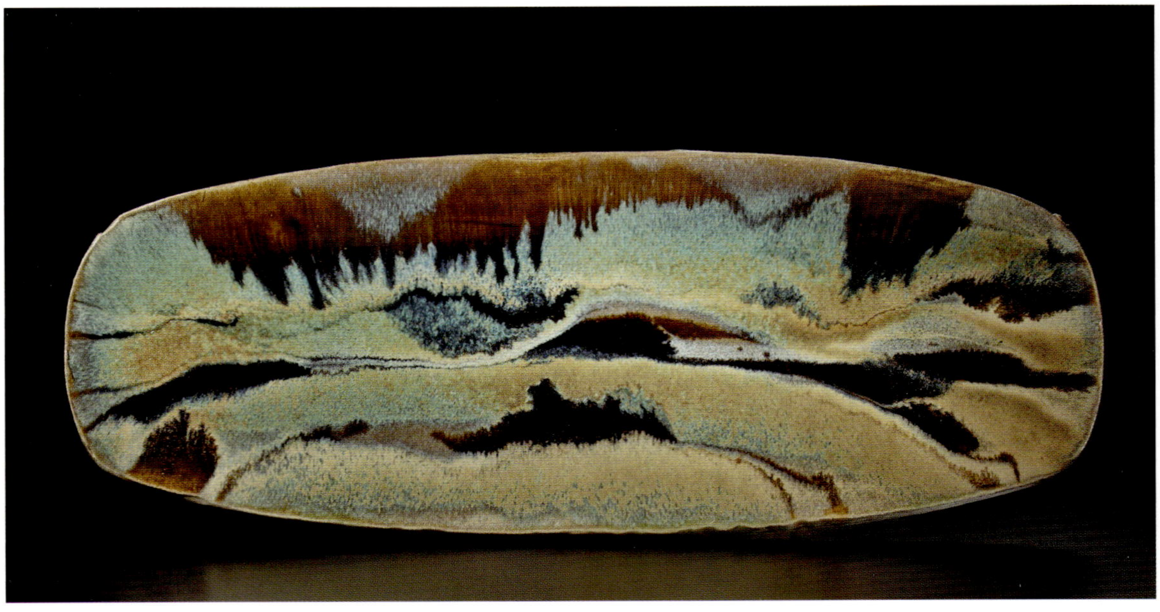

Willi Singleton, *Landscape Slab Form*, 2010. Wood-fired clay, slab construction, corn stalk ash glazes with and without copper, creek clay glaze over, 32" x 12" x 1.5". *Photo credit: Kenek Photography.*

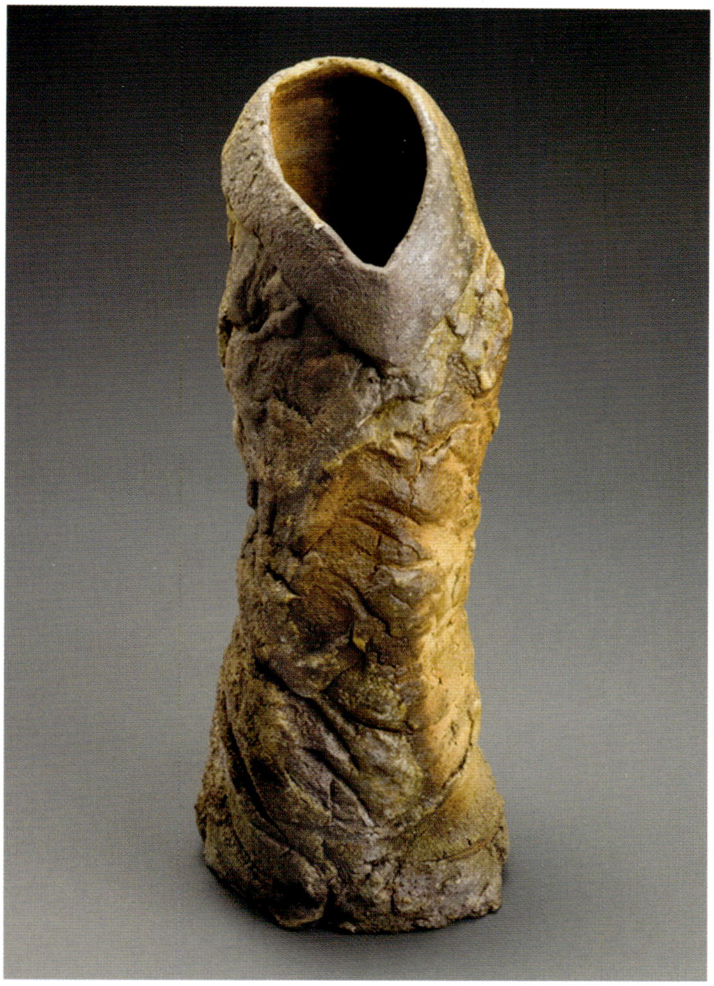

Kristin Müller, *Balzak*, handbuilt stoneware with natural ash, 15.5" x 5.5" x 3". *Photo credit: Douglas Foulke.*

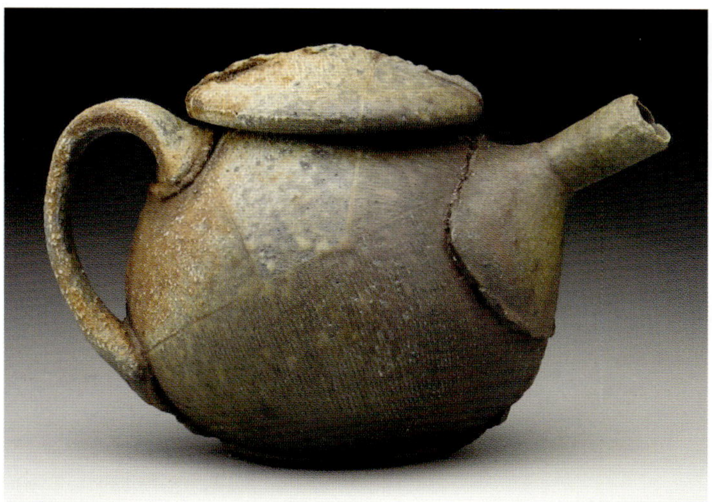

J. Daniel Murphy, *Tea Pot with Back Handle*, 2012. Wood-fired iron rich clay, natural ash glaze, handbuilt and wheel-thrown parts, 9" x 6" x 6". *Photo Credit: J. Daniel Murphy.*

Other artists, such as Joy Brown and Peter Callas, choose to make more sculptural works, and their approach to firing is also very different. Joy's beautifully rounded figures are made from an iron rich clay, which are fired in the back section of her anagama kiln to relatively low temperatures for a week, resulting in a flashing of soft earth tones.

Peter Callas brings over four decades of wood-firing experience to the table. Peter fires his abstract sculptures and wares for eight days, consuming a cord of wood per day to get a totally different effect with his art. The forms emerge from his anagama kiln full of rich surface textures with a depth and complexity from the ash deposits that can't be appreciated without closer inspection. His work illustrates the fuse and hibuse results of heavy ash encrustation and hikidashi, the outcome rapid cooling produces. Peter has turned these techniques into an art form.

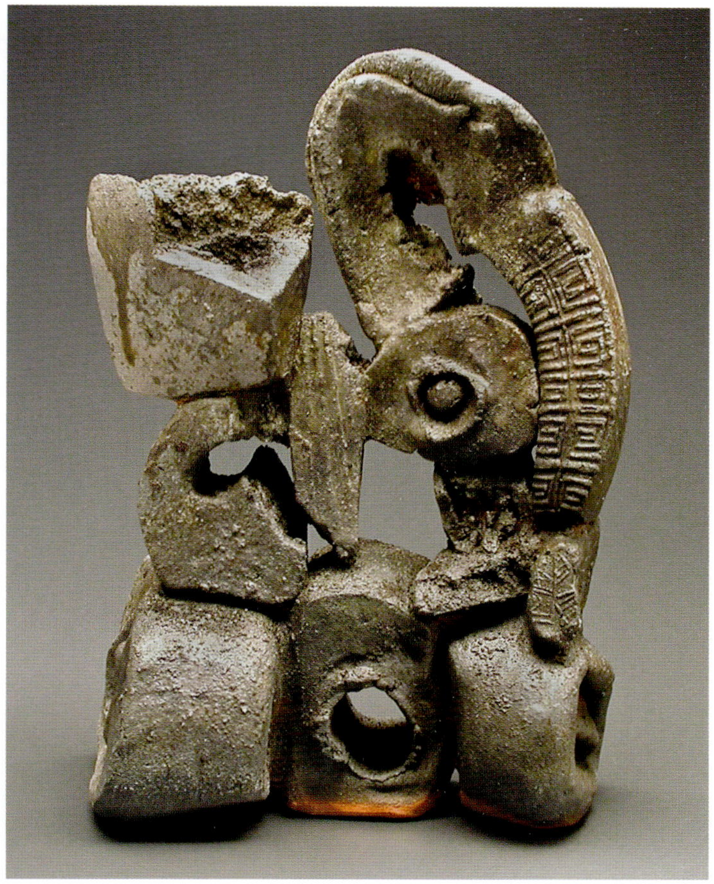

Peter Callas, *Maquette, 2007*. Handbuilt stoneware, 12" x 8" x 19"(American Museum of Ceramic Art). *Photo credit: Peter Callas.*

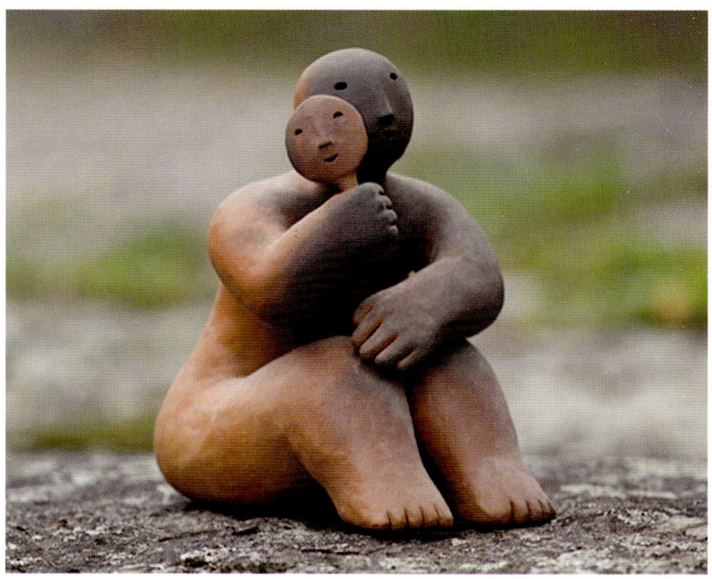

Joy Brown, *Sitter with Mask*, 2011. Wood-fired stoneware, handbuilt, 15" x 9" x 10". *Photo credit: Richard Wanderman.*

I will never forget my first wood-firing experience. It was in August of 1991 at Chester Hewell's Groundhog kiln in Gillsville, Georgia. The kiln was a simple groundhog kiln. There were no gas burners, no electricity, and no cones—only fire, experience, and intuition.

I recall many of the conversations with some of the other potters at the time. I had many questions. Why did they fire with wood? Surely it would have been more efficient to fire with gas or electric. These types of kilns would have melt the tobacco-spit glaze that they used much quicker and more efficiently than using wood. They would not have to cut, chop, and stack countless cords of wood in preparation for each firing. I did not understand at the time what they were doing or why!

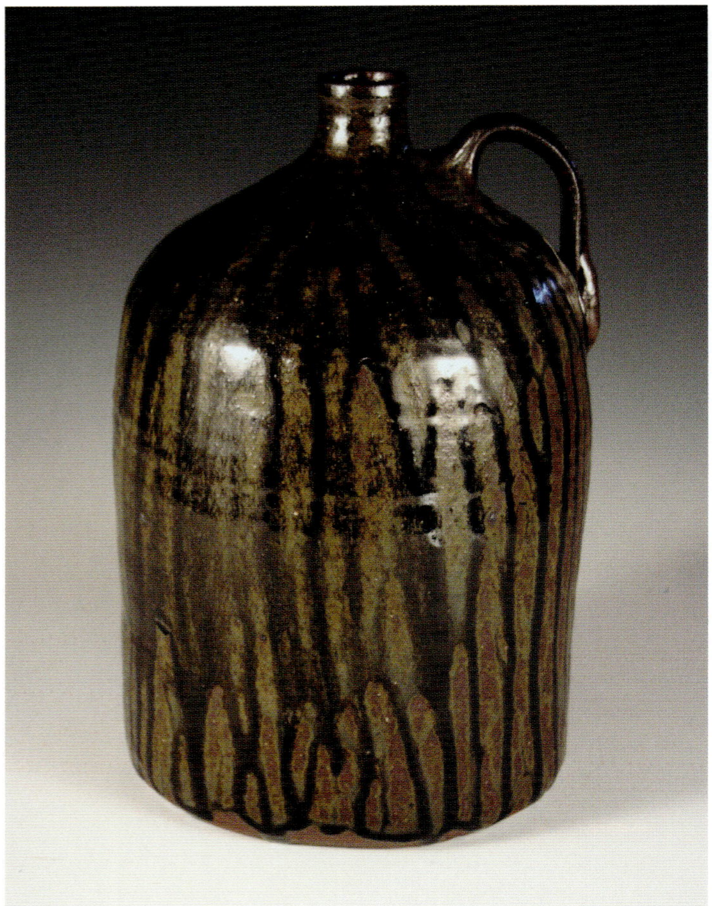

Amedeo Salamoni, *One-Gallon Jug,* 1994. Wood-fired in Groundhog Kiln, Tobacco-Spit Glaze, 12" h. *Photo credit: Amedeo Salamoni.*

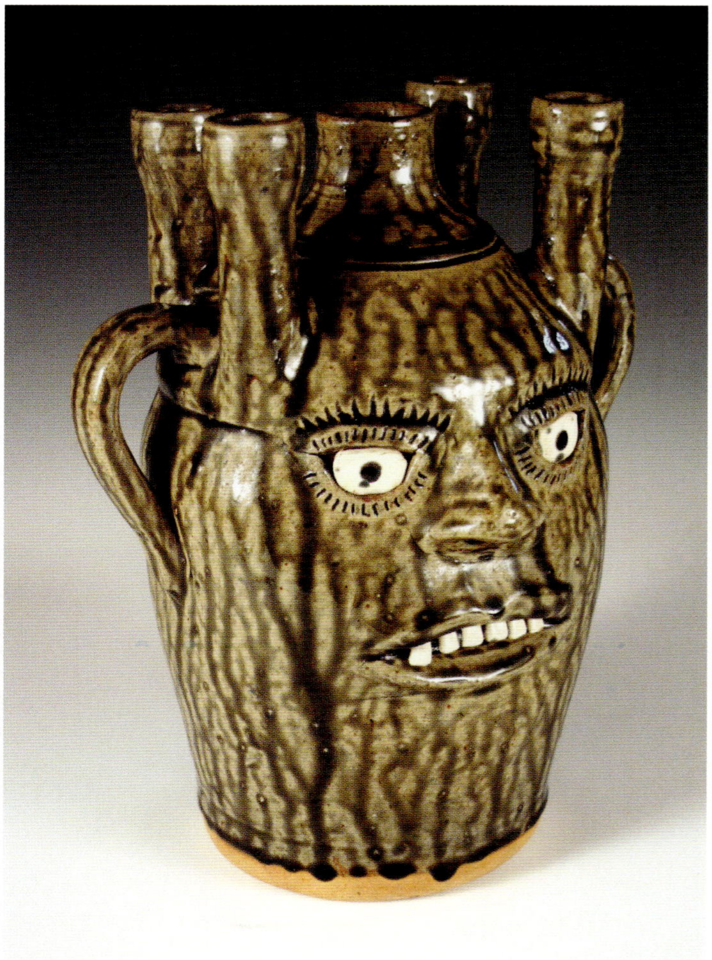

A.G. Meaders, *Candelabra Face Jug,* 1993. Wood-fired in a Groundhog Kiln, Tobacco-Spit Glaze, 11" h. (Collection of Author). *Photo credit: Amedeo Salamoni.*

I became so fascinated with this process, that every time I could be involved in a wood-firing I would volunteer my help. Over time I learned to tell the temperature of the kiln by the color inside the chamber, watching the kiln temperature rise and fall, using the smoke and flame patterns coming from the chimney as a guide to stoking the kiln with wood, learning to watch the pots inside the kiln for ash buildup, and listening to the crackle and the roar of the fire. I was totally mesmerized by the flames, the smells, and the intense heat coming from the firebox. Later, after examining the pots from the firing, seeing the path of the flame on the pots, the ash deposits, and the depth of surface variations on the pots, I was amazed. This is what made me fall in love with wood-firing.

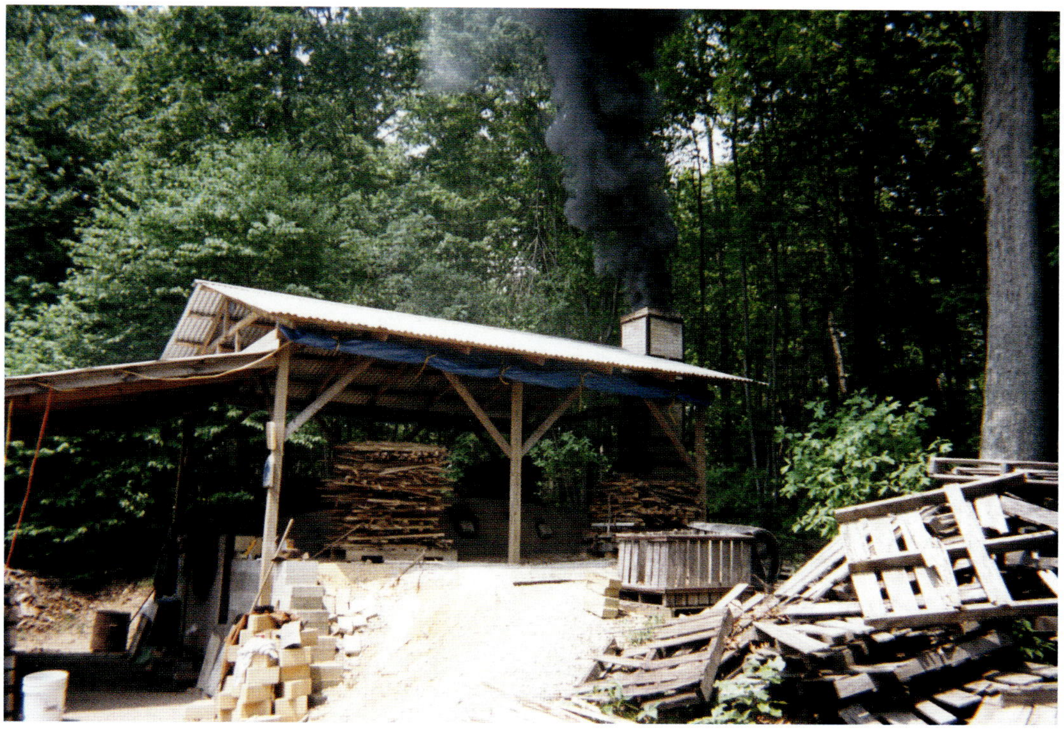

Amedeo Salamoni, anagama kiln firing, 2000. *Photo credit: Amedeo Salamoni.*

Working with my hands and senses is what drew me to clay. I admired the skill and knowledge it took to be a potter. After my exposure to wood-firing, it made perfect sense to me to further explore the process on my own. I went on to build a tube kiln based on an anagama style kiln. This would be the first of many kilns that I would build and fire.

This project has been in my thoughts for the past ten years. With this book, I hope to provide a way for potters, students, collectors, and patrons of the arts to discover where to find these artists and view their work. Ultimately, I want others to see and share my love for this type of art and to expand the understanding of contemporary wood-fired ceramics.

—Amedeo Salamoni

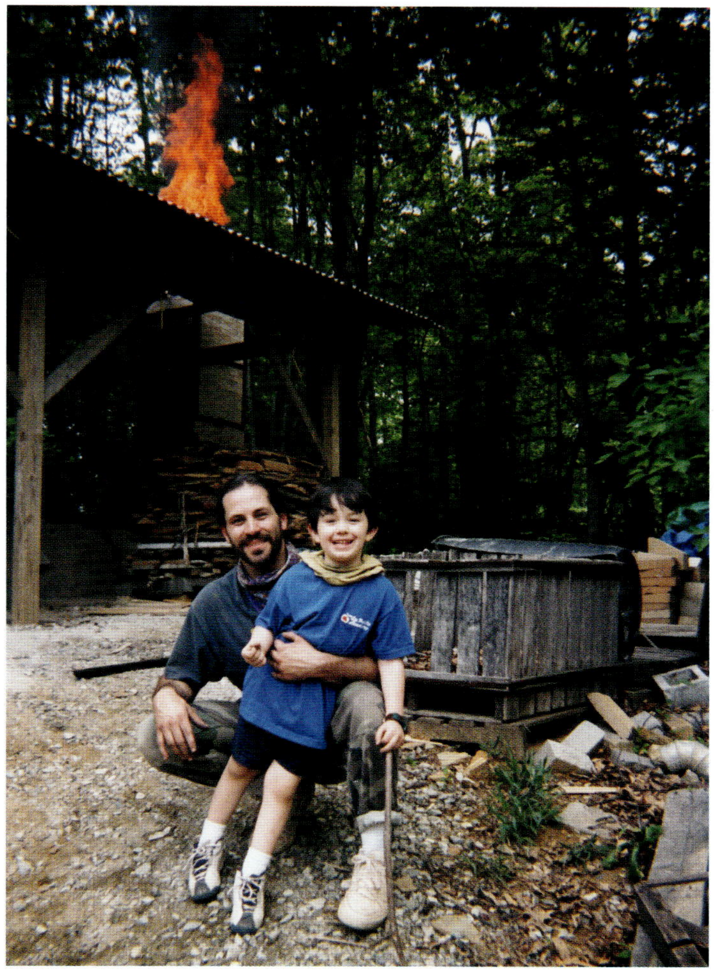

Amedeo Salamoni, author, with his son Ian firing the anagama kiln, 2000. *Photo credit: Diane Salamoni.*

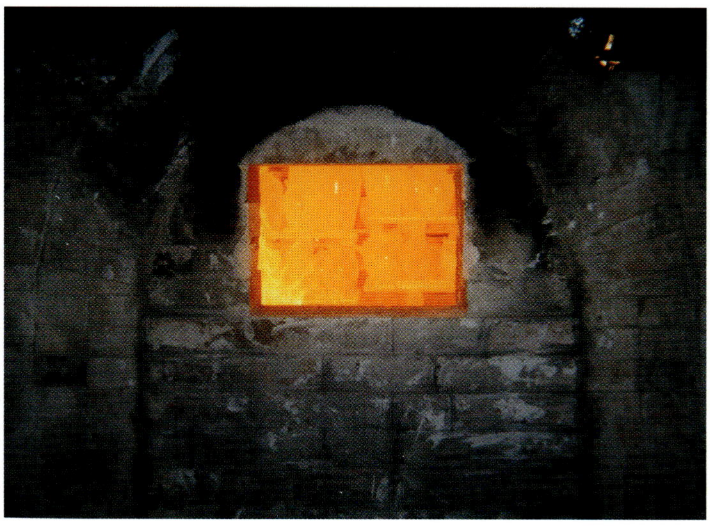

Amedeo Salamoni, view into anagama firebox, *2004. Photo credit: Amedeo Salamoni.*

HADI ABBAS

Orlando, Florida
habbas@mail.ucf.edu, www.autash.com

Artist's Statement

My roots are enriched with culture. I was born in Kerbala, Iraq, to a Persian mother and an Indian father. In my early childhood, I lived in Gorgan, Iran, and in my later teen years moved to Blackburn, England, for a year to study the English language. In addition, I am fluent when speaking in Arabic, Persian, and English. My artistic inspiration stems from a mechanical background and culturally diverse family unit. Many of my forms and designs are influenced by Persian, Indian, and Islamic architecture, such as the domes on the Muslim's shrines and temples. Likewise, the ancient copper containers of Mesopotamia (present day Iraq), stone carved utensils of Persia, and American cast iron ware catapulted my efforts to design and build a device that allows me to intricately cut lids for my jars with precision. Lastly, the study of Ikebana (1999–2003), Japanese floral art, with master artist Mariko Haturi, has led to the creative design on my vases and containers—for my own masterpieces.

My passion for working with clay, and imparting techniques and knowledge to others, is founded on the theory of historical influence. Clay is more than throwing, glazing, and firing. It possesses a mysterious energy that puts us in touch with our past. Clay reflects the maker's attitudes and spirit. Most of all, clay is a natural material that exists nearby and around us. It is an overlooked resource that has great familiarity and yet people seldom take the opportunity to think much about it.

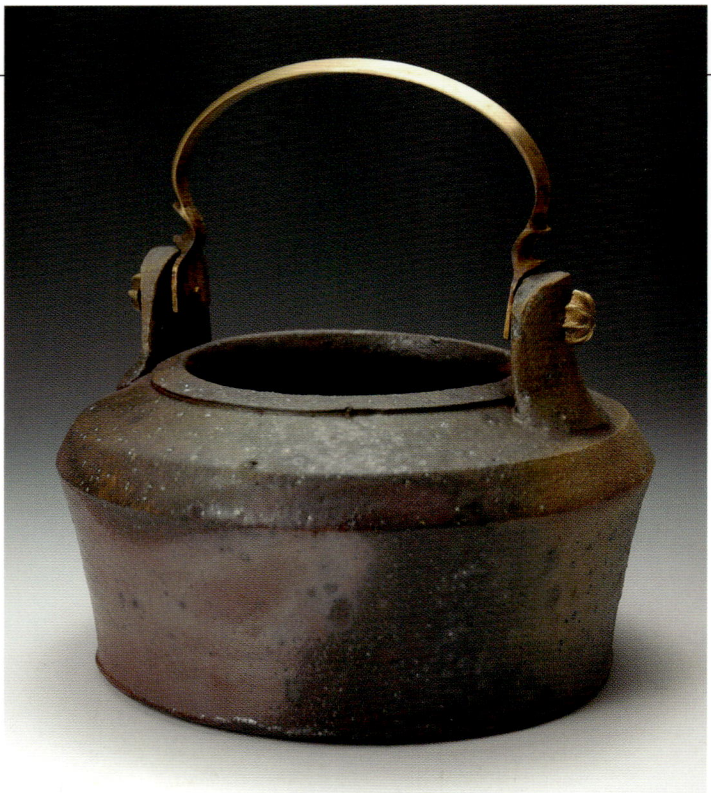

Hadi Abbas, *Basket with Brass Handle*, 2009. Wood-fired red stoneware, wheel-thrown, 9" x 7" x 7". *Photo credit: Joshua Freeman.*

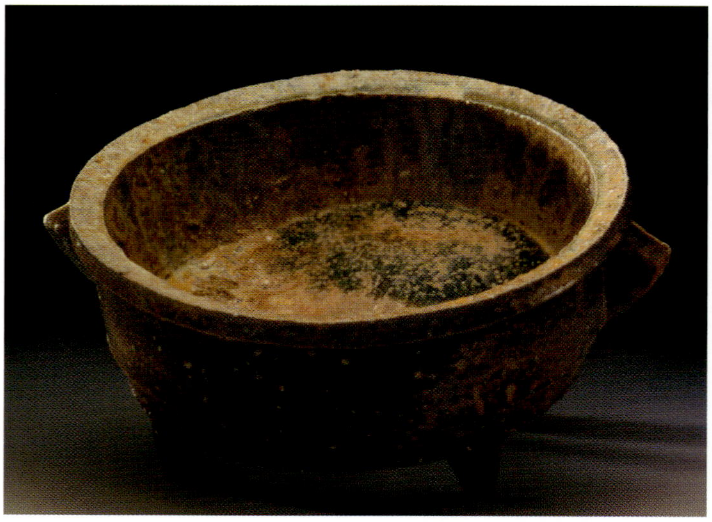

Hadi Abbas, *Bahir*, 2000. Wood-fired white stoneware, wheel-thrown, 6" x 14" x 16". *Photo credit: Joshua Freeman.*

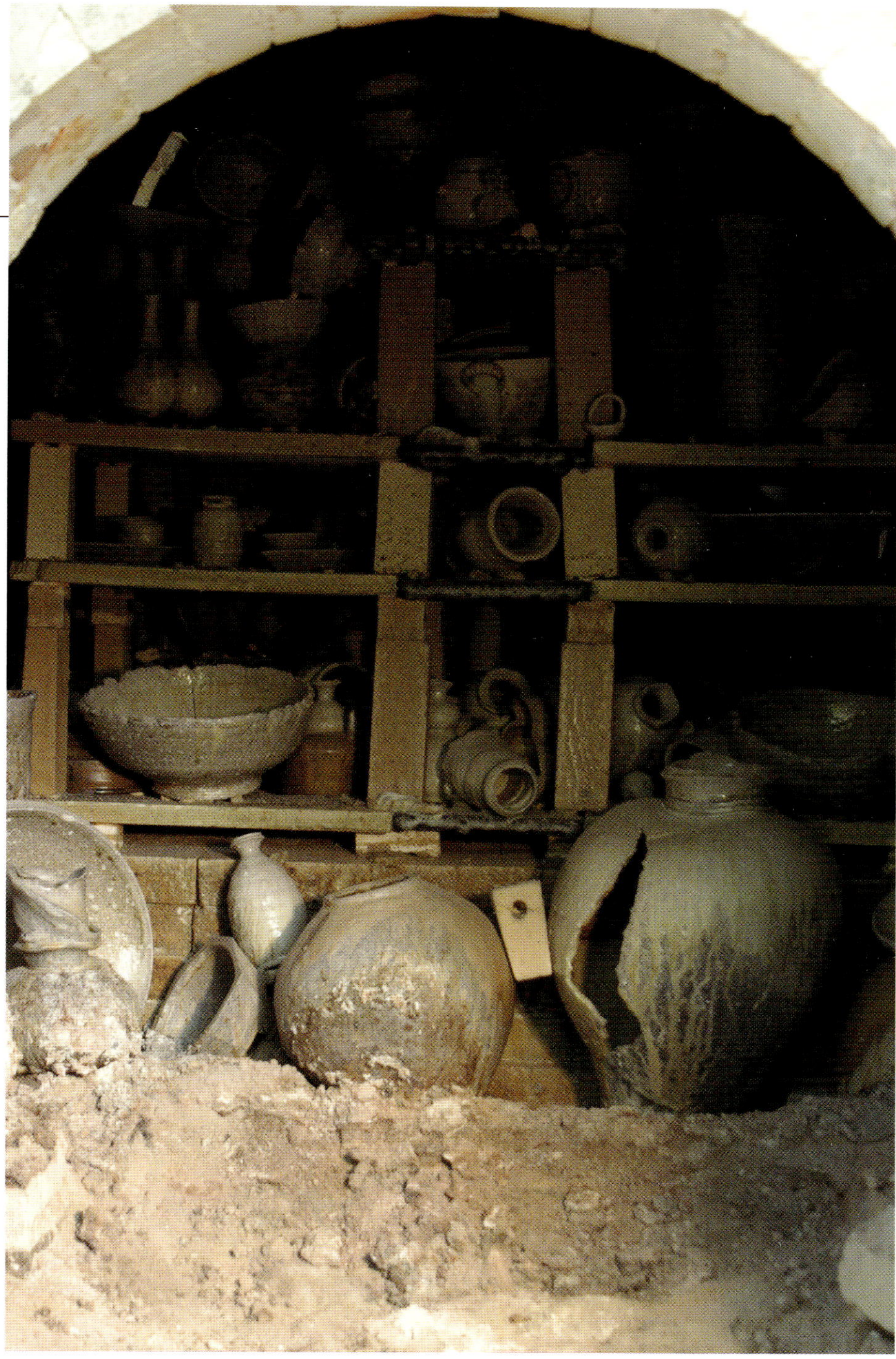

The front main firebox and the first shelf stack in the *Autash anagama kiln*. *Photo credit: Joshua Freeman*.

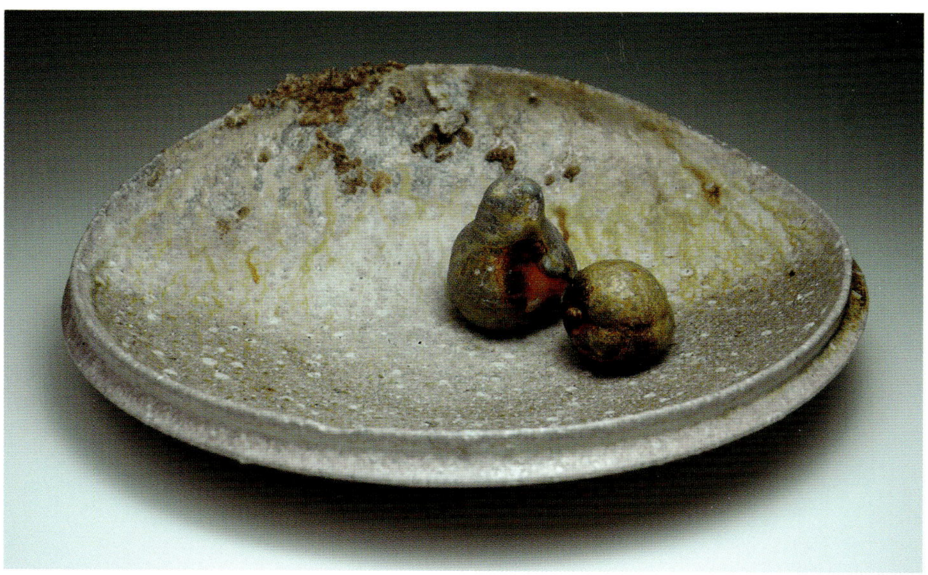

Hadi Abbas, *Platter with Fruit*, 2010. Wood-fired white stoneware, wheel-thrown and handbuilt, 3.5" x 17" x 17". *Photo credit: Joshua Freeman*.

CLAYTON AMEMIYA

Hilo, Hawaii
claytonamemiya@yahoo.com

Artist's Statement

My aesthetic preference leans heavily towards the organic, as do the techniques I use to create my art. Most of my work is utilitarian, made on the potter's wheel to be used in daily life. Even the larger, handbuilt objects are created with a functional intent, specifically to serve as vases and food vessels. I utilize a wood fueled anagama type kiln to fire my pots because wood ash circulating in the kiln at high temperatures causes glazes to form in irregular patterns. Some parts of the clay surfaces remain unglazed, showing how the flames "paint" the pieces on their journey from firebox to chimney. Lines and other carved markings on my larger vessels also contribute to the random patterns, and the extended firing time creates textures reminiscent of stone. In my pots, I see and feel the upland, rainforest environment where I live and work in Hilo, Hawaii.

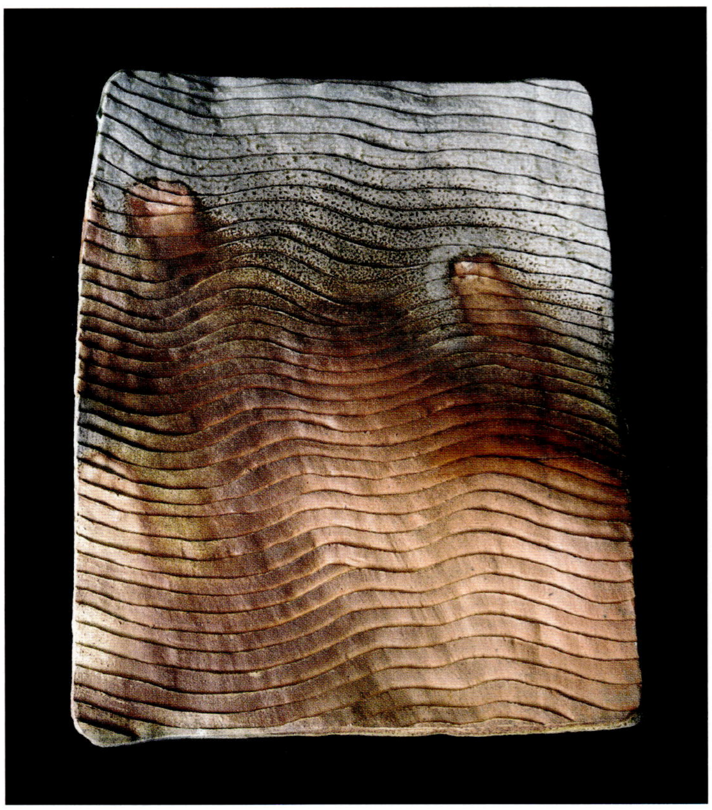

Clayton Amemiya, *Platter with carved lines*, 2004. Wood-fired stoneware, handbuilt, 30" x 23". *Photo credit: Franco Salmoiraghi.* This large platter form was carved with layered, parallel lines throughout the entire surface. Placed on its side, three abalone shells prevented it from sticking to another platter stacked next to it during firing.

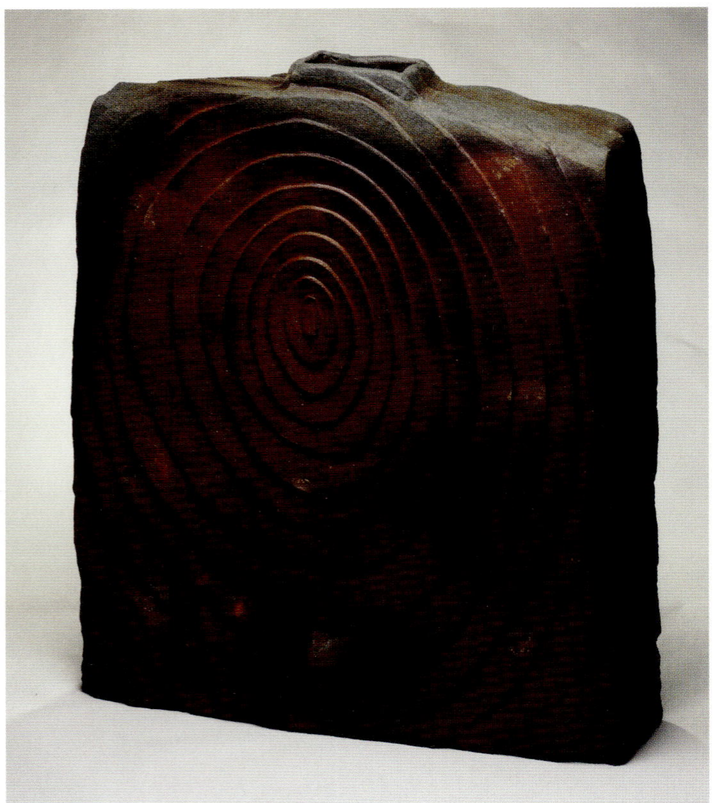

Clayton Amemiya, *Vessel with circular design*, 2006. Wood-fired stoneware, handbuilt, 20" x 18" x 6". *Photo credit: Cory Lum.* This rectangular vessel was handbuilt and carved with a design of circular lines. It was fired face down with the mouth facing the firebox, leaving a dusting of ash on the shoulders. White residue of the limpet shells, from which the piece was separated from the kiln shelf, can be seen.

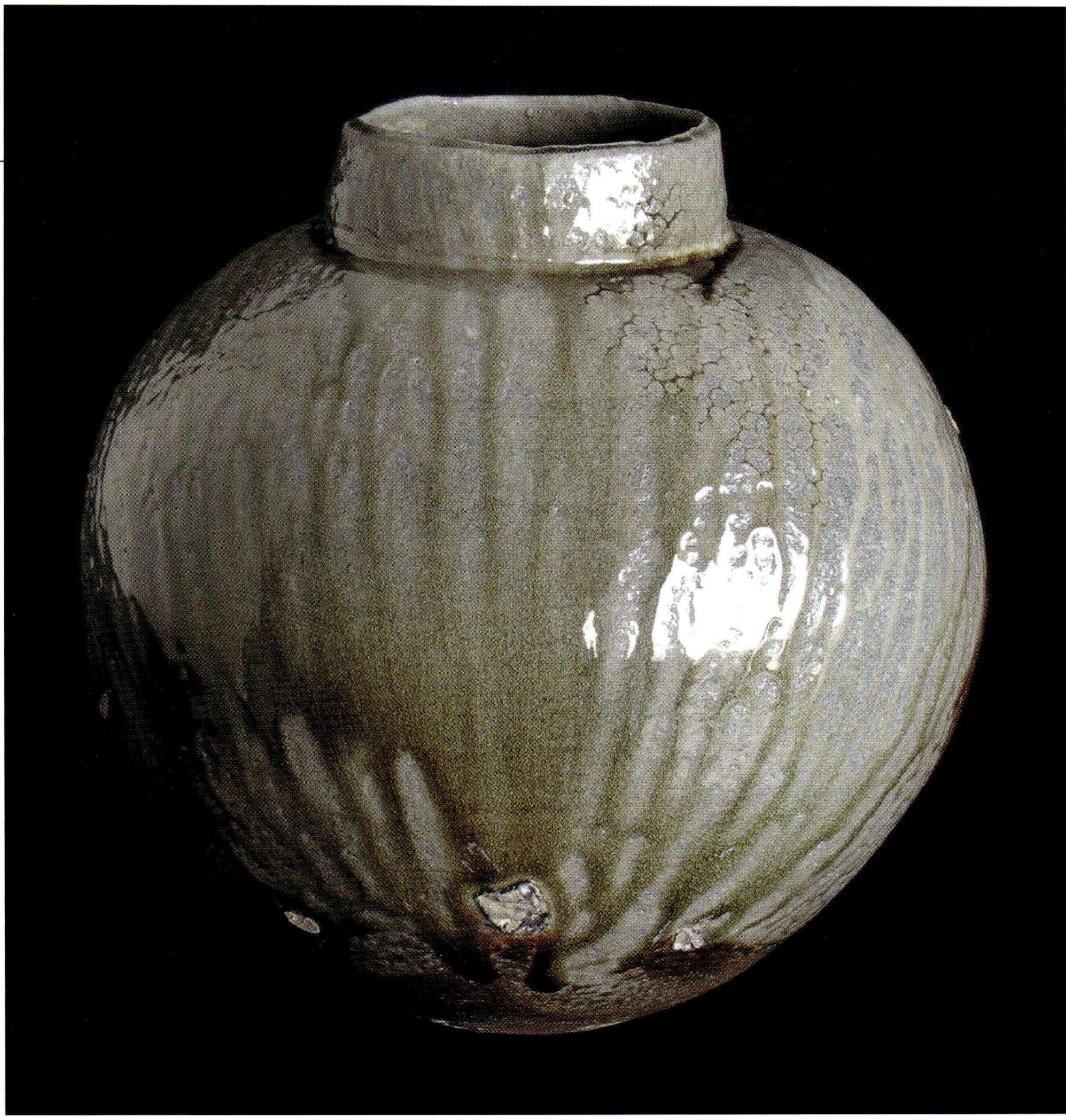

Clayton Amemiya, *Jar with shell scars*, 2005. Wood-fired stoneware, wheel-thrown, 15" x 14" x 14".
Photo *credit: Franco Salmoiraghi.*
This jar was tilted during placement to accommodate it in the limited kiln space. It rested on three limpet shells, which prevented it from sticking to the shelf. Heavy fly ash dripped downwards and accumulated as droplets near the shell scars.

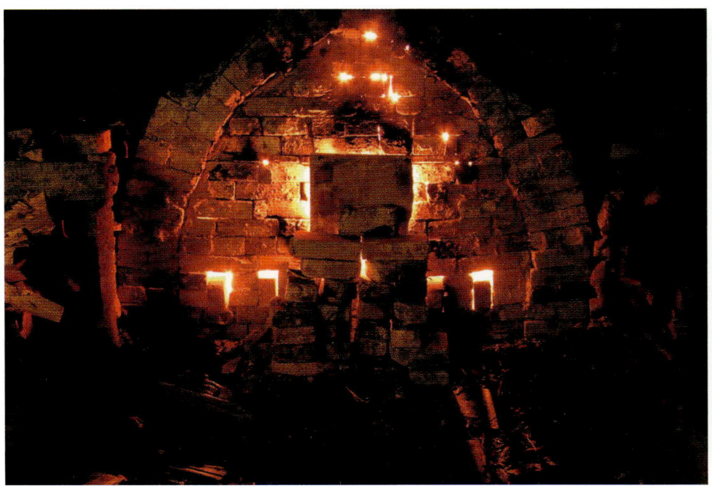

Kiln, night view, immediately after stoking.
Photo credit: Franco Salmoiraghi.

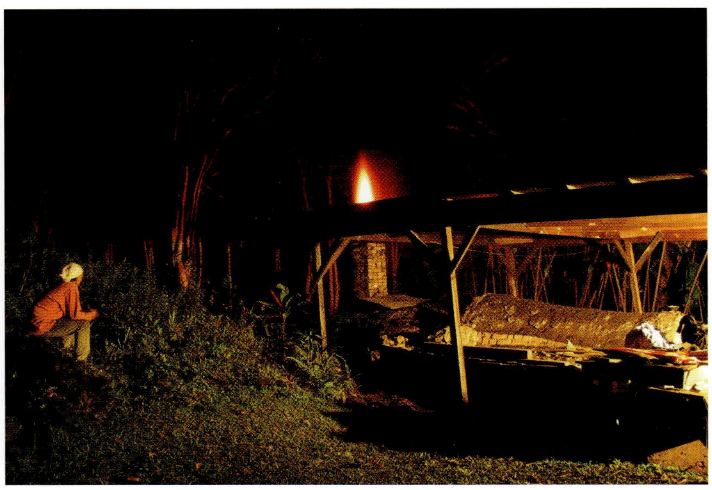

Clayton Amemiya, Kiln, Night view, fourth day.
Photo credit: Franco Salmoiraghi

STEFAN ANDERSSON

Tibro, Sweden
stefan@ceramics.se, www.ceramics.se

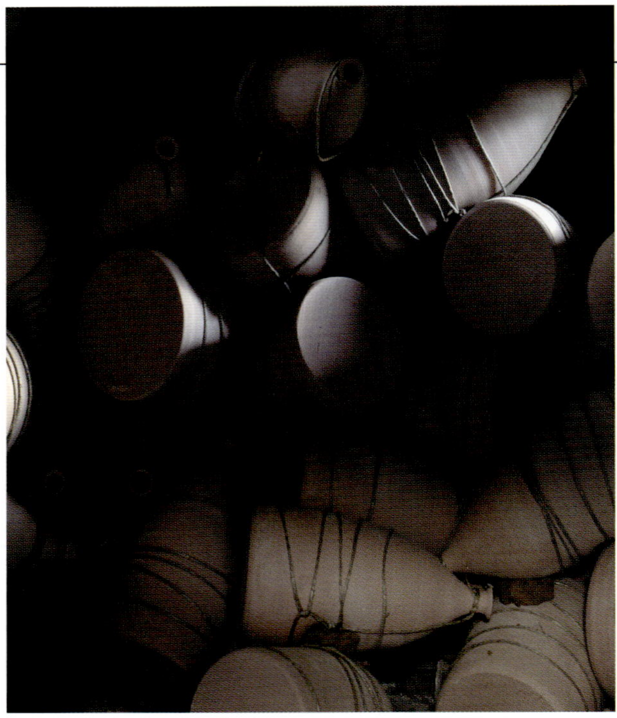

Stefan Andersson, Tumble stack anagama, 2009.
Photo credit: Anabeli Diaz González.

Artist's Statement

I work in small, persistent steps. I take pride in paying meticulous attention to detail, whether on a large vase, teapot, or the tiniest of cups. I freely draw shapes, techniques, and thoughts from history, resulting in a loving study of the potter and his means.

My pots are made on the potter's wheel in a series of steps. Clays, glazes, and slips are studied for use in a wide range of results. Firing in a wood kiln also provides a seemingly endless variety of results—giving the final pots a considerable depth.

I work with many different kilns. I use a modern and a traditional anagama, a salt kiln, and an array of kilns that I have developed over the years. Depending on which kiln I use, the outcomes vary due to the length of time and type of kiln fired. While I may try to control the firing process, I can't help but applaud each kilns' personality and unpredictability.

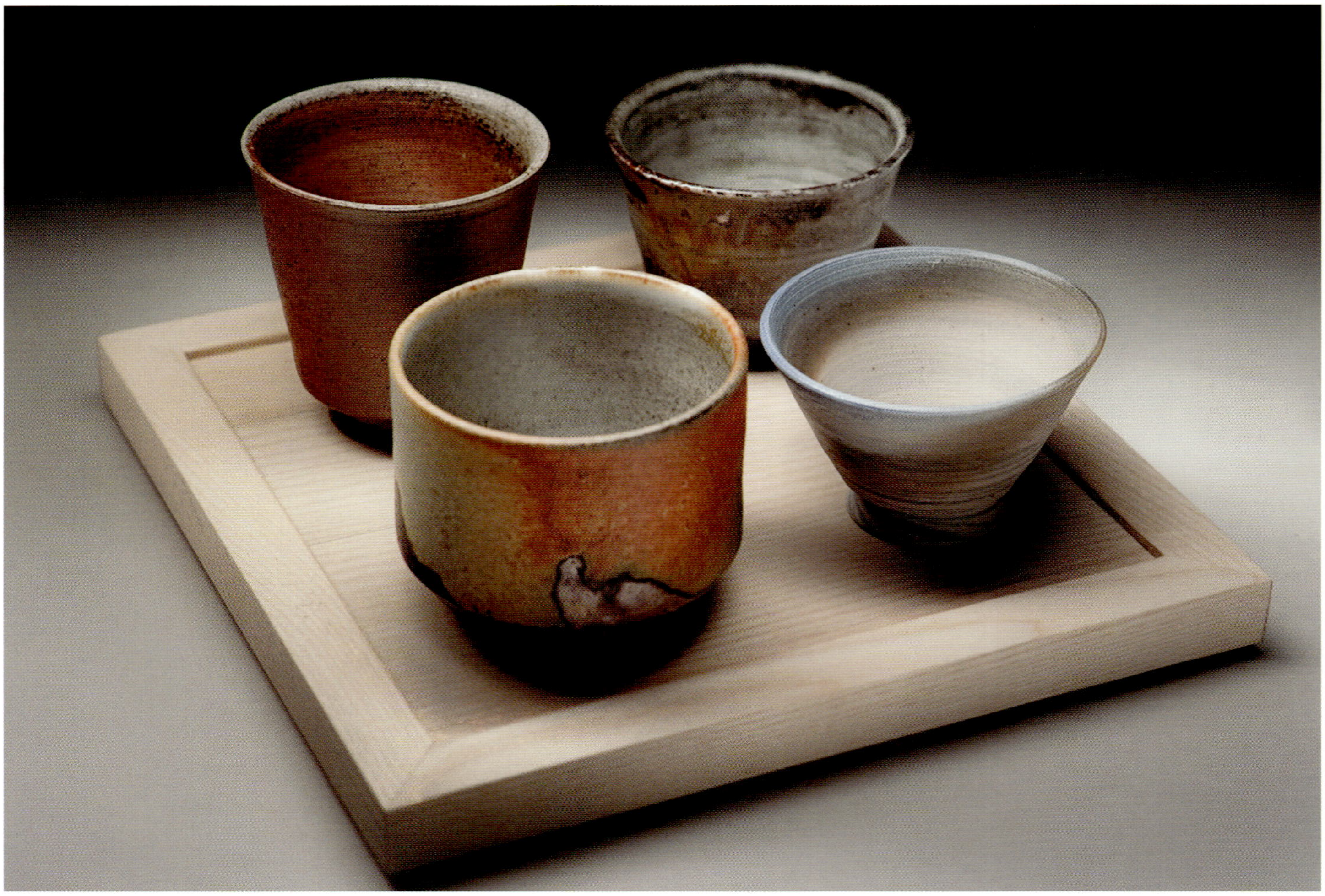

Stefan Andersson, *Quartet*, 2011. Stoneware, porcelain, slip, sinter engobe, glaze, unvarnished elm box, wheel-thrown, 8" x 8" x 6".
Photo credit: Stefan Andersson.

Stefan Andersson, Stoking anagama, 2010. *Photo credit: Matilda Haritz Svenson.*

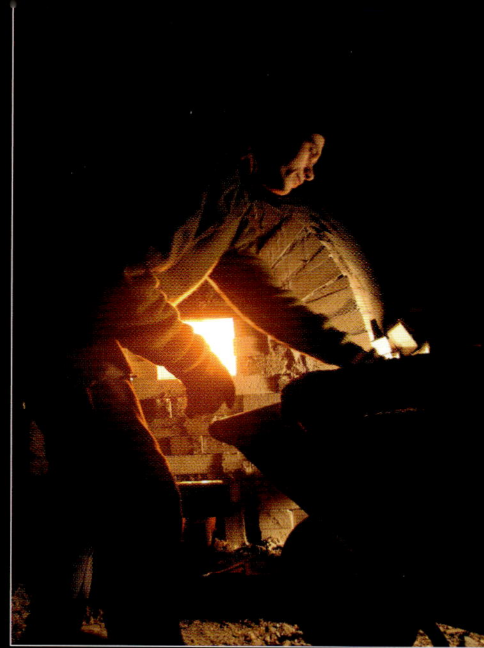

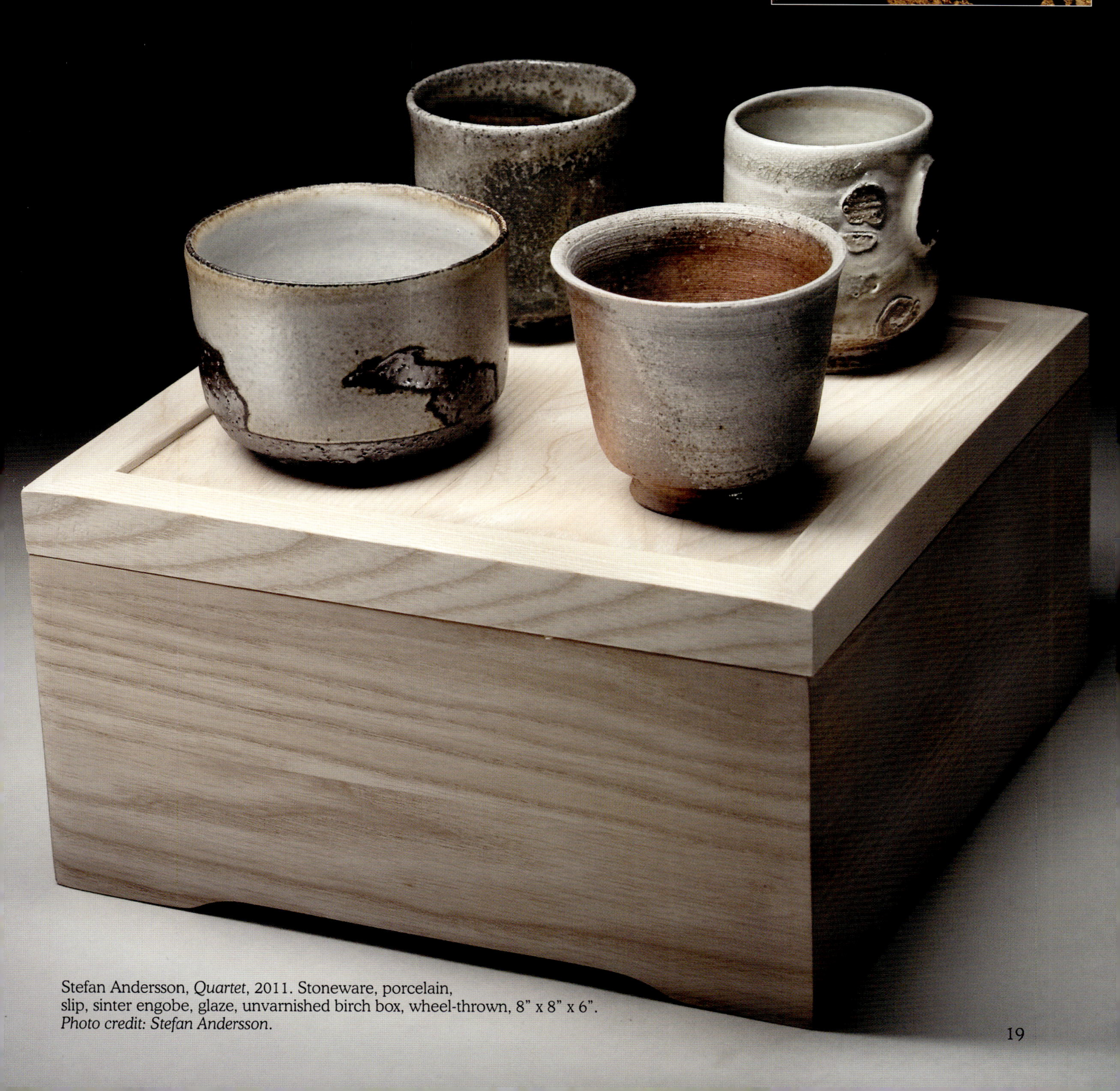

Stefan Andersson, *Quartet*, 2011. Stoneware, porcelain, slip, sinter engobe, glaze, unvarnished birch box, wheel-thrown, 8" x 8" x 6". *Photo credit: Stefan Andersson.*

WILLIAM BAKER

Bakersville, North Carolina
will@williambakerpottery.com, www.williambakerpottery.com

Artist's Statement

My current work reflects a careful balance of form and surface. Strong forms with crisp edges and lines that are carefully refined contrast with the irregular and dramatic markings that comes from wood and soda firing. I choose to create mostly functional work because I enjoy the intimacy with which these pieces can be experienced. For me, knowing that a piece of pottery I created will be held closely on a daily basis adds to the beauty of that piece, and to the attention to detail it deserves. As each piece takes form between my hands one thousand thoughts fill my head—of strong form, proper function, and firing surface. I create the structure, the bones of the pot, already thinking ahead toward the firing chamber. Building on this structure I refine the surface until the taut skin reveals the intentional elements of the seemingly simple form.

I fire my stoneware and porcelain forms in a wood-fired soda kiln, fueled by discarded waste wood. The kiln I currently use is a one chamber sprung arch design with an internal firebox. This kiln has about 40 cubic feet of stacking space, fires in about 18–20 hours using discarded pallets. This kiln produces a nice variety of surfaces with a few areas receiving heavy ash deposits, and other areas showing more soda vapor than wood ash. The constant attention a wood-firing requires is an important part of the process for me, as it allows my complete interaction with fueling the fire. Submitting my deliberate forms to the whim of the wood ash, flame, and soda vapor, demands a relinquishing of control and at the same time invites a wide range of surfaces from the subtle to the bold. The surface I seek for in my work is simply not possible without the combination of all of these factors: the flame and ash and vapors swirling through the kiln, the hours of attention and interaction of a skilled observer, and the acceptance of chance.

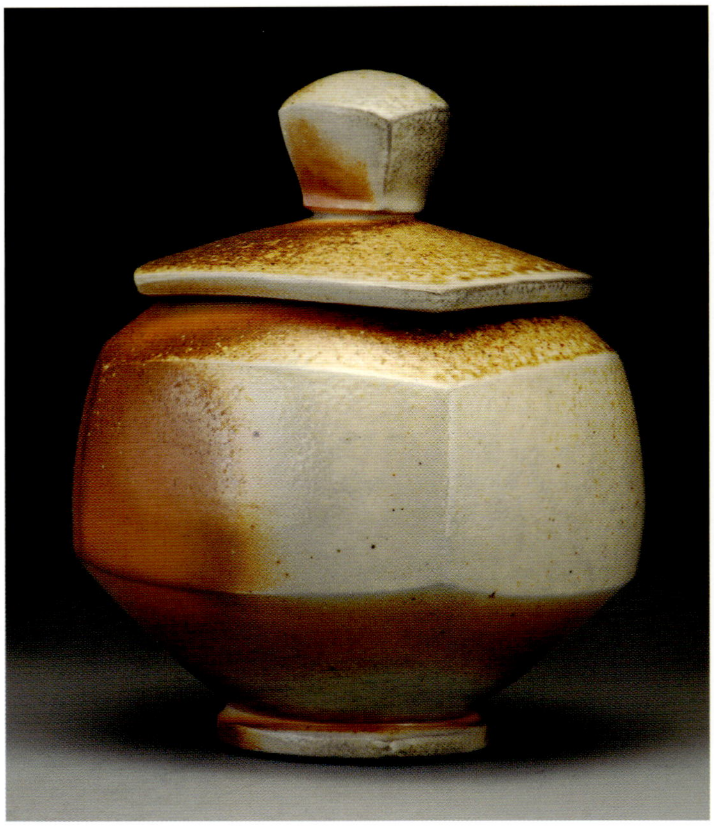

William Baker, *Square Jar*, 2011. Wood and soda fired porcelain, wheel-thrown and altered, 8" x 6" x 6". *Photo credit: Joy Tanner.*

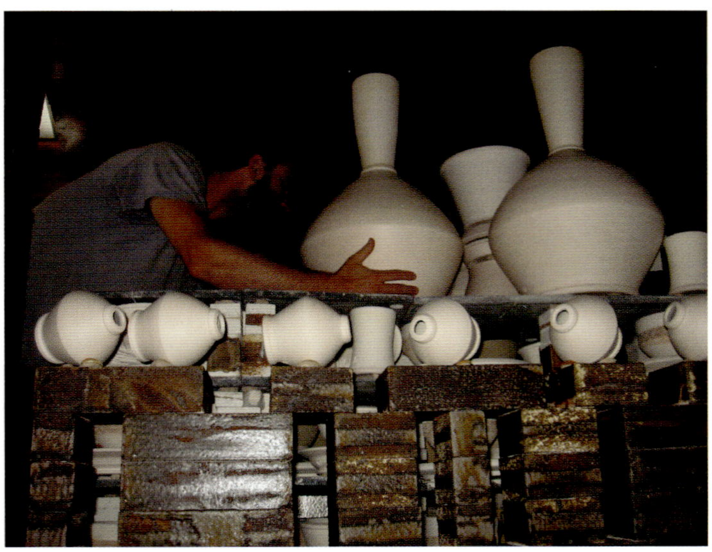

William Baker, Loading wood kiln, 2011. *Photo credit: Joy Tanner.*

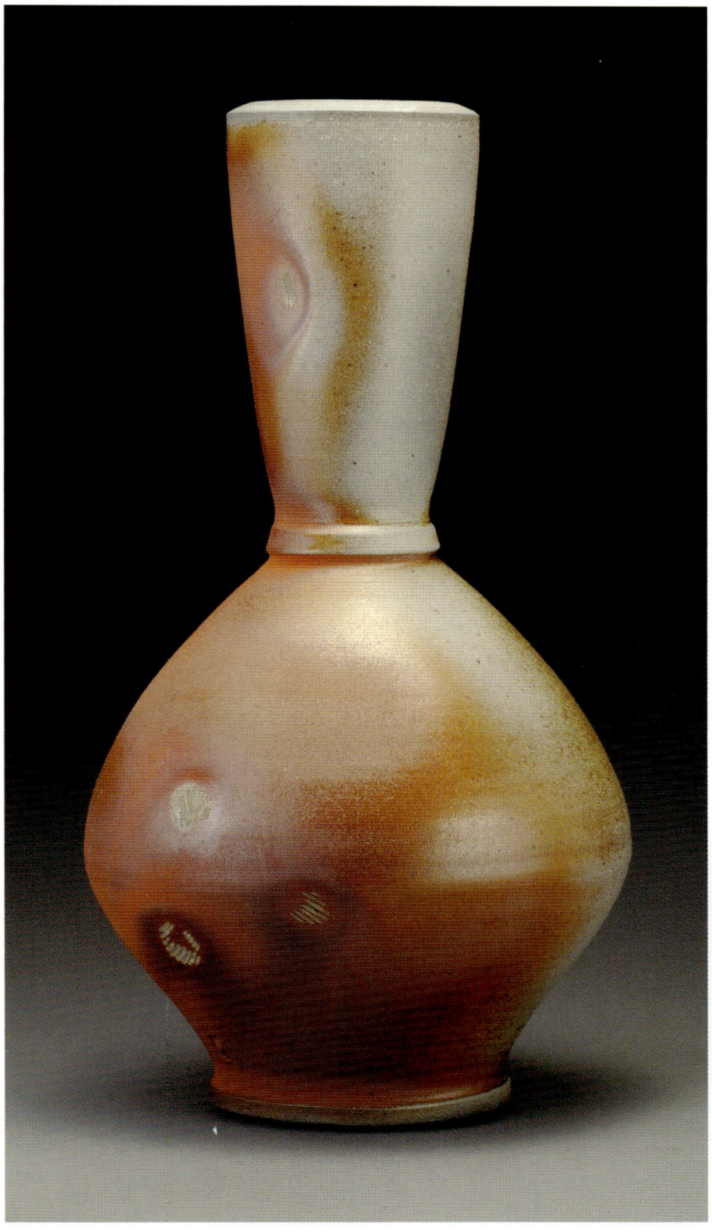

William Baker, *Tall Vessel,* 2011. Wood and soda fired porcelain, wheel-thrown, 20" x 11" x 11". *Photo credit: Joy Tanner.*

All of my pieces begin on the wheel, which was what first lured me into the clay studio. Over the years my style has evolved from one that portrayed a straightforward wheel-thrown aesthetic to one that often employs altering and refining of the wheel-thrown forms to sharpen edges and refine lines. I use stoneware and porcelain clays, forming pieces on the potter's wheel. Some of these pieces are done once they leave the wheel; others are altered using different techniques such as squaring and paddling. Many pieces are dipped in a flashing slip while others might remain bare clay. I bisque fire my work before glazing, if necessary, and then it is on to the final step a wood/soda firing to cone 10–11.

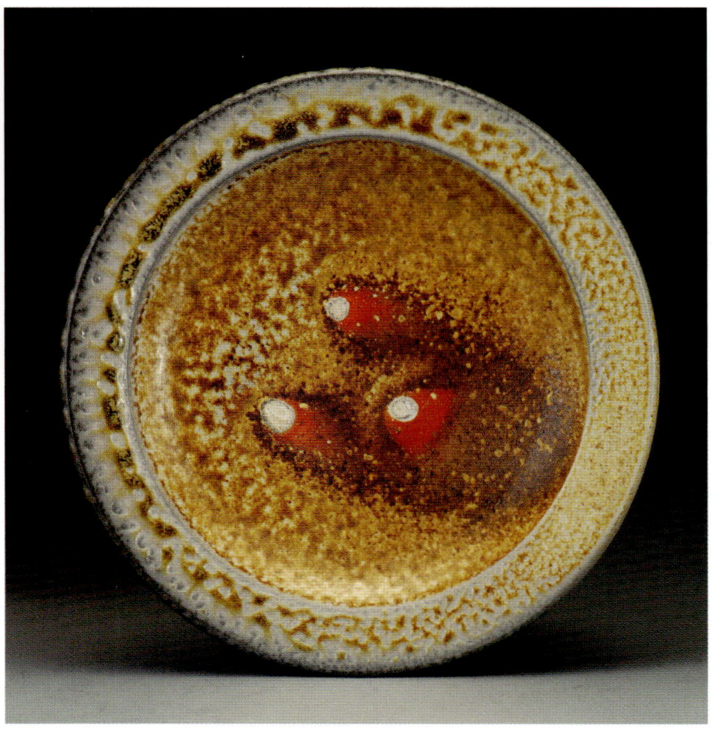

William Baker, *Plate*, 2011. Wood and soda-fired porcelain, wheel-thrown, 2" x 7". *Photo credit: Joy Tanner.*

JOHN BAYMORE

Wilton, New Hampshire
JBaymore@nhia.edu, www.johnbaymore.com

Artist's Statement

Since I began wood-firing in 1969 with the first kiln I built, my commitment to wood-firing has evolved. The origins came from of a blend of youthful enthusiasm and pure necessity, combined with a huge dose of technical ignorance. As a young ceramist, old hard bricks were easy to find, but I knew nothing of how to create gas burners, nor did I have a budget for them anyway. I knew from my studies that for thousands of years people had fired pottery with wood. So building a wood kiln simply made sense from my position of student naiveté. I didn't know what I didn't know.

Later, the process itself got its hooks into me. The primal activity connected me to the work in the kiln in a way that other firings did not. Electric kilns seemed like toasters and gas kilns felt like medical testing devices. A wood kiln gave me the feeling I was firing, not that the pieces I made were being fired. It was an important distinction.

As a young man, there was the "*guy thing*" quality about it all; smoke and flame and danger, and all that potentially "*macho*" garbage. It was so hot that it was "*cool.*" Few people in the USA were wood-firing then, so it was a rarified group of ceramists I joined in pursuing this arcane art. It made what I did seem unique, and more my own.

As I became far more technically savvy I became a professional kiln designer and builder. Because of this, I realized the incredible thermal efficiency of the noborigama-style kiln. It is simply the most efficient type of design there is, without shifting to continuous firings and low mass refractories. I even built some totally gas-fired noborigama. Noborigama already have their roots grown in the environmentally fertile soil of wood as a sustainable and renewable fuel; it simply just made infinite sense to continue to utilize the two paired together.

By 1977 the first "*energy crisis*" had hit America. I knew then that energy consumption, particularly of non-renewable types, was going to be a serious issue in the coming decades. Using re-growing trees to fire my pots, instead of tapping into oh-so-convenient fossil-based fuels, or slightly scary nuclear sourced electricity also made all the sense in the world.

By '77, when gasoline was an astronomical $0.63 per gallon, I knew that a wood-fired noborigama was my future. It was apparent that fossil fuel reserves were not going to increase nor get cheaper. I looked to wood for both economic as well as environmental reasons. With eight years of wood-firing experience

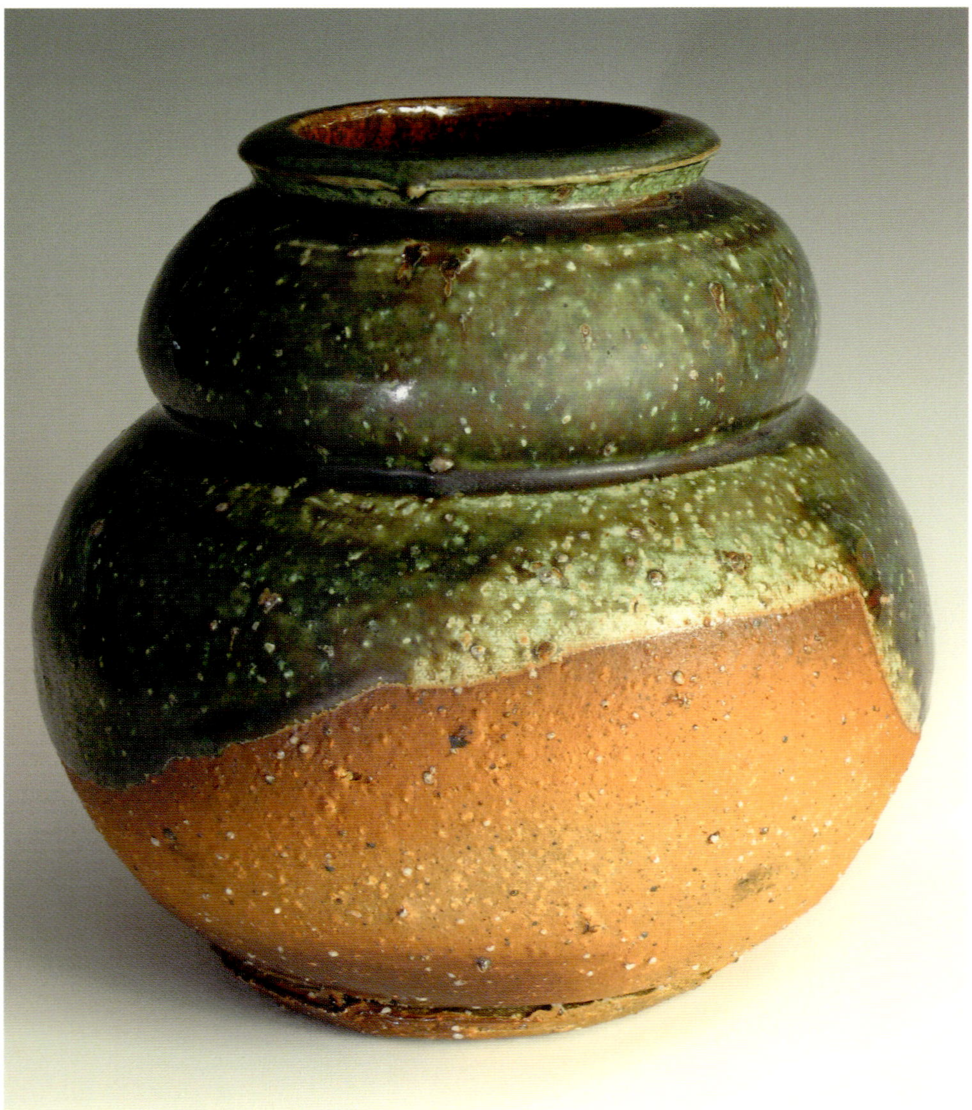

John Baymore, *Gourd-shaped Oribe Vase*, 2011. Noborigama fired stoneware, Orton cone 10; thrown, Oribe-style glaze, 7.4" h x 7" d [19 cm h x 18 cm d]. *Photo credit: John Baymore.*

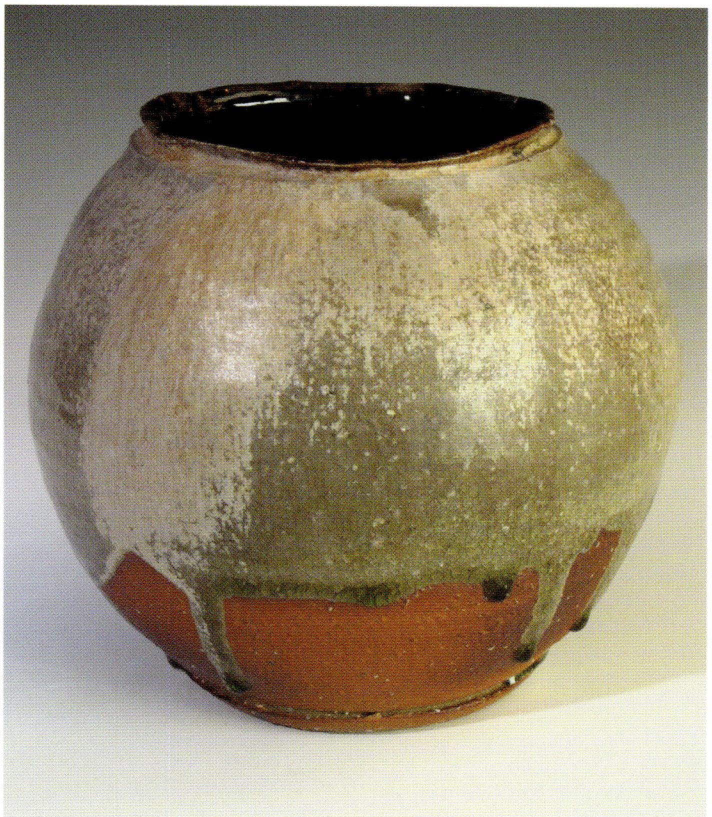

John Baymore, *Vase,* 2011. Noborigama fired stoneware, Orton cone 12; thrown, wood ash, and N.H. granite glaze, 9" h x 7" d [23 cm h x 18 cm d]. *Photo credit: John Baymore.*

by then, I was fully committed conceptually, environmentally, economically, and aesthetically. The noborigama I still fire, named "*Kawagama –River Kiln–*," took shape by 1980.

Over the ensuing years, I have become totally enthralled with the play of wood ash and wood flame on both unglazed and glazed ceramic surfaces. The effect that ash creates on the wares has become a driving force in the genesis of my creations. The touch of the fire on everything I make is very important to me.

Too recently, we've become aware of the damage greenhouse gases are doing to our planet. Conveniently, in addition to being renewable, wood turns out to be just about carbon neutral when burnt. Over about thirty years, trees re-sequester released carbon dioxide into the wood they produce. Thermally efficient noborigama, combined with wood fuel, look very environmentally friendly. Those Japanese potters of the fourteenth through sixteenth centuries who refined the noborigama were onto something well ahead of their time.

I'm getting older, and the heat energy efficiency that the noborigama kiln offers feels very attractive as I view my approaching years. I want to wood-fire for my entire ceramic-producing life, and will wood-fire until I am physically unable. As a professor of ceramics, I hope to pass on this ecologically sustainable pursuit to some of my students. Like growing new trees, I hope to see a new crop of wood-firers continue the tradition where I eventually leave off. Hopefully, just like in burning trees, this will take what I have put out there into the "*ceramic atmosphere*" and channel it back into new growth in the field.

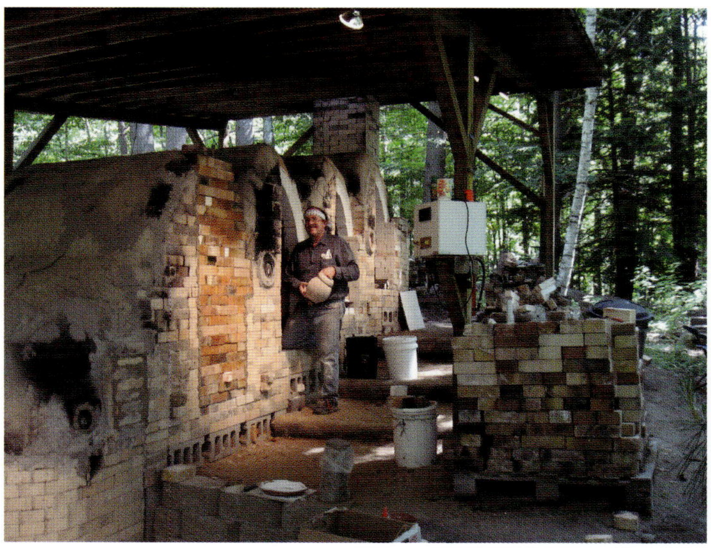

John Baymore loading the noborigama at his studio in Wilton, New Hampshire. *Photo credit: John Baymore.*

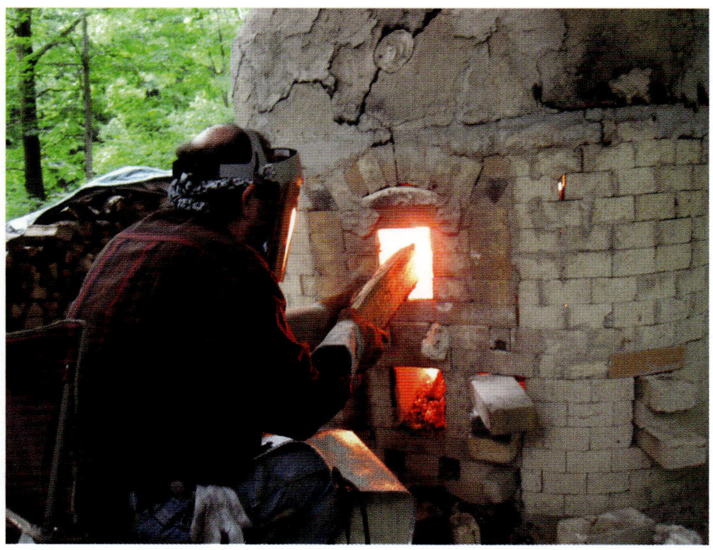

John Baymore stoking the main firemouth of the noborigama at his studio in Wilton, New Hampshire. *Photo credit: John Baymore.*

JOHN BENN

Shelton, Washington
bennpottery@gmail.com, www.benngallagher.com

Artist's Statement

On beautifully forested Harstine Island in Washington State's Puget Sound, John Benn makes wood-fired pottery and sculptural tiles. He digs several local stoneware clays and fires his kilns using trees from the surrounding forest: fir, alder, madrona, and maple. The pots are wood-fired in a twenty-five foot long anagama hybrid design wood kiln, the Mongrelgama. The pots, often large scale, are unglazed, with natural ash effects from the three-day firing.

"The nature of the wood-fire process is compulsive and inefficient, and it connects me closely to the earth. The work is formed from clays beneath my feet. I submit it to my kiln to be reborn as permanent objects that record the touch of my hands and the fire and ash that have transformed it.

My role as a wood-fire potter is to find and make forms that will be transformed by the flame. I want them to be able to carry glaze drips that will roll over the surface and culminate in "dragonfly eyes." Scorch marks from black to red, as well as blue and green areas can be painted by the firing process as well. Textures are also created. Sometimes all these things happen in a single firing. I feel that a successful piece is a two plus two equals five equation, where I have given up some control in return for a surprising outcome that is a little beyond what I could have planned. I do everything I can as an artist to allow these surprised to appear in my work.

The local clays that I dig are integral to the look that I want from my kiln. At the present time I use a total of six different clays in different parts of the kiln to get different colors, textures, and patinas. Although digging my own clay is time-consuming and requires even more effort than chopping wood, the rewards are there in the finished work.

My commitment to wood-firing is obsessive and non-intellectual. I discover my pots in addition to creating them."

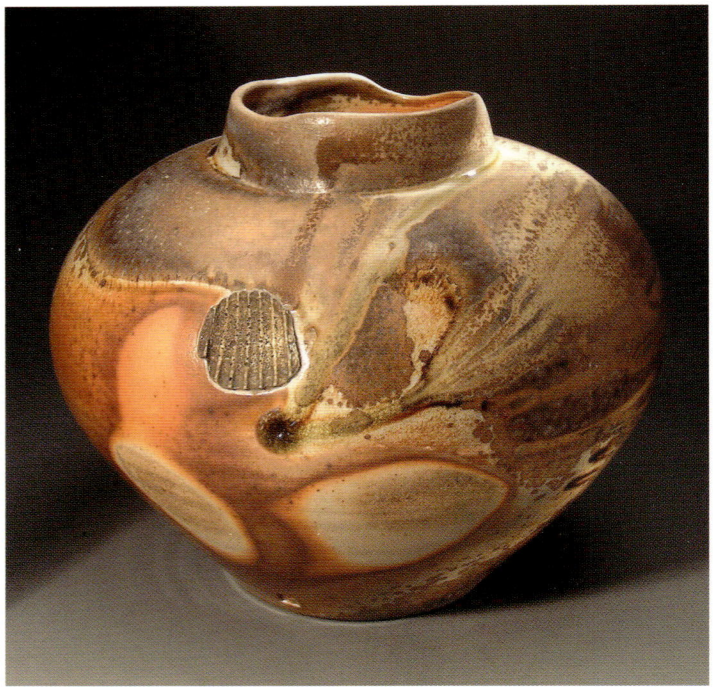

John Benn, *Vase Form*, 2010. Wood-fired stoneware, wheel-thrown, 11" x 10 x 10". *Photo credit: Colleen Gallagher.*

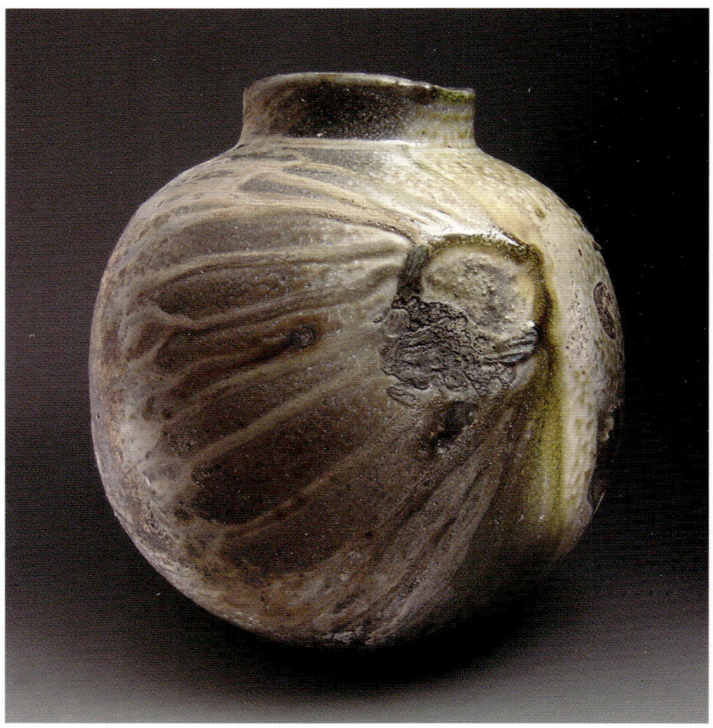

John Benn, *Jar*, 2011. Wood-fired local stoneware, wheel-thrown, 14" x 12" x 12". *Photo credit: Colleen Gallagher.*

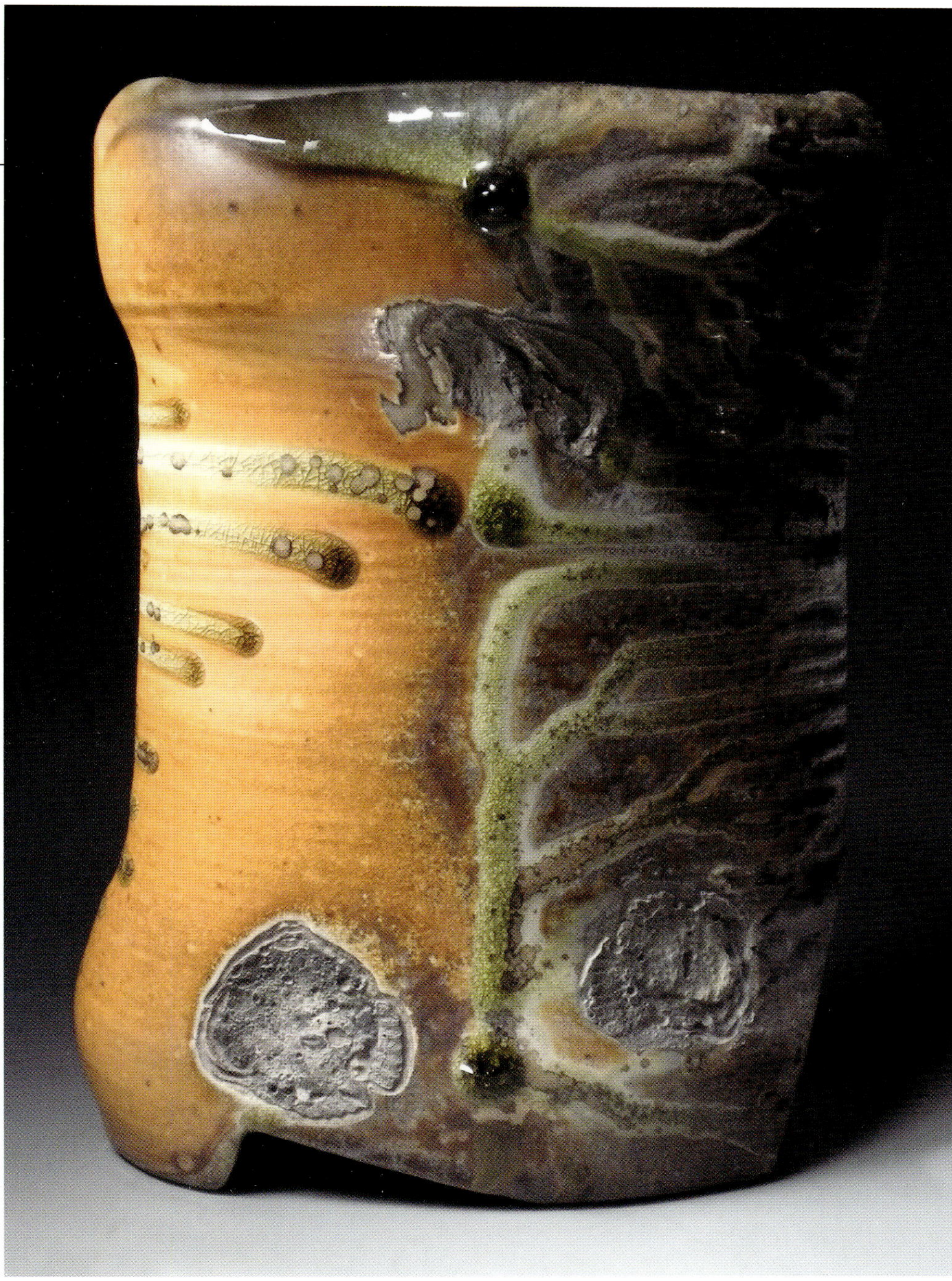

John Benn, *Envelope Vase,* 2011. Wood-fired local stoneware, wheel-thrown, 12" x 9" x 5". *Photo credit: Colleen Gallagher.*

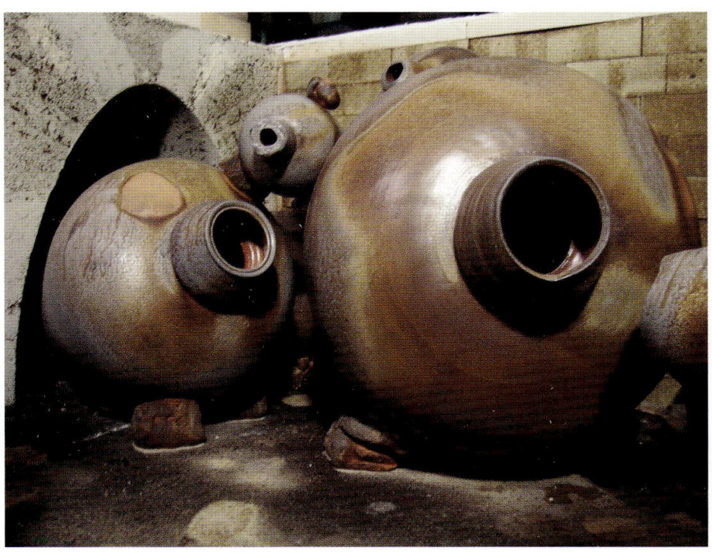

John Benn, Harstine Island Wood Kilns. Tumble stacked pots post firing. *Photo credit: Colleen Gallagher.*

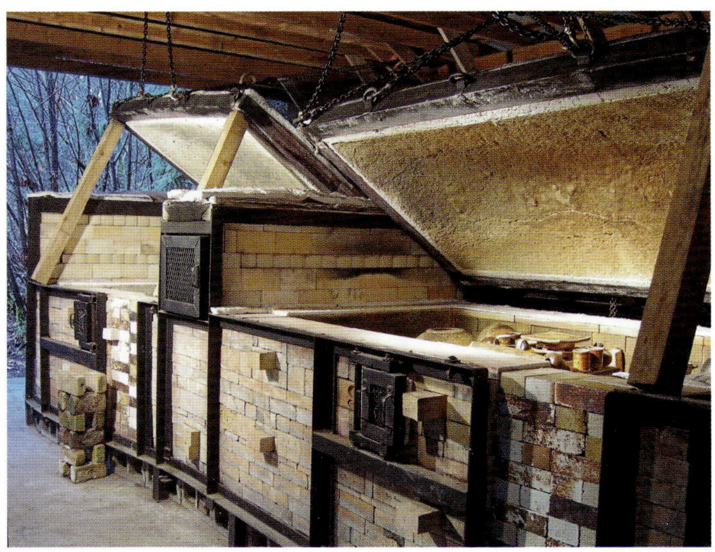

John Benn, Harstine Island Wood Kilns. The *Mongrelgama* with its lids open. *Photo credit: Colleen Gallagher.*

BRETTON SAGE BINFORD

Portland, Oregon
brett@mudsharkstudios.org, www.mudsharkstudios.com

Artist's Statement

My most recent bodies of work encompass the exploration of Americana objects and objects of nostalgia. I enjoy pushing what can be done with mold making and slip casting found objects.

When presented with the opportunity to fire with wood, I felt drawn to the sense of irony of objects realistically cast and rendered but then given a new context in the wood-firing. The wood-firing speaks of old traditions and with contemporary iconic objects there is a lot of conceptual play.

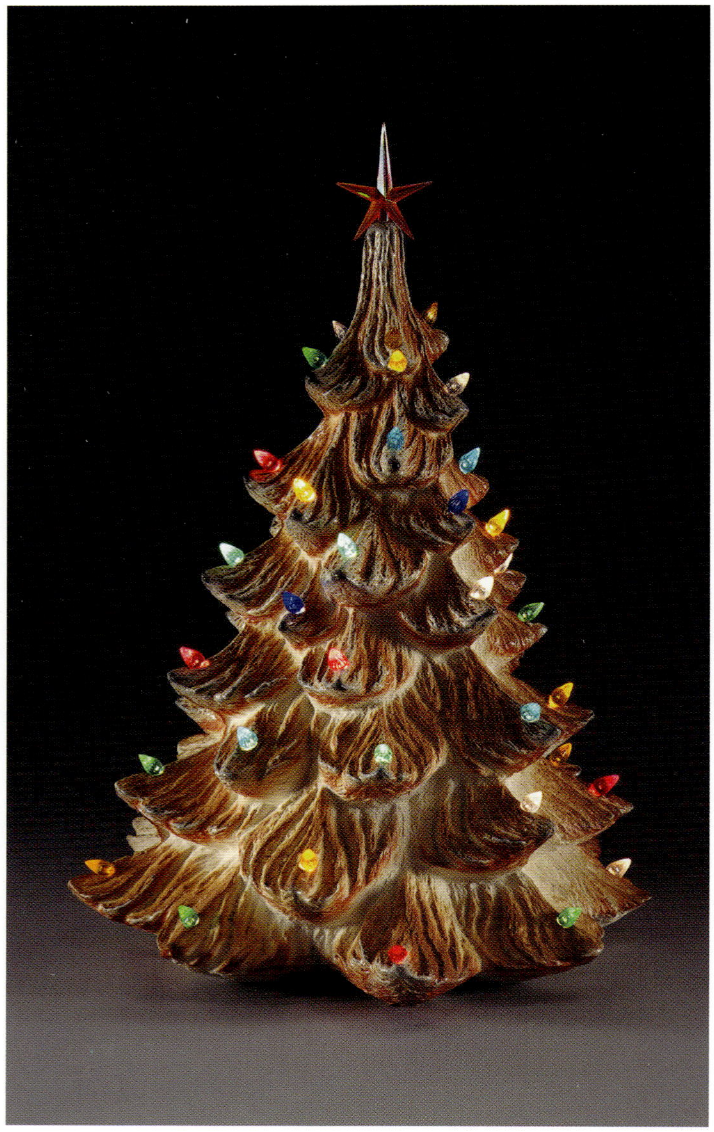

Bretton Binford, Christmas tree, 2010. Wood-fired slip cast stoneware, 18" x 12" x 12". *Photo credit: Courtney Frisse.*

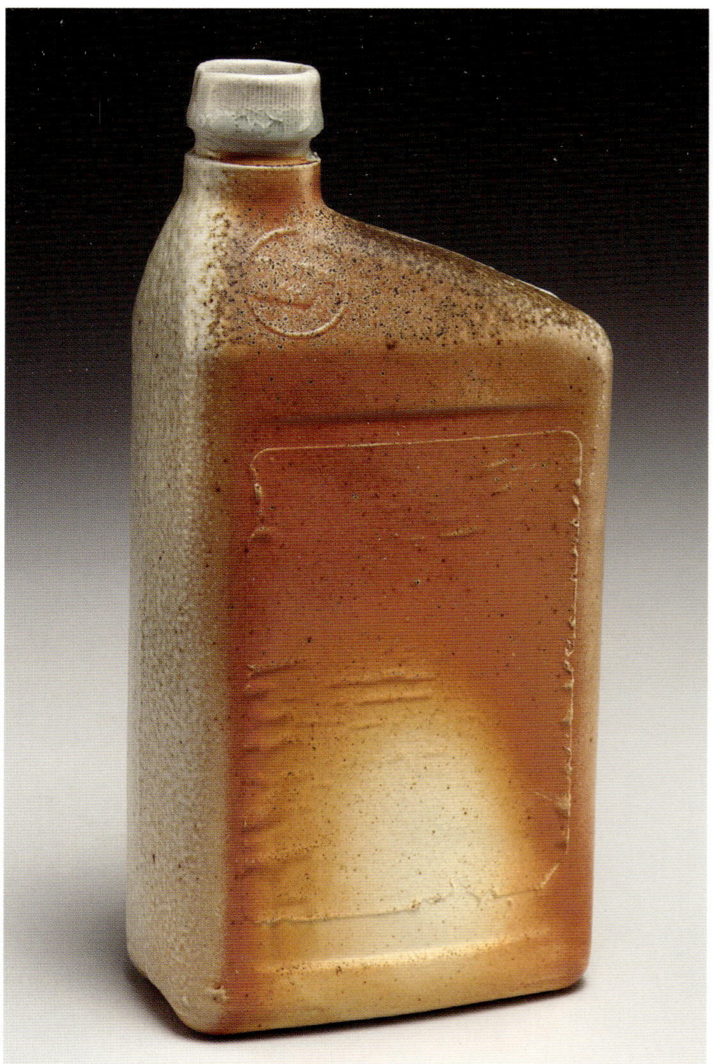

Bretton Binford, *Oil Quart*, 2010. Wood-fired slip cast stoneware, 10" x 4" x 2". *Photo credit: Courtney Frisse.*

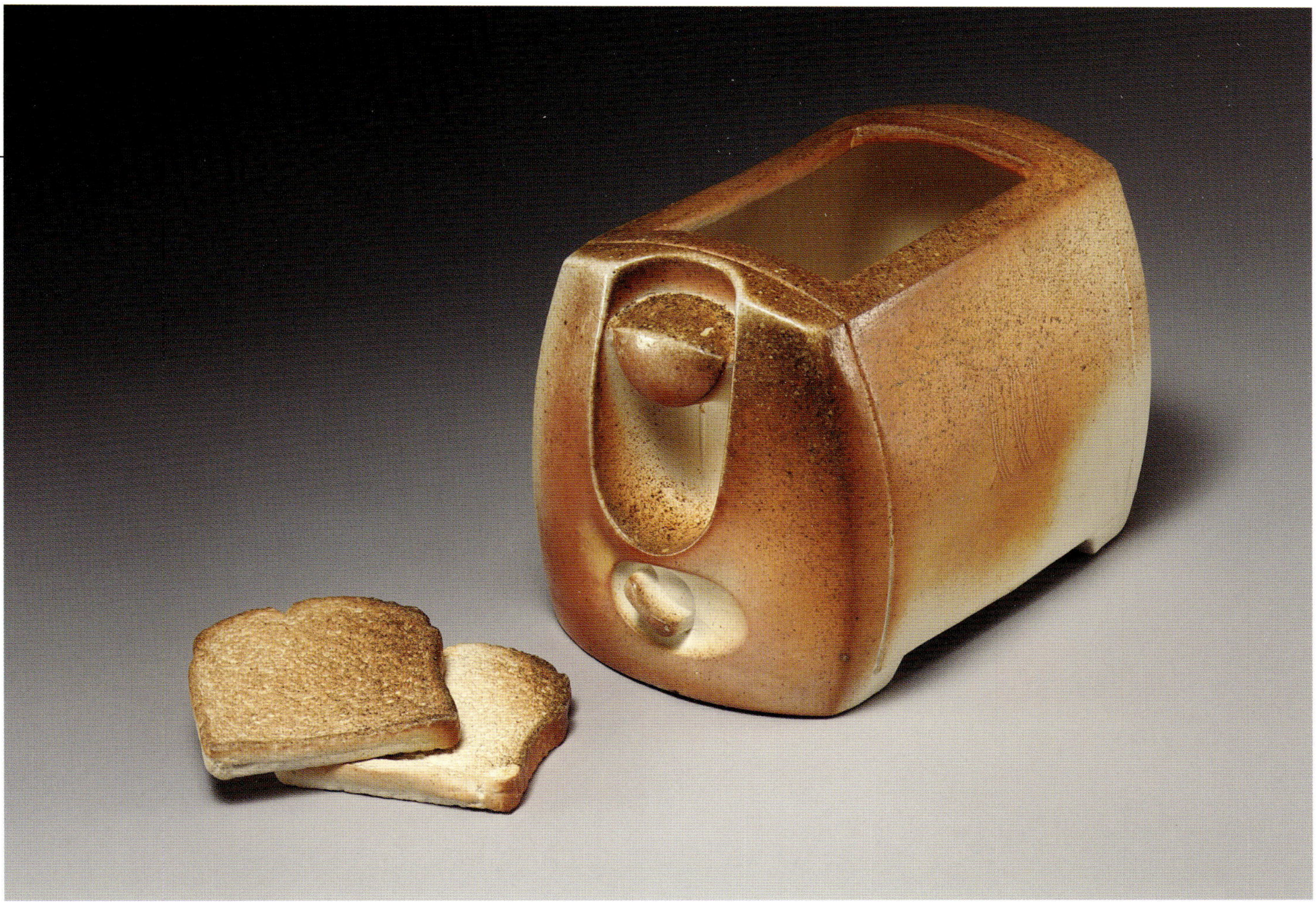

Bretton Binford, *Toast and Toaster,* 2010. Wood-fired slip cast stoneware, Toast, 4" x 4", Toaster 8" x 12" x 6". *Photo credit: Courtney Frisse.*

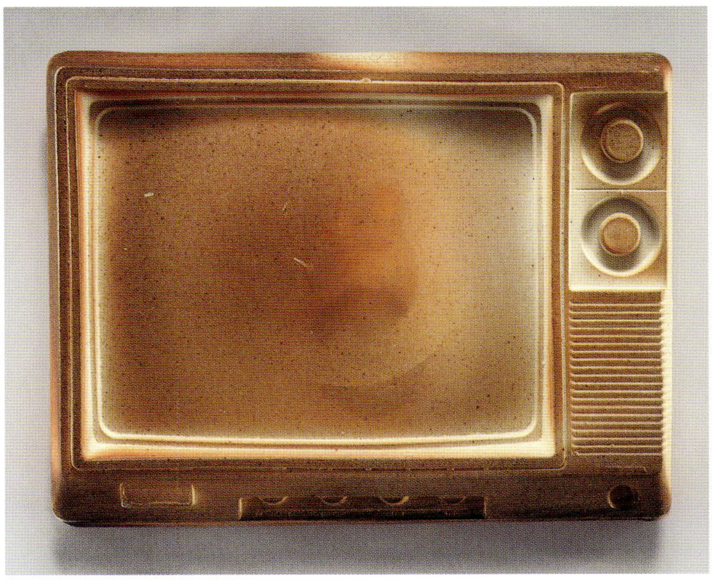

Bretton Binford, *Television*, 2010. Wood-fired slip cast stoneware, 2" x 12" x 18". *Photo credit: Courtney Frisse.*

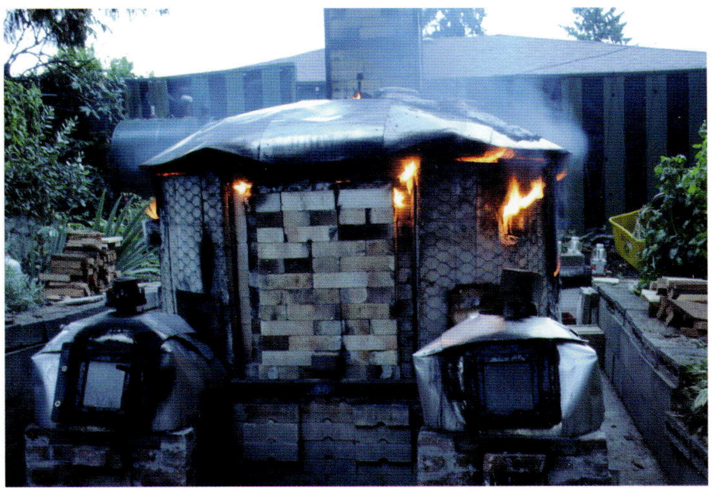

Careen Stoll's "*Tin Man*" kiln. The tin man is a cross draft kiln with oil assist burner system. At the end of our firing we introduced a light amount of soda.

SCHYLER BINKLEY

Kalamazoo, Michigan
schyler.binkley@gmail.com, www.schylerbinkley.com

Artist's Statement

For me, the choice to wood-fire is as much about tradition as it is about concept and aesthetics. Mankind has a very long history with pottery and ceramics. We, as humans, have always made things out of clay. I think it is significant that so many people throughout history and continuing into today love to work with this material. Wood-firing has been a part of this tradition for thousands of years. A community of potters coming together and participating in a wood-firing is something that has changed very little over the years. In this age of technological advancement, it is refreshing to participate in something that is so decidedly low-tech. I feel wood-firing keeps alive a very strong and important connection to our past.

As the maker of my pots, it is conceptually profound for me to relinquish control over the end result of my work. The results of wood-firing are very much affected by the firing process itself, which it is literally impossible to completely control. One can influence the flame, but certainly not control it. Committing my hard work and ideas to a process that may destroy said work is a helpful and profound experience. One has to simply trust and have faith that things will work out. Sometimes they do, sometimes they don't—but therein lies the conceptual value. It is about the process of letting go.

Aesthetically, wood-fired pots are the most precious and beautiful of ceramic wares. Because the results depend so much on the infinite and uncontrollable variables of the firing, beautiful results must be appreciated that much more. It is like seeing destiny played out, as if the best pot in the kiln was destined for greatness. To produce a beautiful wood-fired pot, many variables must come together of their own accord. The surfaces and colors that result from wood-firing are incredibly rich and layered. A wood-fired pot can be read like a book. A sharp and educated eye can identify which side of the pot was facing the firebox, the path the flame took around it, even where the pot was in the kiln. These factors and more are recorded on the pot by the flame itself and contribute to a very interactive viewing experience. The layers of information create a very complex whole. Wood-fired pots can be many things: subtle yet bold; humble yet proud; beautiful yet ugly.

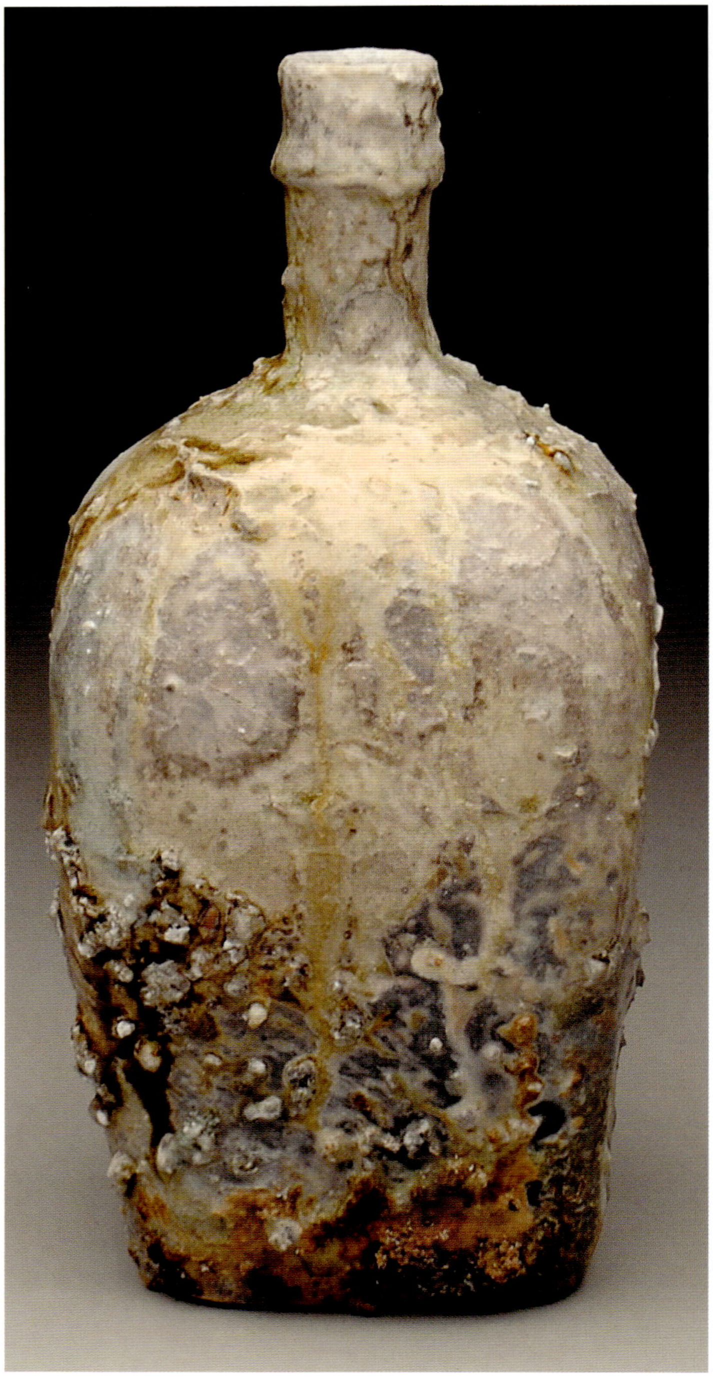

Schyler Binkley, *The Poison*, 2011. Wood-fired stoneware, wheel-thrown; twice-fired in firebox, 12" x 5.5" x 5.5". *Photo credit: Schyler Binkley.*

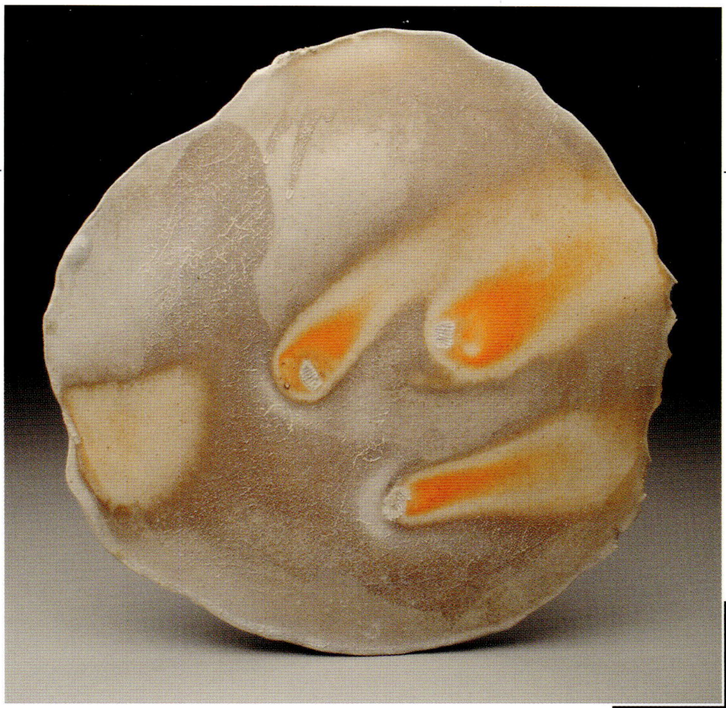

Schyler Binkley, *Plate with Comet Tails*, 2011. Wood-fired stoneware, slab built, flashing slips, 1.5" x 13" x 13". *Photo credit: Schyler Binkley.*

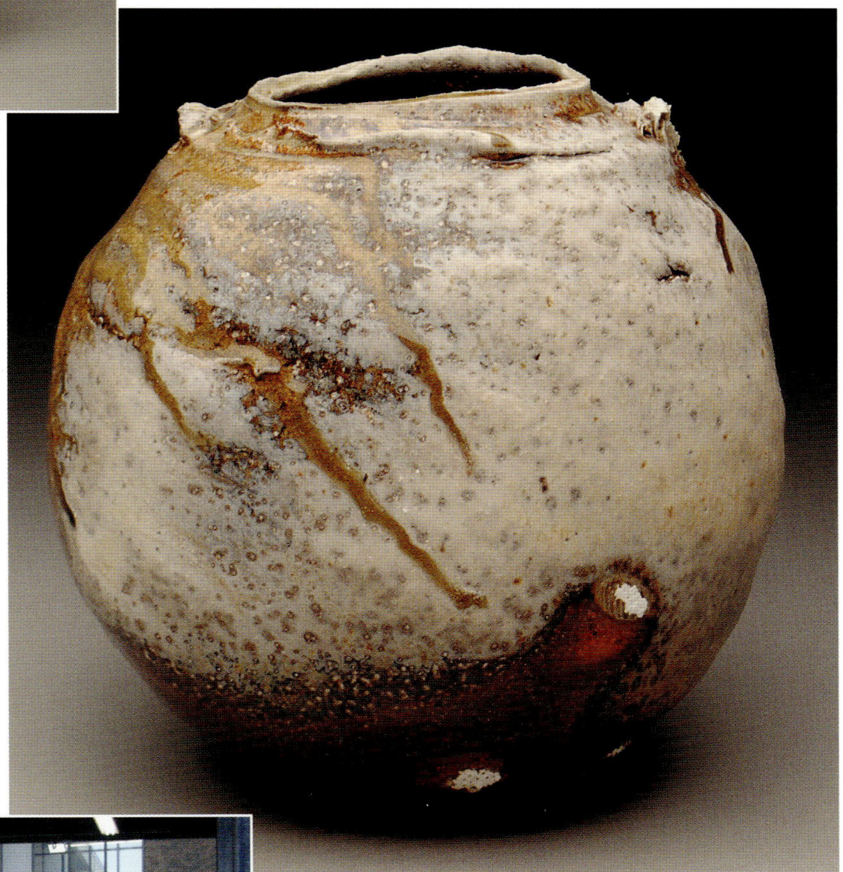

Schyler Binkley, *A Dream For Us*, 2011. Wood-fired iron-rich stoneware, wheel-thrown, 9" x 10" x 10". *Photo credit: Schyler Binkley.*

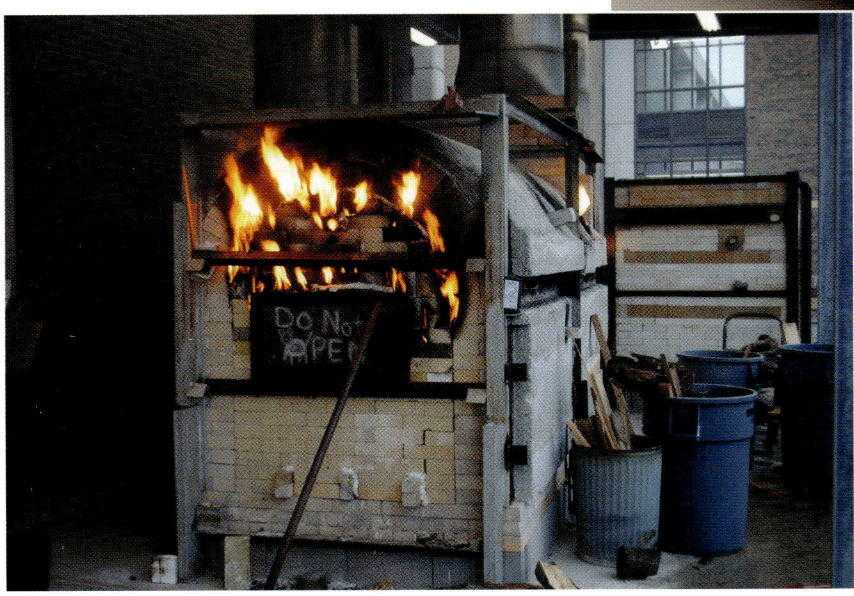

Heavy reduction cycle near end of firing. *Photo credit: Schyler Binkley.*

RICHARD BRESNAHAN

Collegeville, Minnesota
rbresnahan@csbsju.edu, www.csbsju.edu/pottery

Artist's Statement

Richard Bresnahan directs a pottery studio on the campus of Saint John's, a Benedictine abbey and university in the woodlands of central Minnesota. Originally from Casselton, North Dakota, where his father ran a grain-elevator company and his mother raised a family of six, Bresnahan (b. 1953) is unapologetic about his rural Midwestern roots.

Richard Bresnahan secured an apprenticeship with Nakazato Takashi, a thirteenth generation potter in the seaside town of Karatsu. (Takashi was the son of Tarōuemon XII, who was named a "Living National Treasure" during Bresnahan's time with the family.) Bresnahan's apprenticeship with the illustrious Nakazatos lasted nearly three-and-a-half years, and through them he absorbed the traditions, techniques, and reverence for clay that permeated every aspect of their daily lives.

Rather than ordering materials from a catalogue as he had done in Minnesota, he discovered that it was possible—even preferable—to excavate local clays and to wash away their inherent impurities. He also found out how to make glaze from ash and then how to apply the solution quickly and evenly to greenware. As importantly, the Nakazatos encouraged Bresnahan to think critically about kiln design and to consider how certain structural alterations might induce better aesthetic results. "The verve of sixteen-hour days," says Bresnahan, "and of doing everything from sweeping the floor and making tea to digging clay and building kilns provided me with a holistic approach … and a deep respect for indigenous systems."

Indeed, an abiding respect for ecology and natural systems has always characterized Richard Bresnahan's philosophy and

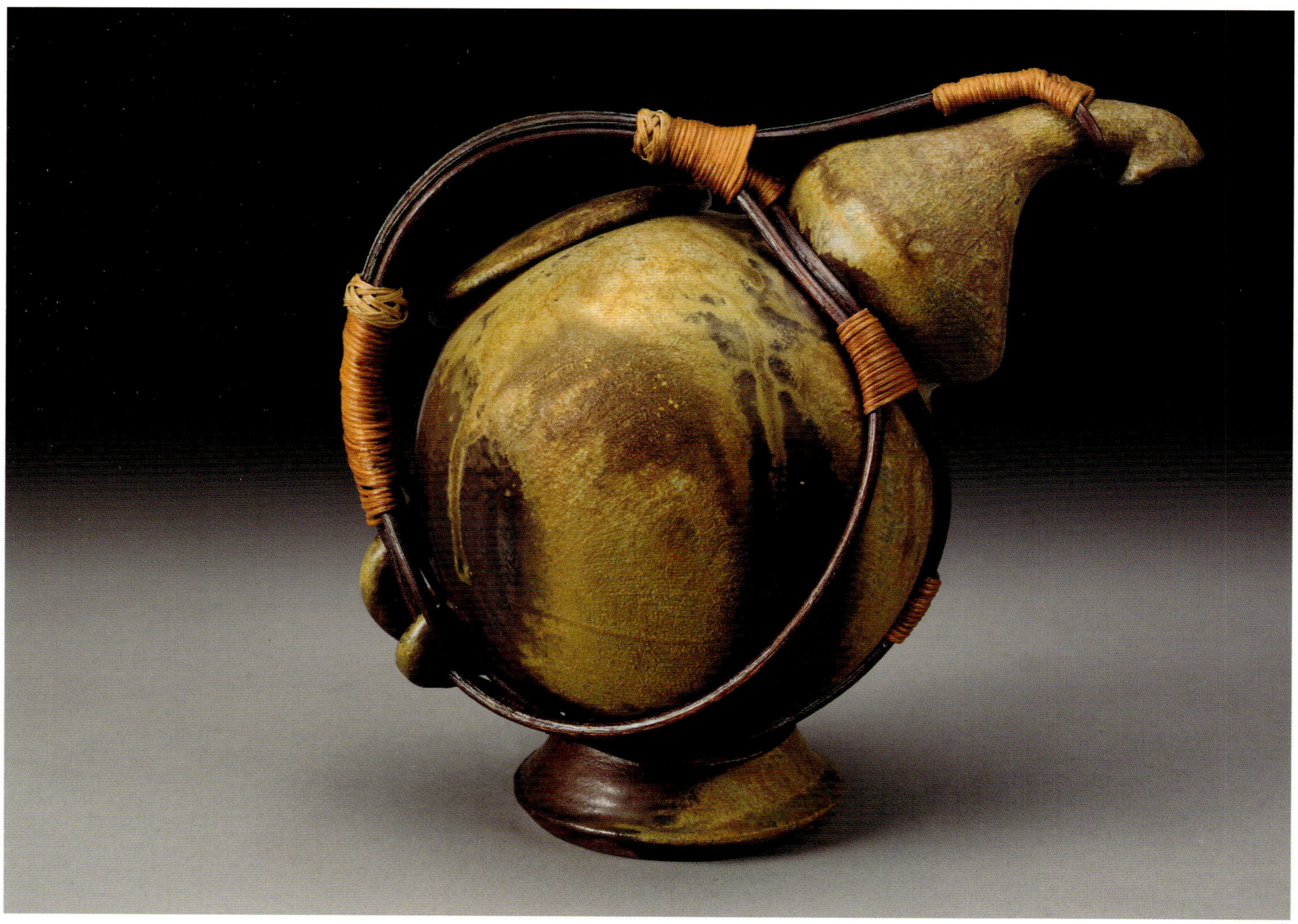

Richard Bresnahan, *Double-gourd Teapot,* 2008. Wood-fired stoneware with reed handle by Paul Krueger, 6.5" x 7". *Photo credit: Brian Zehowski.* Saint John's Pottery, Johanna kiln, front chamber. Melting wood ash settles on the pots in this chamber. For use with green tea and has a pleasant pouring sound.

his ongoing dialogue with fire and clay. Throughout his long career, he has consistently challenged himself to undertake new firing techniques, to test potential glaze materials, and to devise unique and elegant functional forms. As a result, the dynamic pottery Bresnahan creates reflects both the specifics of his physical environment—by utilizing local resources—and his own far-reaching artistic vision.

Dr. Matthew Welch, Deputy Director of the Minneapolis Institute of Art, Stoked: Five Artists of Fire and Clay

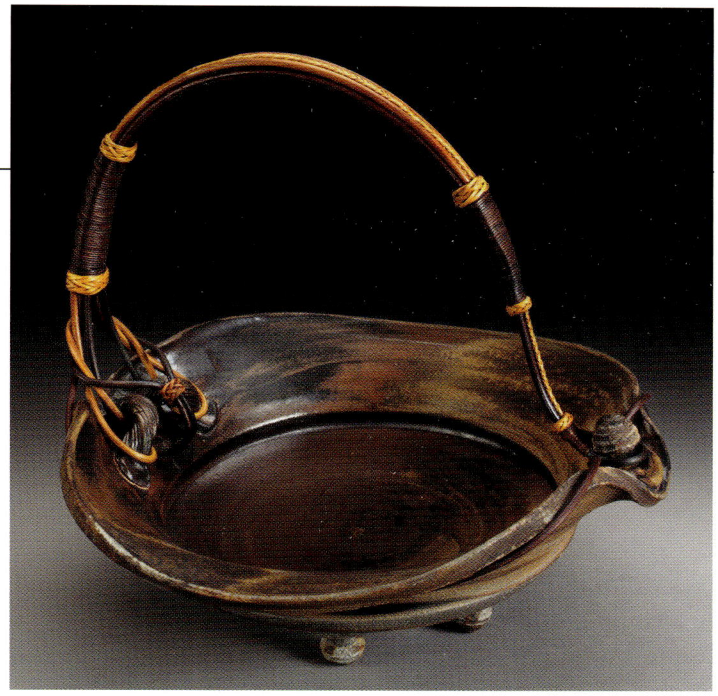

Richard Bresnahan, *Folded Lip Tebachi,* 2008. Wood-fired stoneware with reed handle by Paul Krueger, 12.5" x 12". *Photo credit: Brian Zehowski.*
Saint John's Pottery, Johanna kiln, front chamber. Melting wood ash settles on the pots in this chamber. A *Tebachi* is a handled dish in the Japanese tea ceremony.

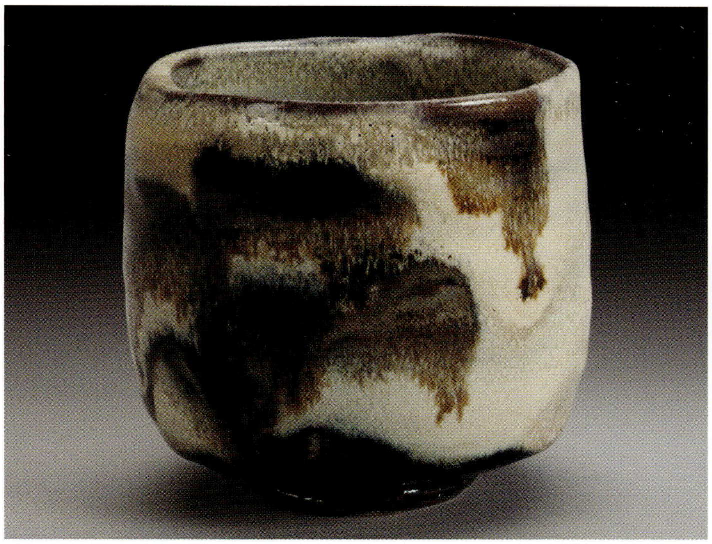

Richard Bresnahan, *Chosen Tea Bowl,* 2010. Wood-fired stoneware with navy bean straw ash, basswood ash, and granite dust glazes, 4" x 4". *Photo credit: Brian Zehowski.*
Saint John's Pottery, Johanna kiln, glaze chamber. Multiple glazes were inlaid into one another. *Chosen* is a subset of a technical glazing process.

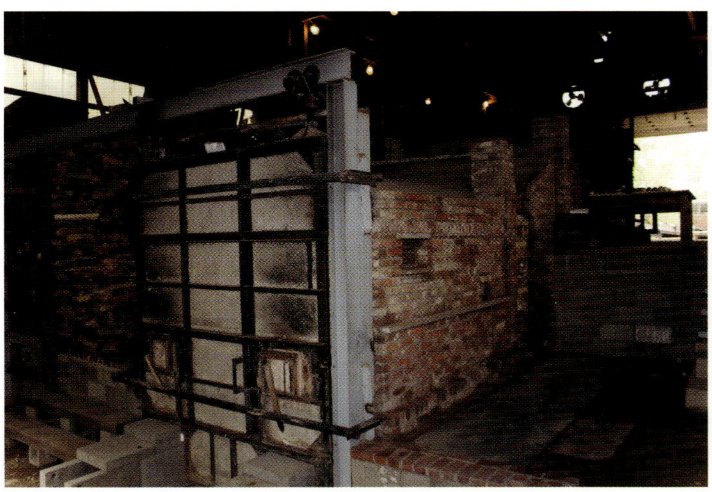

Richard Bresnahan, Front-facing photograph of the Johanna kiln, the largest wood kiln of its kind in North America. *Photo credit: Brian Zehowski.*

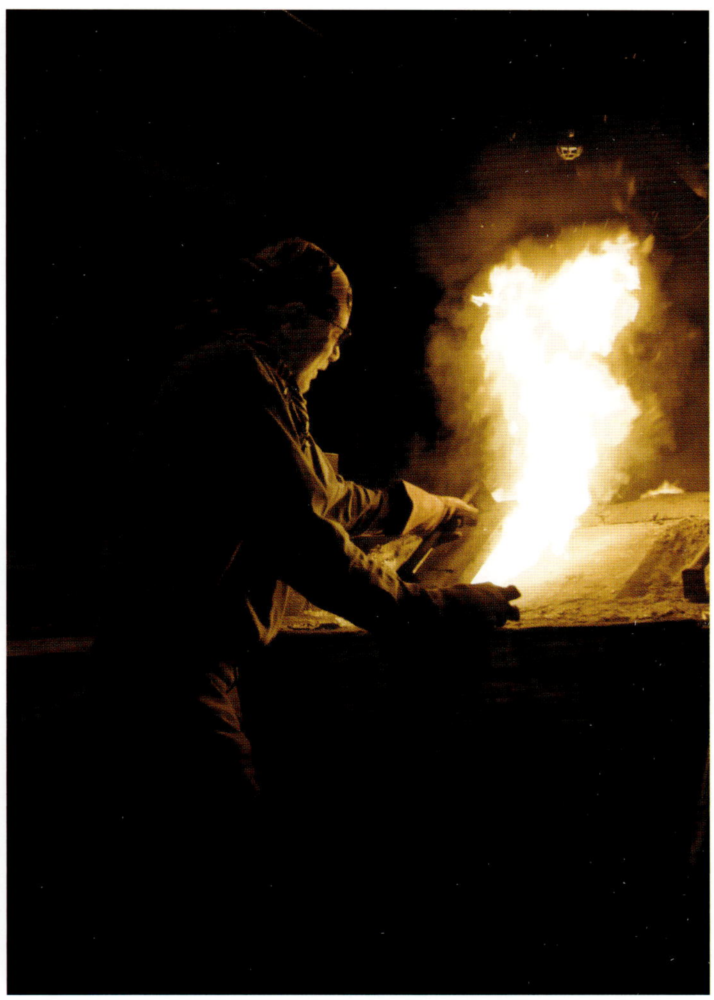

Richard Bresnahan, Richard Bresnahan stoking the *Tanegashima*-style third chamber of the Johanna kiln through the kiln ceiling. *Photo credit: Brian Zehowski.*

JOY BROWN

South Kent, Connecticut
joy@artwithin.net, www.artwithin.net

Artist's Statement

My work reflects the influence of the Japanese aesthetic resulting from my childhood in Japan and apprenticeship in traditional Japanese wood-fired ceramics. I now work in Kent, Connecticut, where I built my studio and thirty-foot long anagama, a Japanese style wood-firing tunnel kiln.

My life with clay began as a potter firing an anagama for unglazed pottery. I have always fired with wood and it is what I know and love best. I was drawn to wood-fired pots; the mystery, darkness, and depth were compelling. I love wood-firing! I love the process, the collaborative effort, and the results. I find it liberating and challenging to explore the relationship between the clay I have chosen, the kiln and the firing, the shapes that have evolved, and my own intention. The process of integrating these elements over many years is reflected in my work and empowers my life.

All my figures are hollow, made by pinching and paddling, building up the forms inch-by-inch. The clay I use is a red brick clay from Georgia. The results from the relatively low temperature weeklong firing compliment, and are an integral part of, the forms that I make. The firing process leaves just a dusting of ash and the soft earth tones of the clay itself.

Over the past forty years my work has evolved from vessels to animal shapes, and then to the human-like forms and abstract wall reliefs of recent years. These rounded figures convey the heavy gravity of stone; the expressions and gestures transcend that weight, suggesting a warm lightness of being.

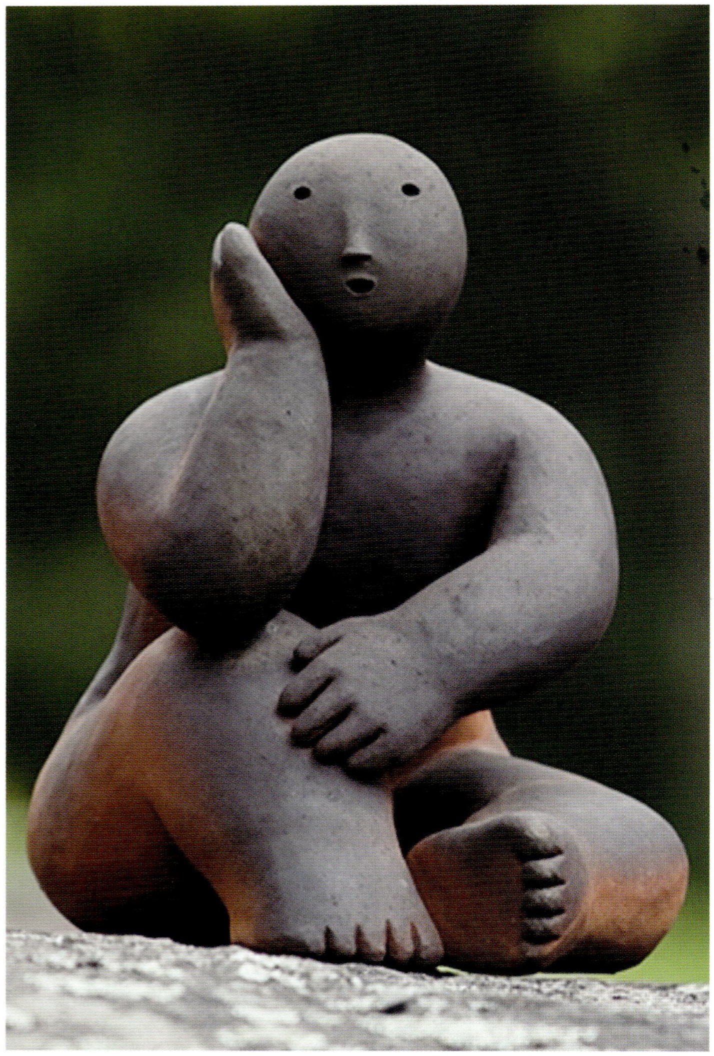

Joy Brown, *Sitter with Head in Hand*, 2008. Wood-fired stoneware, handbuilt, 14" x 11" x 13". Photo *credit: Richard Wanderman.*

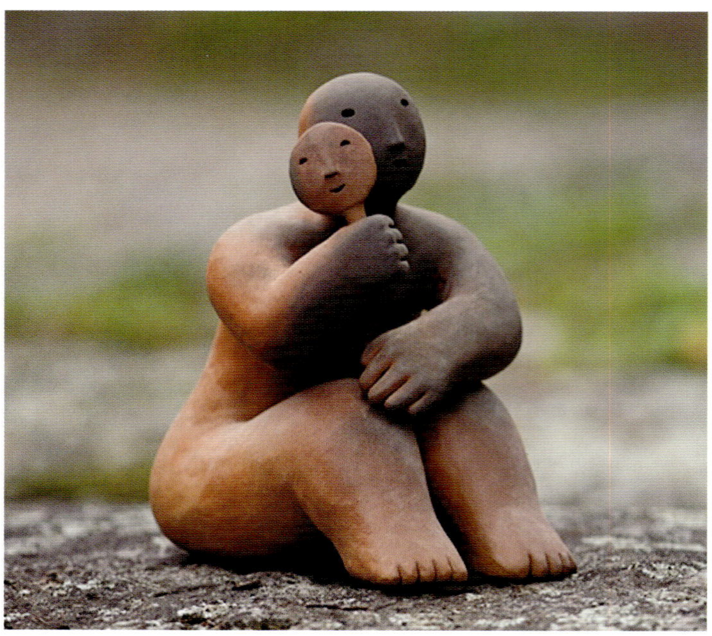

Joy Brown, *Sitter with mask*, 2011. Wood-fired stoneware, handbuilt, 15" x 9" x 10". *Photo credit: Richard Wanderman.*

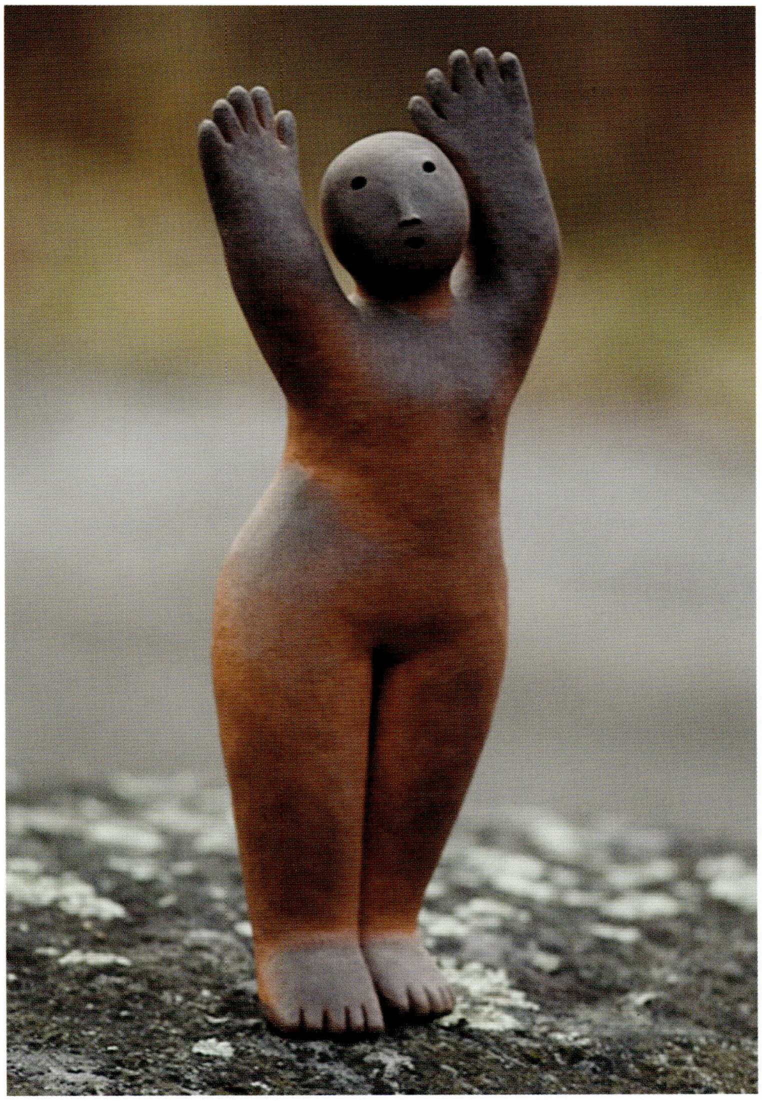

Joy Brown, *Standing figure*, 2010. Wood-fired stoneware, handbuilt, 12.5" x 15" x 16". Photo *credit: Richard Wanderman*.

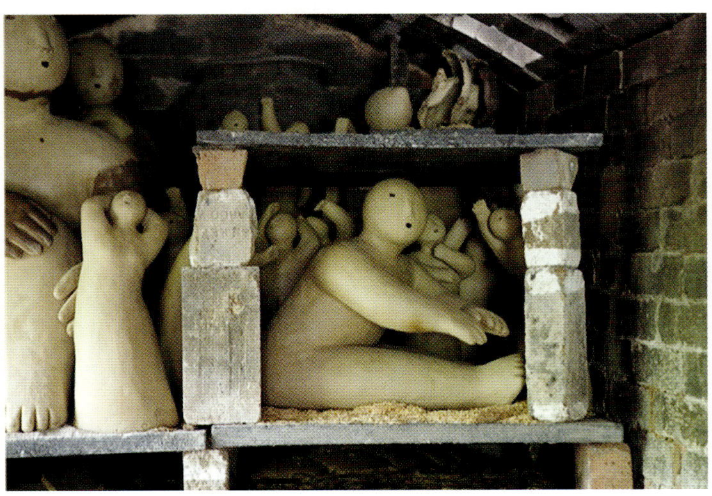

Joy Brown, Kiln loading. *Photo credit: Richard Wanderman*.

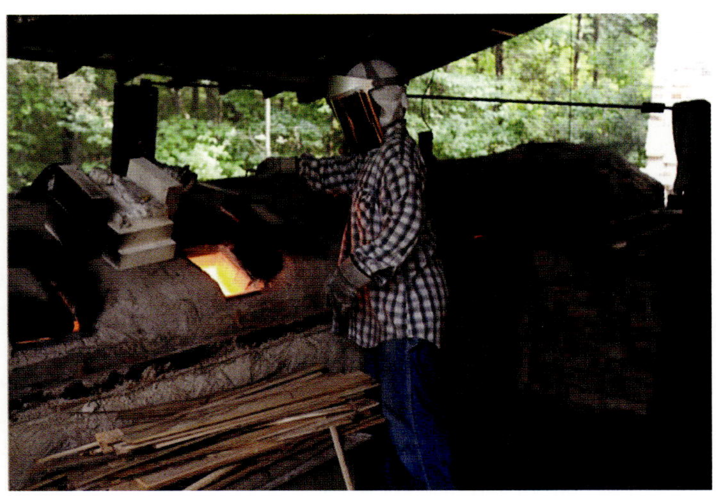

Joy Brown, Stoking a side port. *Photo credit: Richard Wanderman*.

JOE BRUHIN

Fox, Arkansas
joebruhin@gmail.com, www.joebruhin.com

Artist's Statement

My kiln is an anagama. It is half buried in the ground and measures forty feet long. It requires ten days of constant attending and ten cords of split yellow pine to achieve the results I seek and to complete one firing. The works from this kiln have a natural glaze. I do not apply glaze to the pots. The color and texture is caused by the climate, age, and type of wood being used, atmospheric conditions inside the kiln caused by my intuition and experience, placement of the works, wood ash from the burning fuel melting into the clay, fire flashing the work, and coals being maintained on the pots. The blending of these variables results in endless variety of effects. By being in harmony with the elements and forces of nature, the work becomes a sacrificial offering surrendered to the flames and reborn. It aspires towards the ideal of timeless beauty. Complacency is not an option. Each firing gives information to grow from and inspires me to dive deeper, which keeps the work fresh and enables me to express with heart and spirit. My goal and ideal is to make objects of beauty that have a transcending quality that can inspire another human being, bring a little joy, or add a positive presence to one's living environment. For me, working with clay and being an artist that wood-fires, is a vehicle for spiritual growth. One can say that my pots are offerings or prayers and my kiln the temple to manifest them.

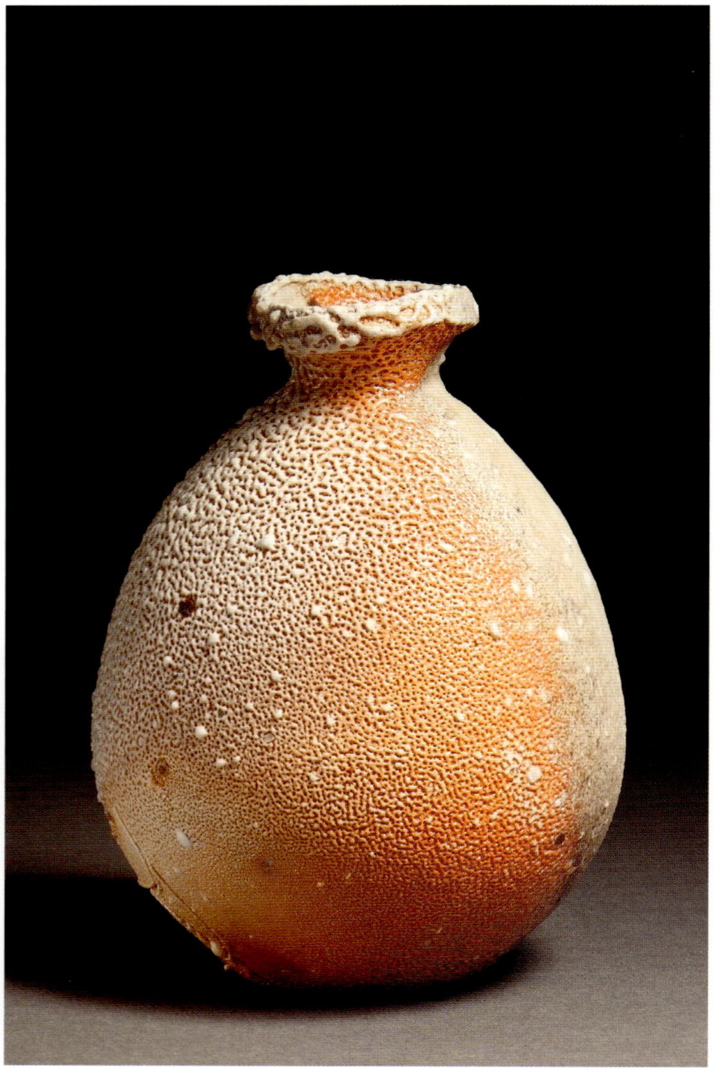

Joe Bruhin, Bottle, 2011. Anagama fired stoneware, wheel-thrown, 7" x 5" x 5". *Photo credit: George Chambers.*

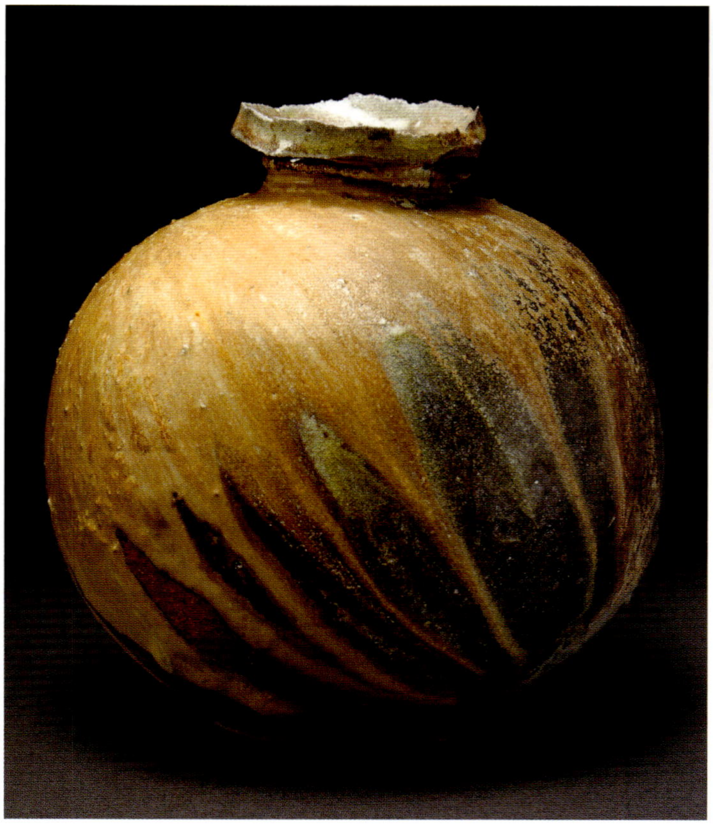

Joe Bruhin, *Iga Jar "Niyaz,"* 2009. Anagama fired stoneware, wheel-thrown, 11" x 11" x 11". *Photo credit: George Chambers.*

Joe Bruhin, *Iga Bottle "Hatuyuki,"* 2010. Anagama fired stoneware, wheel-thrown, 6" x 4" x 4". *Photo credit: George Chambers.*

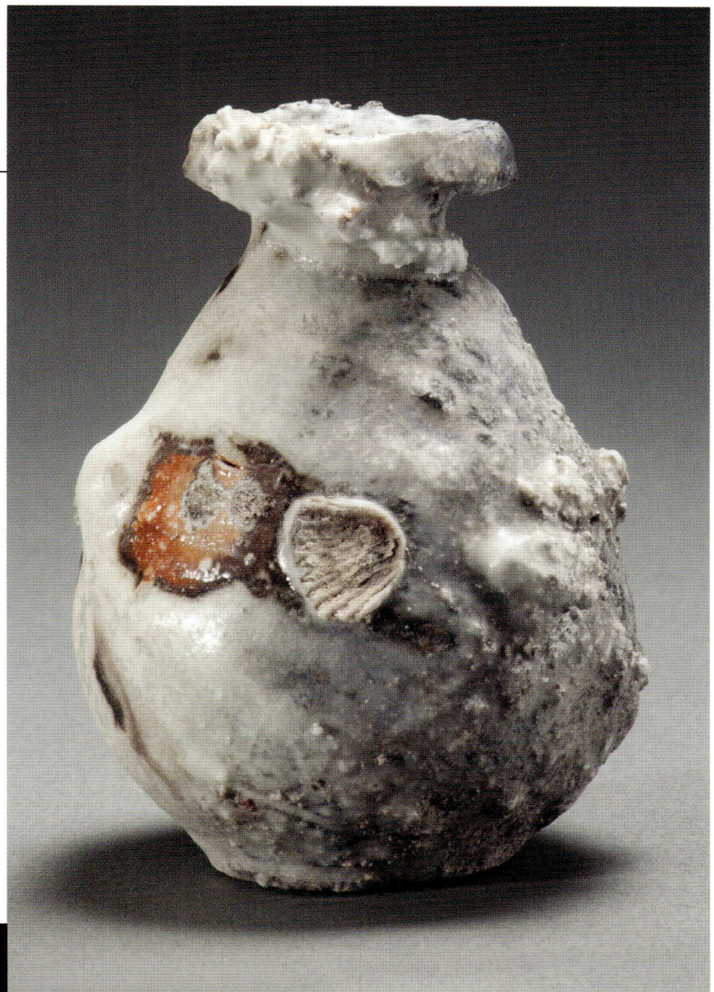

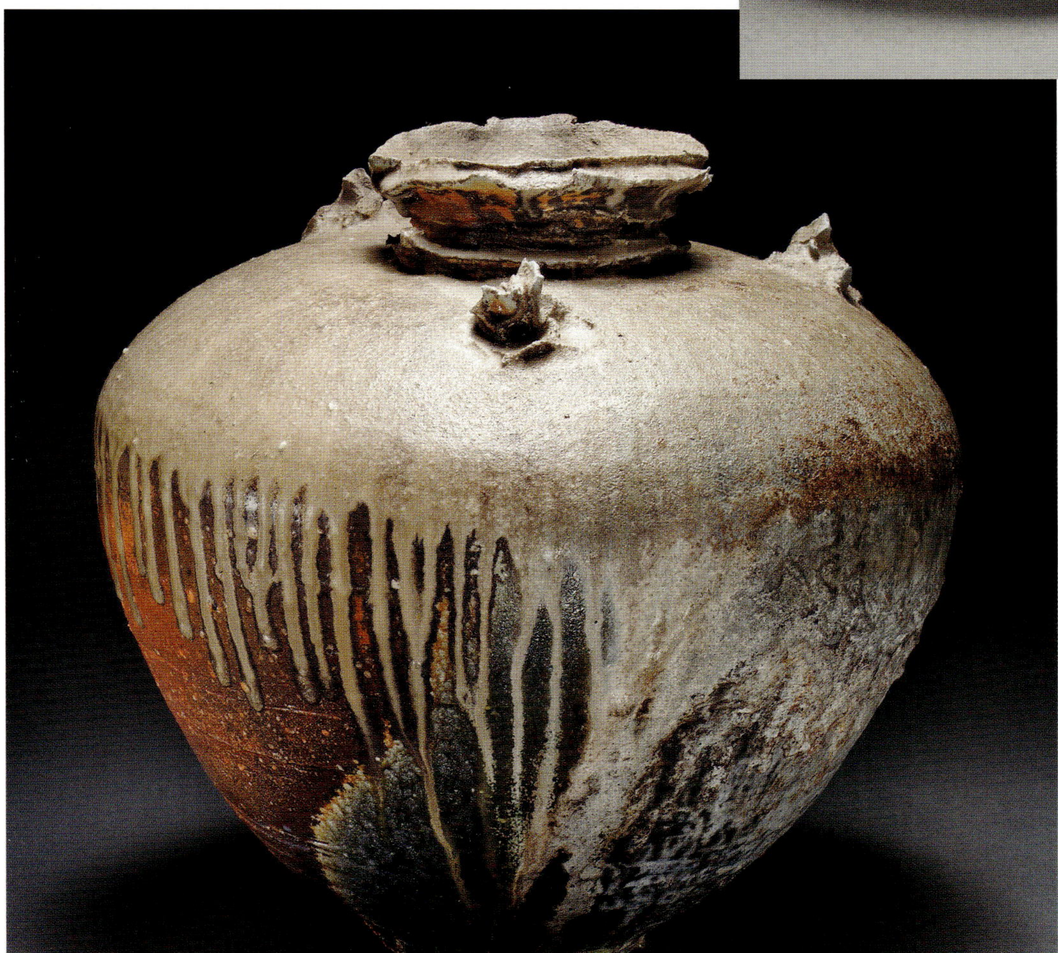

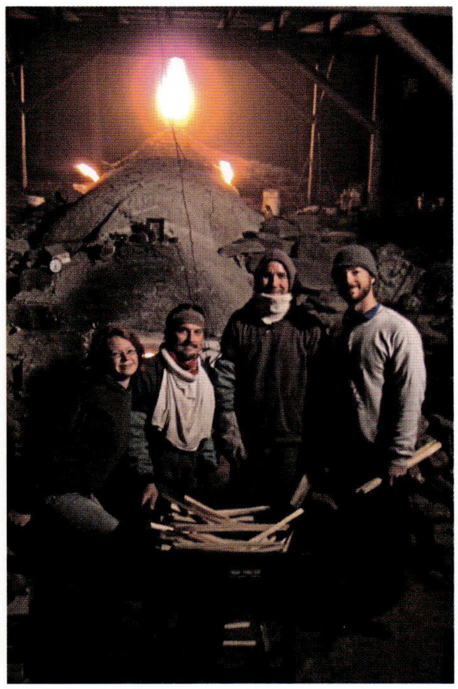

Joe Bruhin, The crew and myself near completion of firing. *Photo credit: George Chambers.*

Joe Bruhin, Iga Tsubo "*Heart Cave,*" 17" x 17" x 17", 2008. Anagama fired for ten days. *Photo credit: George Chambers.*

JIM BUSBY

Iowa City, Iowa
Design1performance@gmail.com,
www.jimbusbyceramics.blogspot.com

Artist's Statement

My interest in clay is focused on form and surface. The clay responds to my hands and form is found predominantly on the wheel. The surface is from fire and melting ash.

I am intrigued by the elusive simplicity within the tactile universe of clay.

I strive, on a daily basis, to remain open-minded, imaginative, and playful in my approach. With serious intent, I am compelled to make work that is honest to this medium of art, that in some way expresses what it is to work with clay.

I am obliged to always question and explore and am prone to frequent "failures." Failures are certainly marvelous teachers; they offer us direct insight into the unexpected. They are the ideas we didn't have, but should have been looking for. After all, art is about motivation—a time of exhaustion and renewed inspiration.

The spirit that moves us should be evident in each piece.

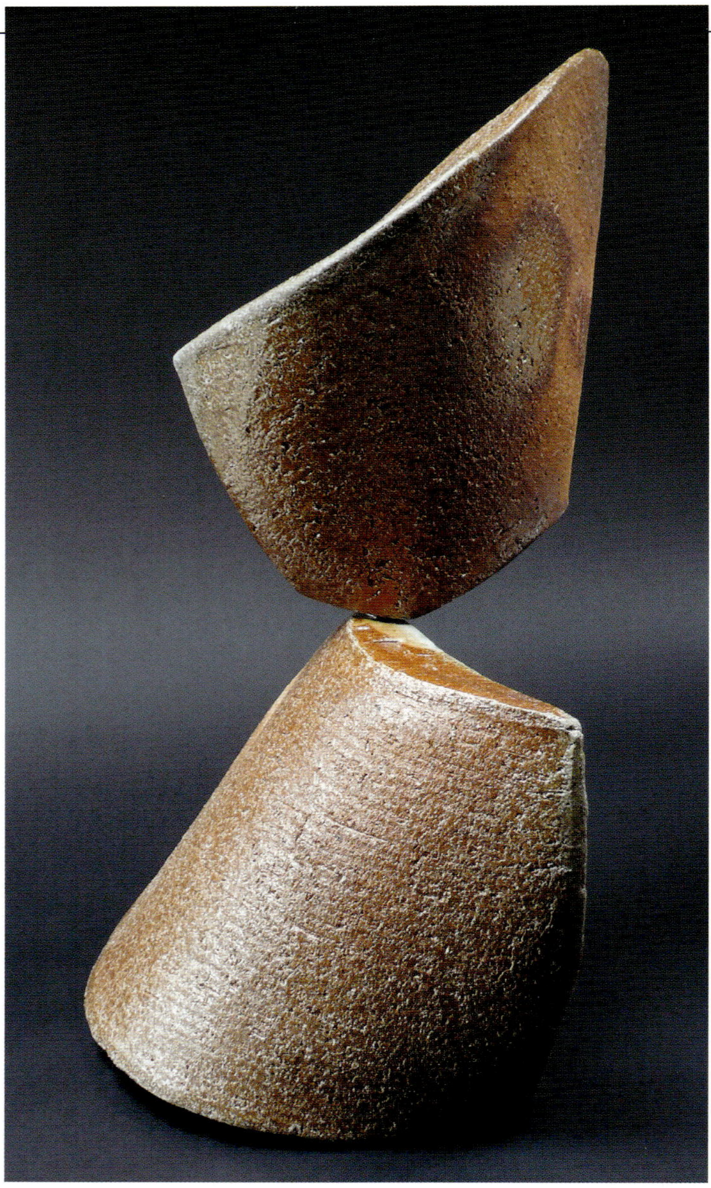

Jim Busby, *Sail Stack*, 2011. Wood-fired stoneware, slab built, 38" x 27" x 20". *Photo Credit: Jim Busby.*

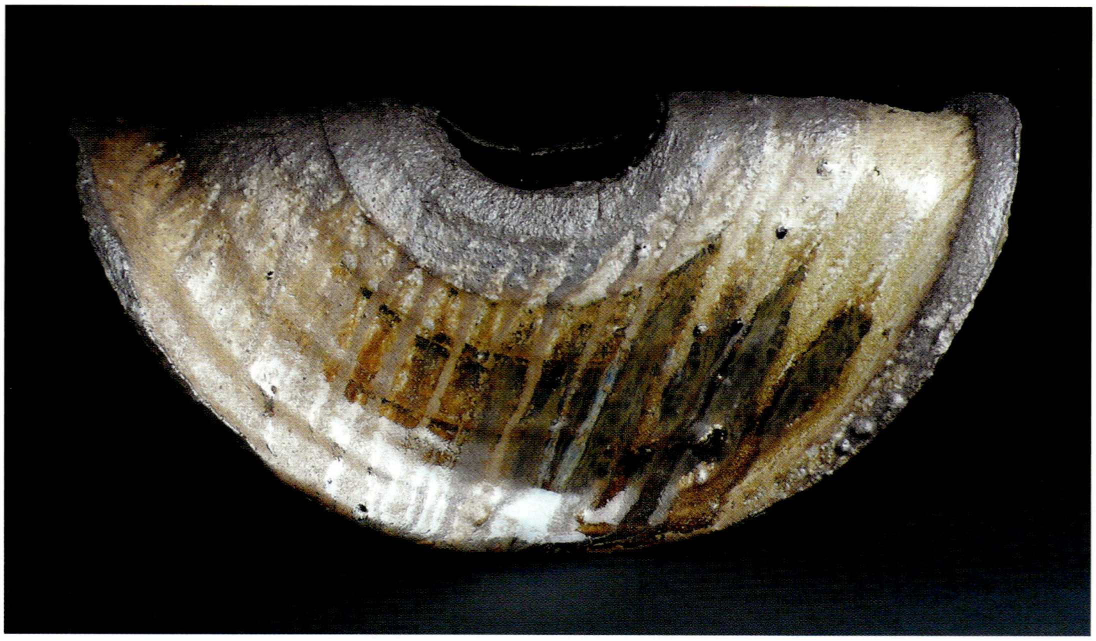

Jim Busby, *Open Vase*, 2010. Wood-fired stoneware, wheel-thrown, 6" x 5" x 12". *Photo Credit: Jim Busby.*

My first wood-fire experiences were with Greg Kennedy at Idyllwild Arts Foundation in the mountains of southern California. I often reminisce about the contentment of sitting out under the "simple" stars on a clear night while firing up a kiln. Fire crackling, pine trees murmuring in the dark, salivating over the "something to munch on" that my partner Nancy made—it was nothing short of spectacular.

We had a small noborigama kiln. I was a beginning ceramic student and was allowed to participate in a firing by some of the advanced students. I feverishly split wood for the whole weekend, feeling honored to place my one little perfect bowl in the kiln. They ended up placing it over to the side of the firebox. The next weekend we unloaded the kiln—what came out was a bowl ripped apart by fire, crusted with ash and nail half melted into it. Not quite my idea of functional.

I was "hooked" ... and still am.

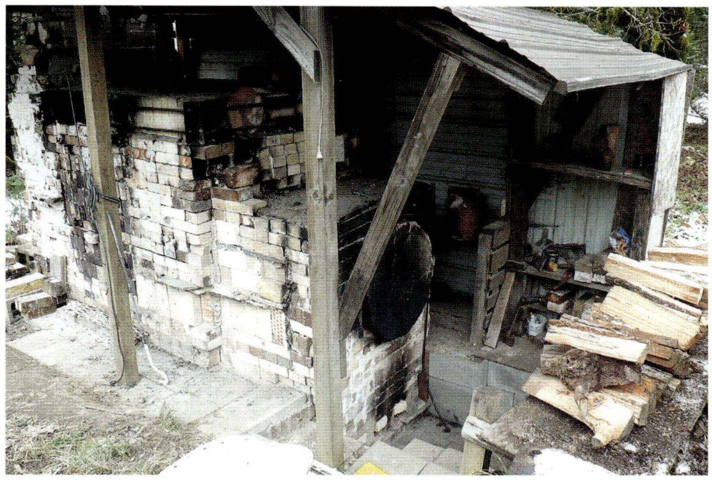

Jim Busby, Caboose Kiln, Firebox, 2008. *Photo Credit: Jim Busby.*

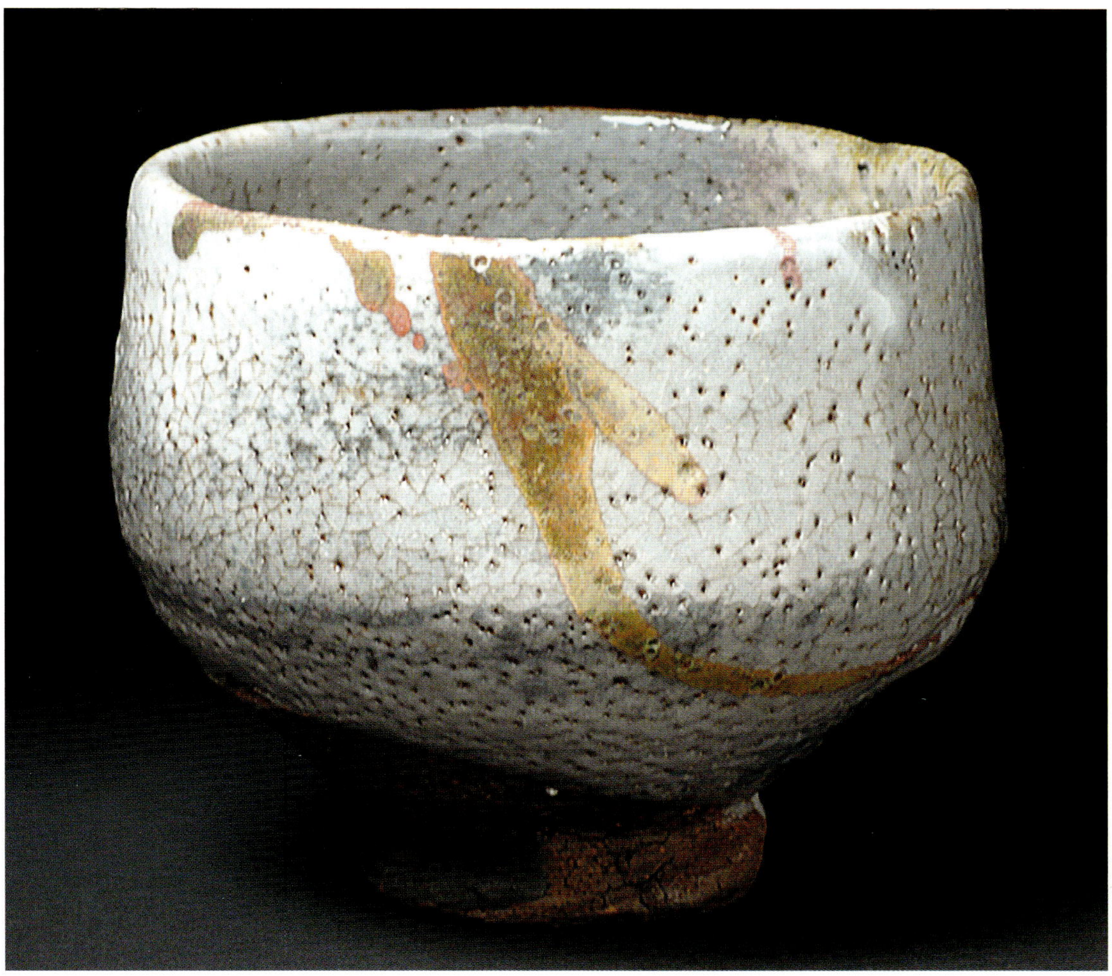

Jim Busby, *Tea Bowl*, 2010. Wood-fired stoneware, wheel-thrown, 4" x 4" x 4". *Photo Credit: Jim Busby.*

CONRAD CALIMPONG

Ferndale, California
ccalimpong@thekilnshed.net, www.thekilnshed.net

Artist's Statement

In 1985, I helped Fred Olsen build a wood-fired cross draft kiln at his home in Pinyon Crest, California. It was a turning point for me. I discovered an immediate and deep affinity to wood-firing. The way the flame moved about the kiln fascinated me, depositing ash and leaving a trail of color as it wove its way through the chamber toward the chimney. Each kiln firing was a learning experience, and the learning continued long after the kiln had cooled. The subtleties of each firing were slowly revealed by the pieces over time. The changes in color and texture fueled a desire to handle pieces countless times. My approach to making pots and why I am drawn to wood-fire can be summarized by the thoughts of an ancient, unknown Chinese scholar when he wrote about the revered art of calligraphy, "The essence of beauty in writing is not found in the written word but lies in response to unlimited change; line after line should have a way of giving life; character after character should seek life-movement." Wood-firing compliments my quest toward this ideal with the pots I make, giving life to the inanimate and static. The results aren't always good, but there is always something to learn. Twenty-six years later, I am still enthralled.

When it came time for me to build my own wood-fired kiln, I designed it based on my experience with Fred Olsen's "Hollyweird Tube." My kiln was synthesized from his design. Instead of two chambers and an interior length of twenty-one feet, my kiln is a single flame-shaped nine-and-a-half-foot long chamber. I felt it would be a more manageable size to fire alone, though this has yet to happen.

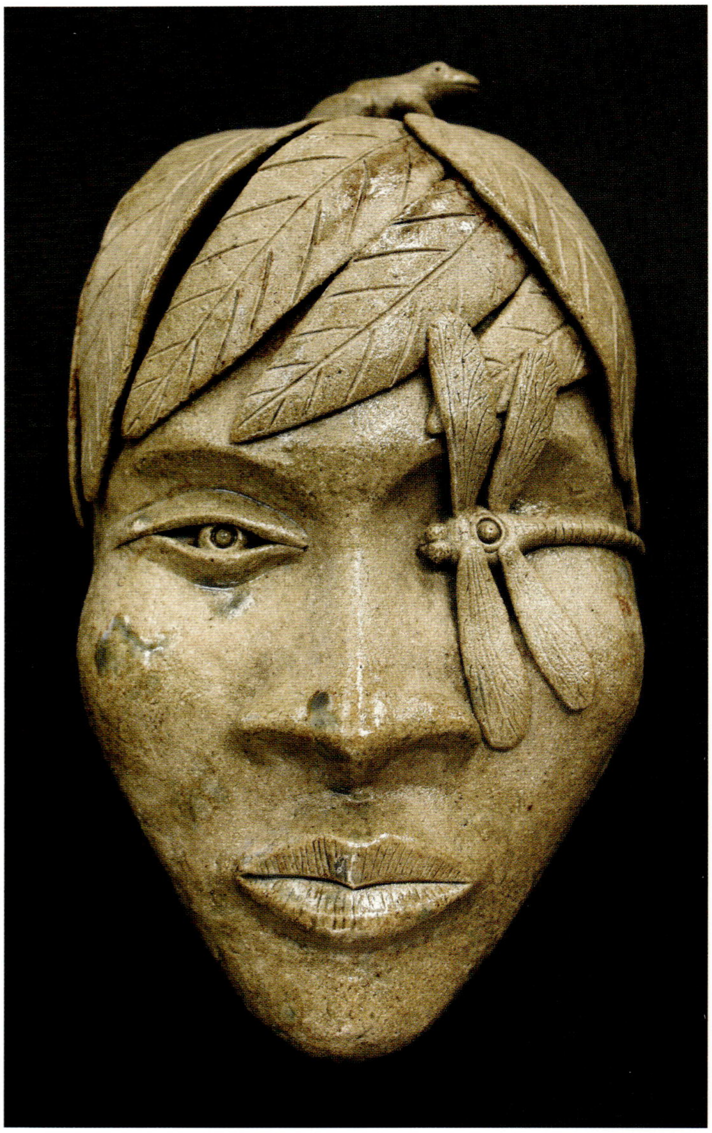

Conrad Calimpong, *Siem Reap*, 2008. Wood-fired stoneware, handbuilt, 12" x 7.5". *Photo credit: Conrad Calimpong.*

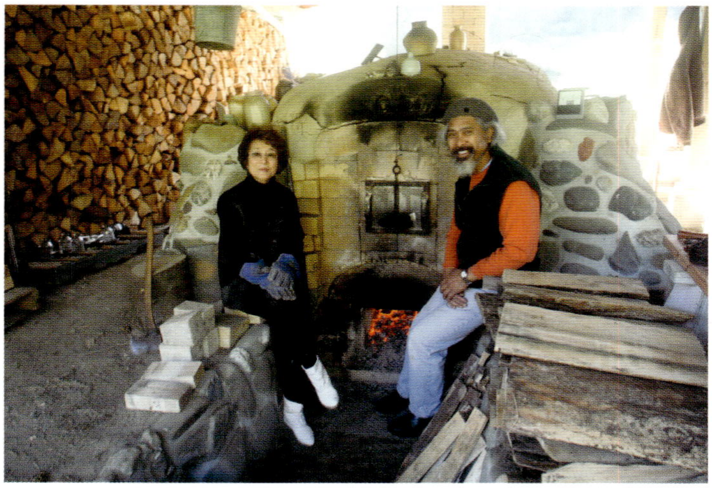

Conrad Calimpong in front of the kiln with Minako Lee, 2006.
Photo credit: Donna Allen.

The work I fire in this kiln is a mix of functional pottery and sculpture. As a young ceramics student, I was infatuated with a photograph of a Sung Dynasty Tz'u Chou vase from the Victoria and Albert Museum shown prominently in Glen C. Nelson's *Ceramics: A Potter's Handbook*. Its strong shape and spontaneous decoration touched me in a way that was very different from other pots I had seen. That pot became my benchmark. In 1985, I started an on-going series of work called "In Classical Shape," using the mei-ping vase shape as a point of departure. When I first came to ceramics, I was more interested in reproducing the shape on the potter's wheel. Years later I've become more interested in the form as sculpture, making it out of clay, wood, found objects, beaver sticks, and other items that have replaced ceramic containers such as cans and plastic bottles. We as potters and ceramic artists have had to come to grips with living in a society where the articles that we make are no longer an integral and necessary part of the larger society's daily lives as they had been for thousands of years.

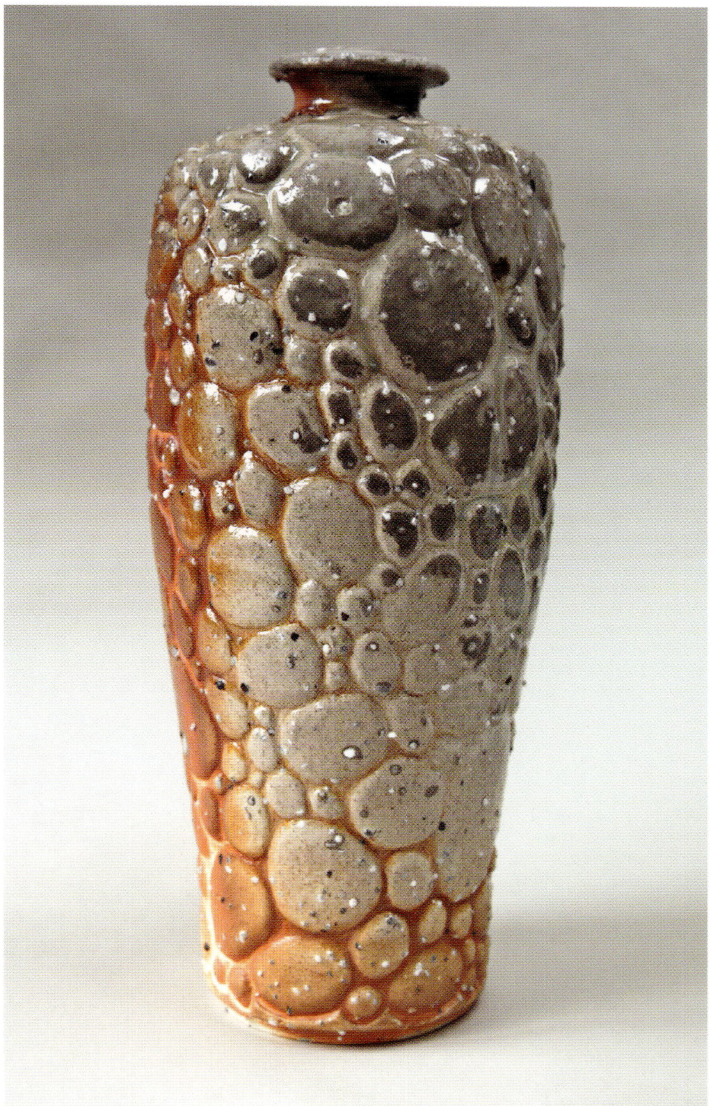

Conrad Calimpong, *In Classical Shape series, River Rock mei-ping vase*, 2011. Wood-fired stoneware, wheel-thrown and hand carved, 16" x 6.75". *Photo credit: Conrad Calimpong.*

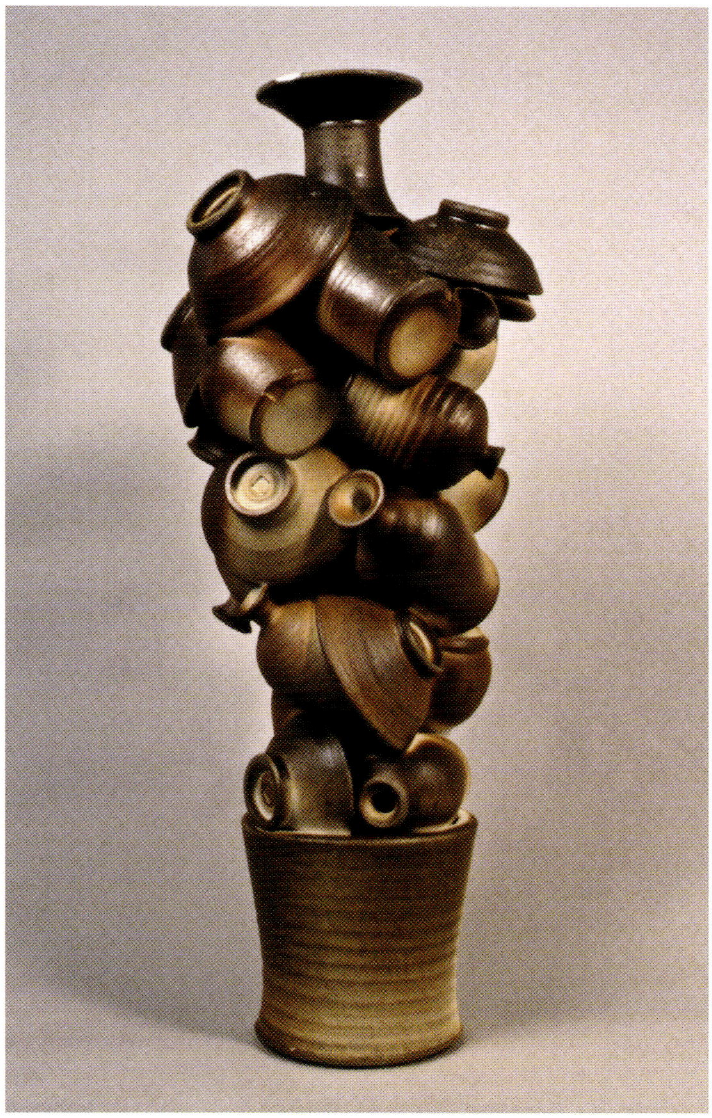

Conrad Calimpong, *In Classical Shape series, Air in Motion*, 1989. Wood-fired stoneware, wheel-thrown, handbuilt, 27" x 9.75" x 10.5". *Photo credit: Conrad Calimpong.*

PETER CALLAS

Belvidere, New Jersey
peter@petercallas.com, www.petercallas.com

Artist's Statement

The process of wood-firing ceramics for over four decades has been the creative touchstone that changed the course of my life.

To fit the forms I choose to make, I fire an anagama kiln. This is a perennially ambitious eight day event using a cord a day, focusing constantly, aspiring to master this mysterious space, create my own universe, and achieve the desired, vibrant subterranean palette.

Elusive by nature, the anagama and its subtle, yet broad color palette forged the collaboration between intuition for the material and the unequaled power of these timeless kilns, allowing me to fire the kind of ceramic work I imagine.

In essence, I am addressing the typical painterly concerns of an abstract expressionist obsessed in evoking characteristic reflections of emotional intensity, unlocking tactile inflections and a unique vision.

I enjoy making spirited creations that emphasize the range and aesthetic sensibilities of forms that fit wood-fire. The work, molded by slab, coil, hand, and/or wheel, is spontaneously interpreted, then executed with a deliberate sense of purpose.

Forms full of inflection, rich in the poetics of clay, using innovative surfaces to express attitude, defiance, power, and the mysterious provocative sensibilities of nature.

This journey, this very indirect route of connecting a volatile mixture of earthiness and refinement, is my hope and mission, creating a fresh look in clay and equalling or bettering past achievements.

A tireless creative challenge that adds up to framing the unpredictable, producing rare objects that do not hit you all at once, but rather stimulate a special emotional narrative.

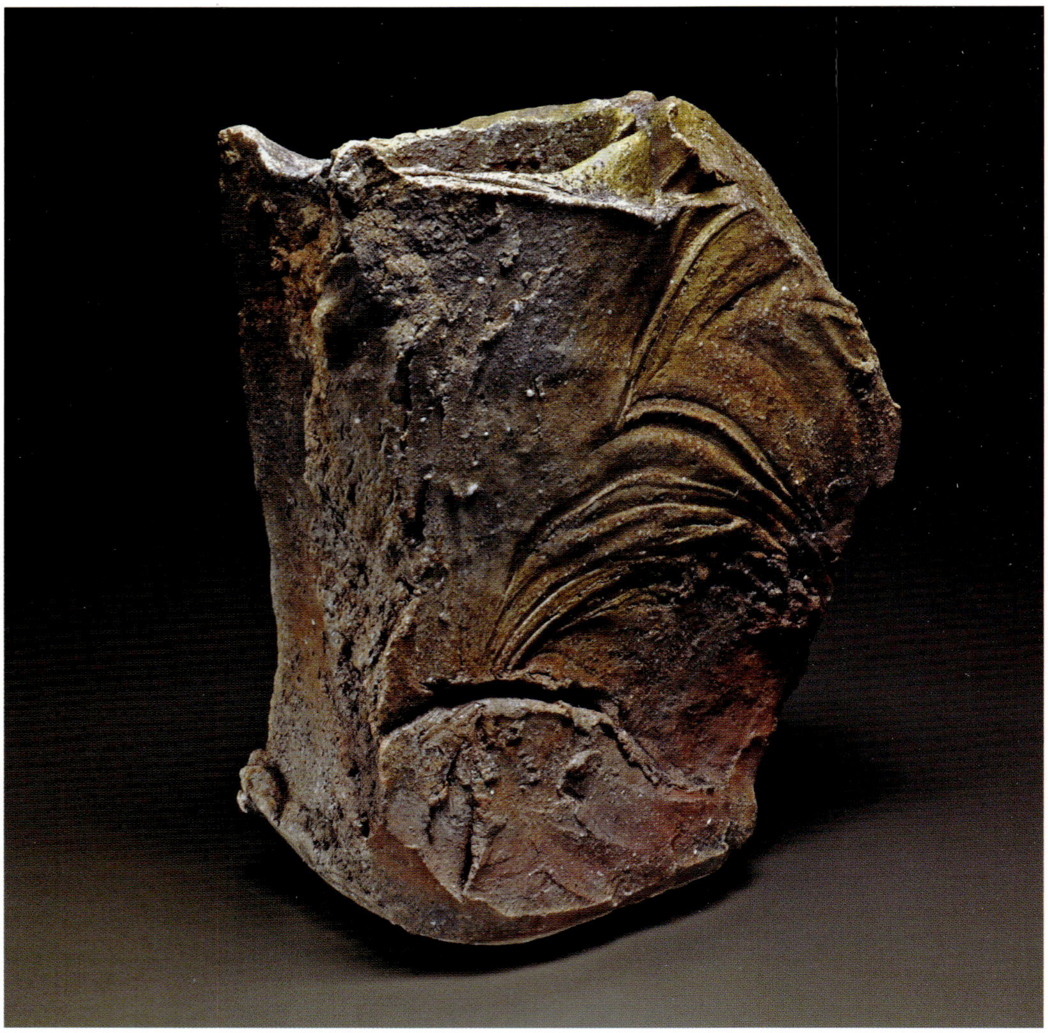

Peter Callas, *Hotpocket*, 2010. Handbuilt stoneware, 15" x 7" x 16", (collection of artist). *Photo credit: Peter Callas.*

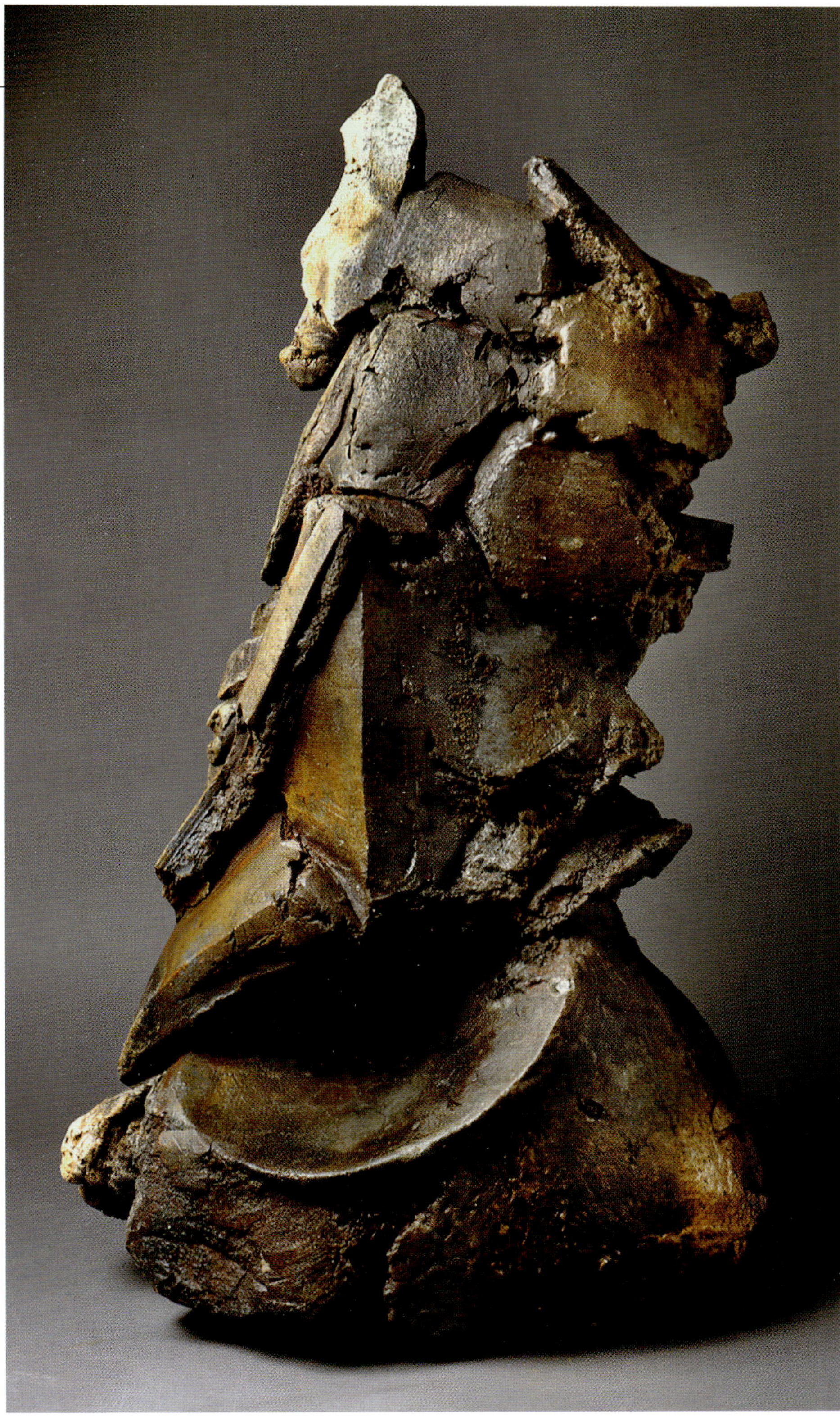

Peter Callas, *Chieftan*, 1995. Handbuilt stoneware, 24" x 18" x 40". *Collection of Daum Museum of Fine Art.*

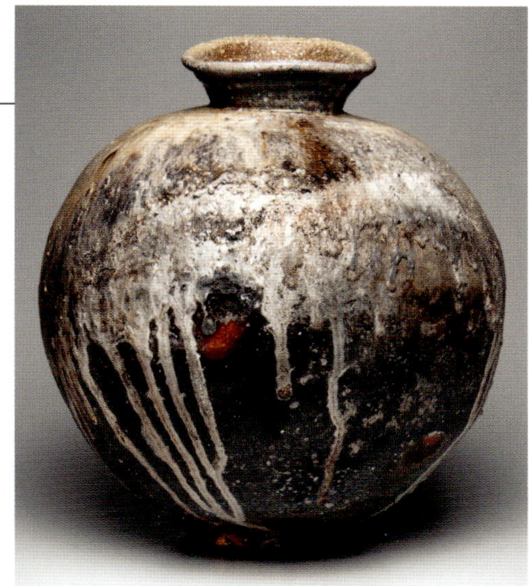

Peter Callas, *Coil Jar*, 2011. Handbuilt stoneware, 19" x 19" x 19" h, (collection of artist). *Photo credit: Peter Callas.*

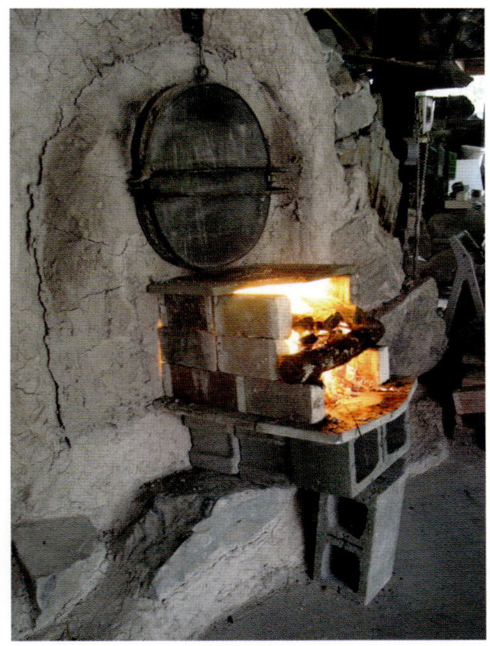

Front firebox. *Photo credit: Peter Callas.*

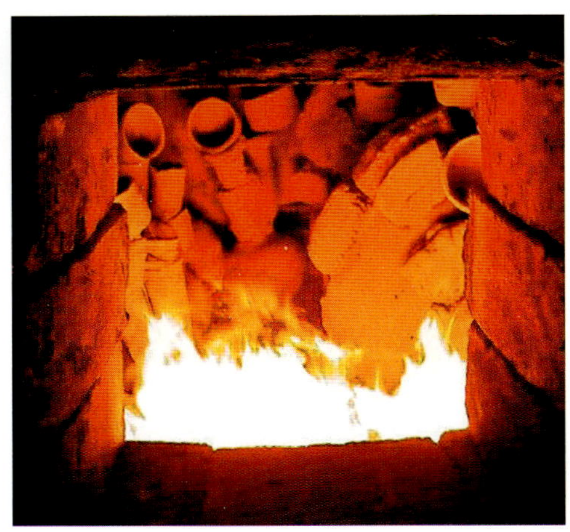

View inside front firebox.
Photo credit: Peter Callas.

LINDA CHRISTIANSON

Lindstrom, Minnesota
linda@christiansonpottery.com, www.christiansonpottery.com

Artist's Statement

The qualities that I search for in my work are fairly straightforward. I am interested in making a pot that does its duty well, yet can stand on its own as a visual object. These pots are not sculpture, and they are not art. They seem to act more like engaging tools than anything else. Wood-firing offers the exterior of the work a surface I find quietly compelling. My current multi-chambered kiln has a Bourry box on the first chamber, an internal firebox on the second, and a collection box as a third chamber. While the firing process is anecdotal to the pot's life, I do enjoy making a woodpile and tending a fire.

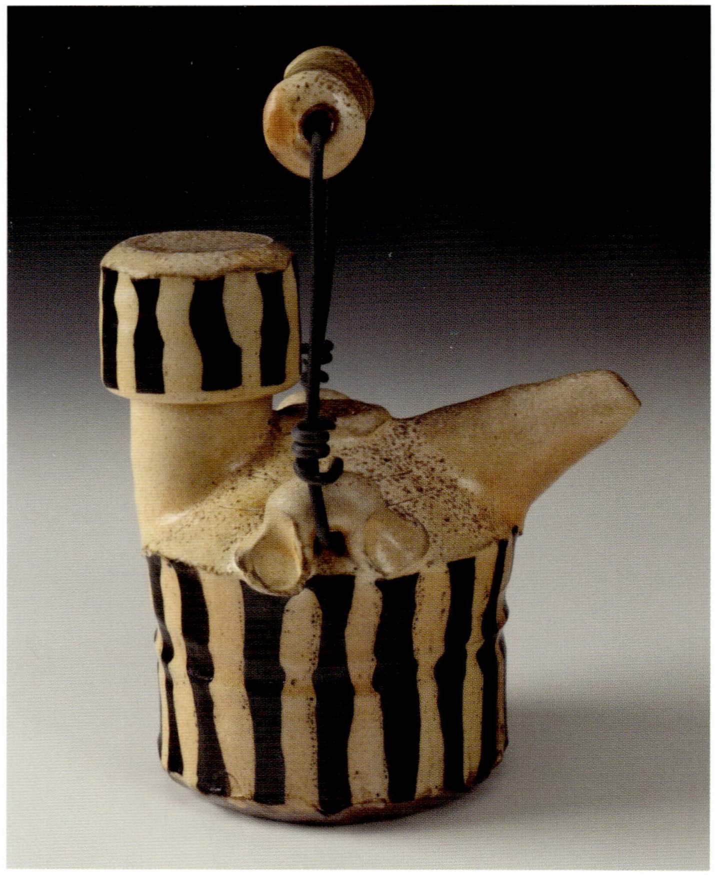

Linda Christianson, *Cooking Oil Can*, 2011. Wood-fired stoneware, wheel-thrown and altered, wire, 11" x 7.5" x 5". *Photo credit: Linda Christianson.*

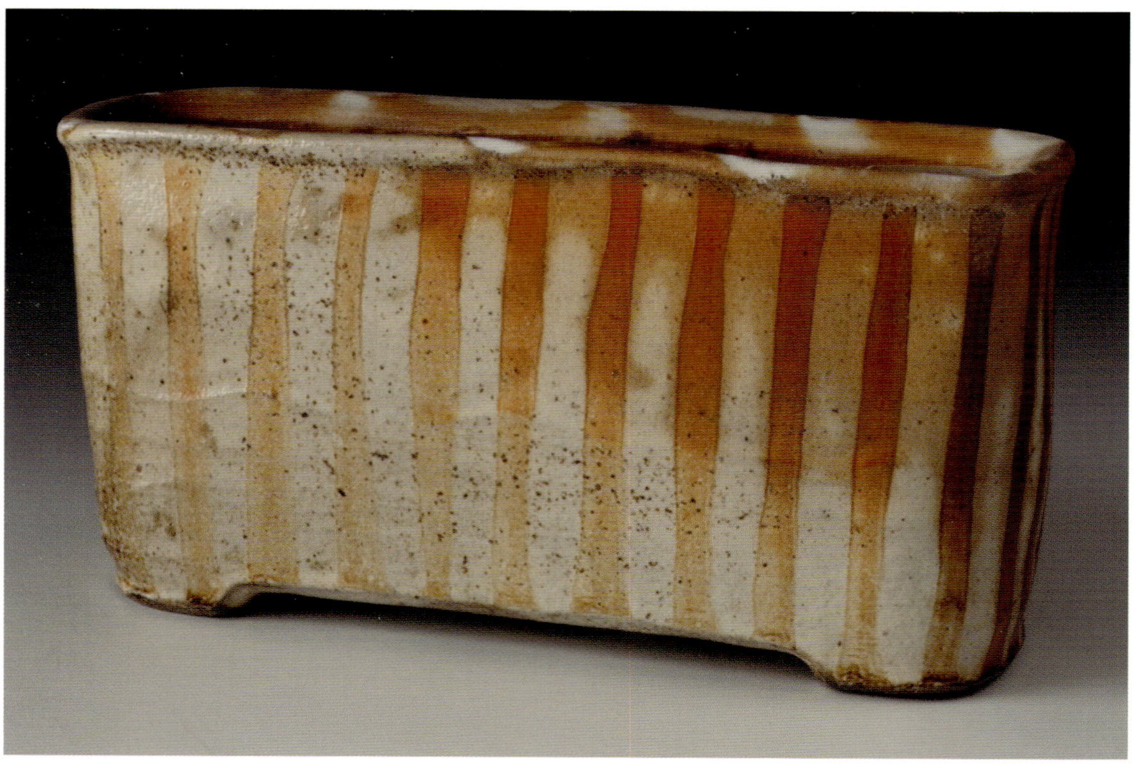

Linda Christianson, *Red Striped Vase*, 2011. Wood-fired stoneware, wheel-thrown and altered, 12" x 7" x 2". *Photo credit: Linda Christianson.*

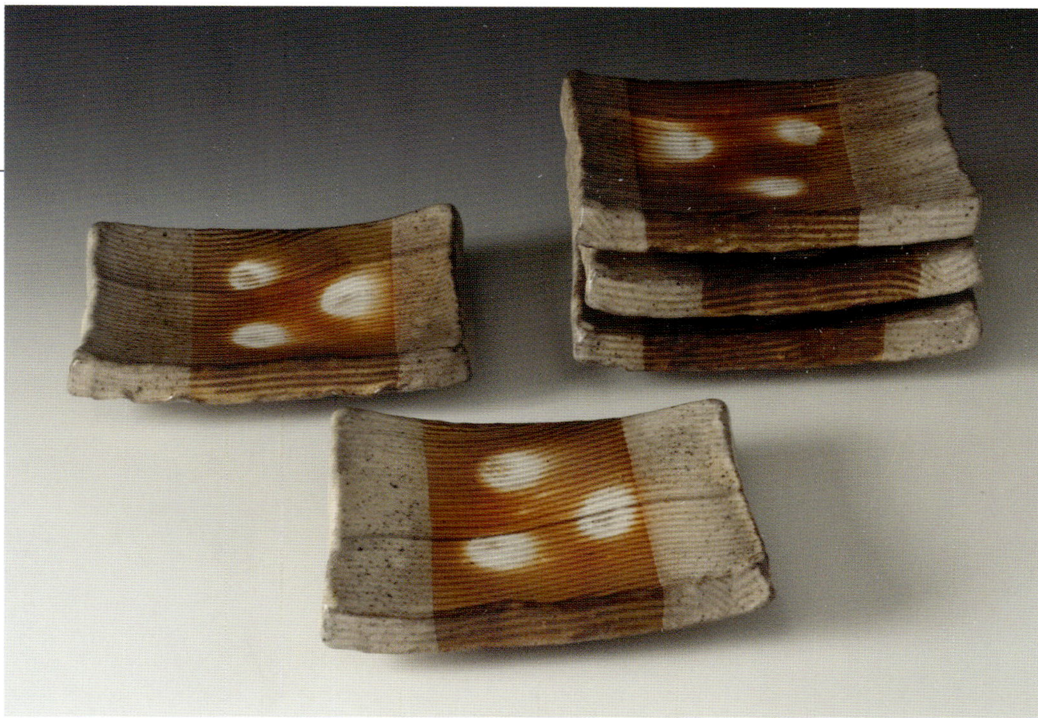

Linda Christianson, *5 Dessert Plates*, 2011. Wood-fired stoneware, slab, 7" x 4.5" x 1" each. *Photo credit: Linda Christianson*.

Most of the pots are made on a Leach treadle wheel, then altered. I dip the pots in one or two slips while still soft, and sometimes use a glaze as a liner or decoration.

I spend a good week loading the kiln, stacking the pots rim to rim and placing them in relationship to the future flame's path. I think of the flame as a flowing river and try to take advantage of its many variables. The firing itself is usually around twenty hours, with a slow heat up and quick finish. A small bit of salt is thrown into each chamber near the end.

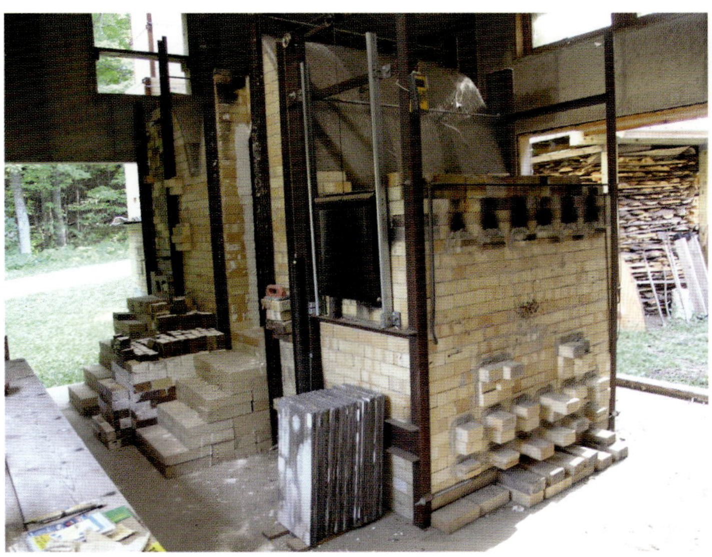

Linda Christianson, Kiln, 2011. *Photo credit: Linda Christianson*.

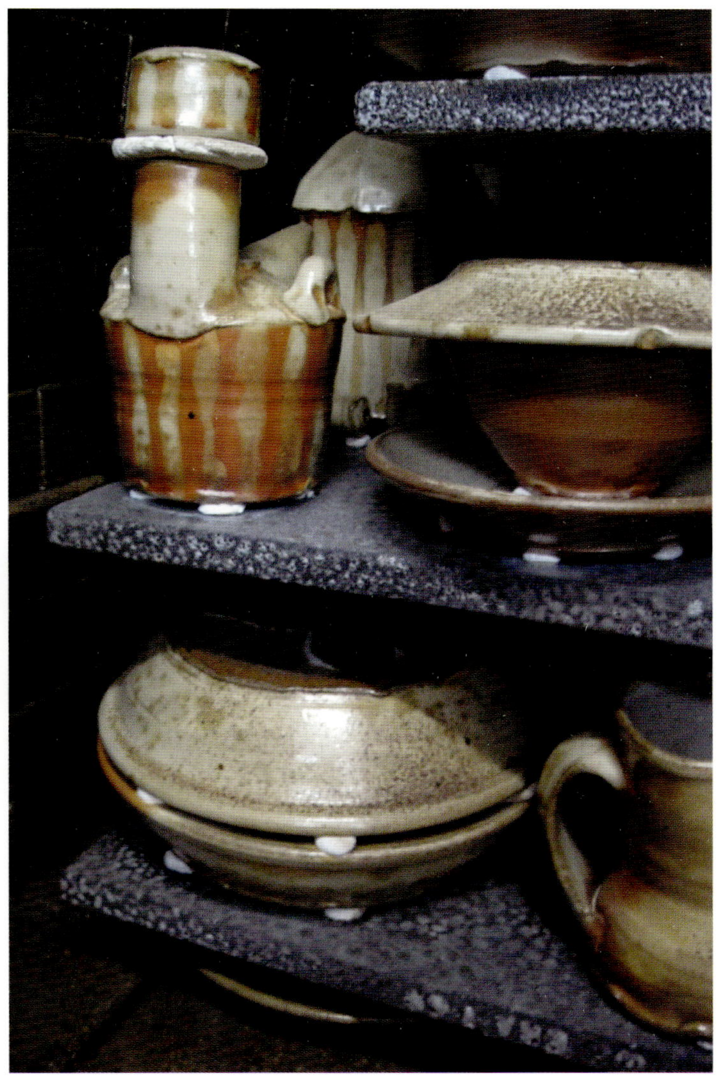

Linda Christianson, Kiln stacking, 2011. *Photo credit: Linda Christianson*.

BEDE CLARKE

Columbia, Missouri
clarkerb@missouri.edu, www.bedeclarkestudio.com

Artist's Statement

My goal in making this work has been to create quiet, simple pots, which like a good meal leave a healthy, full feeling. Pots like these have been my constant companions for forty years. They are forms and surfaces that I never grow tired of. They are rooted in ceramic history; yet, I hope they reflect a personal and individual quality of feeling. Above all, these pots are intended to be good to live with and interact with on a daily basis. They are made with others in mind.

This work has been fired in a catenary wood-fired kiln at my home studio and an anagama at the University of Missouri in Columbia. All of the pieces are wheel-thrown from stoneware clay. Slips and glazes have been applied and then altered through the action of wood-firing.

This interaction of the wood, flame, and ash works to create a rich and complex surface on the pots and so the wood-firing, in no small way, finishes these pots.

For me, good pots spring from compassion. My "technique" amounts to wishing the work well as it moves through the process of forming, articulation of the surface, and firing. I try to find ways of working that respond to the only ability I have ever had, wishing the work well and silently encouraging it to "be good, be good." I keep returning to the studio, simply trying to bring as much sincerity as I can muster to bear upon the work.

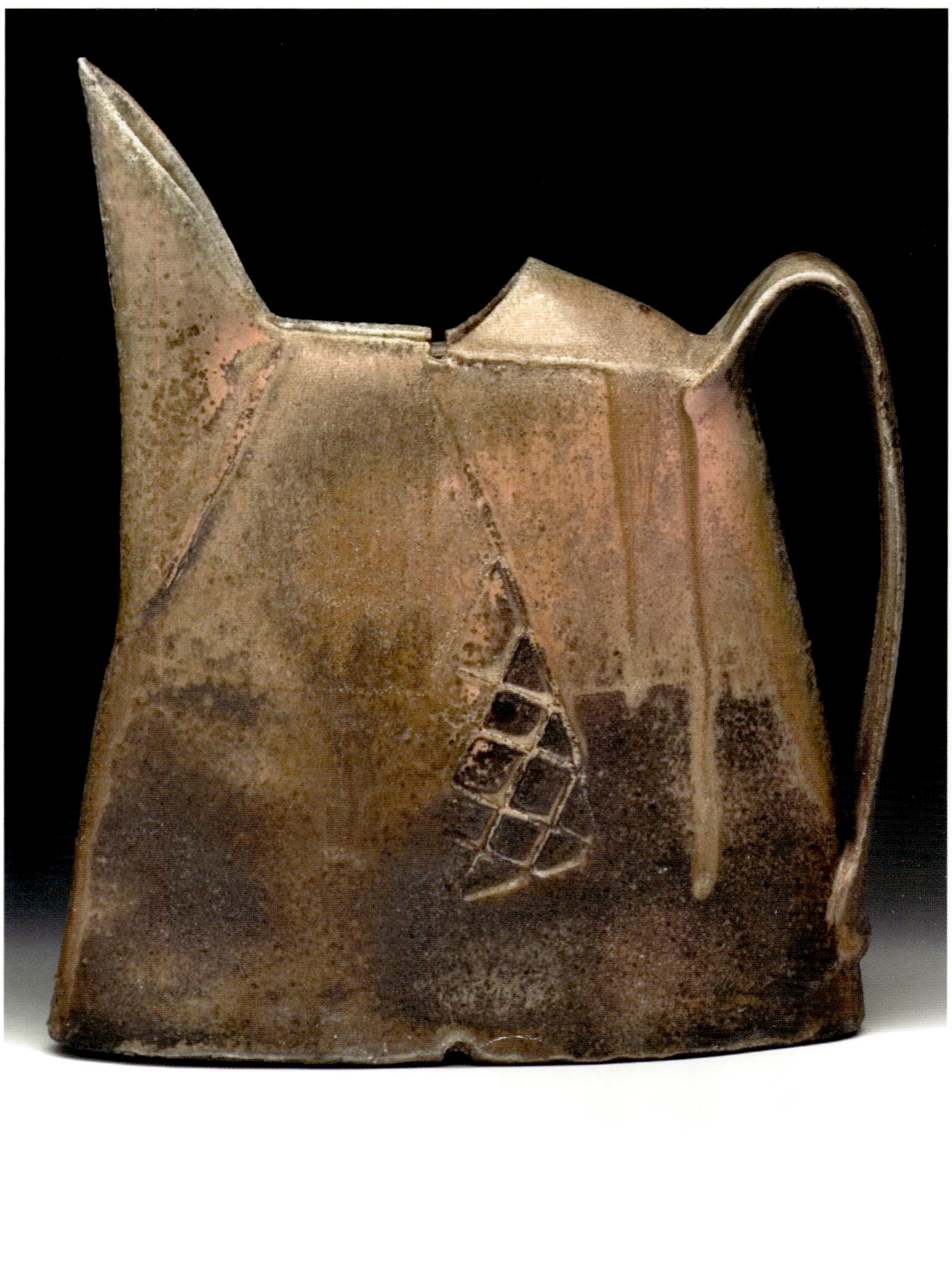

Bede Clarke, *Pitcher*, 2009. Wood-fired stoneware, wheel-thrown and altered, 13" x 11" x 4". *Photo credit: Bede Clarke.*

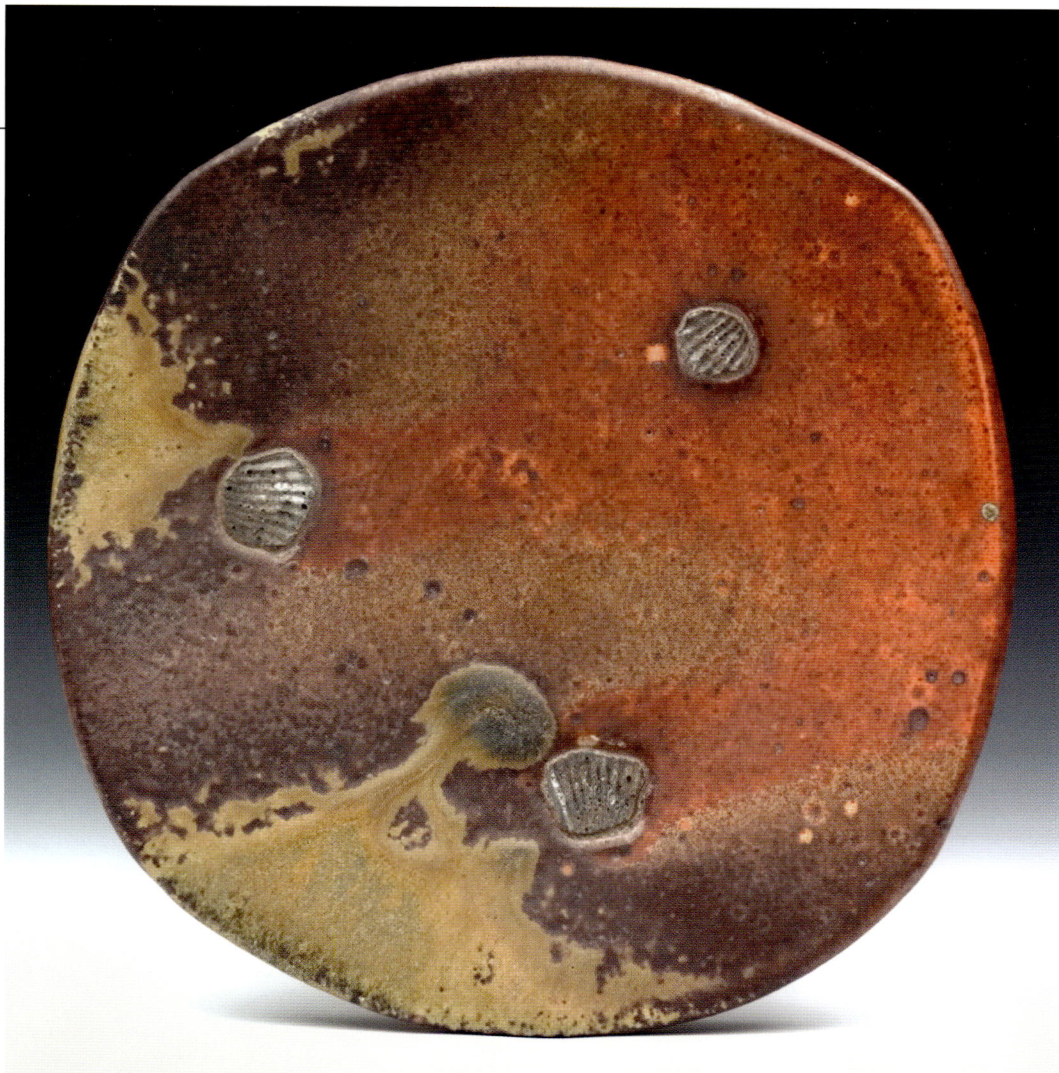

Bede Clarke, *Plate*, 2011. Wood-fired stoneware, wheel-thrown and altered, 10" x 10" x 1". *Photo credit: Bede Clarke.*

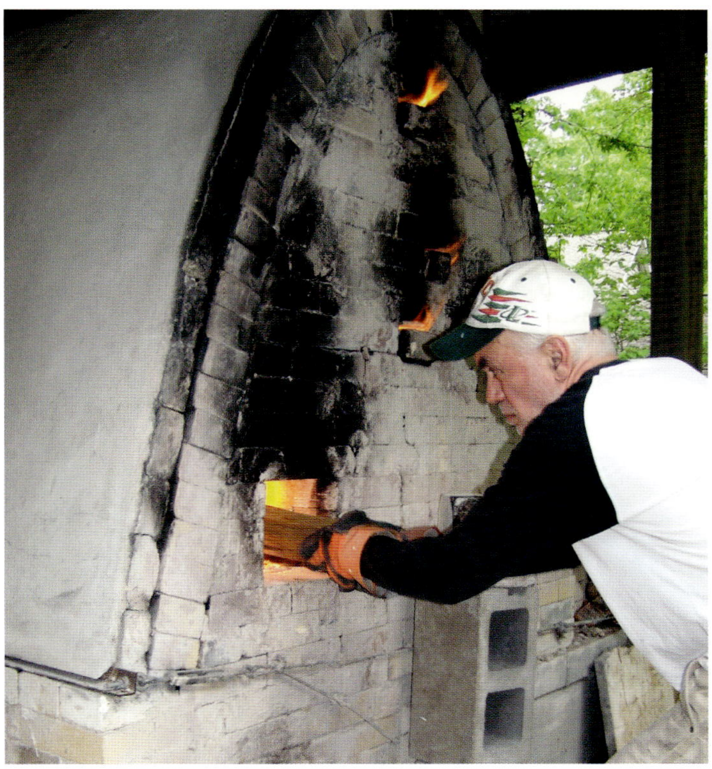

Bede Clarke, Firing Catenary Arch Kiln, 2000. 21 cu. ft. firing area 7' x 7' x 5'. *Photo Credit: Eric Carlson.*

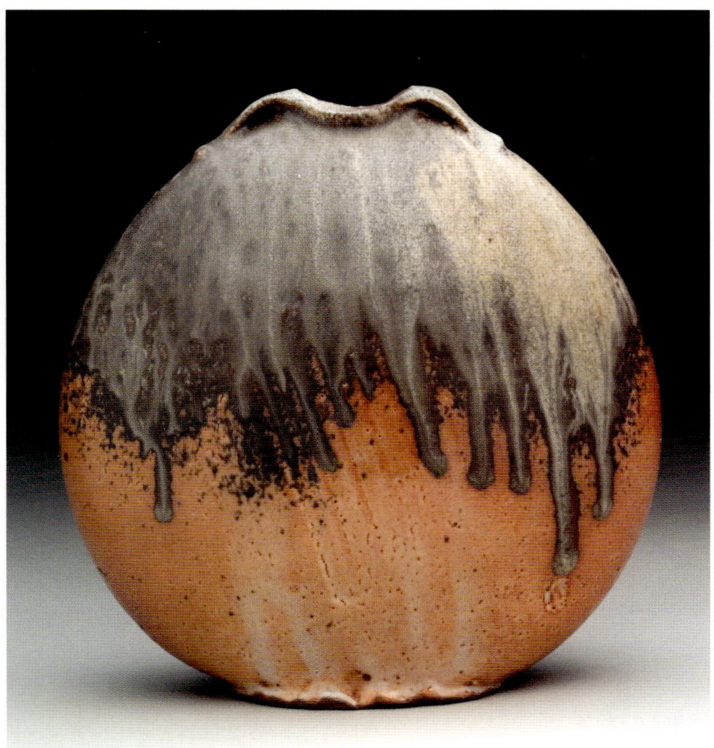

Bede Clarke, *Bottle*, 2011. Wood-fired stoneware, wheel-thrown and altered, 9.5" x 9" x 3.5". *Photo credit: Bede Clarke.*

PAULA COOLEY

Saskatoon, Saskatchewan, Canada
cooleypottery@shaw.ca, www.paulacooley.com

Artist's Statement

I am a full-time ceramics artist with an active studio practice since 1999 when I completed with my Ceramics Diploma. Initially I was interested in functional, domestic pottery, but I became increasingly intrigued by ceramic sculpture and so decided to pursue a B.F.A. in Sculpture, graduating in 2010. I create both functional and sculptural work. I wheel throw, handbuild, and use a wide range of clays, surfaces, and firing methods to achieve particular effects. My one-of-a-kind sculptural work employs traditional technique and craftsmanship to create experimental organic forms that reflect the inspiration of the natural world. Whimsical handbuilt shapes, evocative use of positive and negative space, and rich surfaces characterize my sculptural ceramics. My objects seem to grow or move, shaping and animating the space around them. My sculptural ceramic practice explores the boundaries between art and craft, static and animate, tradition and innovation.

I particularly enjoy firing my work with wood. I am looking for the surfaces that are unique to wood-firing: the subtle colors of reduction cooling, the crusty surfaces from near the firebox, the ash and flash of other parts of the kiln, the haloes from wadding or tumble stacking. I want my work to bear the marks and story of the firing process. I find that wood-firing my sculptural pieces resolves the challenge of trying to satisfactorily glaze a large complicated piece and that the path of the flame adds an additional layer of visual interest to the surface.

I do not own a wood kiln, so instead take advantage of workshops or invitations to fire in other potters' kilns. Over the last ten years I have participated in a number of firings in anagama and bourry box kilns. I usually have the opportunity to wood-fire two to three times a year. Firing in a variety of kilns allows me to experience a range of techniques and atmospheres, but not necessarily to develop a working knowledge specific to one kiln. I do not necessarily consider this a disadvantage as I am interested in experimentation and the different effects different kilns produce.

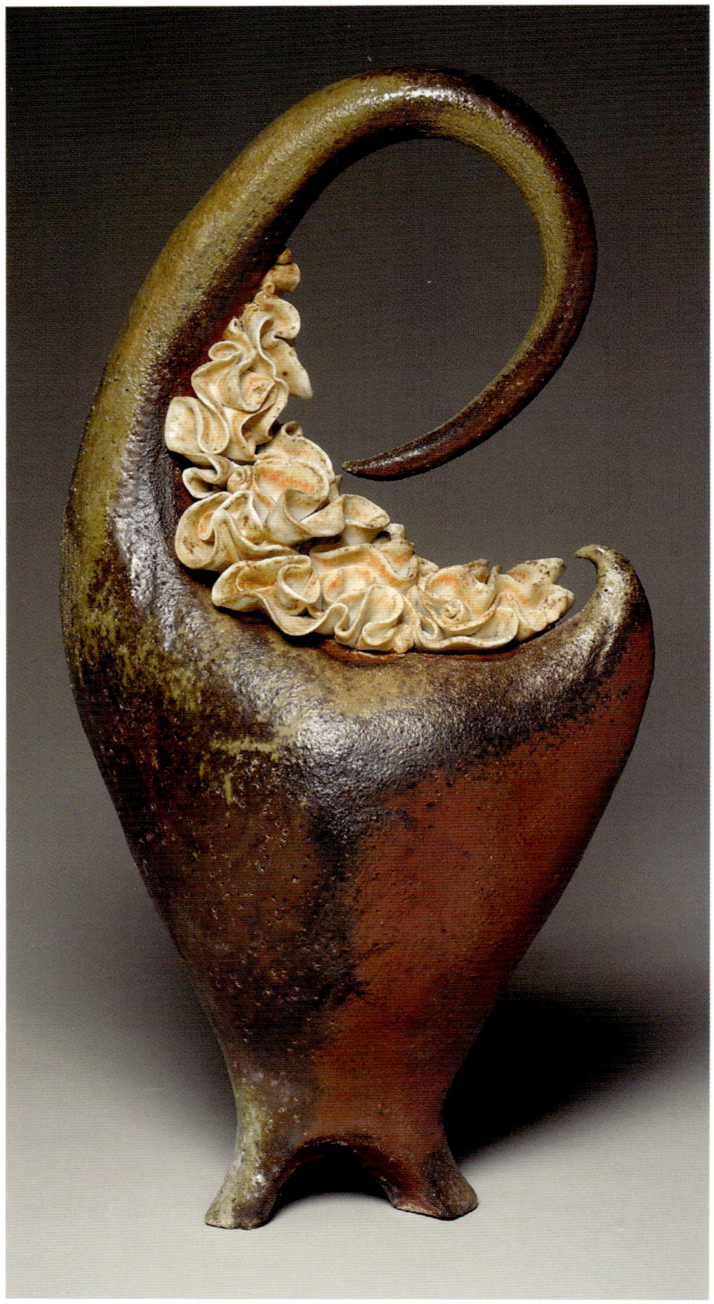

Paula Cooley, *Succor,* 2009. Sculpture, wood-fired stoneware and porcelain, handbuilt, 18" x 9" x 5". *Photo credit: Grant Kernan, AK Photos.*

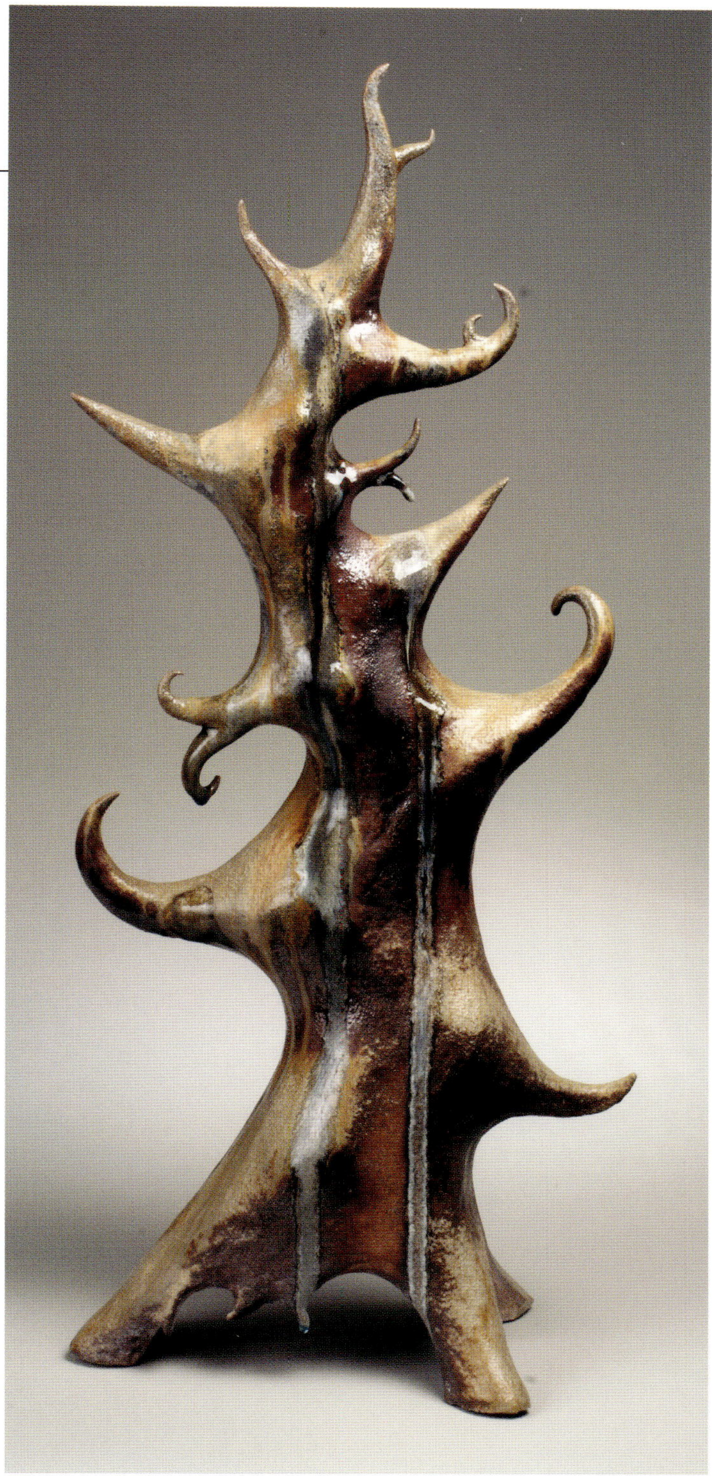

Paula Cooley, *Reach #2*, 2008. Sculpture, wood-fired stoneware, handbuilt, 32" x 16" x 20". *Photo credit: Grant Kernan, AK Photos.*

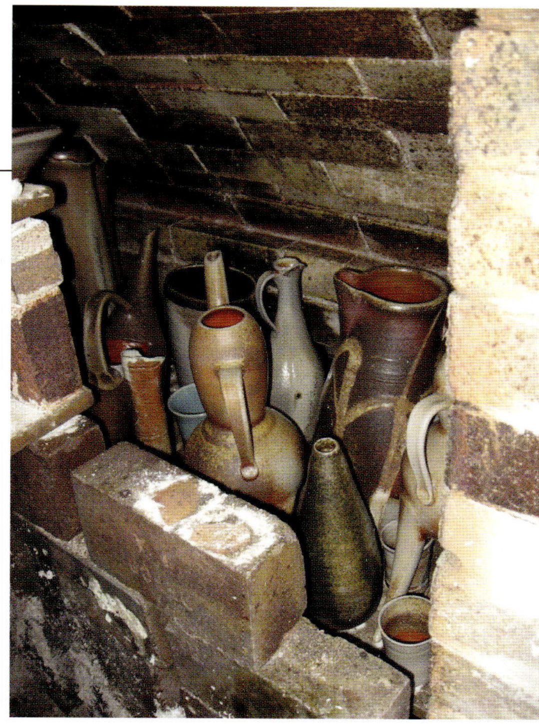

Interior view, unloading wood kiln at Agar's Corners, Saskatoon, Sk., Canada. *Photo credit: Paula Cooley.*

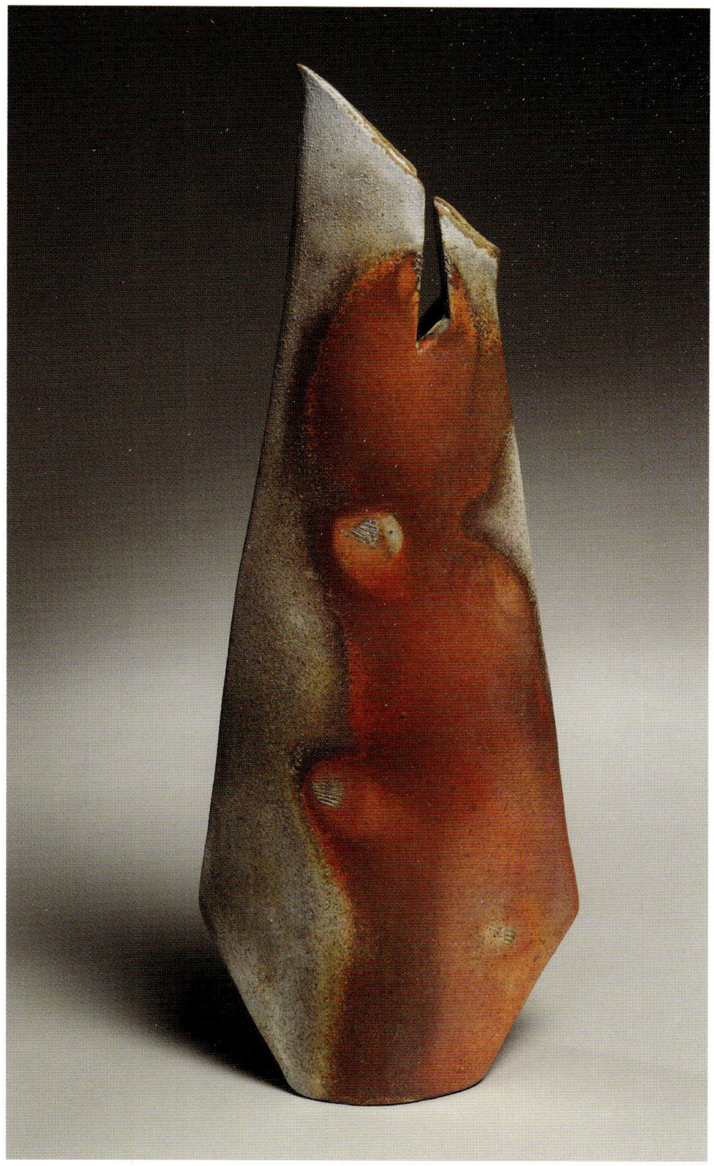

Paula Cooley, *Drift*, 2011. Vase, Wood-fired stoneware, handbuilt, 18" x 8" x 4". *Photo credit: Grant Kernan, AK Photos.*

SCOTT CORNISH

New Galilee, Pennsylvania
cornscott@yahoo.com

Artist's Statement

My choice to use wood as my fuel source is born out of aesthetic and ethical considerations. I create functional wood-fired pottery that is intended for everyday use. Although the role of a potter has shifted in the past 100 years, I think it is more important than ever to create, own, and use handmade objects. There is something quite unique about drinking coffee out of a mug that was created by someone's hands. Surrounding oneself with handmade objects enriches the spirit and provides a welcome reprieve from the sometimes-monotonous nature of daily life. It is my intent to bring art into the home and I hope that my work celebrates the everyday.

The forms that I create are inspired by a variety of different influences. Chinese pottery from the Sung and Ming Dynasty display a remarkable sense of craftsmanship and a subtle yet immense beauty that I can only strive for in my own work. The casual forms and quick execution of early American stoneware has a much different characteristic, yet it can be equally inspiring. Many different aspects of historical and contemporary work, nature, architecture, and firing influence my artistic sensibilities.

In a time when the acquisition of natural gas is clouded in the controversy of hydraulic fracturing and its environmental effects, I wish to reduce my dependency on this non-sustainable natural resource. My fuel source is gathered from downed trees within a mile radius of my studio and kiln. I am fortunate to live in an area that is rich in hardwood forests and it makes sense to utilize this sustainable natural resource as my fuel. While the actual fuel is often acquired free of charge, or perhaps by trading pottery, the labor involved is not to be underestimated. I spend countless hours cutting, hauling, splitting and stacking wood in preparation for a firing. This provides my studio life with a reprieve and allows me to reflect on my ceramic work and previous and future firings. I don't think of the wood preparation as a chore, but as a necessary cog in the wheel of the production of my work.

I designed my double chamber kiln with versatility, aesthetics, and efficiency in mind. It is constructed entirely with reclaimed firebricks that were destined for a landfill. Typically, I fire both chambers together; however, the design is such that I can fire either chamber independently without the other. Before building my kiln, I experimented with a variety of wood and gas fired kilns. I have always been drawn to atmospheric firings, including soda firing. I also enjoy the directional effects and heavy ash deposits of the train style kiln. My design came from the desire to meld these two aesthetics and provide a wide range of effects achievable in one kiln. The first chamber is in the style of a train kiln with a downdraft firebox. The second chamber is connected to the first by a series of flues and forms a catenary arch with an internal firebox. The design has proven to be quite efficient. While firing the first chamber, the second chamber acts as a heat sink, storing heat that would typically go up the flue. This chamber also completes the combustion of the wood, producing a very clean burning kiln. The first chamber is fired off completely before shutting it down and moving to the second. I spray a mixture of soda and water in the second chamber between cone six and cone nine. Although the kiln could fire quicker, I typically extend firing times from two to three days. I am currently preparing for the ninth fining in my kiln. New discoveries come with each firing as I learn more of the intricacies involved with my design and the use of wood as a fuel.

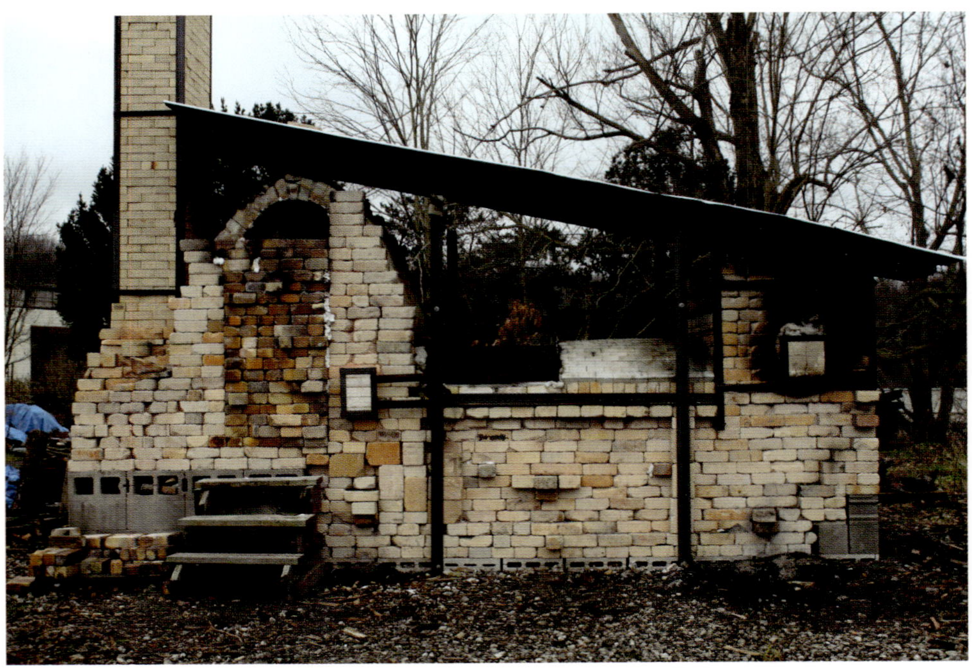

Front of kiln. *Photo credit: Scott Cornish.*

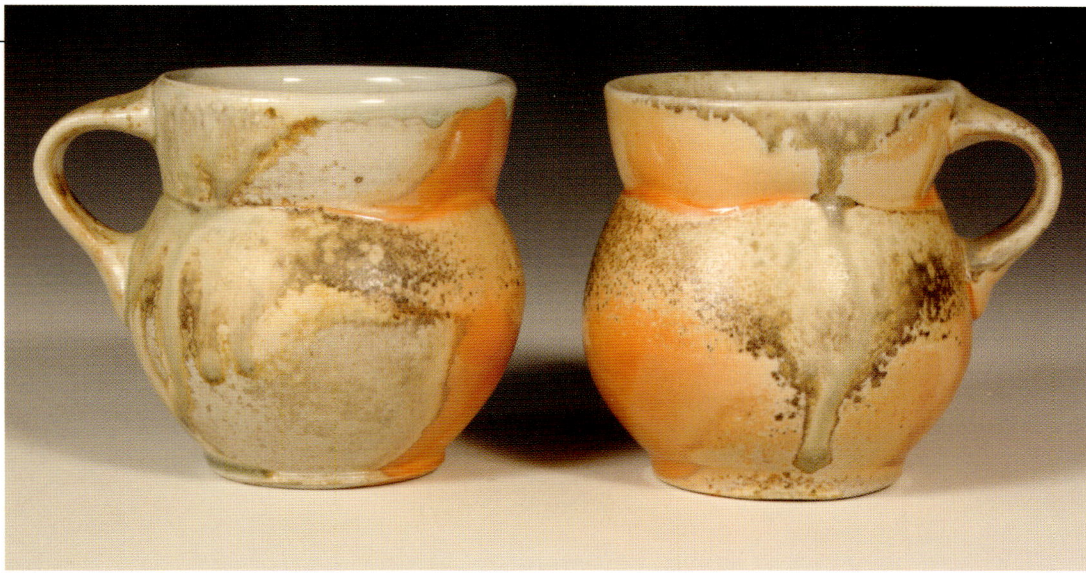

Scott Cornish, *Mugs, 2011*. Wood-fired porcelain, wheel-thrown, 4" x 5" x 4" each. *Photo credit: Scott Cornish.*

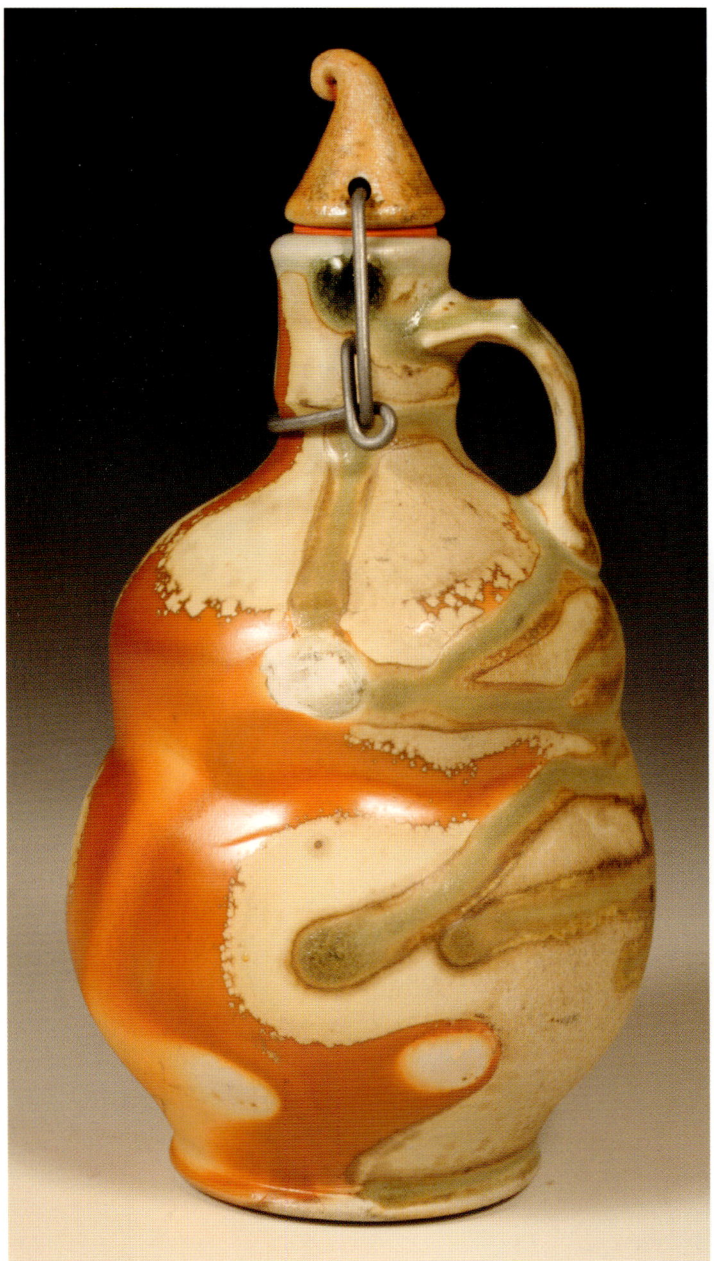

Scott Cornish, *Ale Bottle*, 2011. Wood-fired porcelain, wheel-thrown, 9" x 4" x 4". *Photo credit: Scott Cornish.*

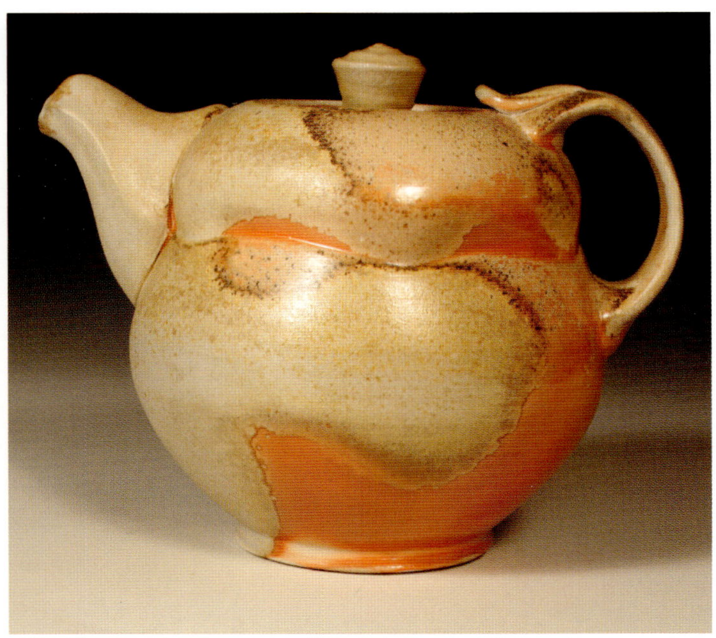

Scott Cornish, *Teapot, 2011*. Wood-fired porcelain, wheel-thrown, 6" x 7.5" x 5". *Photo credit: Scott Cornish*

KEVIN CROWE

Amherst, Virginia
tyeriverpottery@aol.com, www.kevincrowepottery.com

Artist's Statement

My work is strongly influenced by Asian and English forms and surfaces. When I was a young guy, I came across photographs of early Bizen and Shigaraki pots. I had never seen a wood kiln and had no idea how these surfaces were created. But I felt a profound attraction to these pots, for which I had no visual vocabulary. The extraordinary tension between the soft, asymmetrical forms and the directional, fluid, natural ash glaze sunk a deep root in me. Research uncovered the world of wood-firing, its history, its many faces, and its contemporary practice. I was in.

My first wood kiln was a small, three-chambered noborigama. The second kiln was a large, two-chambered kiln with an exterior, catenary arched firebox. My present kiln is a large, 45 cu. ft. combination anagama/noborigama. The anagama chamber is followed by a glaze chamber and a salt chamber. It is a joy to fire.

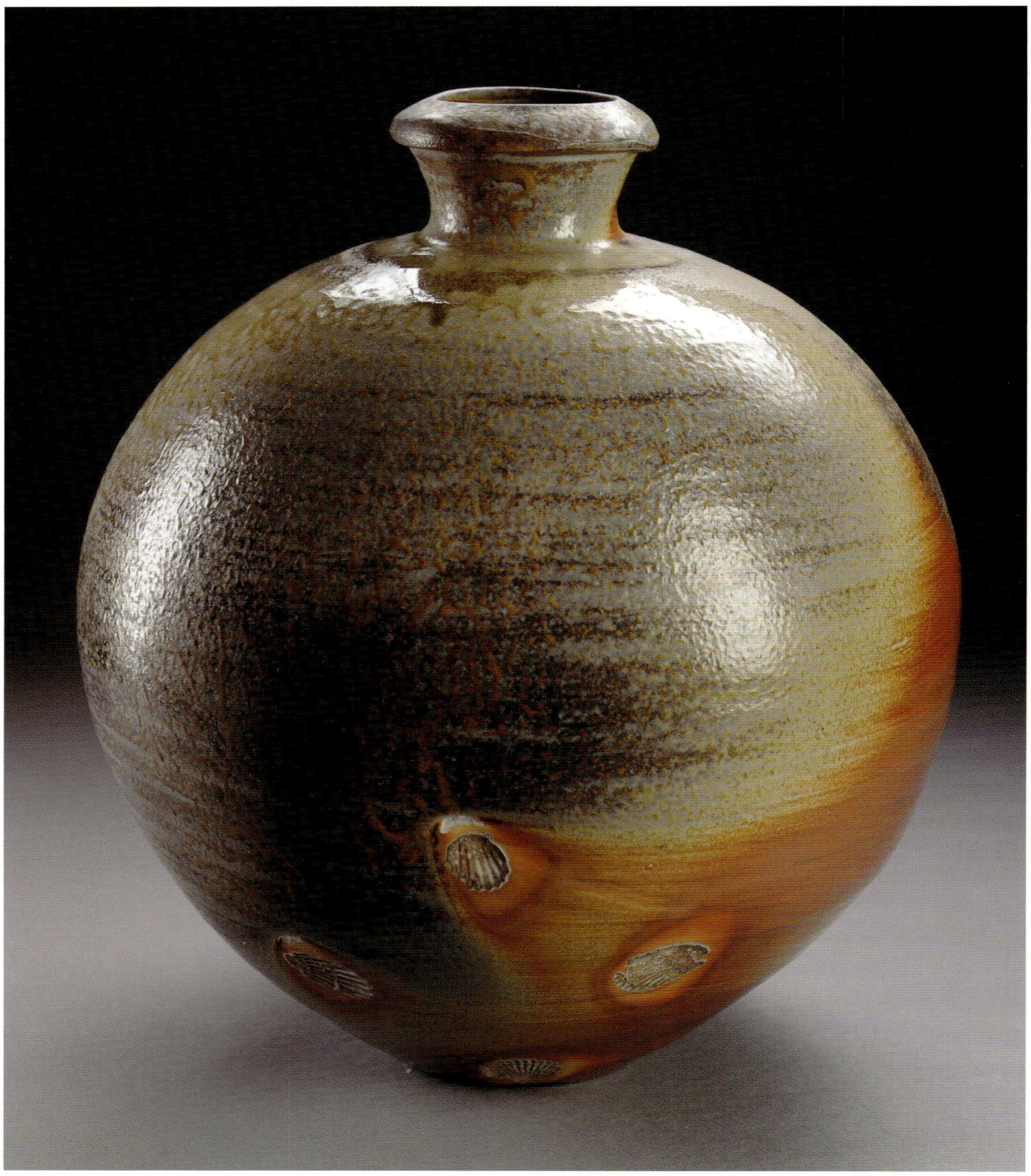

Kevin Crowe, *Large bottle*, 2011. Wood-fired stoneware, wheel-thrown, 15" x 12" x 12". *Photo credit: Tim Carlson, Studio C.*

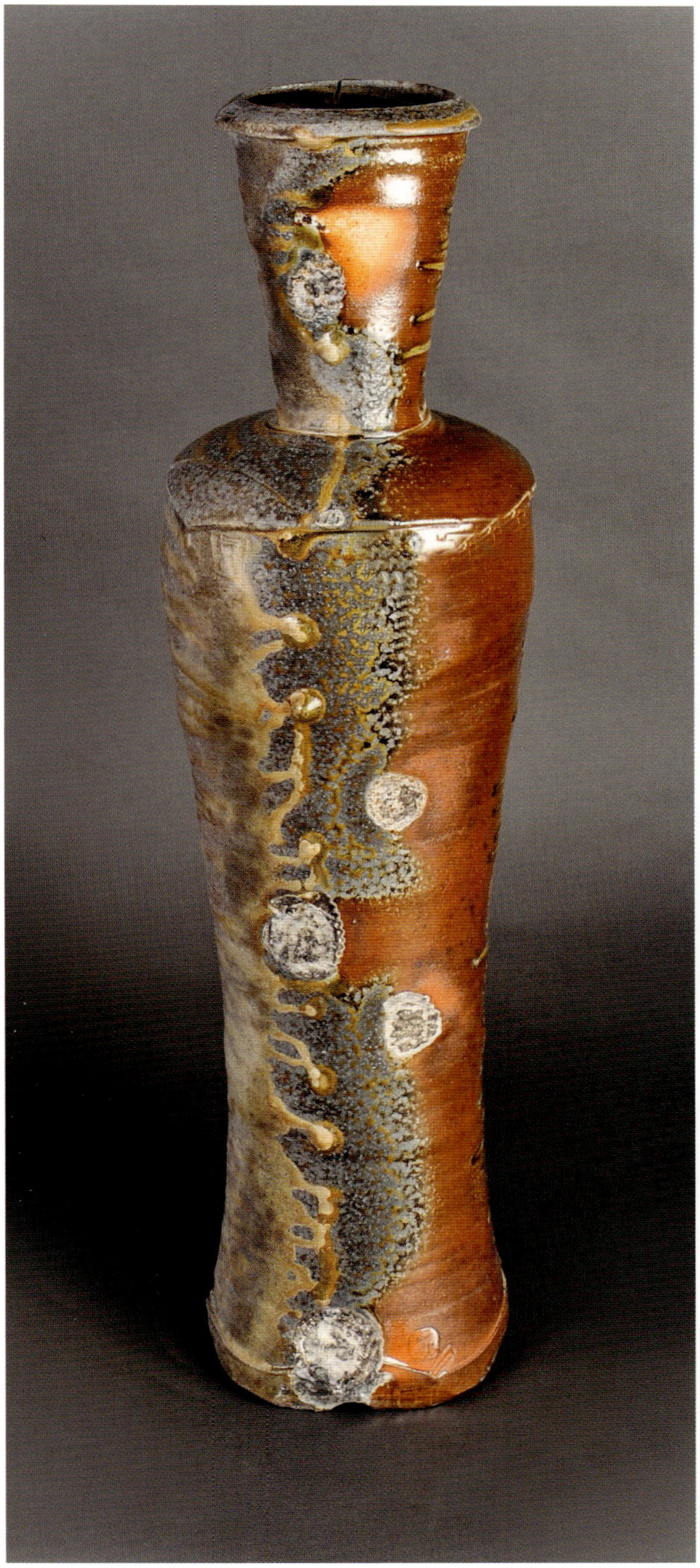

Kevin Crowe, *Tall side-fired vase*, 2011. Wood-fired stoneware, wheel-thrown, 31" x 9" x 9". *Photo credit: Tim Carlson, Studio C.*

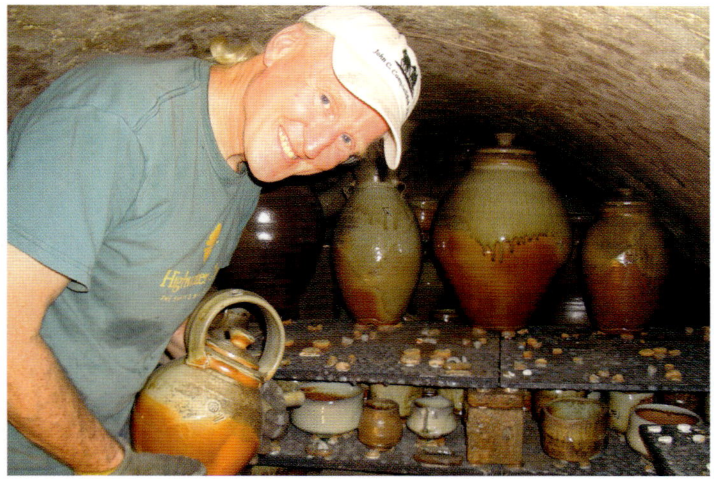

Kevin Crowe, unloading the anagama. *Photo credit: Joy Bridy.*

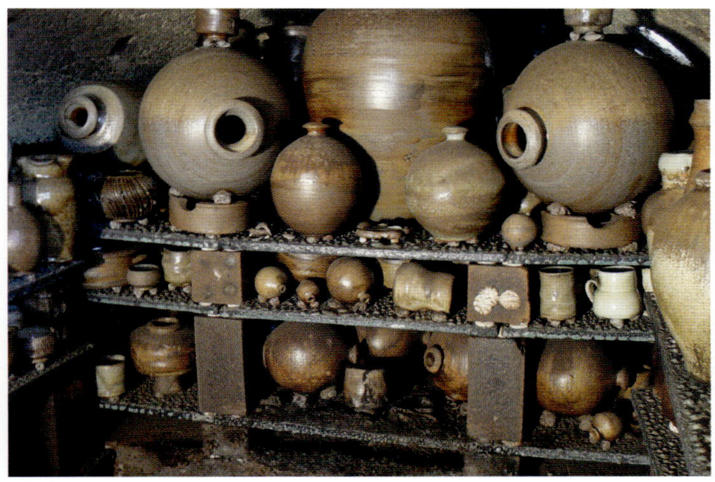

Fired pots in the anagama. *Photo credit: Linda Crowe.*

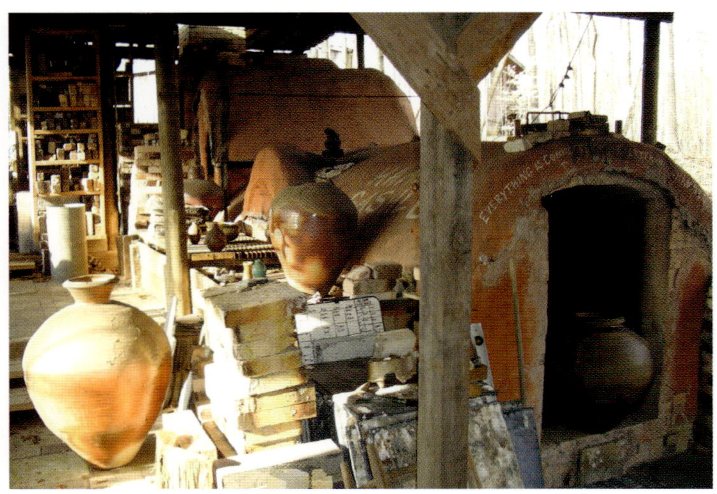

The Tye River three-chambered combination anagama/noborigama. *Photo credit: Linda Crowe.*

DANNY CRUMP

Portland, Oregon
wristrubber@hotmail.com

Artist's Statement

I choose to use clay as medium because of its intrinsic ability to change.

While it can take on any form in its raw state, its fired form is permanent. This permanence of ceramics does not occur in nature, but is brought to us by our own human process. This clay material ties us to our own history through objects and artifacts. It brings associations of other cultures and exists in our contemporary society. My art practice of making sculpture through the ceramic process questions this permanence and fragility, volume, mass, and gravity, and structure to form objects with visual harmonies and strong material dissonance—something, perhaps between wood and steel.

Clay becomes bricks, holes dug in the ground become kilns. Wood is split, objects are made, wood-firing begins, and clay becomes ceramic. I mostly have been taught to wood-fire by a folk tradition passed from experienced teachers, colleagues, students, and by making mistakes. I have fired with dozens of crews, all different styles of kilns, and it is always a combination of dedicated people who are willing to help, and eager to learn to make the firing a success. Wood kilns are the ultimate teaching tools in learning the basics of firing: combustion, atmosphere, temperature, pressure, insulation, etc.

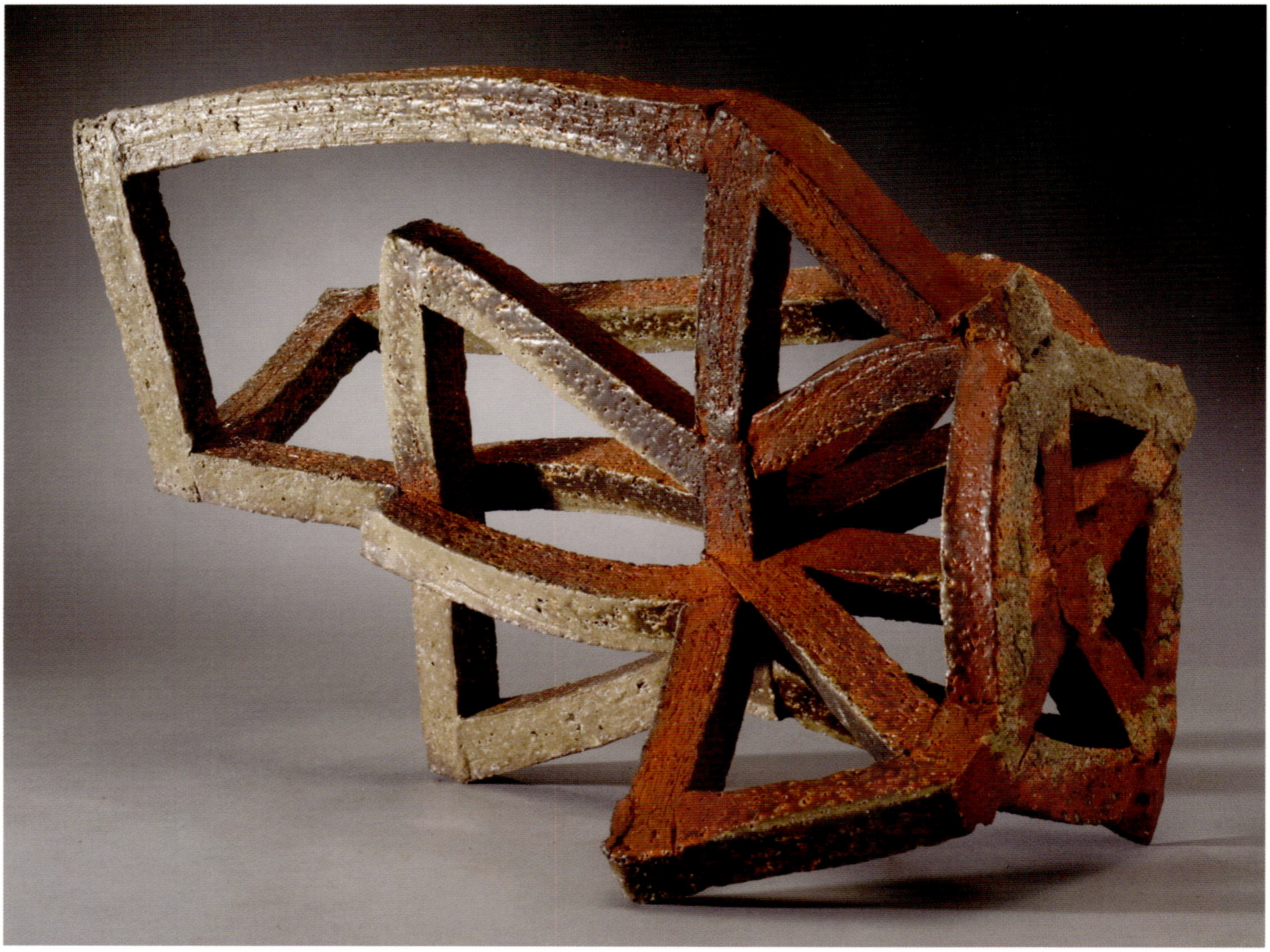

Danny Crump, *Seeing Constructs*, 2010. Wood-fired stoneware, 12" x 24" x 9". *Photo credit: Chris Autio.*

Watching swirling flames has a mesmerizing television-like quality to it. The primal nature of fire's use has a built in comfort for us. Wood as a fuel in the twenty-first century presents an interesting dichotomy of economy: while it is abundant, relatively inexpensive, a renewable and efficient fuel source, it is at the same time incredibly labor intensive to move, process, and plan to set up optimum firing conditions (or learn to compensate for Murphy's law). Not to mention scheduling prep and firing crews, and the actual wadding, loading, firing, cooling, unloading, and grinding that awaits. This allegedly low-tech process can get complicated. I like to keep it basic, and paraphrasing from a mentor, "You put wood in it, it goes up."

I form objects made of simple fireclay, available from the masonry shop, feldspar, and iron bearing clays that are cheap and available from clay suppliers. The clays are blended to different colors and are formed with slabs to create teetering structures, or large solid slabs. The softness of the clay is balanced between its structural capabilities of supporting the weight above it. This gravity is sometimes disturbed mid-construction by rotating the piece in a different orientation. This physical jumbling (and sometimes collapse) metaphorically recalls my own confusion with culture and is my visual response to the technological interface of the world.

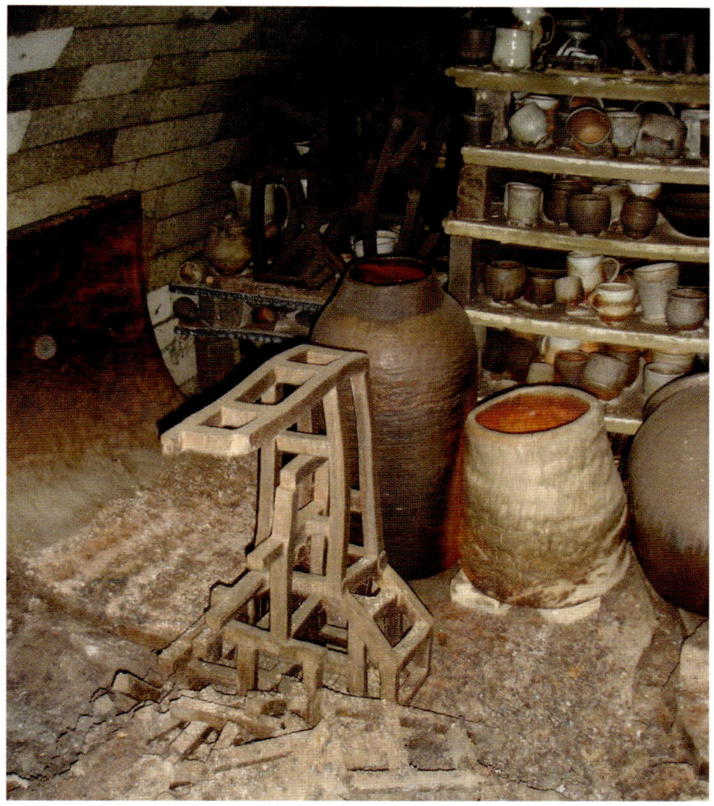

View of the inside of the firebox prior to unloading. Showing pyroplastic pieces, and pieces still entirely submerged in the remaining ash.

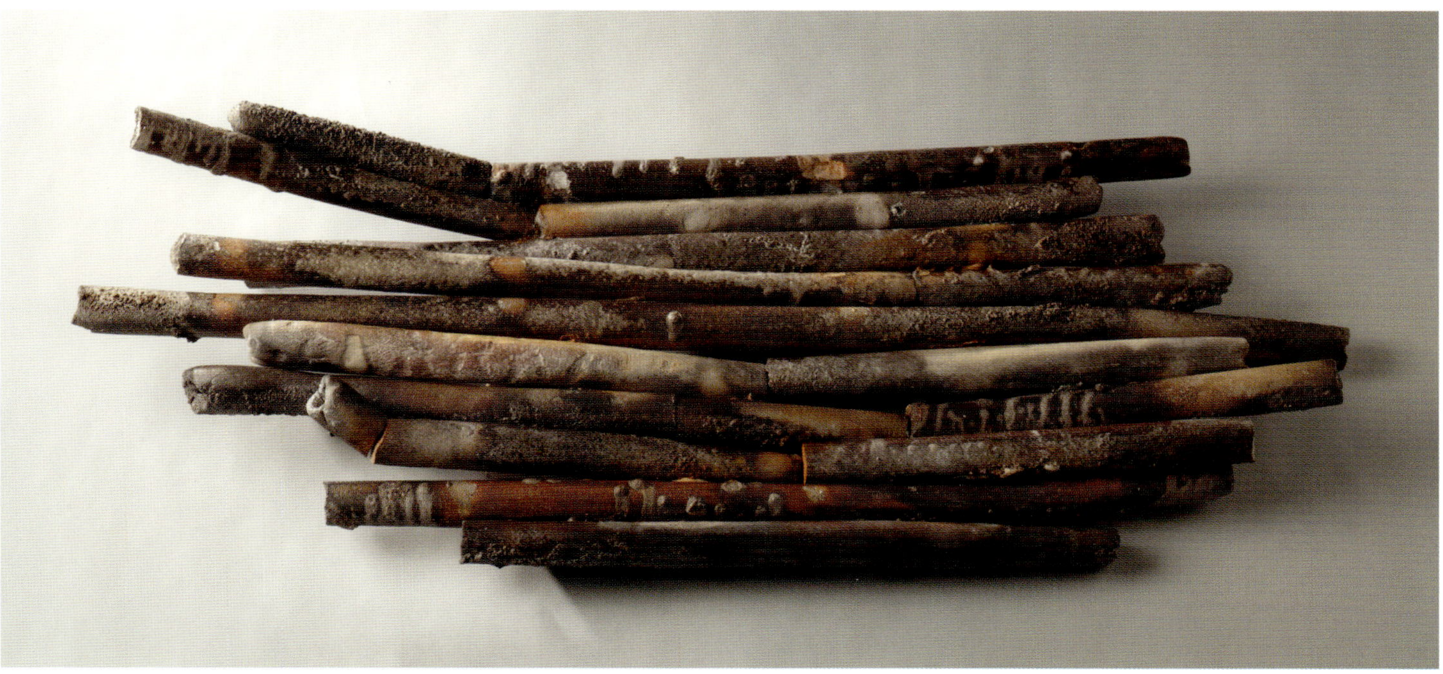

Danny Crump, *Broken Raft*, 2011. Wood-fired stoneware, 42" x 26" x 12". *Photo credit: Chris Autio.*

CLAY CUNNINGHAM

Council Bluffs, Iowa
clay657@yahoo.com, www.claycunninghamceramics.webs.com

Artist's Statement

The focus of my work has developed into the form of thrown and altered clay vessels and sculptures. My research has led me to investigate a variety of interests, including dualities, the nature of clay as a material, and the way we perceive the chaos and clarity of life. I am concerned with using clay as a material capable of recording a powerful gesture in space. These works seek to be heroic in size, form, and sometimes weight.

This research has led to exploration into the area betwixt natural forms and contrived, hand-crafted objects, and the tension and mystery that is achieved in a well-composed balance between the two. My art focuses on broad, simple gestures and fragmented forms as my visual vehicle. The sculptural forms are large and rough, yet often lined with cool, glassy glazes to achieve refined and pristine surfaces. This acts to create tension and set up a dichotomy. This work is explored in vessels of function and more sculptural forms.

I predominantly fire with wood. The way the work is stacked affects the outcome of the firing, leaving a permanent record of the fire's path. Though I enjoy the patina left by the firing process, I choose to fire with wood because of philosophical reasons. I split all my wood by hand with an axe, and find the work invigorating for the body. During the firing, my mind must be in tune with the kiln to fire efficiently and to control the firing to the best of my abilities. I also find it good for the spirit to know that though I have been a part of the process from harvesting the fuel to controlling the fire, there are many serendipitous factors far beyond my control at play.

I see this path of research to be intriguing and ongoing.

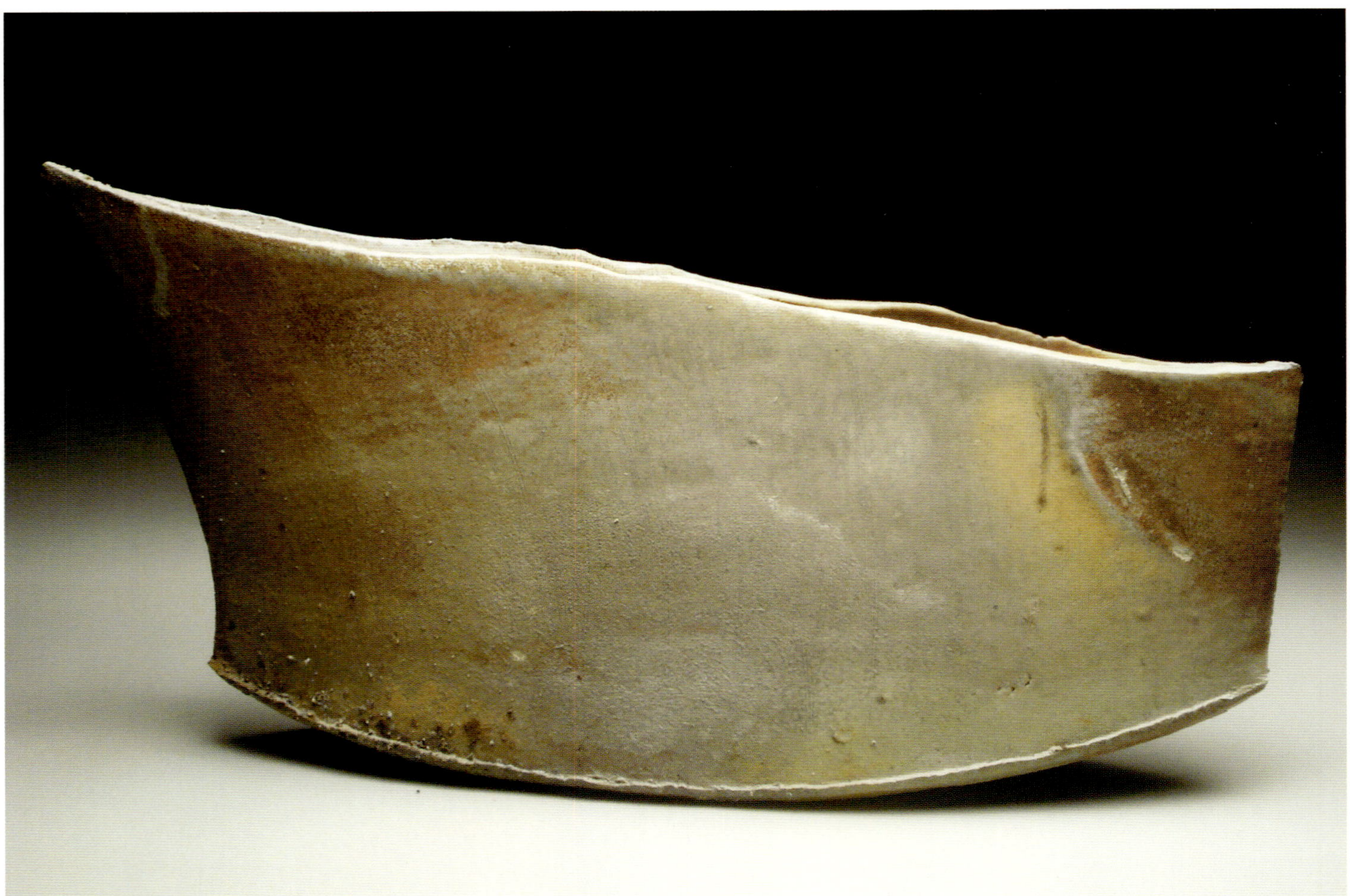

Clay Cunningham, *Ellipse*, 2010. Wood-fired stoneware, handbuilt, 9" h x 20" w x 4" d. *Photo credit: Clay Cunningham.*

Clay Cunningham, *Ellipse with Arrangement*, 2009. Wood-fired stoneware, handbuilt, 7" h x 22" w x 4" d. *Photo credit: Clay Cunningham.*

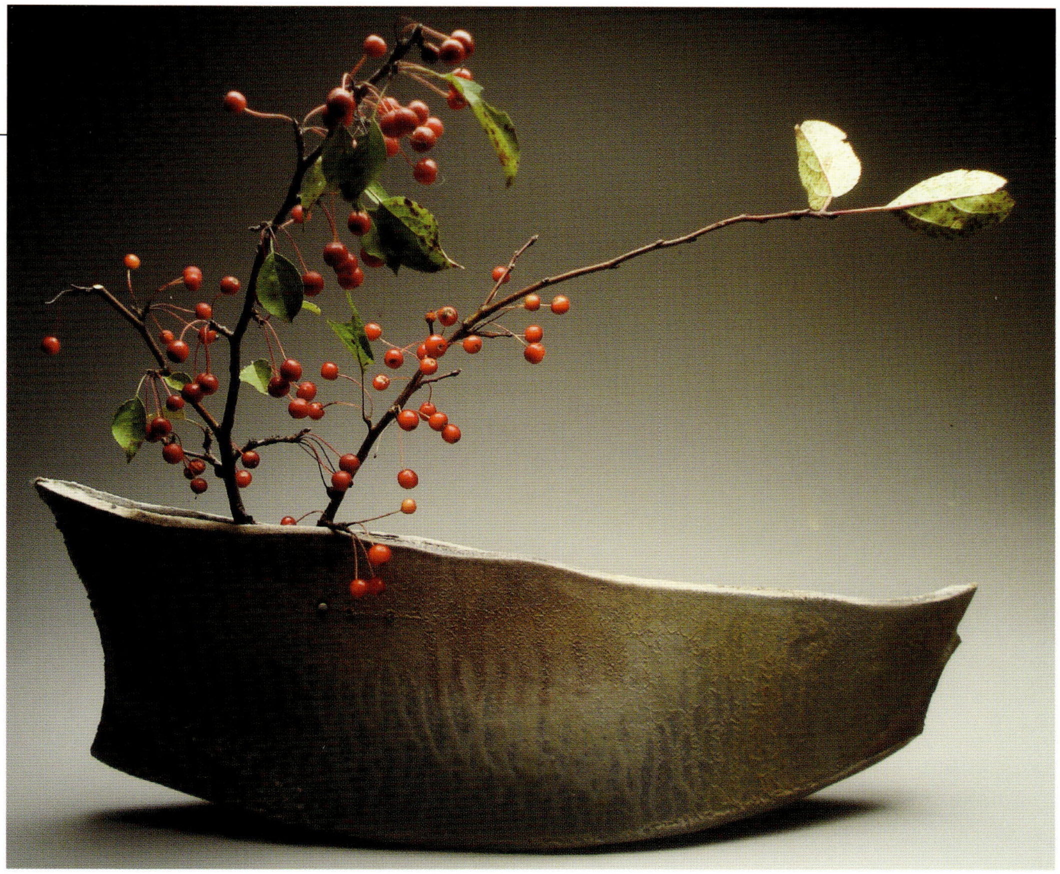

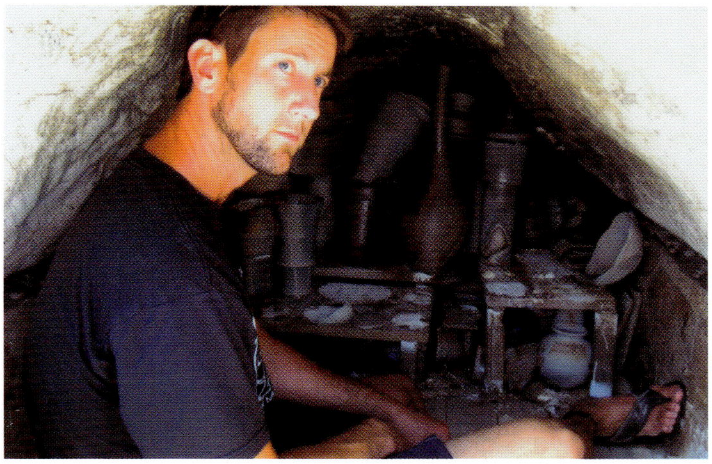

Clay Cunningham unloading a finished kiln from his single chambered kiln, the *Iowagama*, after a summer workshop.

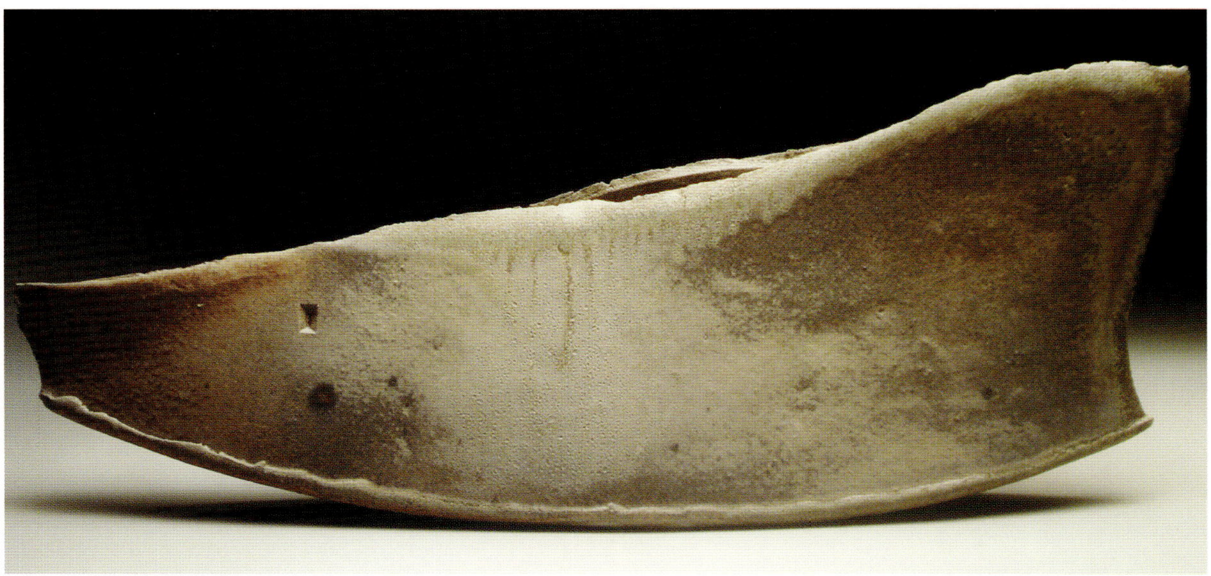

Clay Cunningham, *Ellipse,* 2012. Wood-fired stoneware, Handbuilt, 7" h x 22" w x 4" d. *Photo credit: Clay Cunningham.*

ISRAEL DAVIS

Grand Rapids, Michigan
israel@israeldavis.com, www.israeldavis.com

Artist's Statement

The objects that I make are based upon a series of memories and childhood dreams. At their basic core they represent the freedom and possibilities of youth. Through process and contemplation, these recollections have become a physical vision of past, present, and future.

These cups are created in response to and in honor of both traditional and contemporary practices.

They are based upon traditional Japanese *Yunomi;* a casual drinking vessel meant for daily use that is taller than it is wide.

The images are screen-prints, using slips and underglazes, of drawings derived from photographs of family, friends, and myself. The photographs are drawn over and manipulated in Photoshop. These drawings are treated to reference sketches and writing that might be found in an artist's journal or sketchbook.

Although I make other objects and forms, these yunomi (cups) serve as an intimate entry into the imagery on them. This interaction, being able to hold, turn, and use the vessel, supplements the sentiment that is imbued in the drawings.

Wood-firing imparts an atmospheric quality to the work that just can't be duplicated in any other way. I consider the way I load each piece into the kiln, seeking what I consider a "soft" firing. These cups are stacked in saggars, perforated with holes and slits, and guarded with shields in an attempt to compose with the flame around the imagery. I don't want to obliterate the drawings. Rather, I seek the unison of ash, atmosphere, and imagery. For this reason I often opt for the back of the ware chamber away from the immediate contact of the flame.

My work is fired at Ox-Bow School of Art and Artist's Residency. A one-hundred-year-old art school in Saugatuck, Michigan, that was founded by a group of painters from the School of the Art Institute of Chicago. The kiln is a sixty cubic foot catenary arch cross draft design built by master kiln builder Donovan Palmquist. I have been firing the kiln twice a year since 2009 through a workshop that I teach and organize. I began these workshops as a way to teach students more about firing kilns. As I have continued to make work for these firings I have

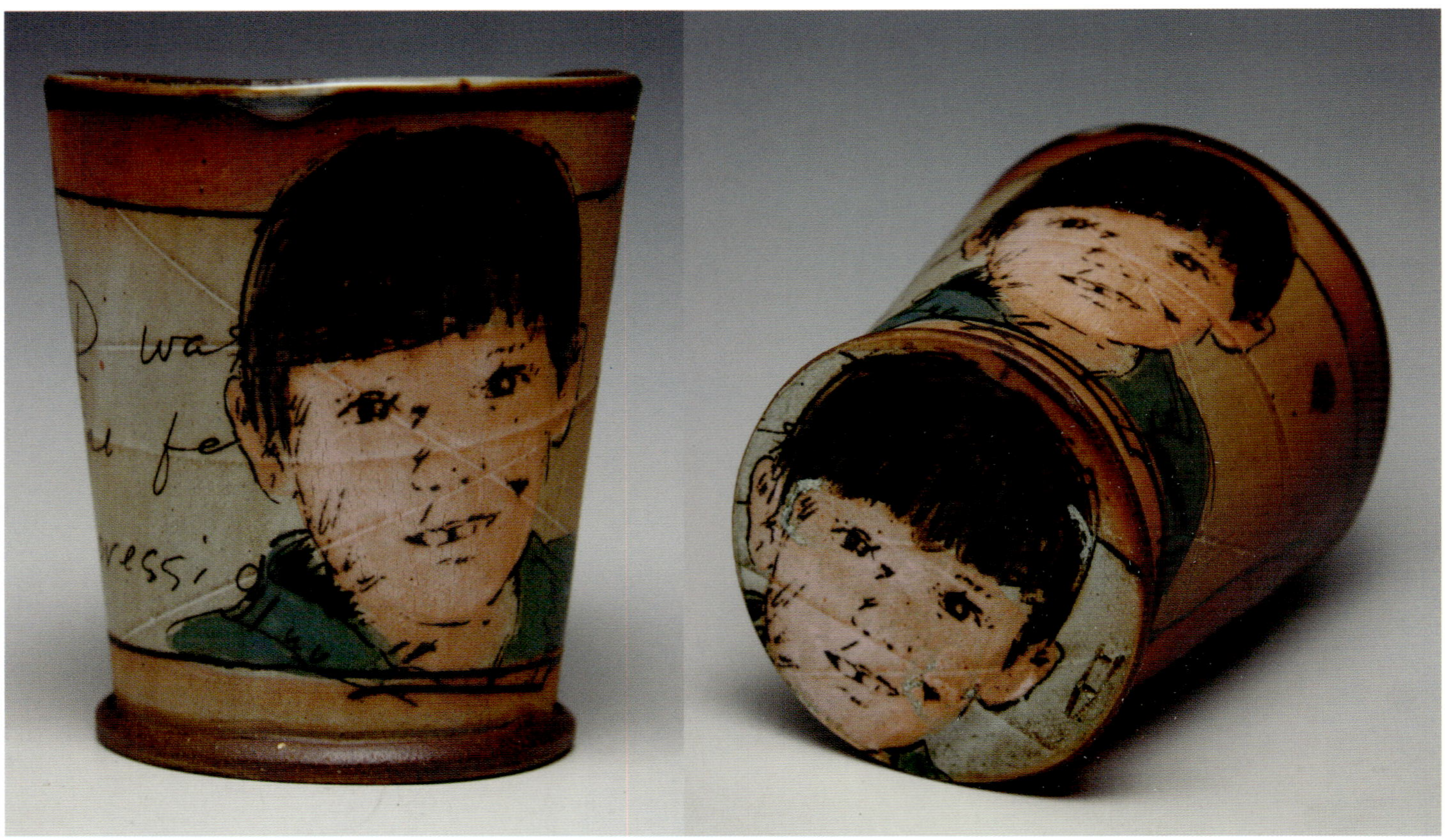

Israel Davis, *Big Izzy D Dreaming,* 2011. Screen-print on clay with slips and underglazes, wood-fired to cone 10, slab built, 4" x 3" x 3". *Photo credit: Israel Davis.*
This piece is a self-portrait based upon a photo at about ten years old.

rediscovered a point of investigation in learning to paint with fire. Because of the frequency of the firing and the nature of the workshop, I don't focus on the numbers of pieces that I get into the kiln. I am more interested in finding those select few works that epitomize the coalescence of imagery and atmosphere. Our firings last for about thirty hours, based upon a firing schedule that I have developed over the years. With good dry wood the kiln fires and is kept on schedule easily.

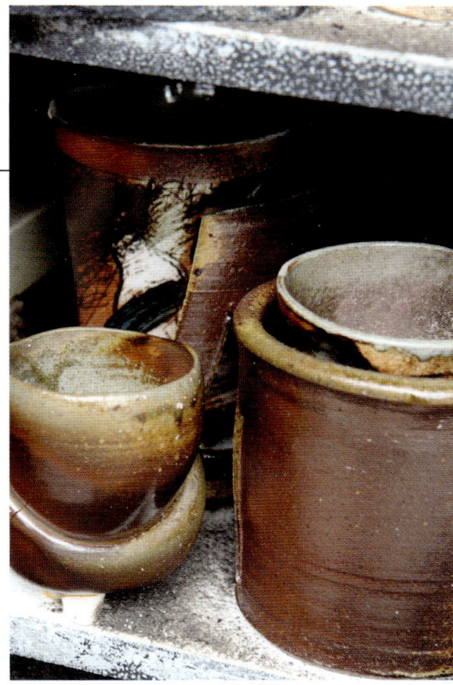

Israel Davis, this is an example of how I load my pieces in saggars and behind shields in an effort to control ash and atmosphere.

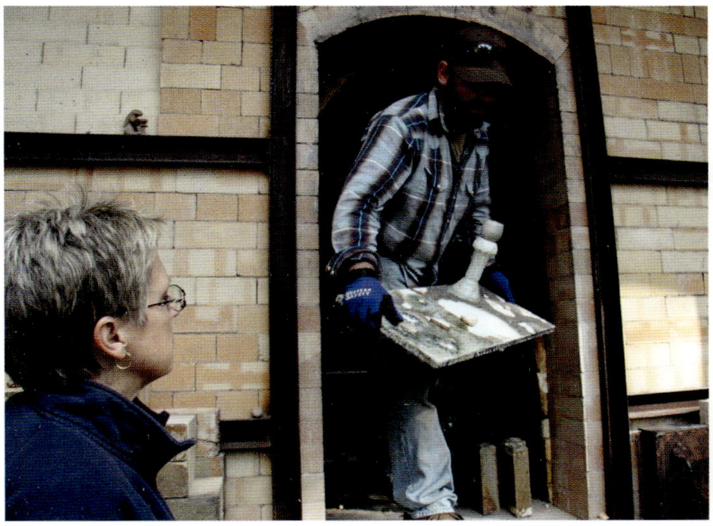

Israel Davis, a picture of myself unloading the kiln after a hot firing that reached cone 13 in some spots.

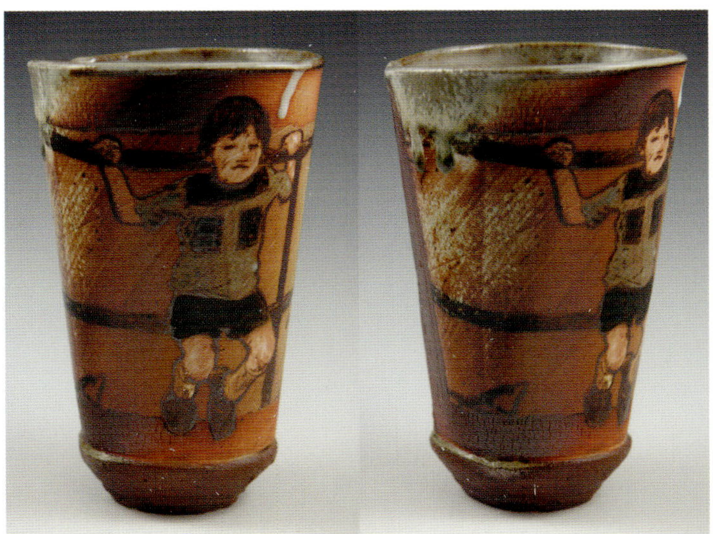

Israel Davis, *Waiting for the Stars Dreaming*, 2011. Screen-print on clay with slips and underglazes, wood-fired to cone 10, slab built, 6" x 3" x 3". *Photo credit: Israel Davis.*
This piece is a self-portrait based upon a photo at about ten years old. I envision myself has having stars in my eyes dreaming of being a pilot.

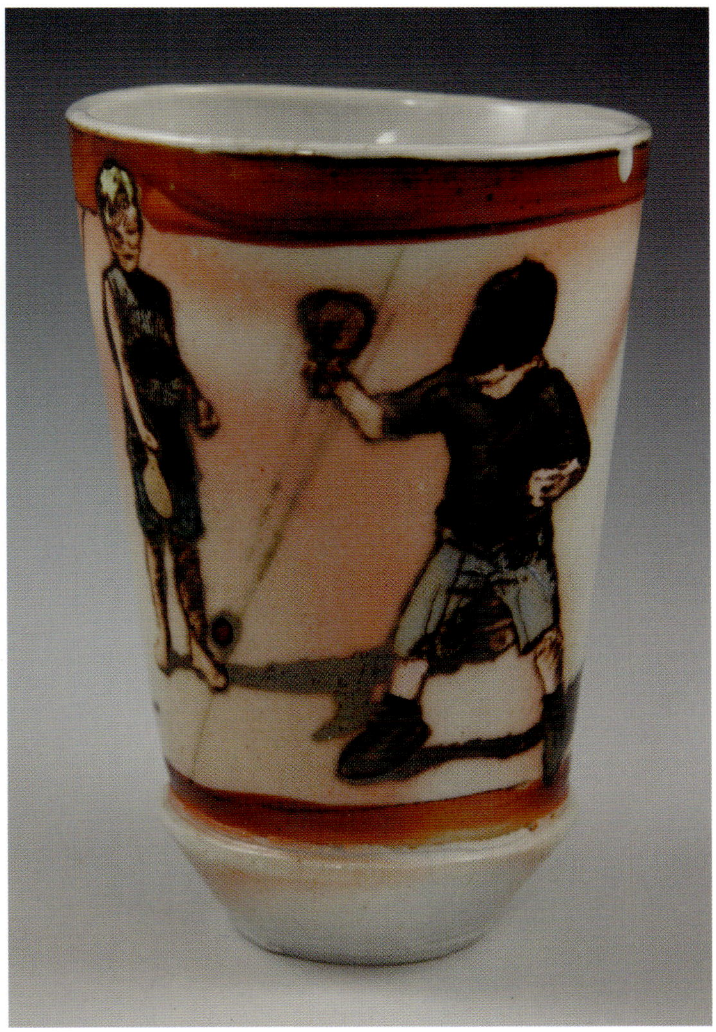

Israel Davis, *Paddle Ball Dreaming*, 2010. Screen-print on clay with slips and underglazes, wood-fired to cone 10, slab built, 6" x 3" x 3". *Photo credit: Israel Davis.*
The drawing on this cup is derived from a photo of my nephews playing paddle ball.

JOANNE DEKEUSTER

Taos, New Mexico
dekeusterclay@gmail.com

Artist's Statement

I enjoy working with white stoneware clay to create my sculptural forms. The contrast between rough and smooth textures play a big part in my slab built work. As I construct my pieces my focus is on form, design, and composition. I then turn my artwork over to the kiln, allowing the flames and ash to enhance my creation. During creation of my work, I am constantly thinking about how the fire will affect each piece.

Although I was introduced to clay as an undergraduate, I was a fiber artist. In 1985 my husband and I started our masters study at Northern Arizona University. At that time I was introduced to wood-firing. I fell in love with the firing process and the outcome. My husband, Kevin, and I built our first one chamber kiln near our home in Southern Arizona. After four long years of firing kilns, and not having any art in any of them, I decided it was time to return to making pots. When I started making pots again, it was because I loved the wood-fire. I truly believe it was an important evolution for me. My pots are made with wood-firing in mind. They are made for the wood kiln. I cannot imagine my artwork fired any other way. If not for those large kilns in Flagstaff, and the fact that I am married to a potter, I would probably still be a weaver.

In 2003, we moved to Taos, New Mexico. Together, my husband and I built a large train style kiln. Since that time all of our artwork is fired in the train kiln.

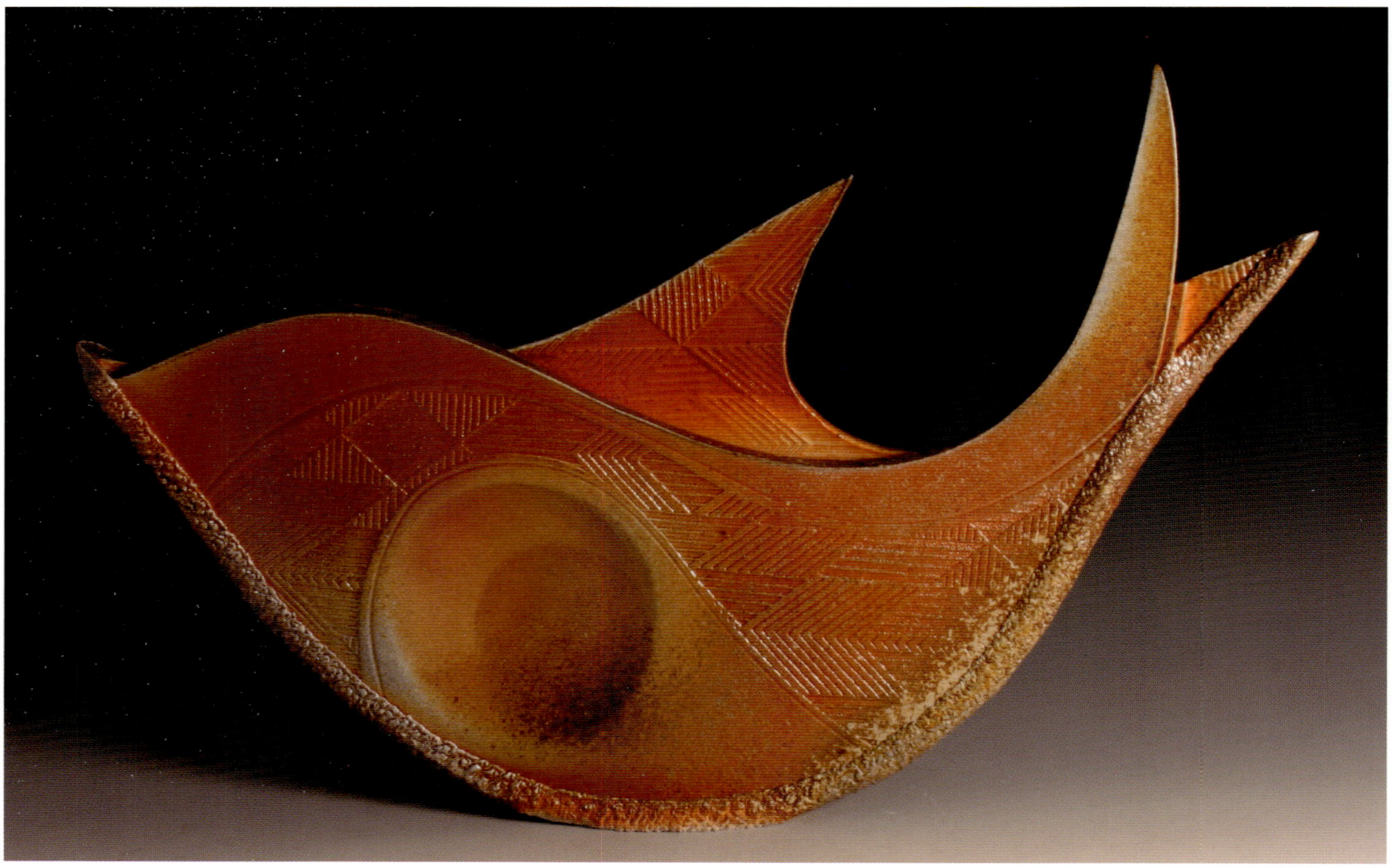

JoAnne DeKeuster, *Untitled*, 2012. White stoneware, slab construction, train kiln, shino glaze, H 15.5" x L 24" x W 4". Photo *credit: JoAnne DeKeuster.*

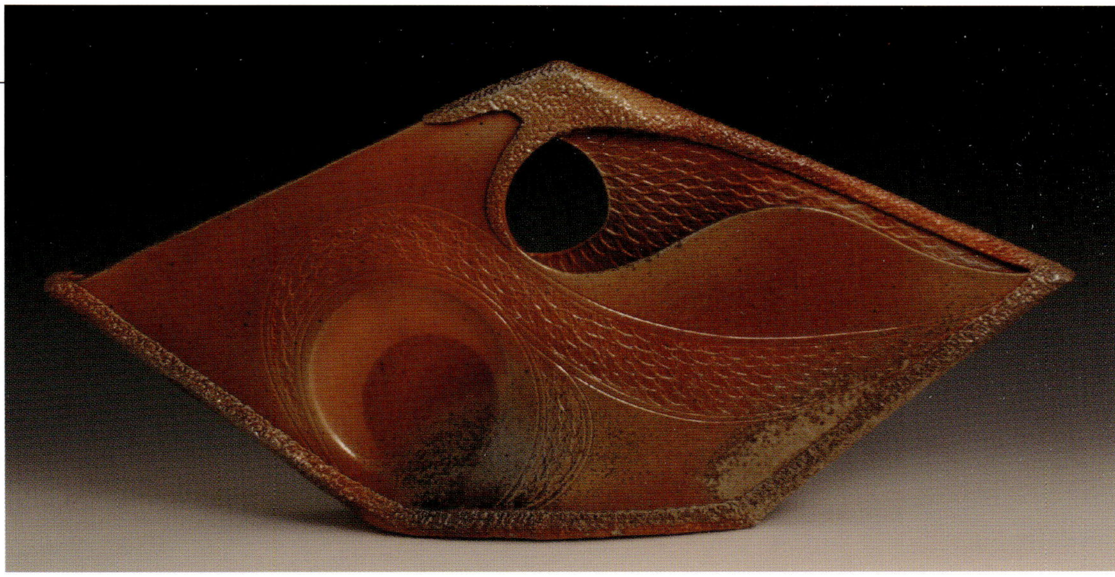

JoAnne DeKeuster, *Untitled*, 2012. White stoneware, slab construction, train kiln, shino glaze, H 12.5" x L 27 x W 5". *Photo credit: JoAnne DeKeuster.*

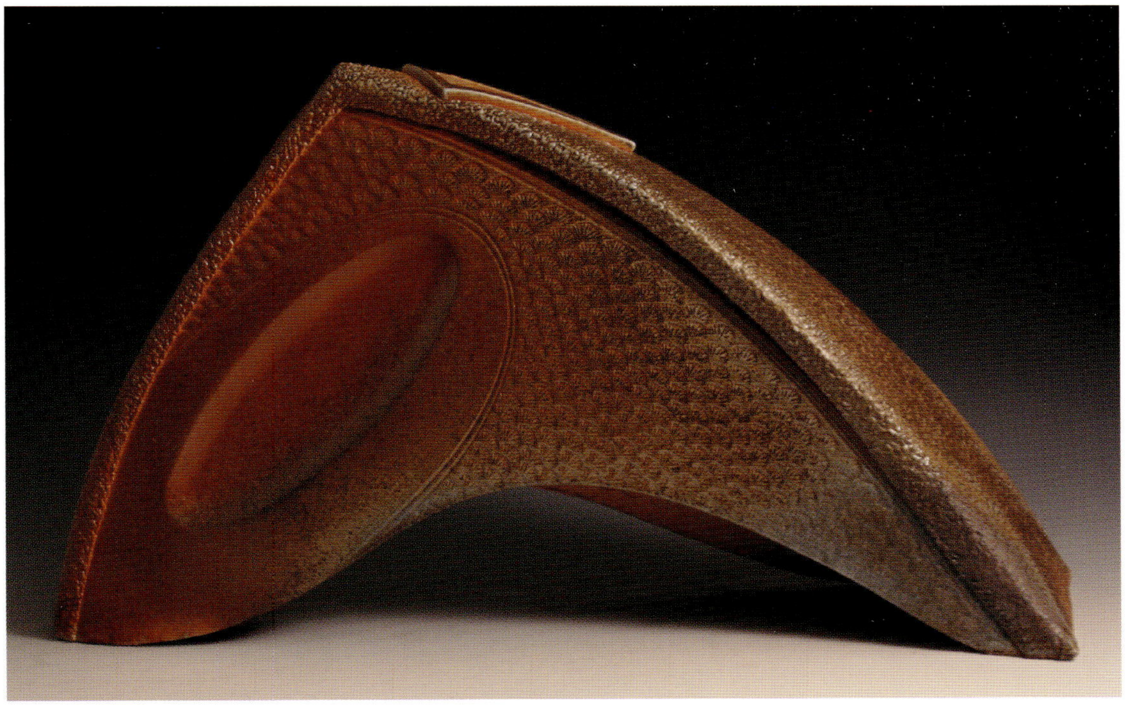

JoAnne DeKeuster, *Untitled*, 2012. White stoneware, slab construction, train kiln, shino glaze, H 13.5" x L 24" x W 6". Photo *credit: JoAnne DeKeuster.*

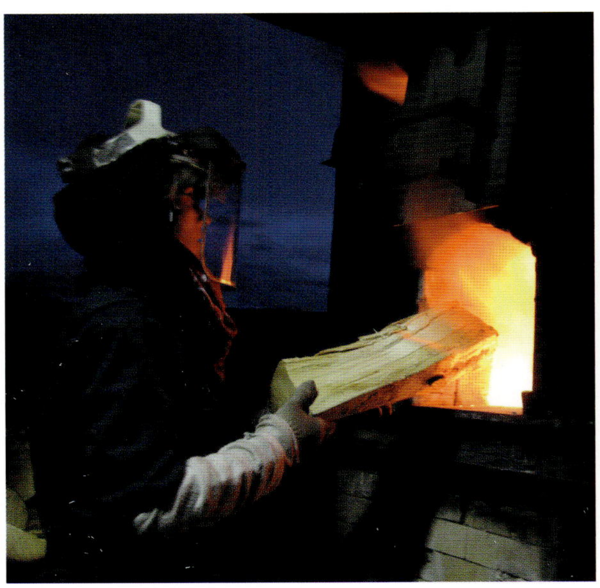

JoAnne stoking kiln. *Photo credit: Kevin DeKeuster.*

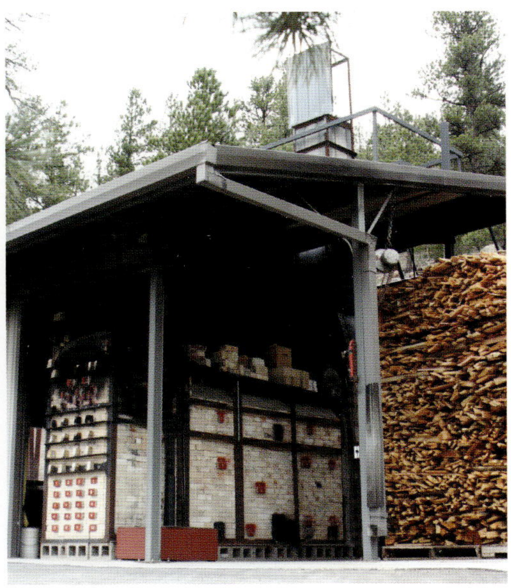

Train kiln. *Photo credit: JoAnne DeKeuster.*

JOSEPH DELPHIA

Pittsburgh, Pennsylvania
jdelphia@gmail.com, www.josephdelphia.com

Artist's Statement

I strive to create surfaces that offer tonal and tactile variations. I want my work to be as enjoyable to hold in the hand as it is interesting to the eye. For the past five years I have been paying close attention to how I stack and place my work in the kiln. I wad the work to create channels, which will compress and direct flame through and around the work to create rich deposits of ash, dynamic flashing patterns, and vivid color. It is the endless possible variations of tone and surface that continue to draw me to wood-firing.

I was originally drawn to wood-firing by the process. I have always appreciated the labor of craft, those hours of work that the customer or appreciator may never consider when he views or uses the final product. This involvement in my craft and the resulting surface makes the work more authentically human. I would be remiss if I did not mention the role that community plays in my desire and love for firing wood kilns. Just as the development of my craft and aesthetic sensibility has informed me as an artist, the communal element of every firing leaves its mark on my work, my process, and my life. For the past five years I have fired my work in a sprung arch, single chamber, anagama kiln.

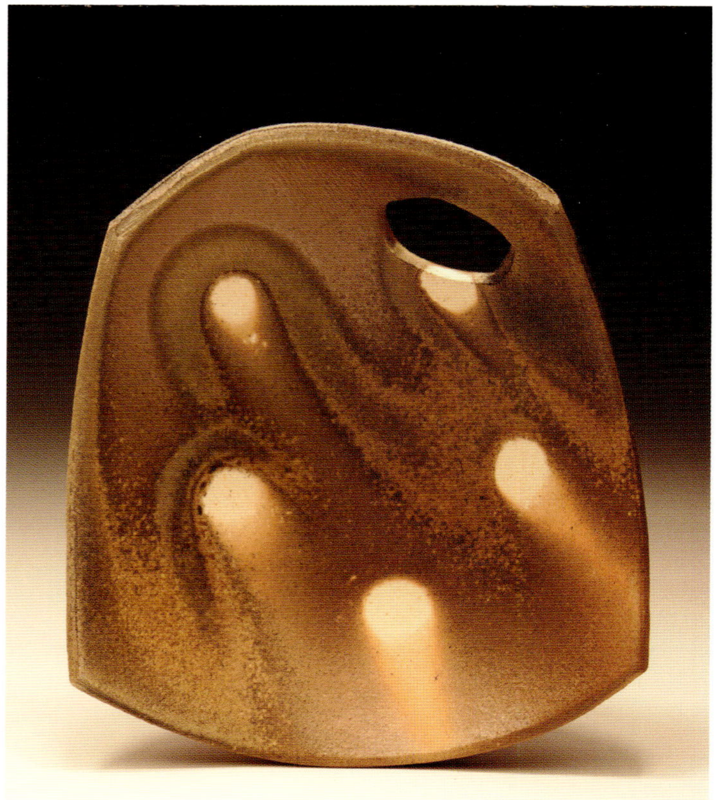

Joseph Delphia, *Plate*, 2011. Wood-fired stoneware, slab built, 8" x 7.5" x 2.5". *Photo credit: Joseph Delphia.*

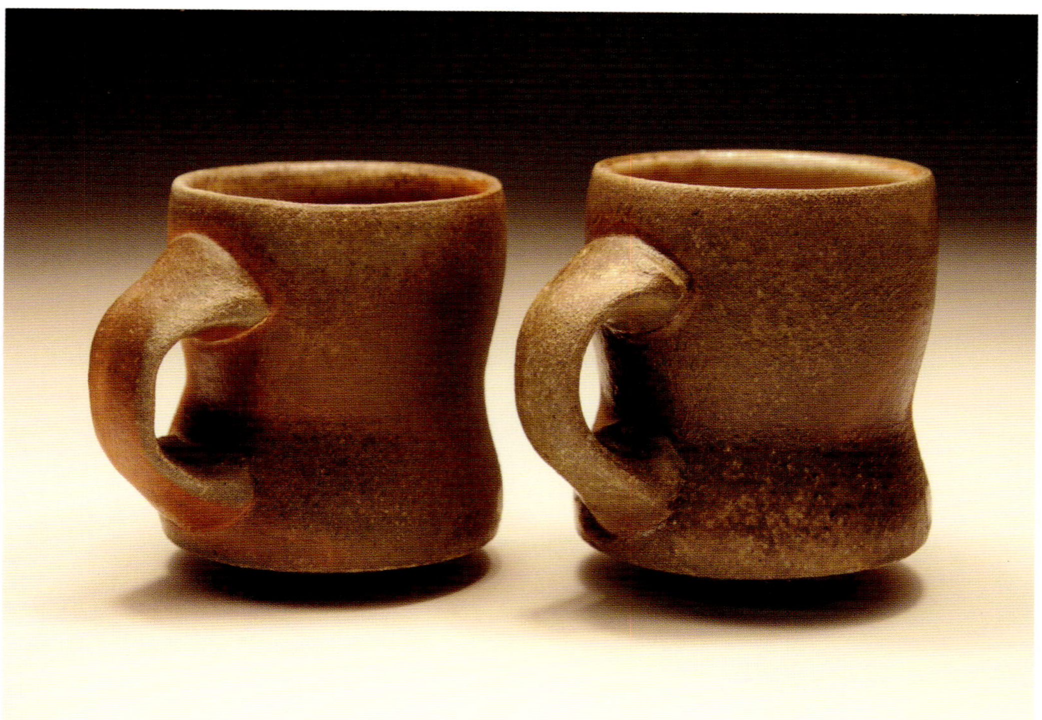

Joseph Delphia, *Mugs*, 2011. Wood-fired stoneware, wheel-thrown, 3" x 4" x 4". *Photo credit: Joseph Delphia.*

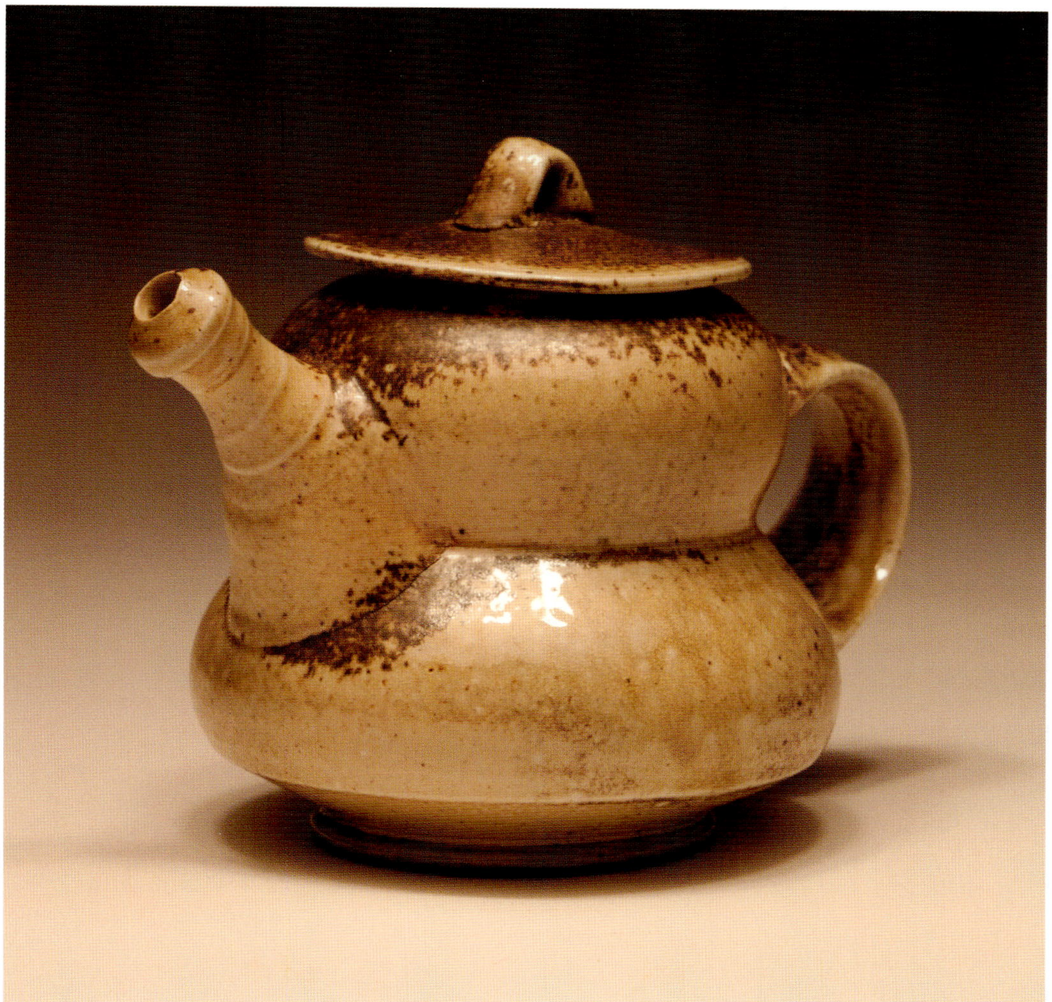

Joseph Delphia, *Tea Pot*, 2011. Wood-fired stoneware, wheel-thrown, 5" x 7" x 5". *Photo credit: Joseph Delphia.*

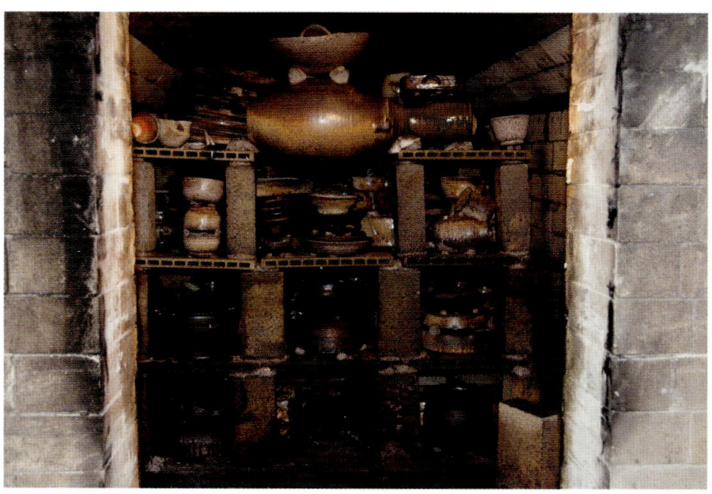

Joseph Delphia, unloading kiln. *Photo credit: Cara Rufenacht.*

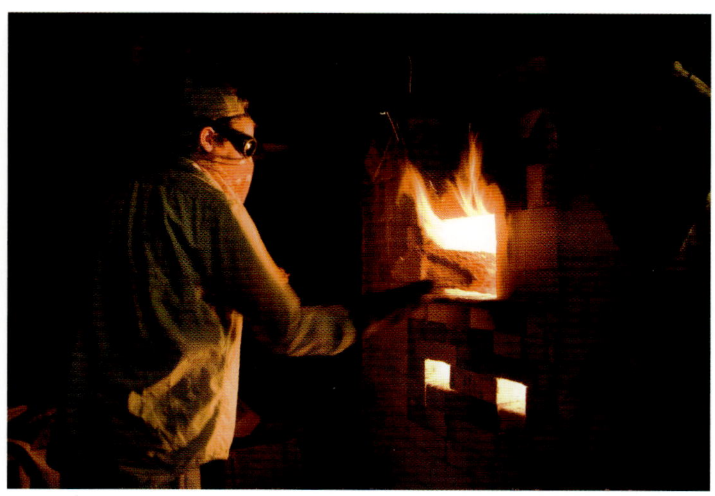

Joseph Delphia, night stoke. *Photo credit: Cara Rufenacht.*

STEPHEN DRIVER

Little Rock, Arkansas
smdriver@ualr.edu, www.stephenmdriverpots.com

Artist's Statement

I started my wood-firing career in Grass Valley, California, at the first Earth, Air, Fire, Water workshop in 1971. Several years later, in 1975, I studied with Peter Dick, a wood-fired earthenware potter in York, UK, and became a confirmed wood-fired potter at that time. I have been wood-firing almost exclusively since 1984. Initially I wood-fired because I liked being so intimately involved with the firing of the work. Soon I came to the conclusion that my pots were just better when wood-fired and have come to prefer painting with the fire; making work with the intent of having the surfaces and form interact with the fire in ways that help the pots come to life. The possibility of the effects of fire and ash on the pots continues to be an addictive experience for me. For years I only fired work in my own kilns (currently a 65 cu.ft. modified fast fire). In the last decade or so, I have been firing in with other potters in their kilns; Joe Molinaro's anagama at Eastern Kentucky University, Kevin Crowe's large three chamber kiln in Virginia, Judith Duff's train and anagama kilns in North Carolina, and most importantly Ron Meyer's anagama in Athens, Georgia. My work at the moment is divided between the softer, more subtle effects of my kiln and the more dramatic surface effects from sixty hours of firing and reduction cooling from Ron's small anagama (fondly nicknamed by Tony Clennell as the "Ratagama"). I have plans to build an anagama kiln at my studio in the future.

My current kiln is a 65 cu.ft. modified fast fire kiln. By modified I mean I have made several changes to Fred Olsen's design. The significant changes are: soft brick construction above the firebox, steel hinged door, positive sealed dampers at the inlet flues (a feature that allows a lot of flexible firing options), side stoking ports, a venturi exit flue (of sorts) with an oversized stack, and active and passive dampers. This kiln is very efficient and fires to cone 13 with very little effort. The most problematic aspect of the kiln is keeping it from over-firing. Works from this kiln have what I call soft ash effects.

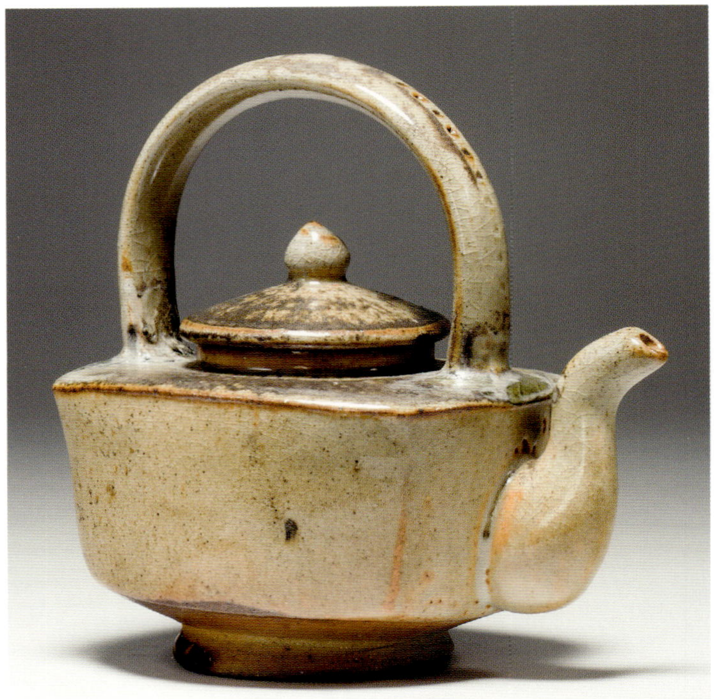

Stephen Driver, *Shino Teapot*, 2009. Wood-fired stoneware, thrown and altered, 8" H x 7" W x 4" D. *Photo credit: George Chambers.*

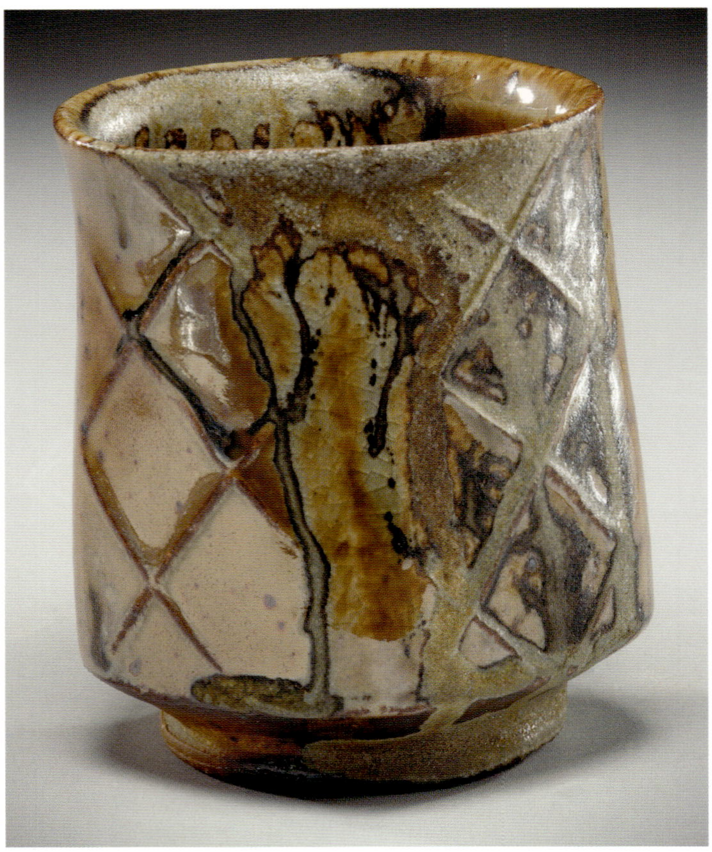

Stephen Driver, *Paddled Cup*, 2006. Wood-fired stoneware, thrown and paddled, 4" H x 3" W. *Photo credit: George Chambers.*

By this I mean the flame, fly ash, and the length of firing (twenty-four hours) creates an environment where the ash lands on the shoulders and tops of pots and interacts with the slips and glazes; enlivening the surfaces in a beautiful, soft manner. The functional pots are quite usable, the ash, slips, and glazes getting along in a civilized manner. In this kiln I use a stoneware body, semi-porcelain stoneware, and porcelain. I apply flashing slips and glazes on all the ware.

Ron's firings are sixty hours with a reduction cooling segment that adds another six or seven hours to the process. In this kiln I use porcelain only, sometimes flashing slips or a laterite wash, but mostly I fire the pots naked. The effects of a long firing and the reduction cooling leave the surfaces of the pots rich in color and texture that have a real sense of mystery to them. As we came to understand how to fire the kiln, the results have varied considerably from firing to firing, but we have always gotten something remarkable from every one.

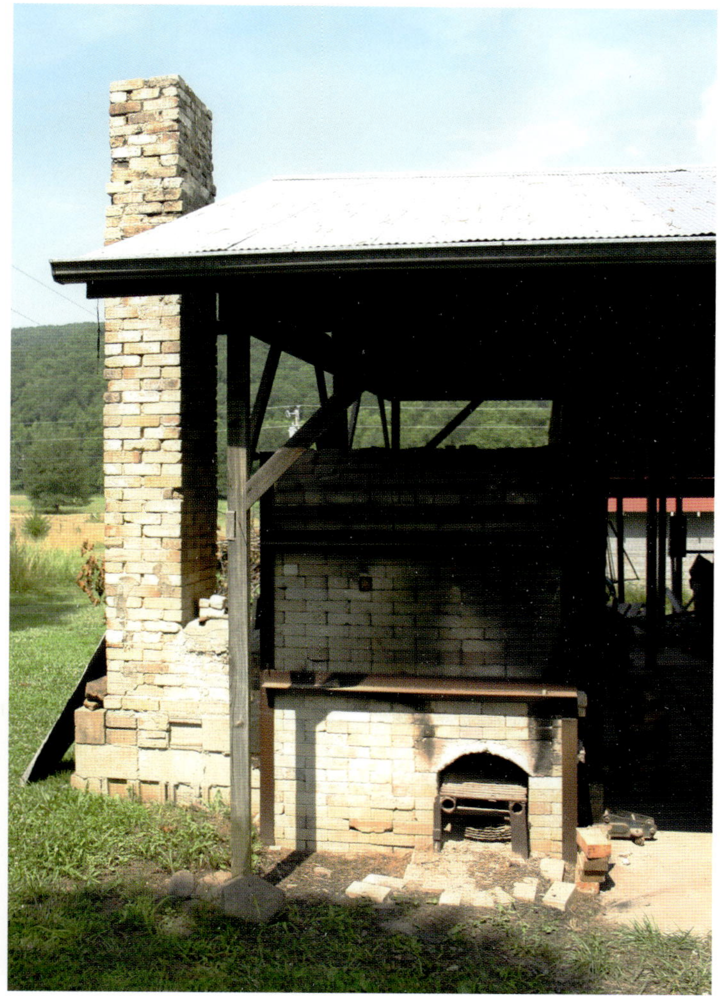

West view of kiln, 2005. Yale, Arkansas, modified fast fire kiln.

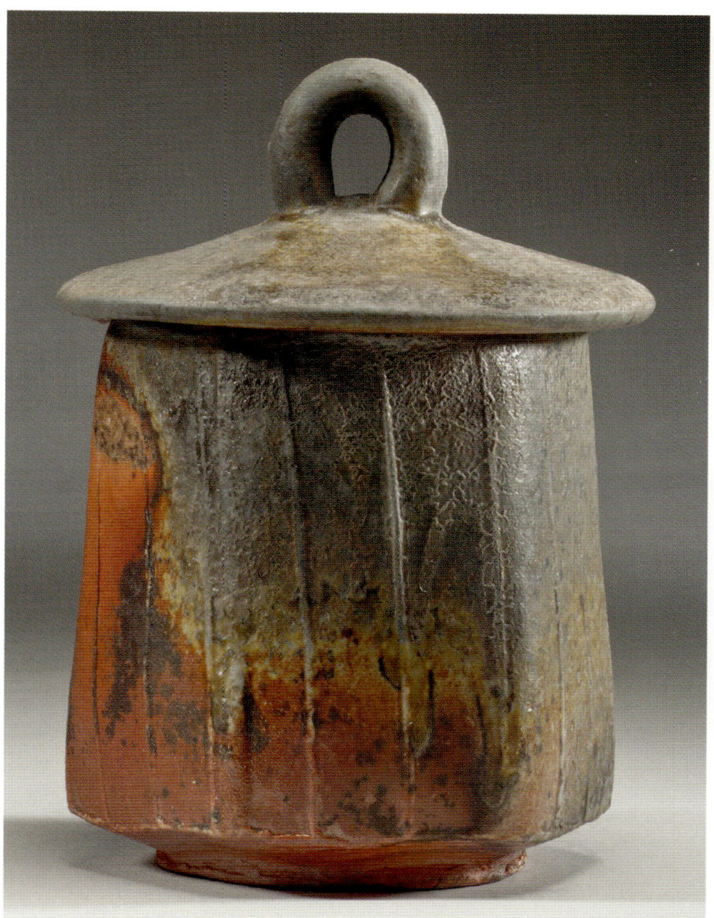

Stephen Driver, *Paddled Covered Jar,* 2010. Wood-fired stoneware, wheel-thrown and altered, Size 8" H x 6" W. *Photo credit: George Chambers.*

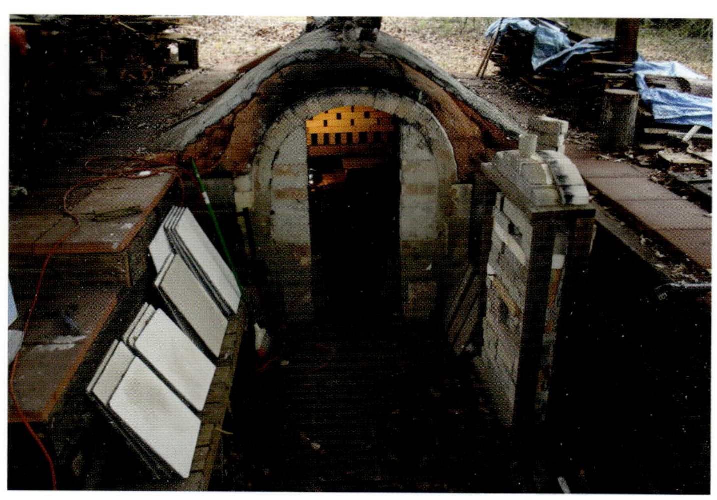

Ron Meyers kiln ready to start loading, 2010. Athens, Georgia.

CHRIS DROBNOCK

Mount Union, Pennsylvania
drobnock@gmail.com, www.drobnock.com

Artist's Statement

I would describe what I do as functional. I use hand-building techniques and the wheel to make objects that can be used in daily life in place of more mass-produced commercial ware. The daily rituals that exist like drinking coffee, using a butter dish, holding a plate, and arranging flowers all give me the opportunity to influence the aesthetics of a user's experience and how they experience it. This is the greatest joy: to know that my work is enhancing a person's life beyond that of the functionality of the polystyrene cup.

I alter the work at different stages during the drying process by stamping, paddling, and using whatever is around me at the time the work is being made to give the work visual interest. With my markings, glaze, and kiln choice, I hope to keep the wares fresh looking. A constant thought always in my mind is "do with intention," staying away from arbitrary marks and trying to preserve the nature of the tool marks, glaze splashing, and wadding marks (spots left by the protective clay used to keep the pots from sticking to the shelves in a wood kiln). Intuition and the free flowing nature of the clay also play a part, but unlike drawing you can't neatly trace in clay, so each mark should be made like it is the last.

I choose to use the anagama (Japanese-style wood kiln) to fire my wares as much as possible. It can, at times, create challenging surfaces to call functional but in the end, with a little creative thought, function can be achieved. There is also this connectedness, much like what deer or turkey hunters must feel (alas, I do not participate in these activities) when they are sitting in the woods waiting for their game to wander by, to all those individuals that have done a very similar thing throughout time. Yet, if those potters saw my work I am sure they would know it was not of their time. It is an aesthetic decision much like choosing to use pencil instead of oil paints. I choose to fire over a period of days with wood to form glaze and surface marks that could really be made no other way.

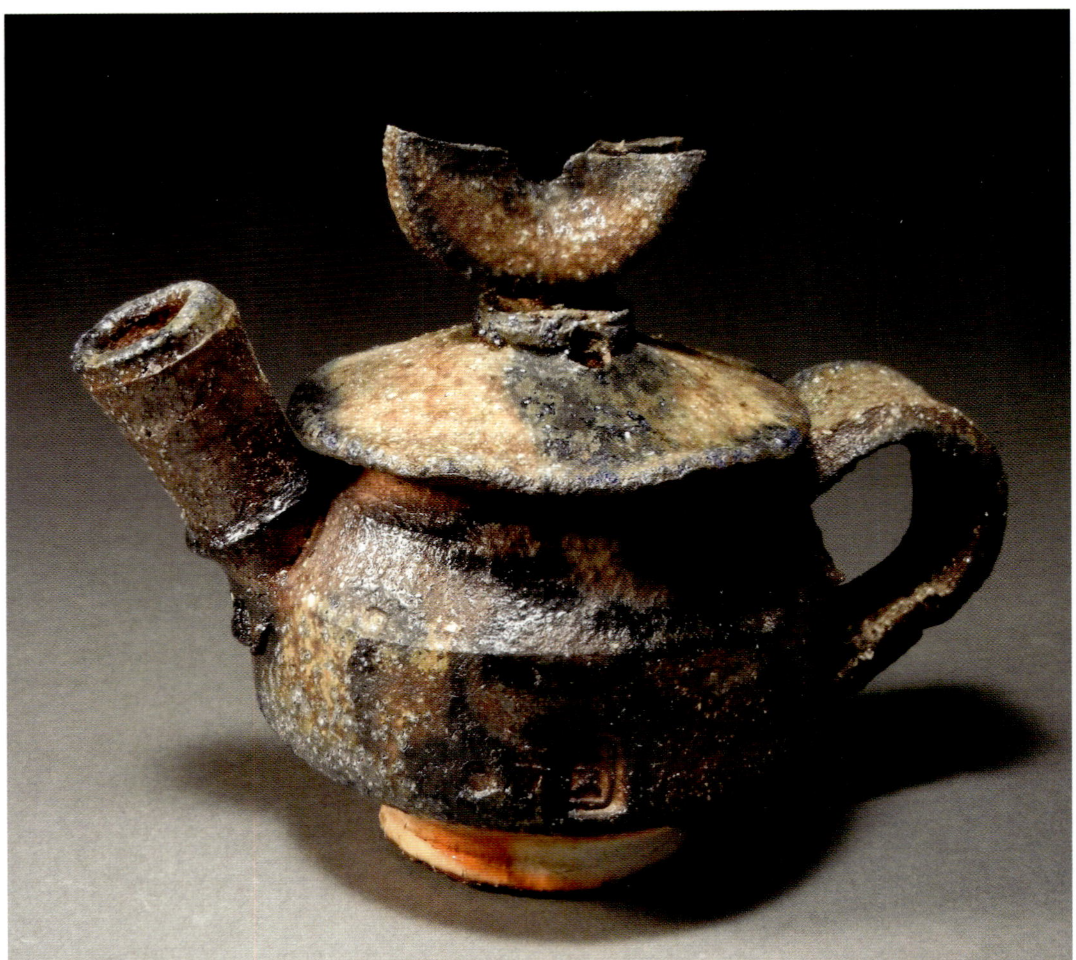

Chris Drobnock, *Teapot with Cobalt*, 2010. Anagama, stoneware, sand, cobalt stain, 6" x 4.5" x 6". *Photo credit: Chris Drobnock.*

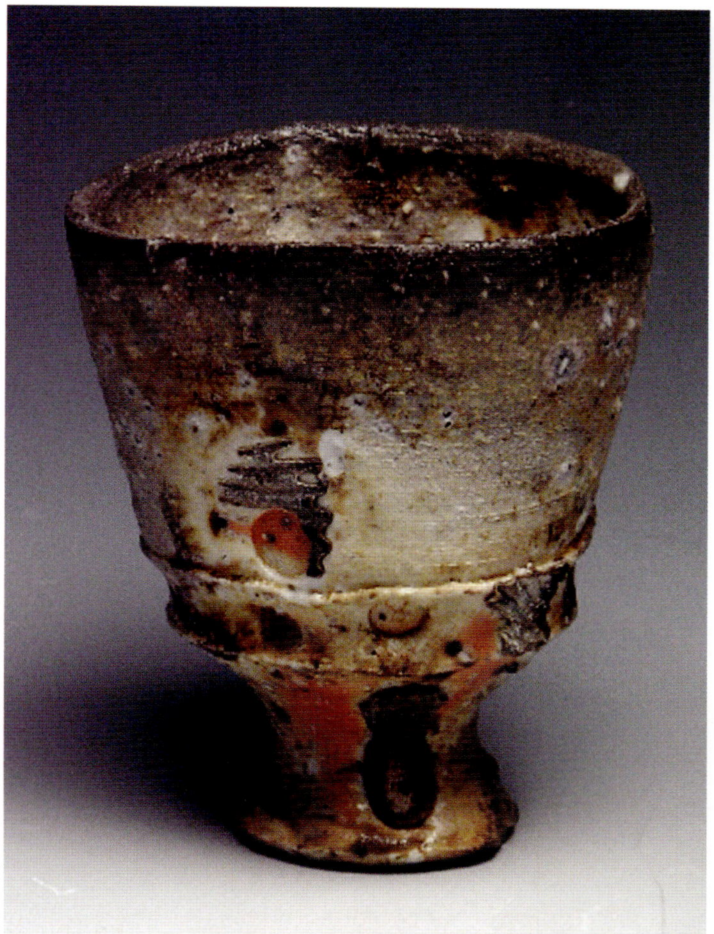

Chris Drobnock, *Porcelain Cup with Granite Inclusions*, 2009. Anagama, porcelain, unglazed, 5" x 3" x 3". *Photo credit: Chris Drobnock.*

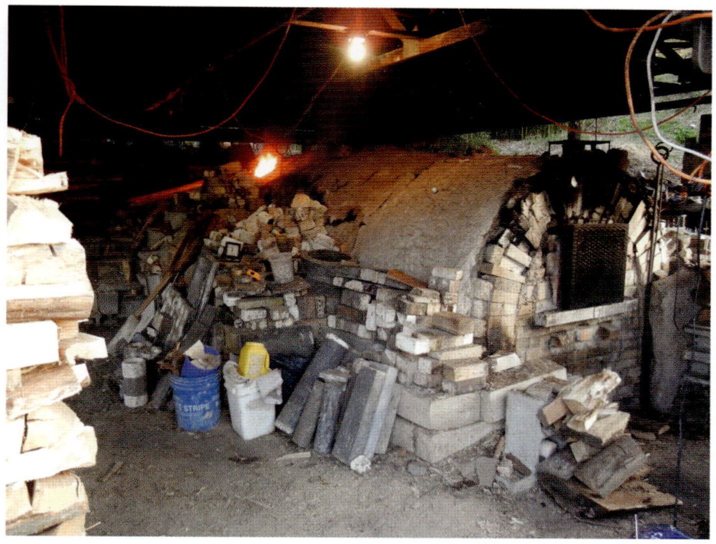

Chris Drobnock, Juniata anagama kiln, 2012. Juniata College anagama firing at cone 10. Huntingdon, Pennsylvania. April 2012. *Photo credit: Chris Drobnock.*

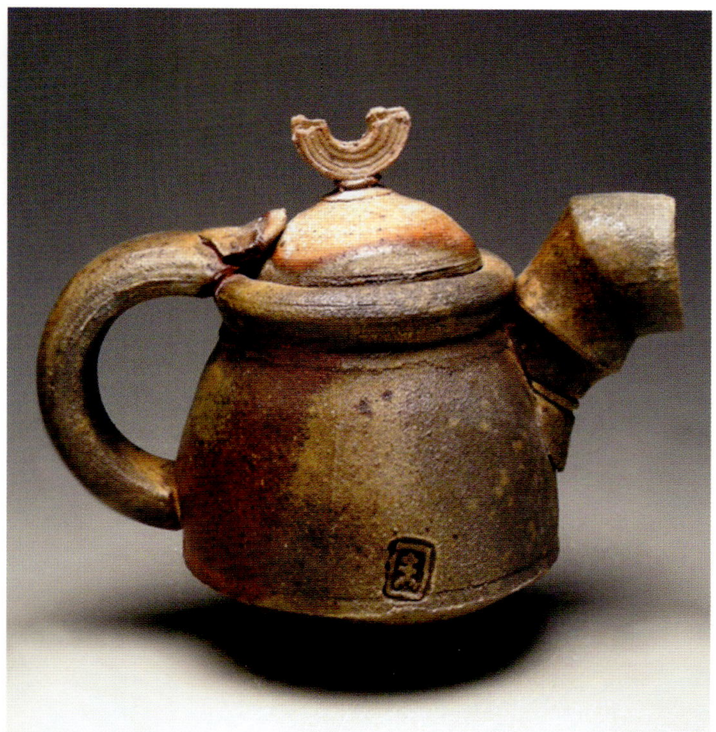

Chris Drobnock, *Teapot*, 2009. Anagama, stoneware, unglazed, 5" x 4" x 6". *Photo credit: Chris Drobnock.*

JUDITH DUFF

Brevard, North Carolina
JDuff@Comporium.net, www.judithduff.com

Artist's Statement

The pots I create are a combination of wheel-thrown and slab construction. They are altered in various ways to produce interesting and organic shapes and are fired in either a wood or gas kiln.

Surface decoration, whether from natural wood-ash or glaze applications, has been a consistent theme in my work. My love of nature and my background in biology and painting have influenced the forms, line-quality, and textures found in my pottery.

My work took on an exciting new dimension in 1998 when I built my first wood-fired kiln, a *Train Kiln*. Later, in 2005, I built an *anagama*. I am drawn to the unpredictability and one-of-a-kind nature of wood-firing and the resulting surface textures and ash deposits. Though labor-intensive, I find the process of wood-firing energizing and satisfying.

Every wood-firing creates a unique opportunity for fresh expression. Each pot has a history revealed only when it is removed from the kiln. Although each piece is carefully designed, I have learned to value the cracks and distortions that emerge as an aspect of what I call *unplanned art*.

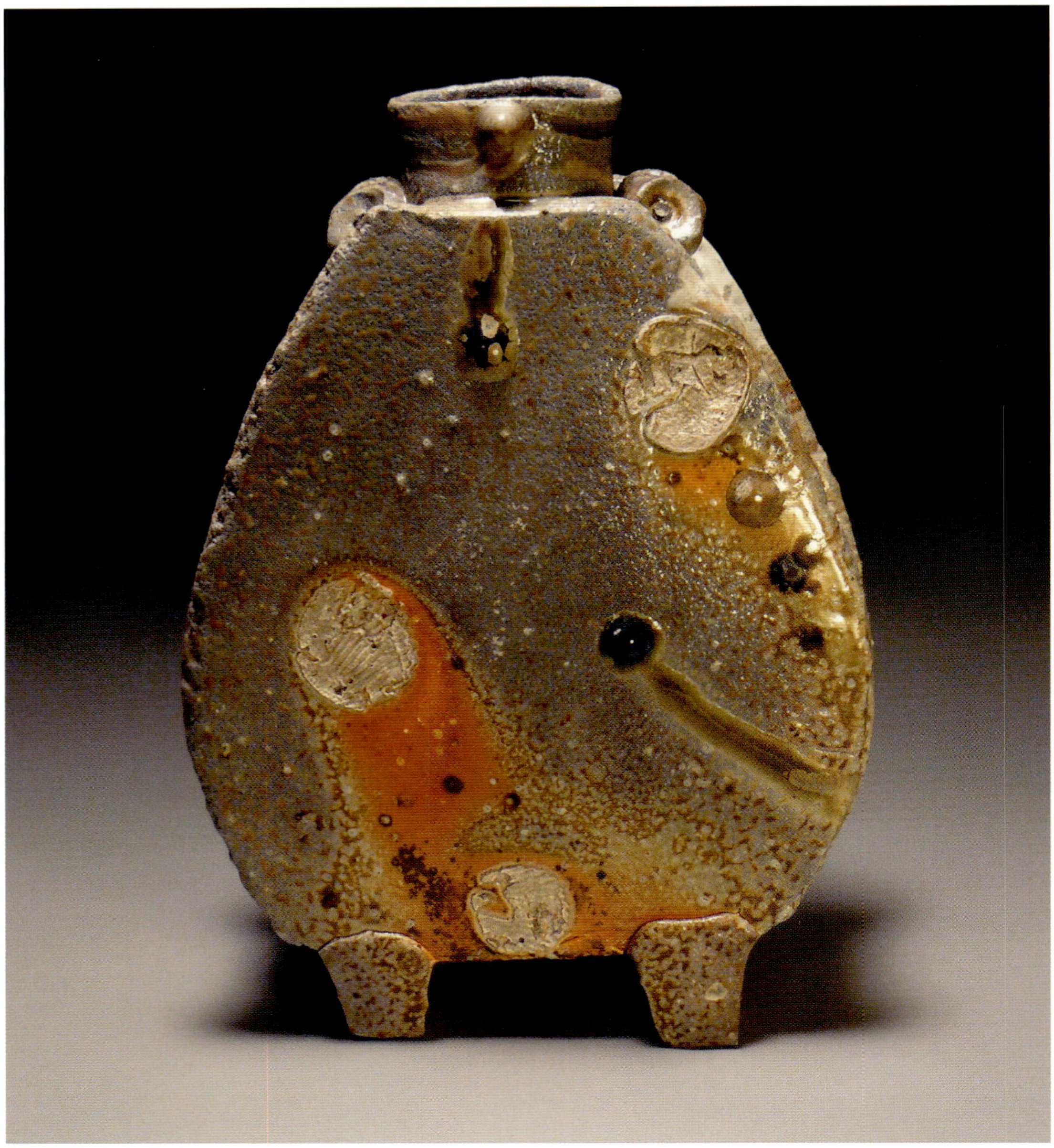

Judith Duff, *Tumble Stacked Bottle*, 2009. Shigaraki-style clay, wheel-thrown and slab construction, wood-fired to cone 12, natural ash surface, 13" x 8.5" x 2.5". *Photo credit: Judith Duff.*

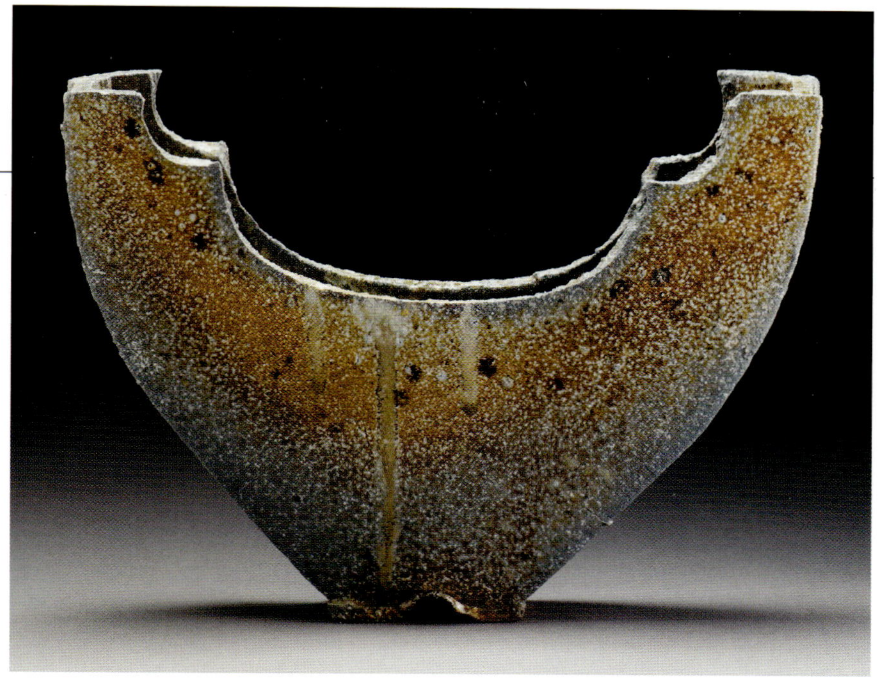

Judith Duff, *Scallop Vase*, 2010. Shigaraki-style clay, wheel-thrown and slab construction, wood-fired to cone 12, natural ash surface, 9.5" x 13" x 3". *Photo credit: Judith Duff.*

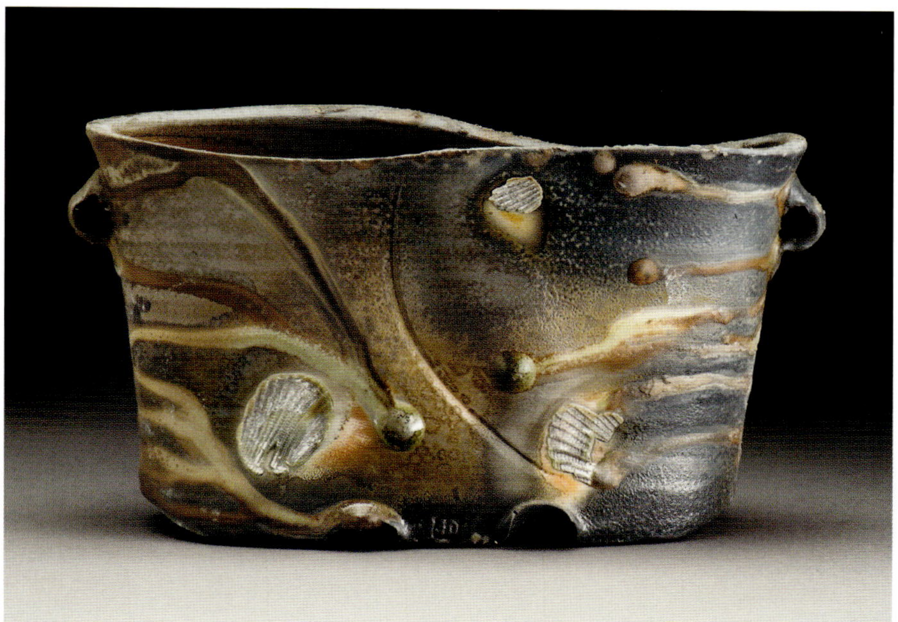

Judith Duff, *Oval Vase*, 2009. Porcelain clay, wheel-thrown and slab construction, tumble-stacked, wood-fired to cone 12, natural ash surface, 7" x 12" x 4.5". *Photo credit: Judith Duff.*

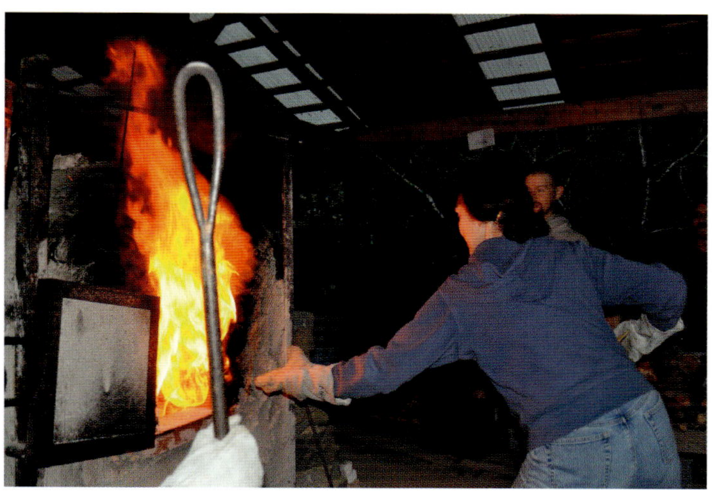

Firing anagama kiln.

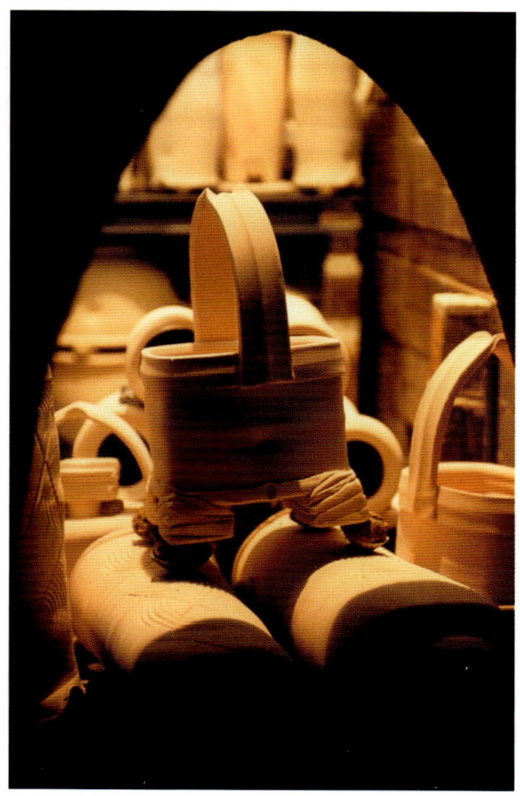

View of stacking through throat arch of Train Kiln.

TREVOR DUNN

Jacksonville, Florida
trevor.dunn@unf.edu, www.trevortdunn.com

Artist's Statement

My work in ceramics encompasses three different genres of the medium: utilitarian, ceremonial, and sculptural forms. For the past several years I have focused on reliquaries that reference ceremonial altars and discarded artifacts. The reliquaries function as a pedestal for supporting mezcal cantaros. Mezcal cantaros are simple earthenware vessels that were traditionally used to store and transport mezcal. These bottles became obsolete during the 1940s. By making pieces that refer to cast off functional objects and placing them in context with a truly antiquated relic, I am encouraging the user to ask questions of this relationship. Some of the questions may cause one to ponder objects they themselves use on a daily basis. What rituals do I employ in my daily life? What is my relationship with the past? How do my interpersonal relationships function? What is my purpose, and how will I avoid becoming obsolete?

I am interested in creating a ritualistic setting for consuming mezcal. Keeping the traditional form of the round-bodied mezcal cantaro necessitates an altarpiece to keep the mezcal from spilling. By creating a reliquary that functions on a basic level to hold the cantaro upright, I am grounding the dialog between the users to a certain location in hopes to mandate intimacy. Users of the cantaro and reliquaries will hopefully incorporate a respectful reverence for the dialogs that ensue.

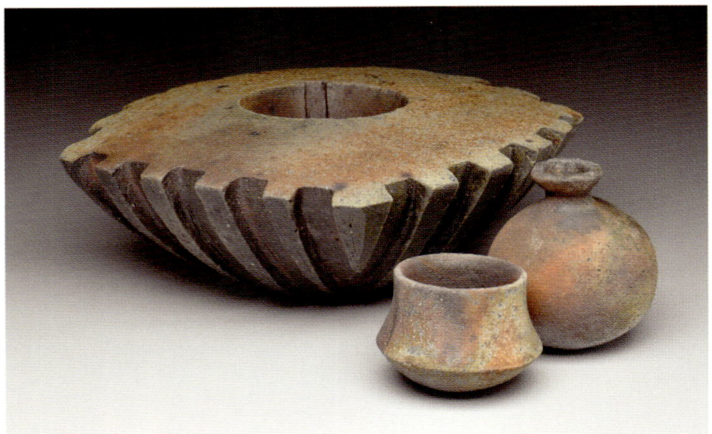

Trevor Dunn, *Square Cam Gear Reliquary with Cantaro and Cup*, 2010. Iron-rich stoneware, wood-fired to cone 8, cooled in reduction, 9" x 11" x 11". Photo *Credit: Trevor Dunn.*

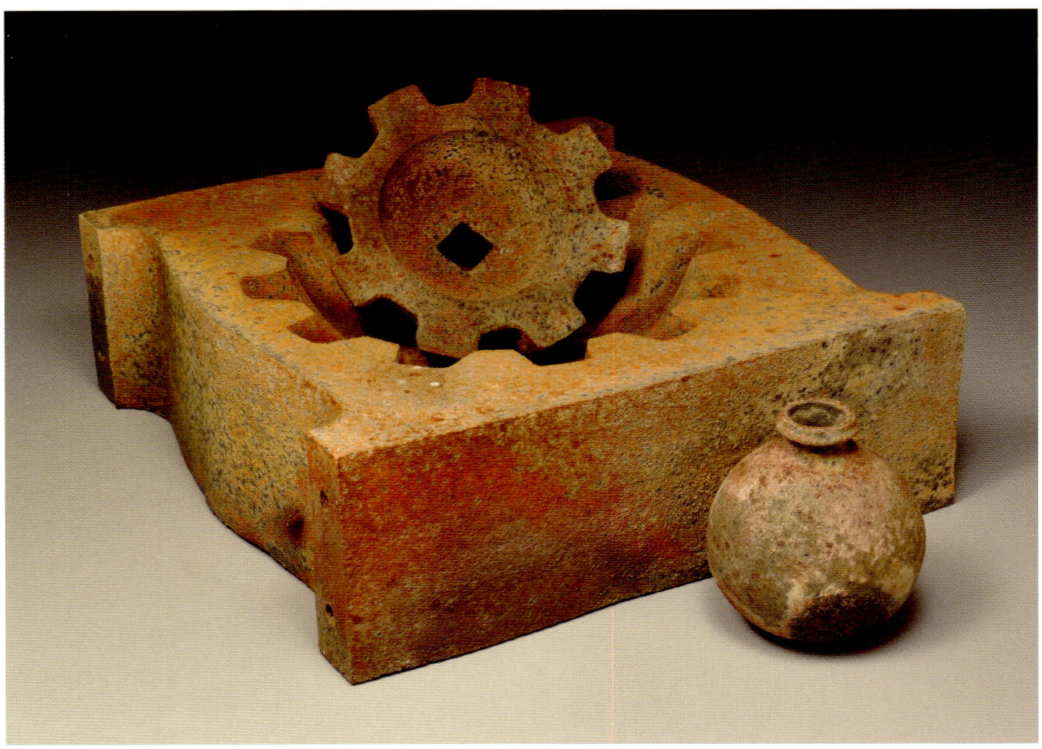

Trevor Dunn, *Vestige of Modernity*, 2009. Iron-rich stoneware, wood-fired to cone 8, cooled in reduction, 9" x 16" x 16". *Photo Credit: Trevor Dunn.*

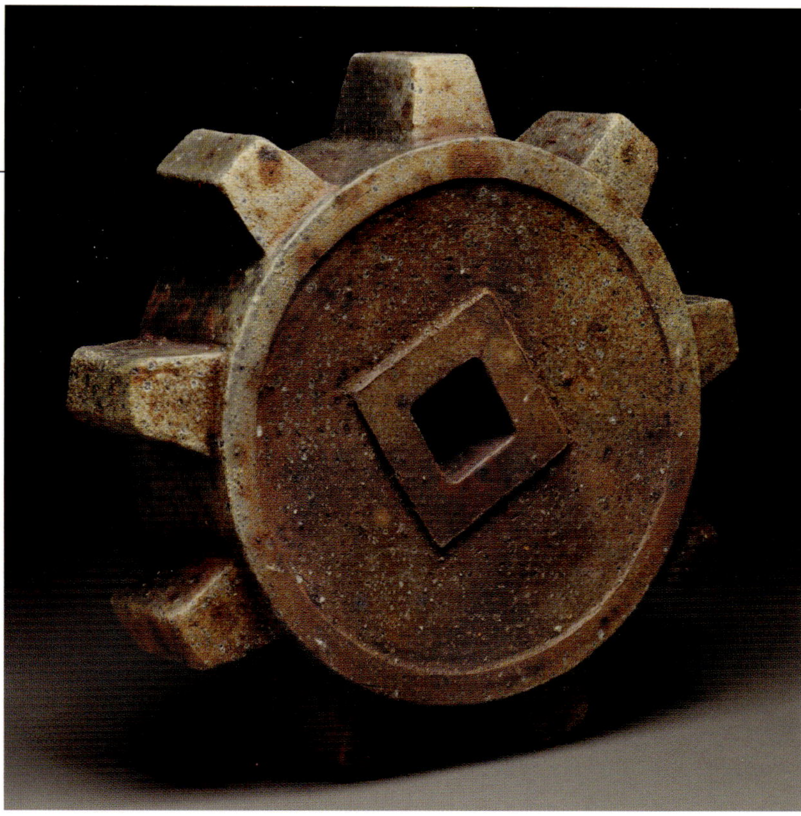

Trevor Dunn, *Hemisphere Block,* 2010. Iron-rich stoneware, wood-fired to cone 9, cooled in reduction, 9" x 11" x 11". *Photo Credit: Trevor Dunn.*

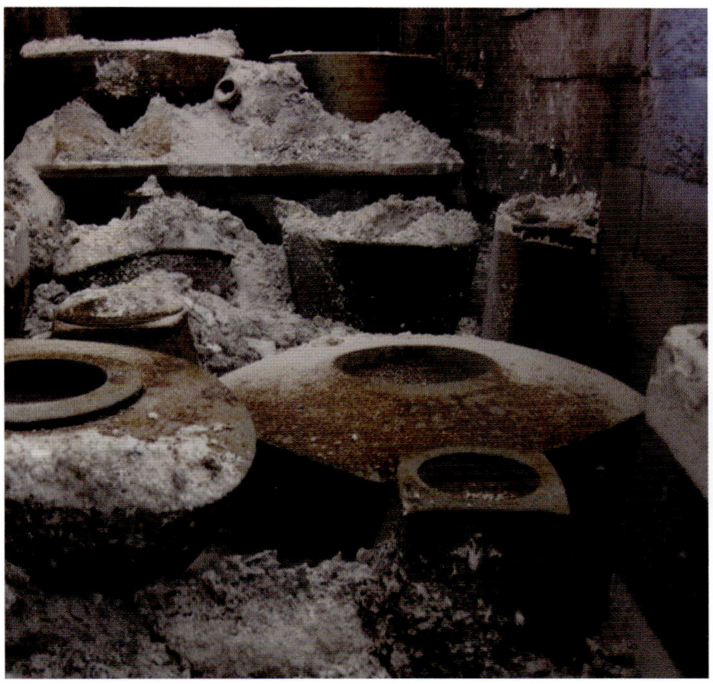

View of ember-buried pieces during unloading. *Photo Credit: Trevor Dunn.*

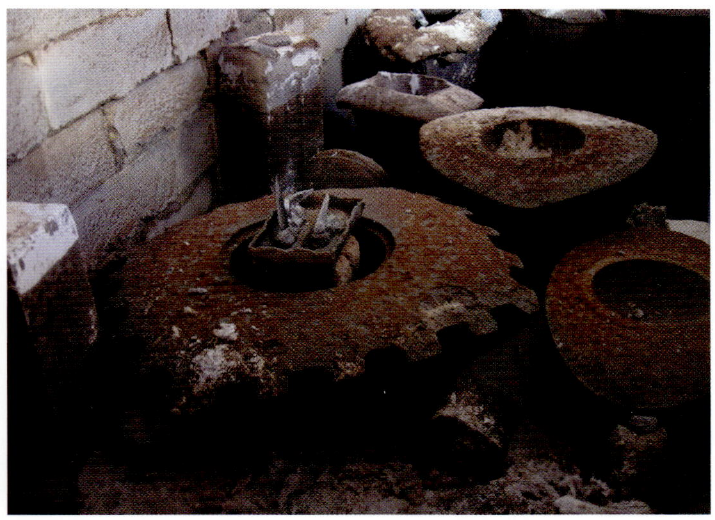

Firebox view during unloading. *Photo Credit: Trevor Dunn.*

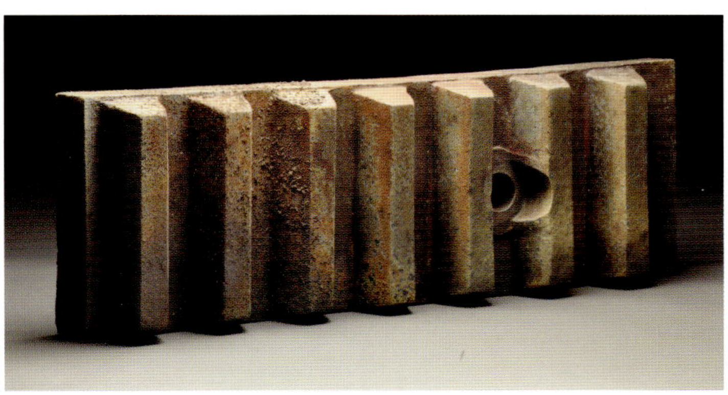

Trevor Dunn, *Rack,* 2010. Iron-rich stoneware, wood-fired to cone 8, cooled in reduction, 11" x 33" x 6". *Photo Credit: Trevor Dunn.*

I prefer to fire in an anagama style kiln or in a train kiln because of the dynamic effects that can be attained by the strong cross-draft of the flame. The firing and cooling technique that I use for my work generates pieces that look like artifacts. The reduction cooling is a critical part of this aesthetic. I use wet or frozen bark to cool the kiln along with a steady stream of water that is introduced into the firebox. I cool the kiln with water and bark down to 1800 degrees Fahrenheit, at which point I stop introducing water into the firebox. Another important factor in my process is to repeatedly bury my pieces in embers while soaking the kiln around cone eight. My anagama has a large firebox and oversized side-stoke troughs to accommodate pieces that are to be stoked upon.

DARREN EMENAU

Central Greenwich, New Brunswick, Canada
www.darrenemenau.ca, www.mnopottery.com

Artist's Statement

During the past fifteen years, my work has been focused on developing a balance between the natural tendency of clay and conscious control over its form and function. Some works are made from unwashed local clays containing rocks, twigs, and other matter, which contribute to unpredictable clay surfaces.

Local materials such as granite feldspars, slate, calcium, quartz, limestone, potash, and wood ash are utilized to create unrefined glazes.

The work is fired in a modified Olsen Fast-Fire wood/salt kiln. The firings are very labor intensive, but can reach 2300 degrees Fahrenheit within five hours and consume less than a cord of one inch hand split cedar. I usually have a group of people to share in the firing, but it is manageable as a solo endeavor.

This wood kiln design allows an array of firing conditions, and is the natural step in working so closely with local materials, and lends the forms an essence of age and history.

This is a local kaolin thrown vessel. After bisque, Ball clay is applied in various layers. The vessel was fired on the side upon wadding, and wood/salt fired to cone 10, with both a body and glaze reduction. Interior glaze made from wood ash, earthenware, granite, limestone, kaolin, and potash.

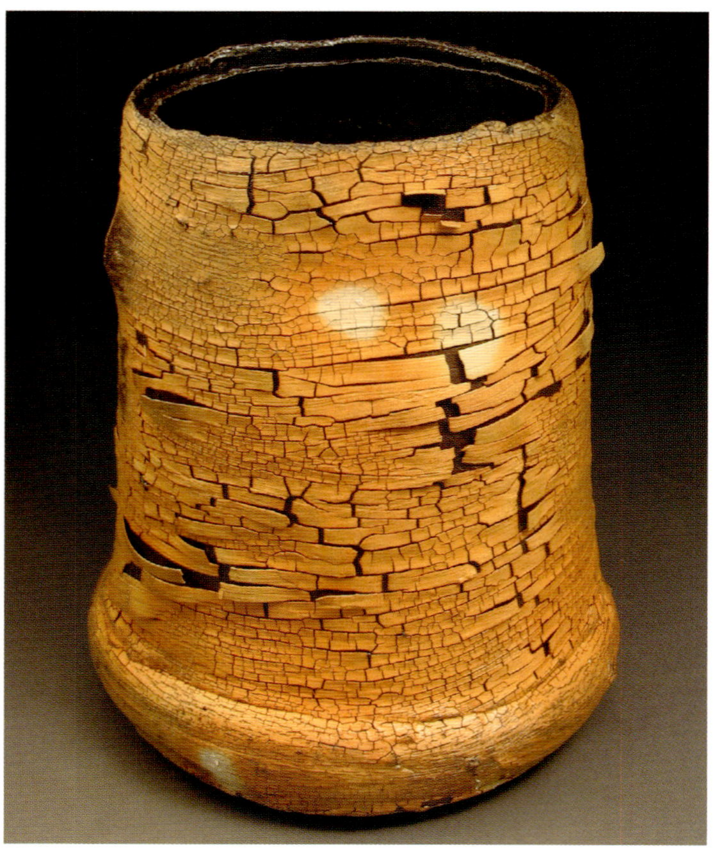

This is an altered extruded vessel with kanthal wire. After bisque, a glaze made from local feldspar, silica, limestone, and kaolin was applies. The vessel was fired on the side upon shells, and wood-fired to cone 10, with both a body and glaze reduction. Interior glaze made from wood ash, earthenware, granite, limestone, kaolin, and potash. A cork was glued as a stopper, and Jeneca Klausen soldered a Sterling silver chain to the kanthal wire.

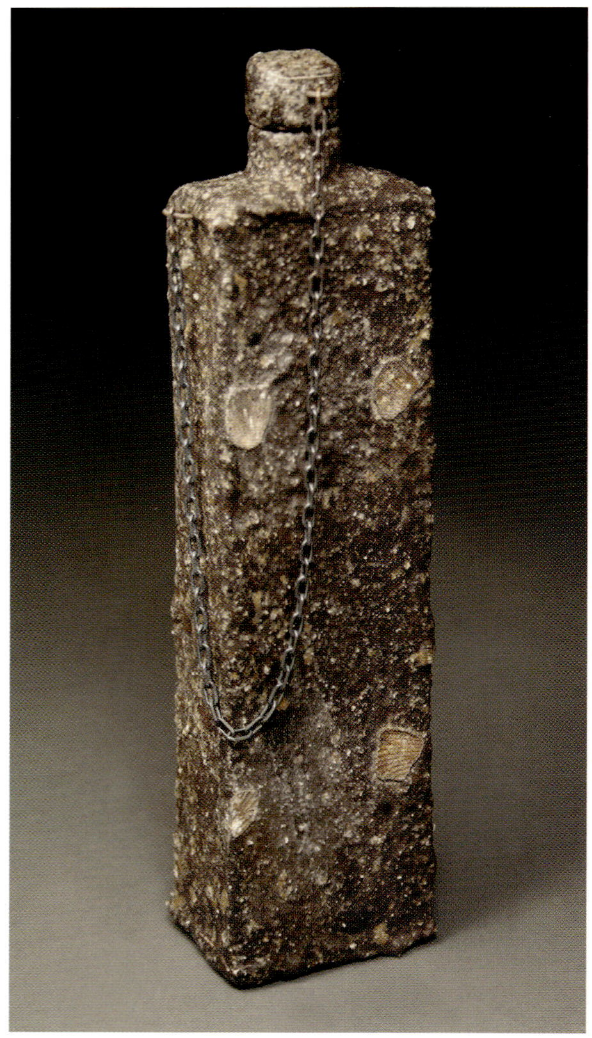

Darren Emenau, *Baron Flask,* 2009. Cone 10 wood-fired porcelain, extruded, altered, Carbon trap shino glaze, Sterling silver chain (Jeneca Klausen), 9" x 2" x 1.5". *Photo credit: Darren Emenau.*

Darren Emenau, *Buckshot*, 2009. Cone 10 wood-fired local Kaolin, wheel-thrown, altered, local granite/limestone/ earthenware, 6" x 4.5" x 4.5". *Photo credit: Darren Emenau.*

This is an altered porcelain-thrown vessel. After bisque, a shino glaze was applied with a coat of sodium silicate on top of the glaze. The vessel was fired on the side upon shells, and wood-fired to cone 10, with both a body and glaze reduction. Interior glaze made from wood ash, earthenware, granite, limestone, kaolin, and potash.

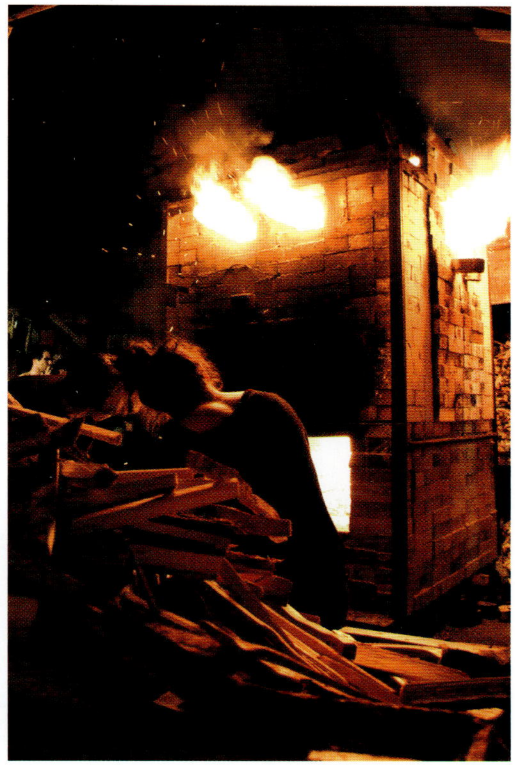

Darren Emenau, Troy Sand with Nora Kennedy. 2011.

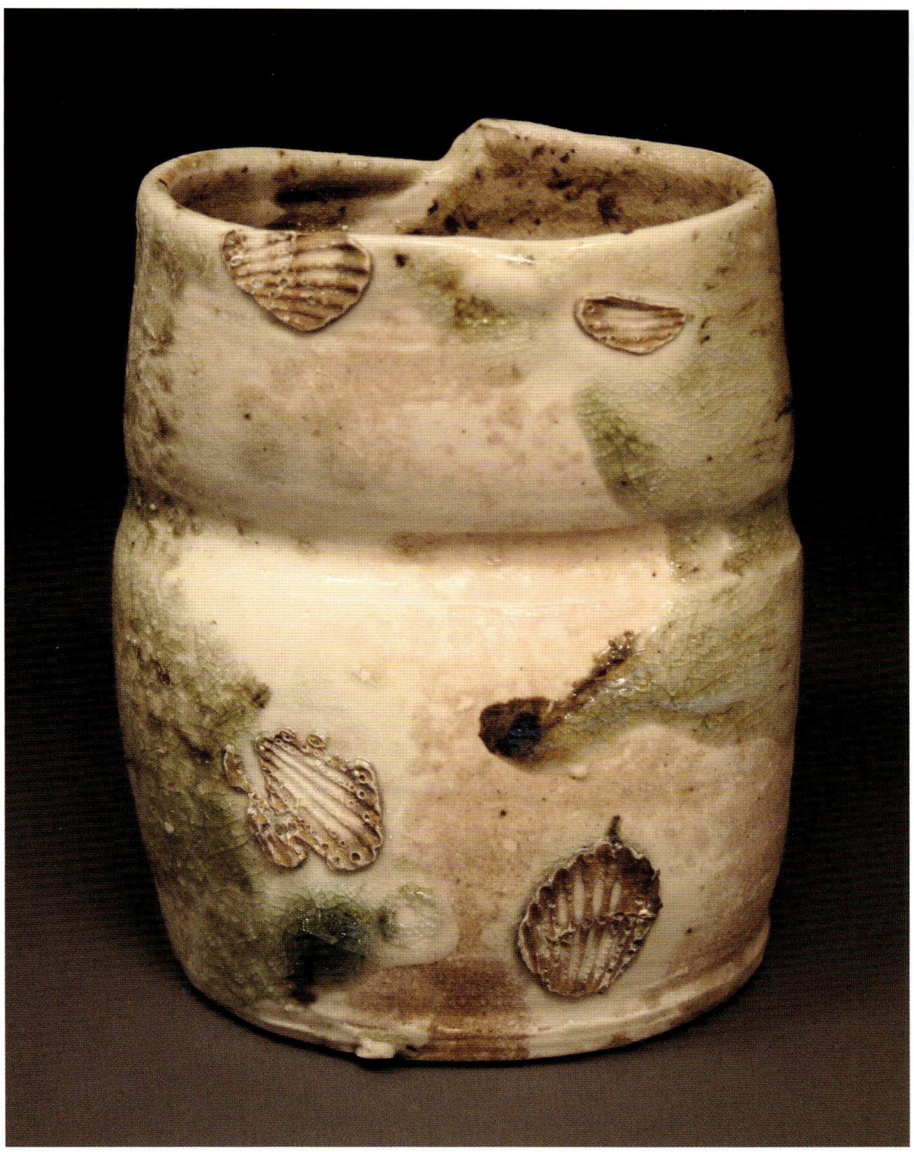

Darren Emenau, *Binky*, 2009. Wood-fired porcelain, wheel-thrown, altered, Carbon trap shino glaze, natural ash glaze, cone 10 wood/salt fire, 7" x 6" x 4". *Photo credit: Darren Emenau.*

BEVERLY FISHER

Philadelphia, Pennsylvania
beverlyfisher11@gmail.com, www.woodfiring.com

Artist's Statement

Some things that are compelling about wood-firing include being an active participant in the finishing of my work and getting to spend time outside in all seasons and weather conditions. Part of it is getting to observe the nature of flow dynamics and fire. To start with a couple of twigs and a match, and end with a river of flame is fascinating and demands complete attention. By working with it's nature, I get a little of what I want, letting it do what it does best. Flame, like water, follows the path of least resistance.

The years spent learning to fire kilns has opened into designing and building kilns. Taking the first kiln from a pencil drawing to a finished form was exciting. The first kiln that Kenton Baker and I built was a double- chambered catenary arch kiln, for Perkins Art Center in Elmer, New Jersey.

In recent years my home kiln is at Still Mountain Center in South Kent, Connecticut, with Joy Brown. We fire once a year in an anagama. I am part of the nomadic tribe that descends every July to load and fire the kiln. The scale of this type of kiln is much larger than most—it can be an intimidating experience to walk up and into a kiln, yet the interior space is covered in secrets, the surface reveals the passage of flame, and the residue of use. My work is influenced by the architecture of kilns as much as by the finish of the firing.

The underlying motivation for making art has always been the desire to make work that was quiet, in the sense that it reveals itself slowly over time. Using a formal language that places the viewer in familiar territory, the work does not depict directly but rather traces a vital definitive gesture suggested by nature, triggering a state of reverie and stillness. Making concentrates my attention in a way that is close to meditation—the eye taking in and the hand giving back.

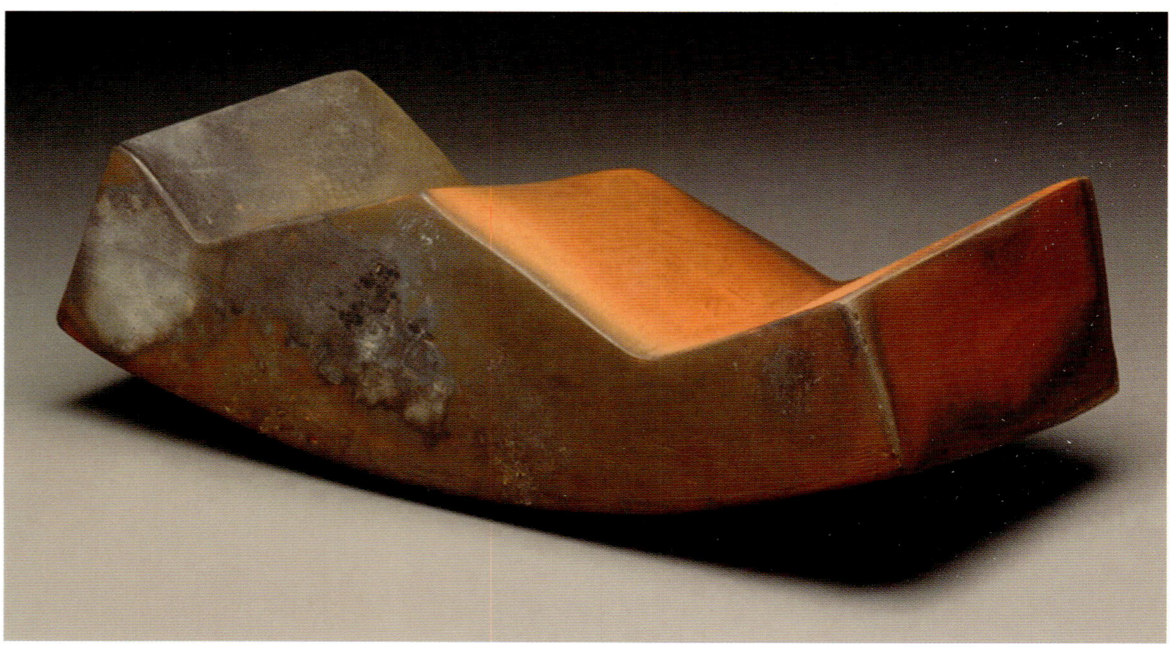

Beverly Fisher, *Orange Zig*, 2007. Unglazed stoneware, wood-fired, 6" x 18" x 10".
Photo credit: John Carlano.

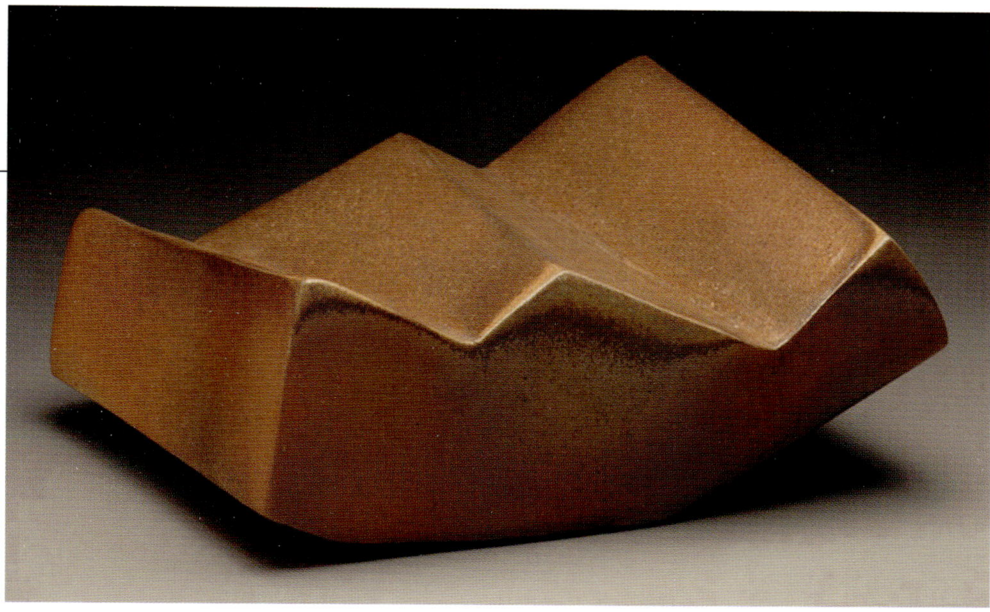

Beverly Fisher, *Original Zig #1*, 2007. Unglazed stoneware, wood-fired, 6" x 20" x 12". *Photo credit: John Carlano.*

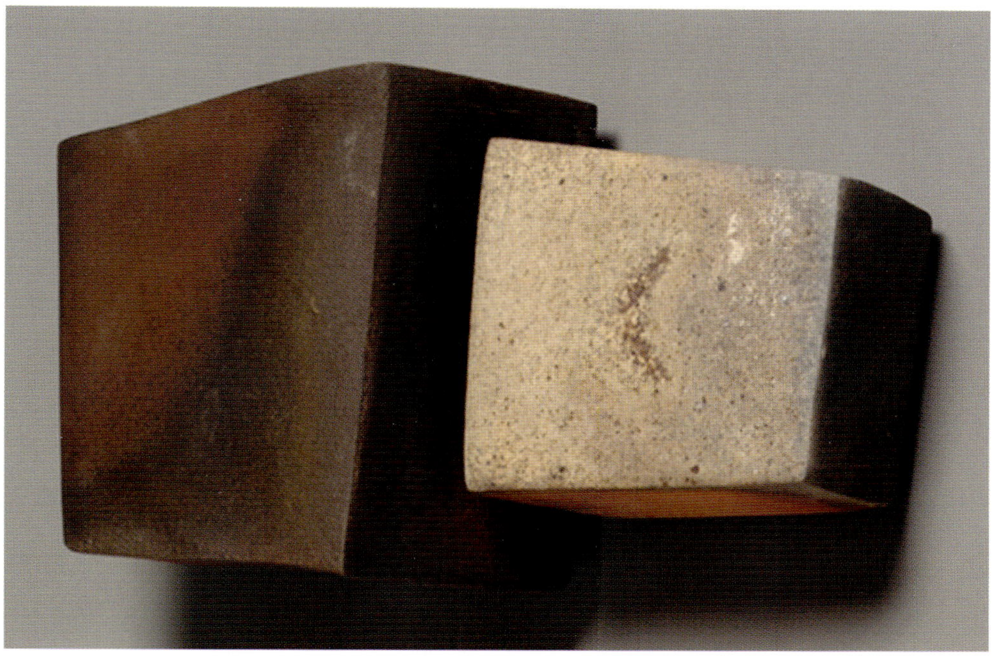

Beverly Fisher, *Untitled*, 2008. Unglazed stoneware, wood-fired, 6" x 9" x 3". *Photo credit: John Carlano.*

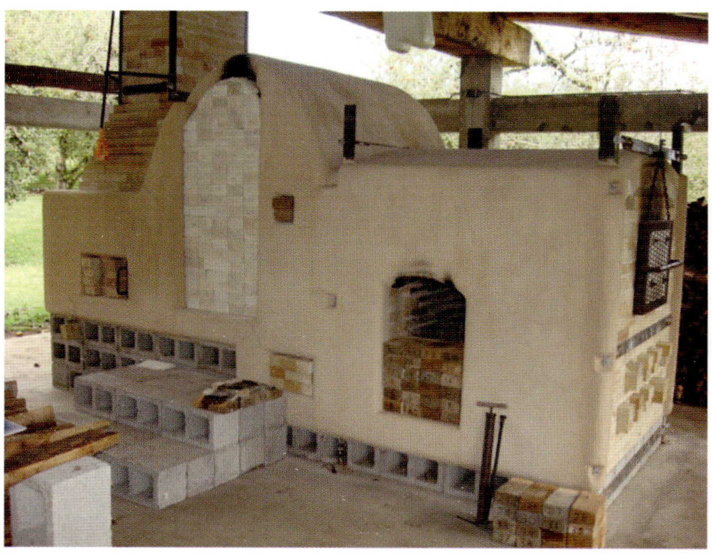

Beverly Fisher/Kenton Baker, 2008. Carlos Vizcaino Kiln Project Mexico City, Mexico, design and consultation on a single chamber cross draft kiln with a step down grate system. *Photo Credit: Carlos Vizcaino.*

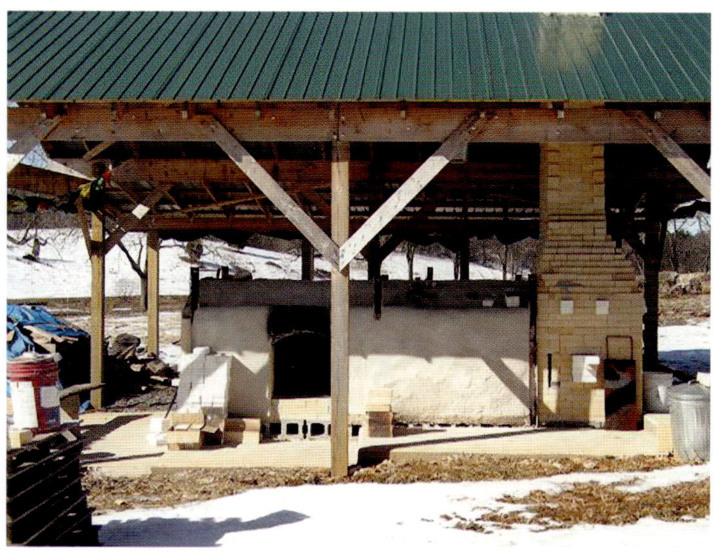

Beverly Fisher/Kenton Baker, 2000. Gil Mallinckrodt Kiln Project Kirkwood, Pennsylvania, design and build a step grate anagama. *Photo credit: Kenton Baker.*

DOUG FITCH

Devon, United Kingdom
www.douglasfitch.co.uk, www.slipware.blogspot.com

Artist's Statement

Wood-firing allows me to engage with the pots throughout their journey, from wet clay, dug from the earth, to finished, fired ceramic.

The flame, licking the surfaces of the pots, creates variations in temperature within the chamber of the kiln. This, combined with gasses from the burning wood, generates a diverse range of tones and colors. I employ a limited palette of materials. I use three slips: black, white, and green. These slips are covered with either a yellow or honey glaze. The ever-varying circumstances within the chamber ensure that, in spite of such a basic selection of raw materials, no two pots are ever exactly the same.

My kiln is a downdraft design, with a chamber capacity of 63 cubic feet. It usually takes me six weeks to make the pots to fill it. It has two fireboxes, fuelled with a blend of soft and hardwood, which I salvage locally. The hardwood is used for a short, slow flame and the softwood for a longer flame when a sharper rise in temperature is required.

Some of the pots are fired raw and some bisque fired. A typical firing will take between fourteen and twenty hours, but the length of the firing depends upon variables within the pack, affecting the path of the flame and the weather conditions on the day.

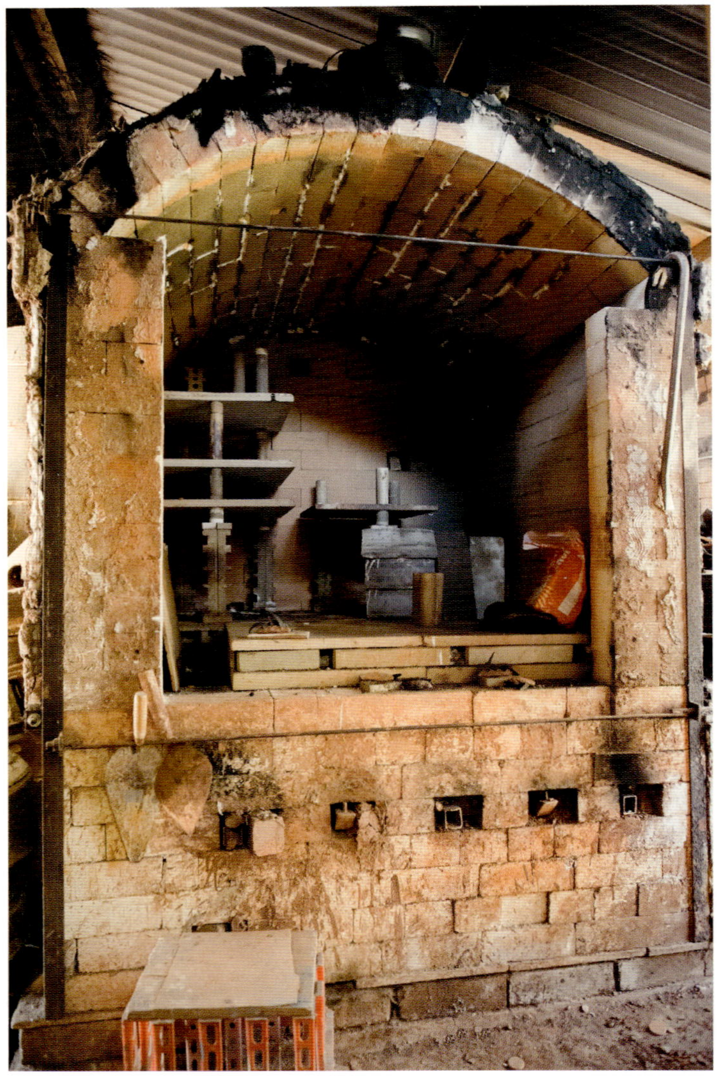

The kiln chamber, 2012. *Photo credit: Claire Borlase.*

My work draws influence from an ancient tradition of English earthenware. I use locally sourced materials wherever possible—as similar as I can find to those that were employed by many generations of potters before me. I am particularly inspired by the work of medieval English potters and my forms often make reference to their work.

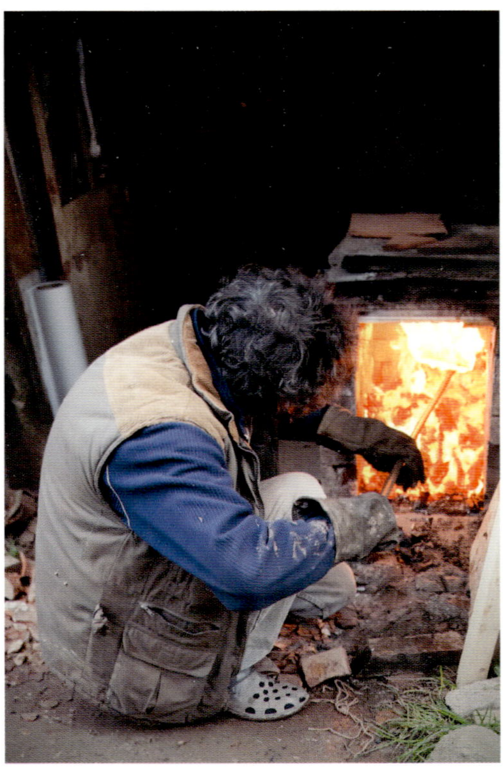

Doug Fitch raking embers, 2012.
Photo credit: Claire Borlase.

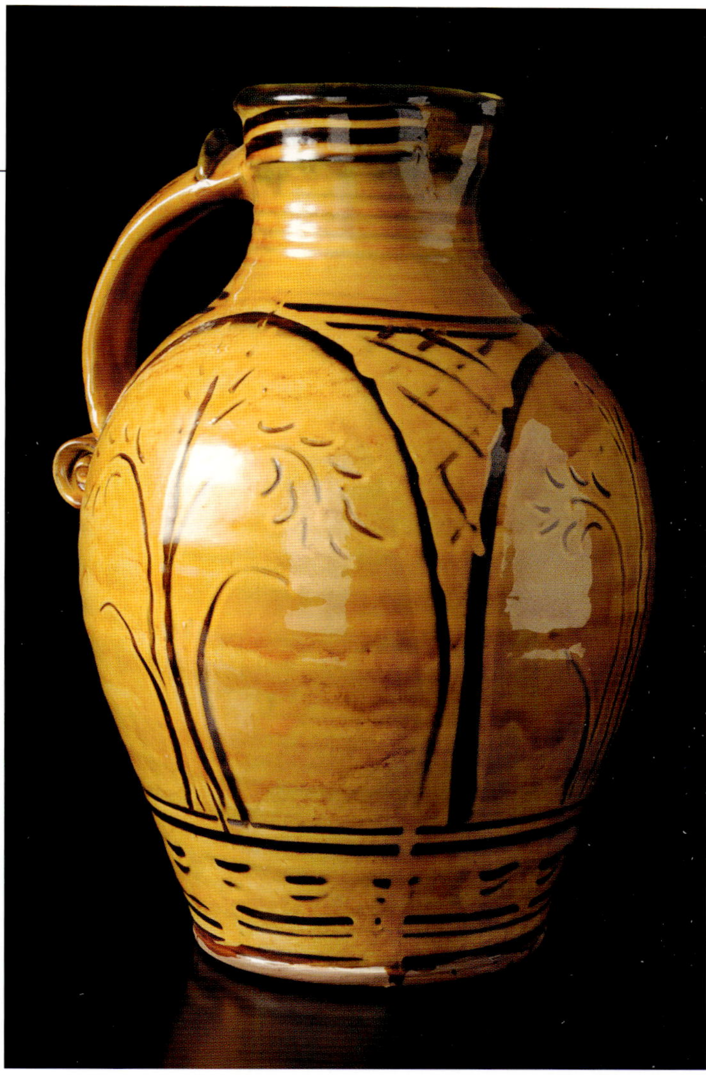

Doug Fitch, *Jug*, 2011. Wood-fired earthenware, wheel-thrown, sgraffito slip decoration. *Photo credit: Jonny Thompson.*

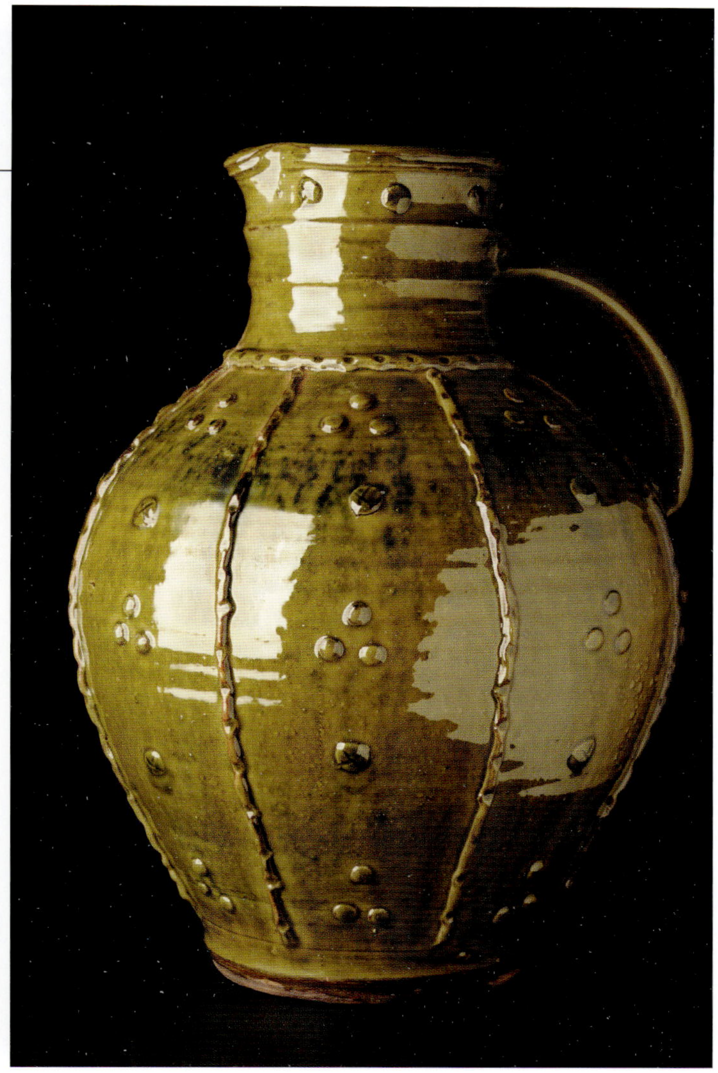

Doug Fitch, *Jug*, 2011. Wood-fired earthenware, green slip with applied decoration, 13" x 9.5" x 9.5". *Photo credit: Jonny Thompson.*

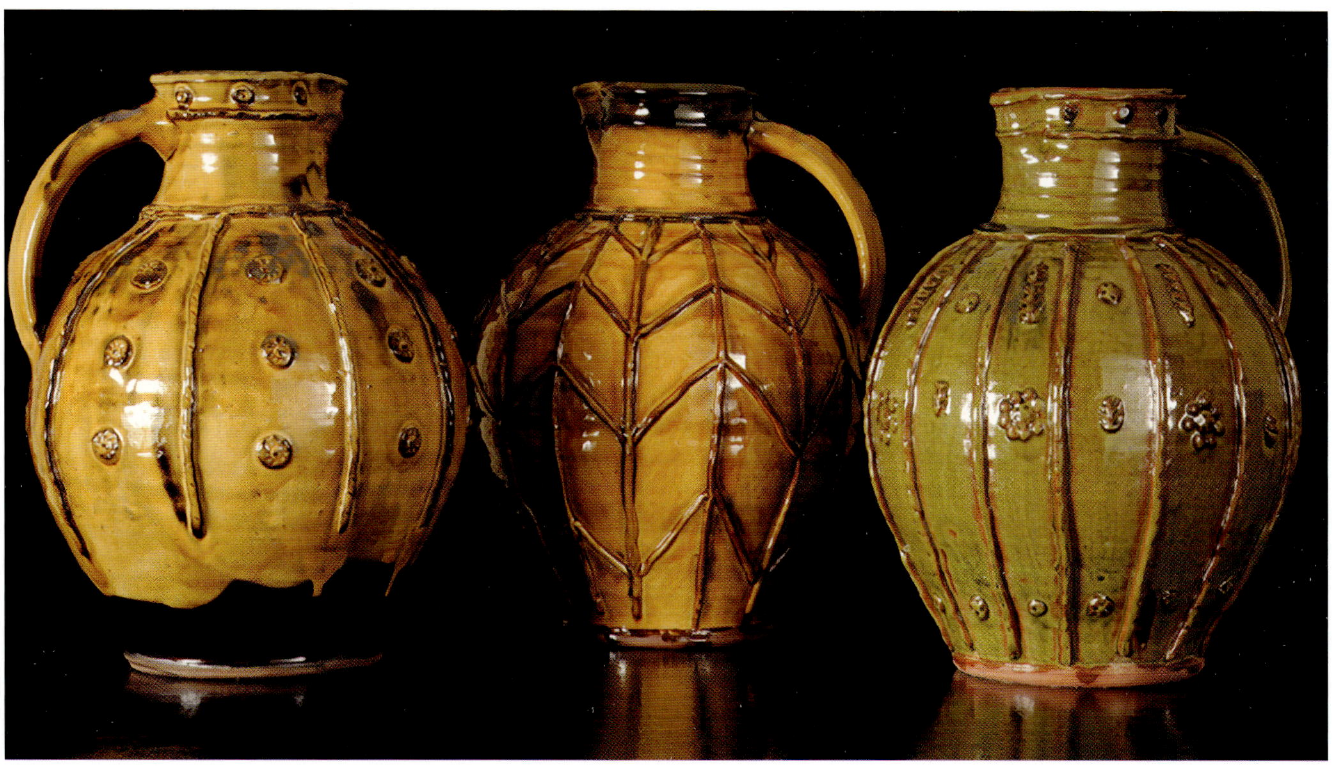

Doug Fitch, *Jugs*, 2011. Wood-fired earthenware, slip decoration with applied sprigs, 13" x 9.5" x 9.5". *Photo credit: Jonny Thompson.*

MATTHEW GADDIE

Bardstown, Kentucky
meadows@bardstowncable.net, www.meadowspottery.com

Artist's Statement

My studio is located at "The Meadows," a 315 acre family farm. The land was purchased in 1864 by my stepfather's great-great grandfather. Moving from a subdivision onto this farm at the age of nine had a profound impact on my life. I remember days of sitting under the old oak on the hill near the barn and wondering what tales of yesterday this old tree retained in its branches. How would it describe the horse drawn carriages that passed by on their way to Bardstown so faithfully following what was then the Wilderness Trail? I would also ponder what could only be described as the spiritual way in which my stepfather was so faithfully the steward of the land. I took note of how intimate his relationship was with the land. My works are an attempt to reflect the ebb and flow of the simple life I grew to admire as a child. My work is primarily functional pottery, straightforward in form and intent. Inside the frame and philosophy of practicality is the struggle to communicate the underlying connection I see between the rural, hard-working days of my youth and the life I want to live as a potter. The cyclical pattern of throwing, trimming, glazing, and firing are comfortable for me. I am a farmed-raised laborer who has wonderful memories of planting, weeding, and harvesting. In the age of quick marts, high speed connections, fast food, and one-stop shopping, I try to live my life slowly and with purpose. In a time of mass produced uniformity, I try to create works that are unique, each piece having recorded its own tale of creation—a tale of struggles, of successes, of failures, of imperfections, and of hopes ... it is a human tale. It is that familiar story that clay can so honestly record. Wood-firing is the most logical and honest way to finish my personal expression. The act of firing brings to harmony many aspects of the work and my life. The labor-intensive processes of wood-firing are a small price to pay for unifying the balances that must be struck between form, function, surface, and concept. I hope that the quality of my work is subtle and simple and takes time to appreciate. Maybe the first cup of coffee in a handmade mug reminds people, in the hustle of their everyday grind, that things of real value take time. It is my hope that everyone who uses handmade objects will slow down for just a moment to talk, to catch a glimpse of dark soils, to view large open skies, and to connect.

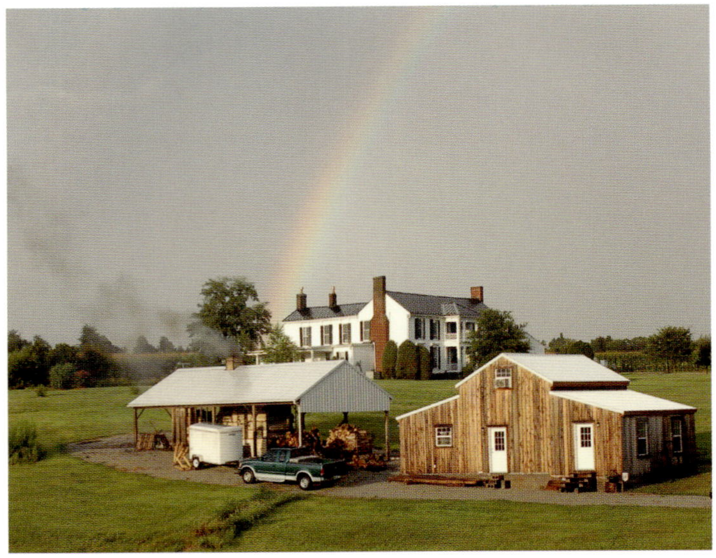

Full view, The Meadows Pottery during firing and after a storm had cleared. *Photo Credit: Matthew Gaddie.*

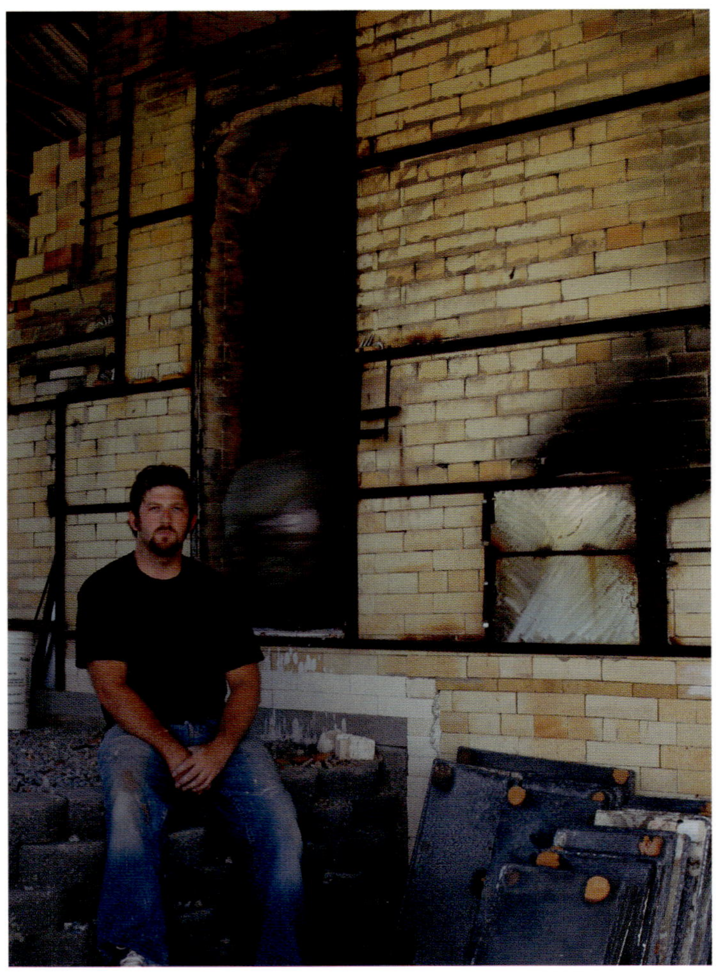

Matthew Gaddie and his wood kiln.
Photo Credit: Patricia E. Hart.

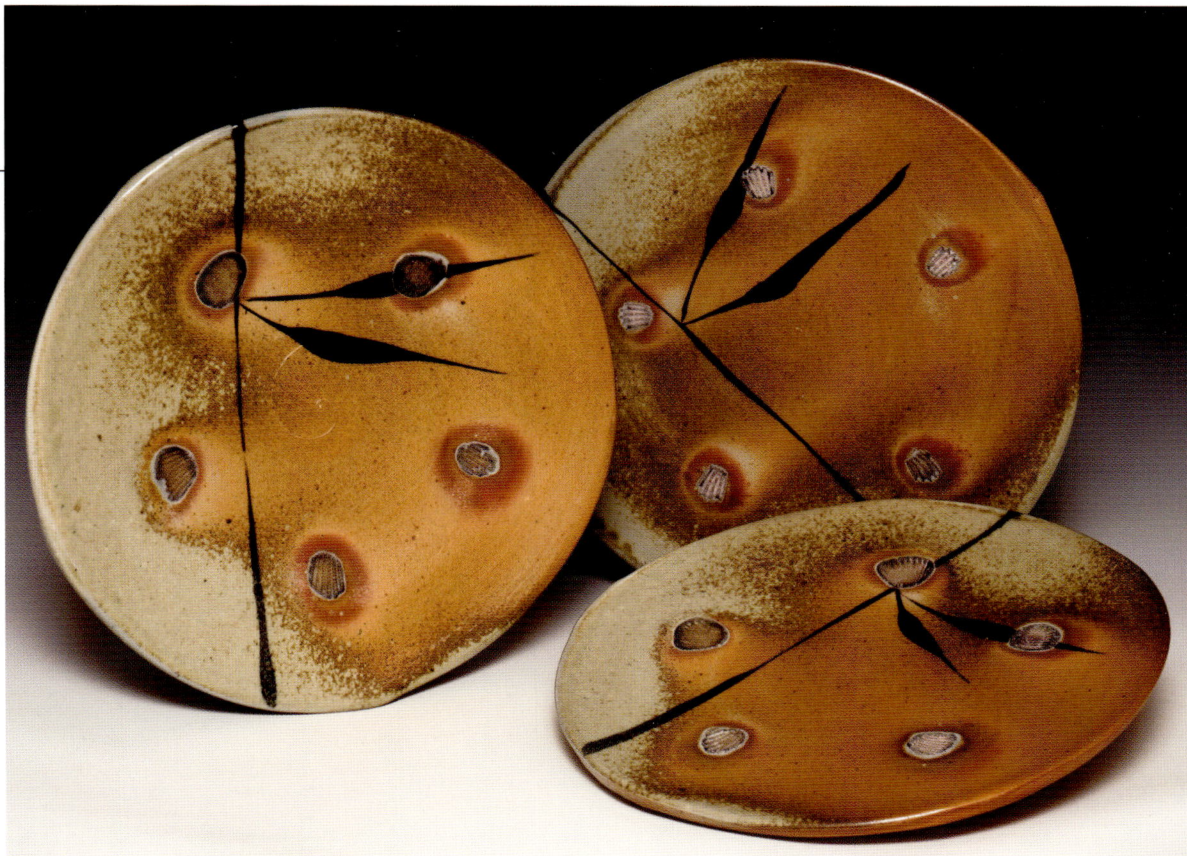

Matthew Gaddie, *Plate Trio,* 2012. Wood-fired B-Mix with black slip brushwork and shells, Wheel-thrown, 2" x 10" x 10". *Photo Credit: Rob Perrin.*

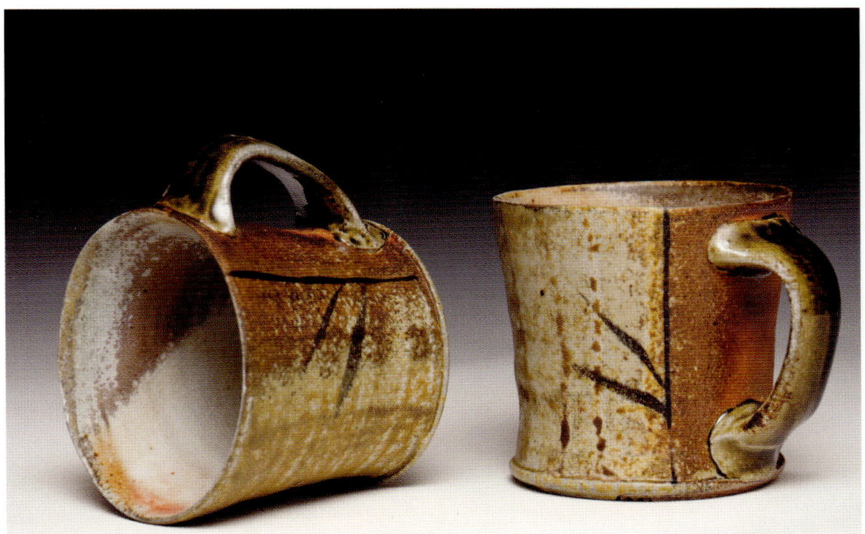

Matthew Gaddie, *Mug Duo,* 2012. Wood-fired stoneware, 6 tile slip with black slip, brushwork, glazed temmoku handle, and Laura's fat white shino liner, wheel-thrown, 5" x 4" x 4". *Photo Credit: Rob Perrin.*

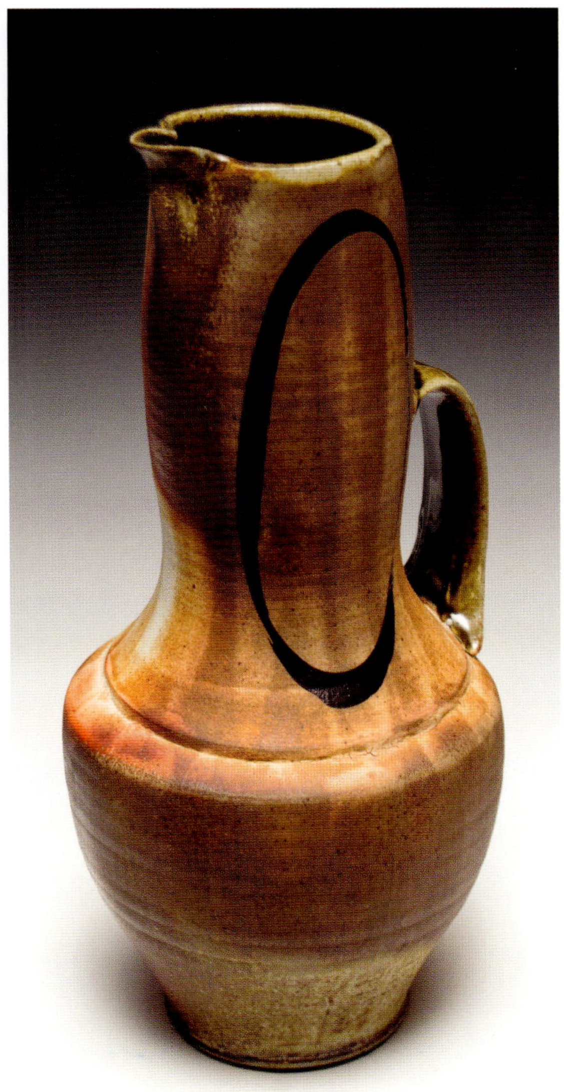

Matthew Gaddie, *Pitcher,* 2012. Wood-fired stoneware, 6 tile slip with black slip, brushwork, glazed temmoku handle, wheel-thrown, 14" x 6" x 6". *Photo Credit: Rob Perrin.*

COLLEEN GALLAGHER

Shelton, Washington
bennpottery@gmail.com, www.benngallagher.com

Artist's Statement

On beautifully forested Harstine Island in Washington's Puget Sound, Colleen Gallagher makes wood-fired pottery and sculptural tiles. She digs several local stoneware clays and fires her two wood-kilns using trees from the surrounding forest.

"The sometimes difficult quest for elusive ash-glazed and vapor-flashed clay surfaces demands hard work, long hours, and a willingness to be a partner to the fire. The fire and the kiln are tools but also teachers.

I do a lot of surface decoration on my pieces before they enter the kiln. I draw with flashing slips and oxides, and also work with textured surfaces and finely detailed relief sculpture. My pieces don't usually like a lot of ash, or an excess of heat. A desirable location in the kiln gives me a balance point in the fire; gathering just enough ash and vapor to give color variation and a polite sheen but not so much built-up ash that the decoration is obscured or obliterated.

Although I control processes, the wood-kiln's ash deposits and color flashes are always a combination of physics and serendipity. Drawings on the successful pieces seem effortless and familiar, yet as mysterious and surprising as a leaf or a feather."

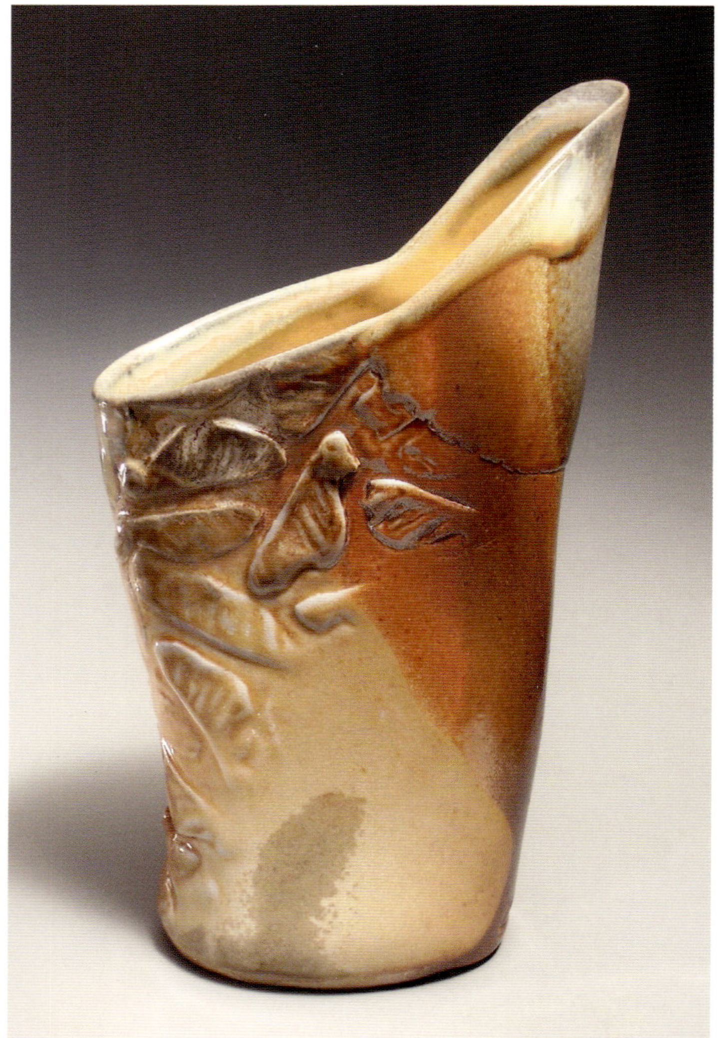

Colleen Gallagher, *Pitcher with Maple Helicopters*, 2011. Wood-fired stoneware, handbuilt, 6" x 3" x 2". *Photo credit: Colleen Gallagher.*

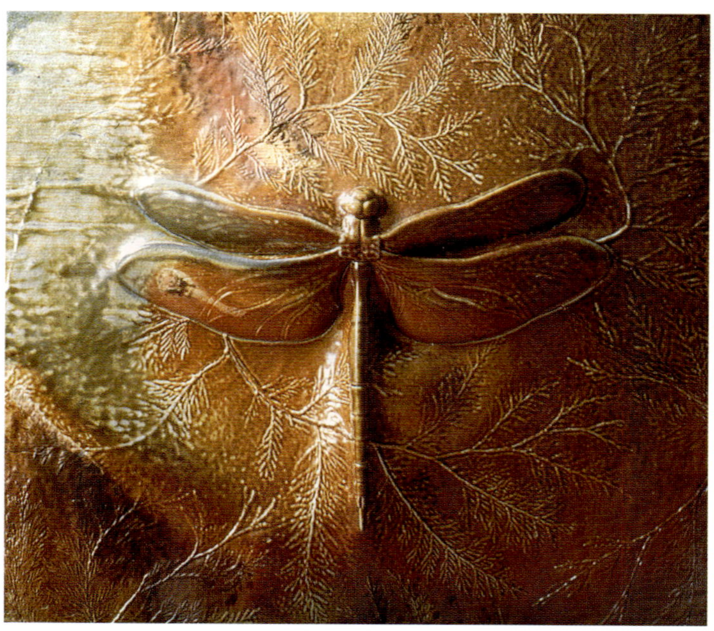

Colleen Gallagher, *Wall Tile*, 2011. Wood-fired stoneware, handbuilt, light salt, 18" x 22" x 3". *Photo credit: Colleen Gallagher.*

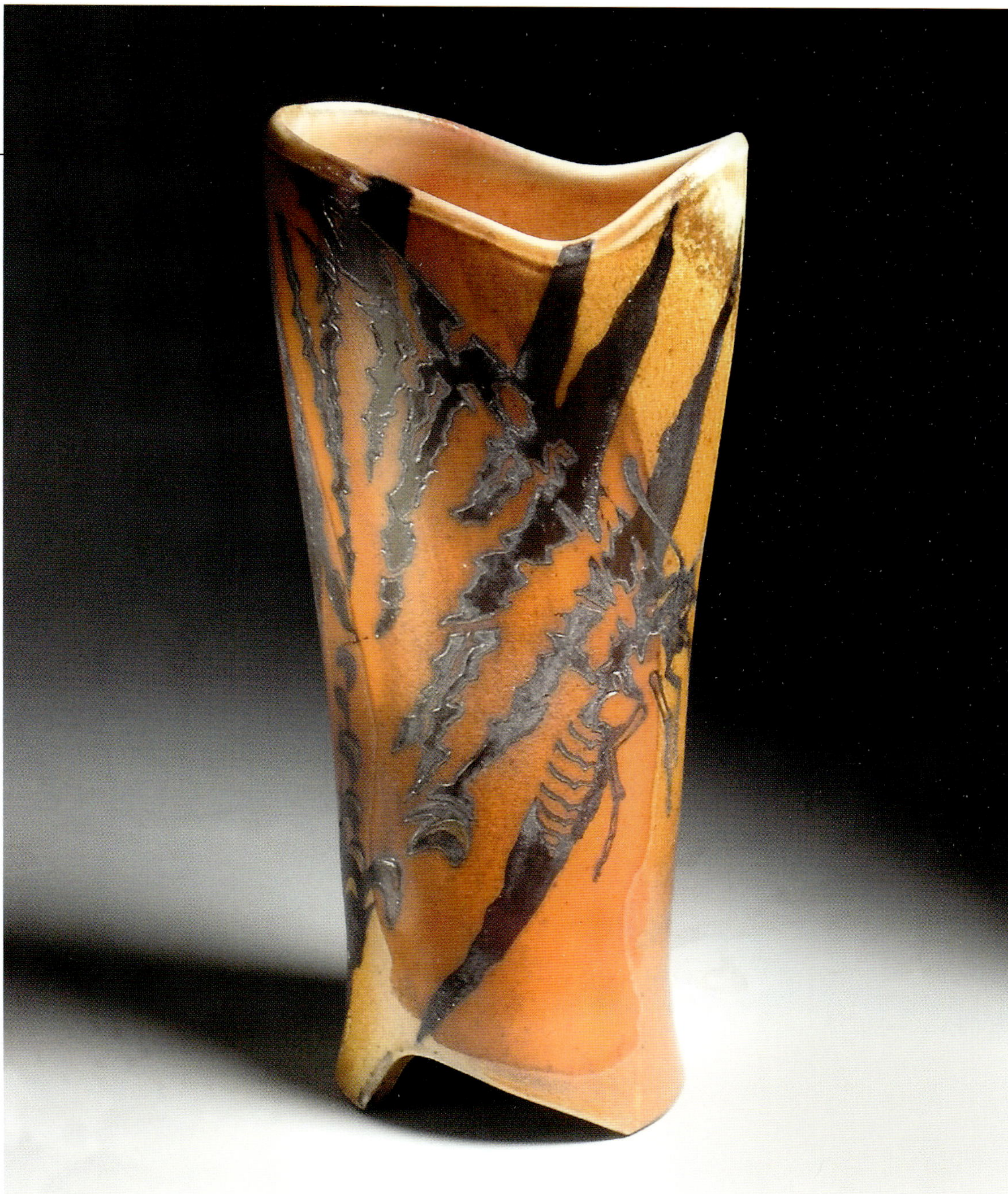

Colleen Gallagher, *Swallowtail Vase,* 2011. Wood-fired stoneware, handbuilt, with oxides and flashing slips, 6" x 3" x 2". *Photo credit: Colleen Gallagher.*

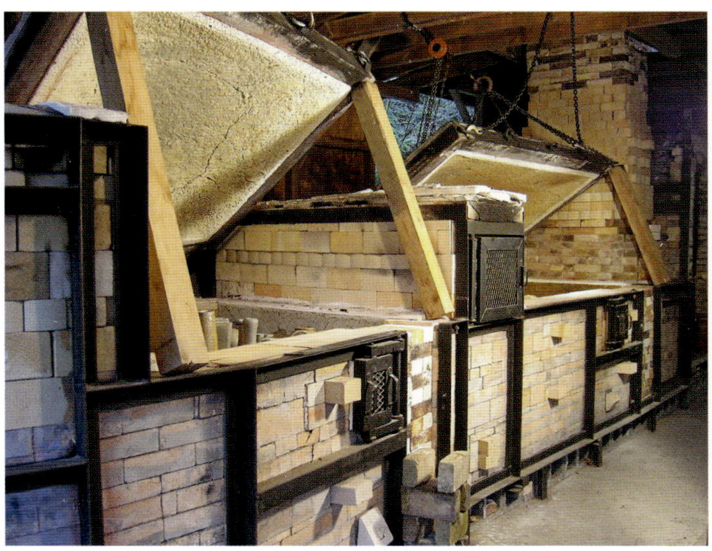

Colleen Gallagher, Harstine Island Wood Kilns, The Mongrelgama with its lids open. *Photo credit: Colleen Gallagher.*

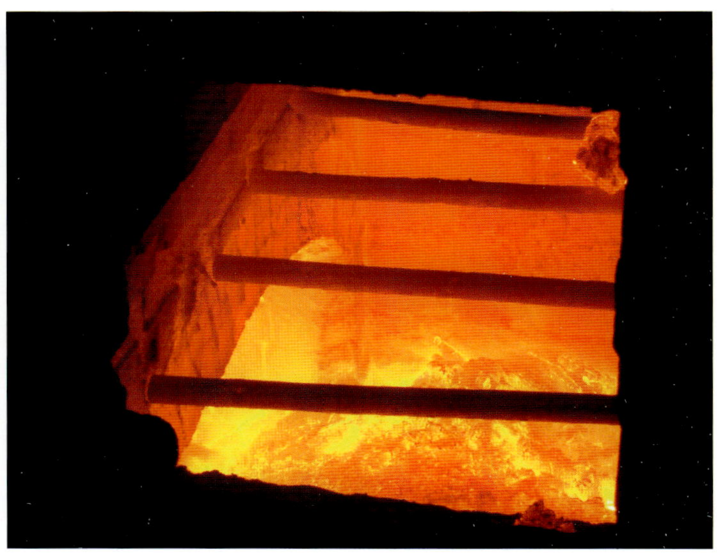

Colleen Gallagher, Harstine Island Wood Kilns, Firebox grate in Bourry box kiln. *Photo credit: Colleen Gallagher.*

STEPHEN GRIMMER

Winnipeg, Manitoba, Canada
grimmer@cc.umanitoba.ca

Artist's Statement

My current studio work investigates the architecture of the Islamic world as a source of structural and surface information for functional pottery. So much of what I find fascinating about form and surface in pottery is embodied in this great architectural tradition: from the stacking, nesting, and juxtaposition of geometric forms, to the obsessive articulation of surface with color and texture. My mathematical side delights in the complex repetitions and permutations of form, pattern, structure, and volume. I am particularly interested in the use of multiple geometries, be they three-dimensional, two-dimensional, or non-Euclidean. Wheel-thrown and cast parts are stacked, inverted, and everted to create pots that have a strong sense of structure. Surface geometries play on geometric abstraction, tessellation, symmetry, fractals, pixilation, and the interplay between negative and positive space.

My stubborn insistence on making functional pots stems from my belief in the value of domesticity, a love of cooking and serving good food and drink, and the power that beautiful, functional objects have to add aesthetic enrichment to and become an integral part of daily life. In addition, pots have a deep and ancient connection to human life, a connection that still resonates with us today. As a potter, I am a part of that tradition, and I feel a responsibility to share it with others. Because of our historical association with pottery, people understand pots on an intuitive level, different from their understanding of many other art forms, and bringing pottery into their homes connects them to this long tradition.

The surfaces achieved at high temperatures in the wood-burning kiln speak to me on many levels. Most important is an ancient, timeless characteristic the firing gives to the work. This technique has been appreciated and utilized by ceramic artists since 1000 AD and is still a vital part of the contemporary idiom. Also important is the way wood kilns record the process of the firing. The fine ash drifting through the ware chamber collects on the pottery early in the firing and in the blast furnace-like conditions of the final hours to form natural glazes and surfaces that are unique to each firing. They are dependent on stoking patterns, wood type, how the kiln is loaded, clay compositions, and even weather conditions. Each

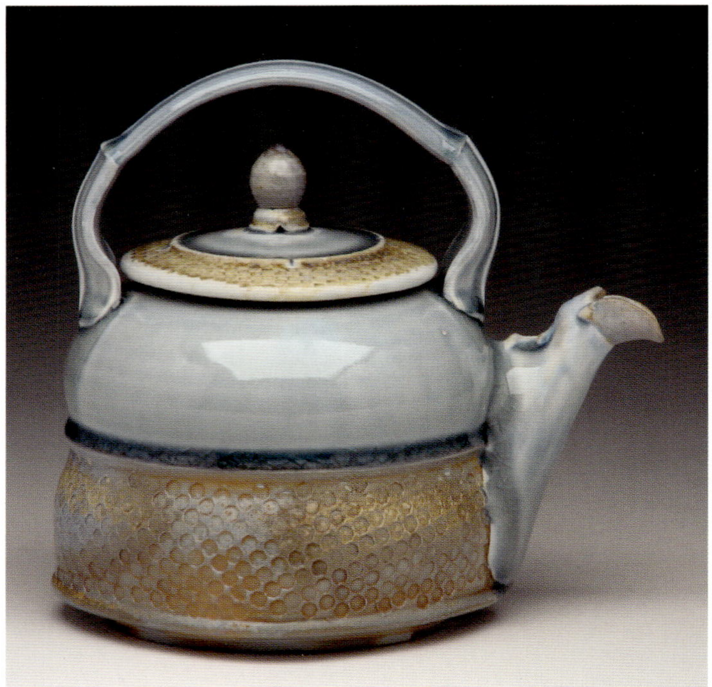

Stephen Grimmer, *Blue Dome Teapot with Octagon-Grid Pattern*, 2009. Wood-fired porcelain, wheel-thrown, 6.5" x 6.5" x 5". *Photo credit: Stephen Grimmer.*

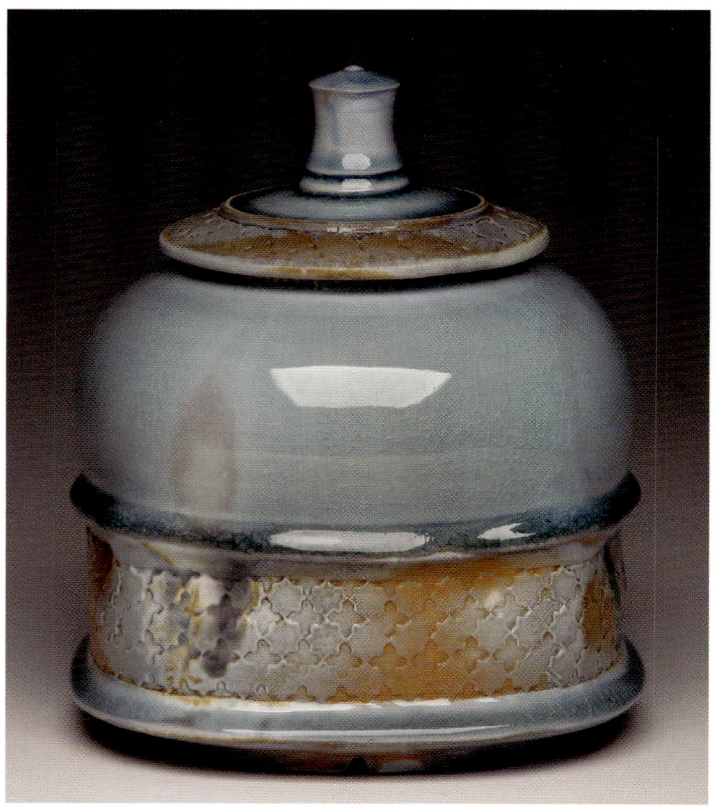

Stephen Grimmer, *Blue Dome Jar with Cross-Star Pattern*, 2010. Wood-fired porcelain, wheel-thrown, 7" x 5" x 5". *Photo credit: Stephen Grimmer.*

firing leaves a unique narrative trace on the ware. I attempt a certain level of control and predictability in the process, but in the end, I give my work up to the kiln. What it gives back to me is often more profound than anything I could have realized on my own.

Our "Train" kiln at the University of Manitoba is based directly on those designed by John Neely at Utah State University. The chamber has space for two stacks of three 12" x 24" shelves, with a 9-inch side stoke between the front and back stacks. Typical firings last between thirty-six and sixty hours, depending on effects desired and wood used. For about five years our primary fuel was tamarack with a little poplar mixed in for ash and color— the more poplar, the better. Our wood supplier is getting out of the timber business and into eco-tourism, so we have found another source that specializes in pine and ash. I'm eager to see the results.

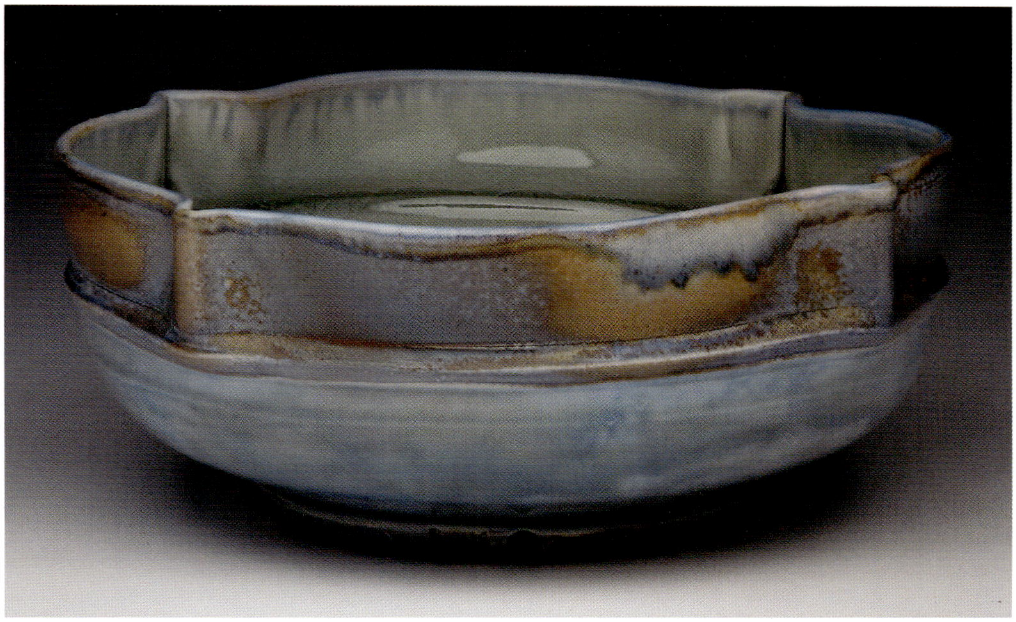

Stephen Grimmer, *Large Bowl with Blue Dome*, 2011. Wood-fired porcelain, wheel-thrown, 4.5" x 10.5" x 10.5". Photo *credit: Stephen Grimmer.*

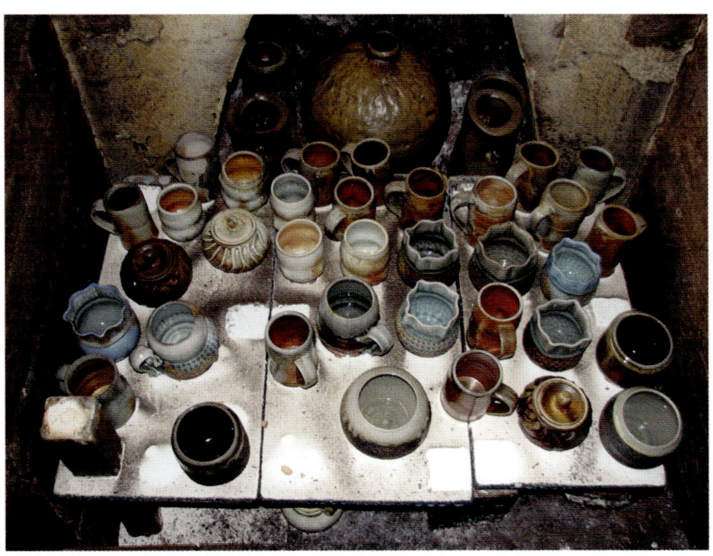

Unloading The University of Manitoba Train Kiln, January 2010. The throat arch holds pots by Tamura Shizuo of Shigaraki, Japan. *Photo credit: Stephen Grimmer.*

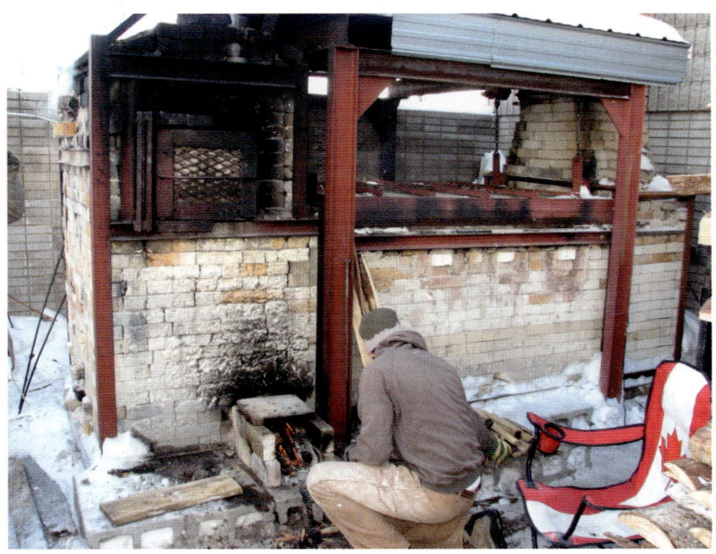

University of Manitoba Train Kiln, "*Warm Wind*" shift, January 2010."*Warm Wind*" is the first six hours of the firing that brings the temperature, in this case, from -18C/0F to about +10C/50F. *Photo credit: Stephen Grimmer.*

PERRY HAAS

Logan, Utah
p.haas3011@gmail.com, www.perryhaasceramics.com

Artist's Statement

The human body and my attraction to geologic formations influence the forms that I develop. The movement of the flame enhances curves and subtle movements of the clay while wood ash is added over time on the flowing curves. A geological approach in my work exposes the raw material of clay, by resembling folds in the earth's crust, to the thrusts of pressure coming from within the earth. This allows the material to expose itself to its most natural state. Both function and form is deeply considered when choosing to fire my work in an atmospheric wood-firing process. The firing method of wood kilns has captured my interest because of its ability to record the symbolic elements of the fire and clay relationship. A minimalist approach to the glazing process allows the ware to form a reflection of textured skin, a combination between humanity and the movement of the landscape.

Perry Haas has been working in ceramics since 2003. He holds a BFA in ceramics from Utah State University. Ceramics has taken him to China and Korea where he stayed and studied the history and process. His work consists of functional wood-fired ceramics and is now the Wood-fire Resident at the Clay Studio of Missoula. The kiln he is in charge of is an anagama, designed by Ryan Mitchell and Jason Lawfer. The kiln is located outside of Missoula in the mountains, overlooking the Mission Mountains. The anagama kiln is approximately 720 cubic feet, and usually fires for six to eight days. We tend to burn about seven to nine cords of wood during the firing and have a firebox that is eight feet wide in the front of the anagama with three stoke doors side by side.

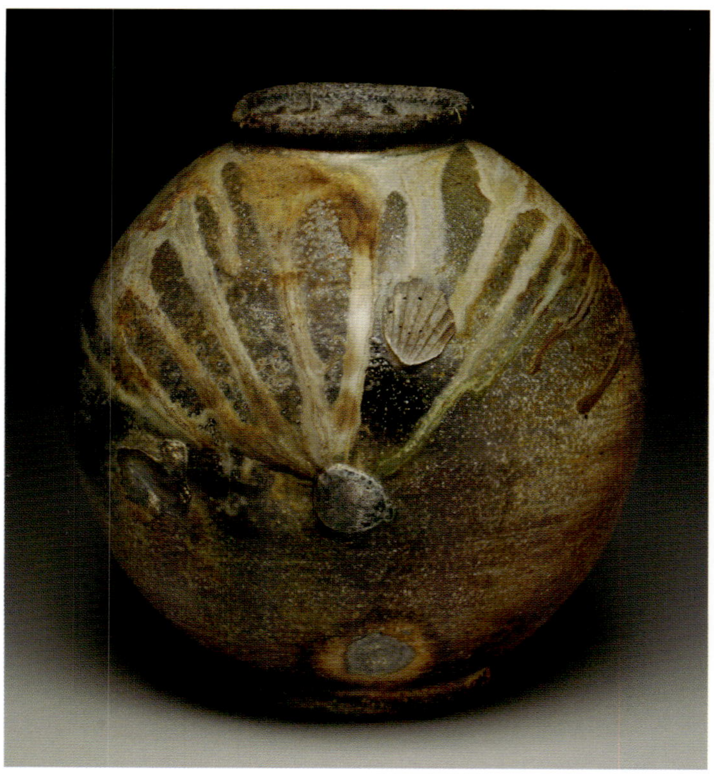

Perry Haas, *Jar,* 2011. Porcelain with Sand, wood-fired with pine and cotton wood, cone 12, thrown and altered, 20" x 18" x 20". *Photo credit: Perry Haas.*

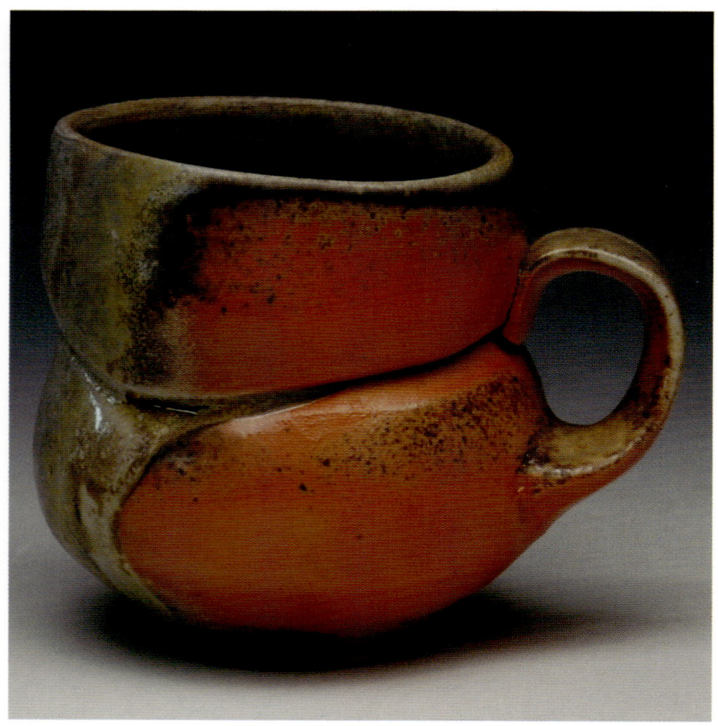

Perry Haas, *Mug*, 2011. Stoneware, wood-fired with pine, cone 11, thrown and altered, 4" x 3" x 4". *Photo credit: Perry Haas.*

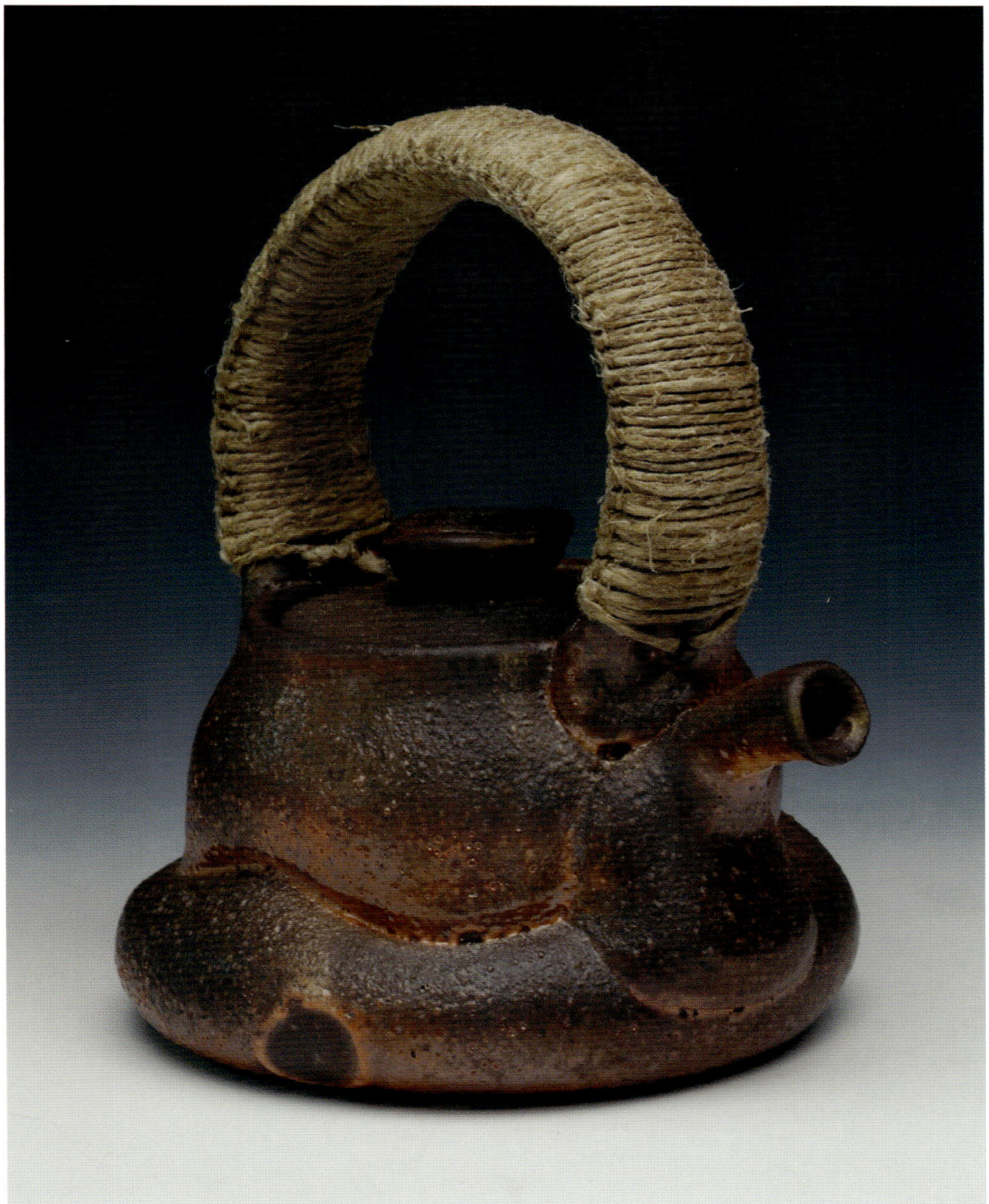

Perry Haas, *Teapot*, 2011. Stoneware, Shino glaze, wood-fired with pine, cone 11, hemp handle, thrown and altered, 8" x 6" x 6". *Photo credit: Perry Haas.*

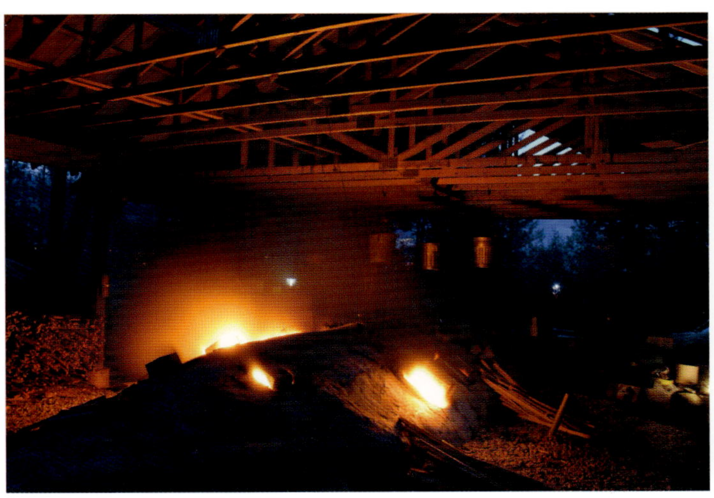

Kiln at night, looking toward firebox. *Photo credit: Perry Haas.*

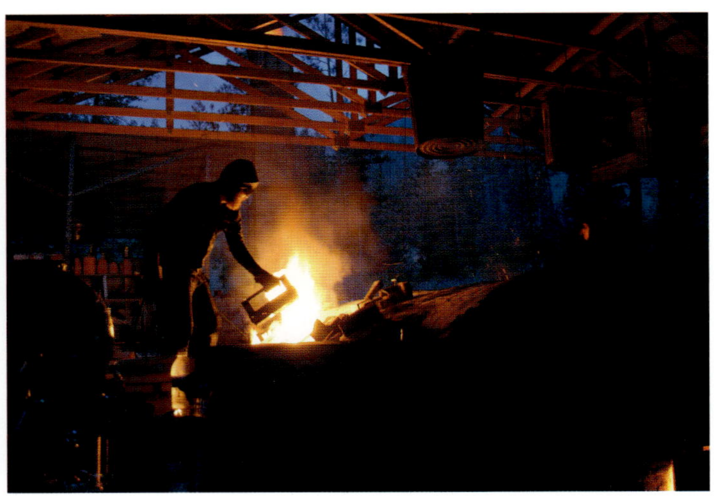

Side stoking kiln at night. *Photo credit: Perry Haas.*

HADDIE HADACHEK

Brainerd, Minnesota
blackbear@brainerd.net, www.blackbearpottery.biz

Artist's Statement

The intent of my art is driven by my fascination with the play of form and color. Wood-firing amplifies this with its flashing and ash accumulation. In addition, I find joy in the gathering of fellow wood-firing potters for a wood-firing. The process, the shop talk, and the sharing of food makes for a strong clay community and that energy sustains me for months.

My wood kiln is a catenary arch downdraft with two stoking ports, one on each side of the ware chamber. It was designed and built by Ken Omundson and me. Two people can fire it but we usually have a crew of about eight to spread out the workload and fun. Many of my former apprentices return again for these wood-firings.

I use Continental Clay's B Clay and also fireclay with iron. My work is usually wheel-thrown or slab made, often in combination. Some work is functional and some is more sculptural. Glazing and raw surfaces are found together on most pieces

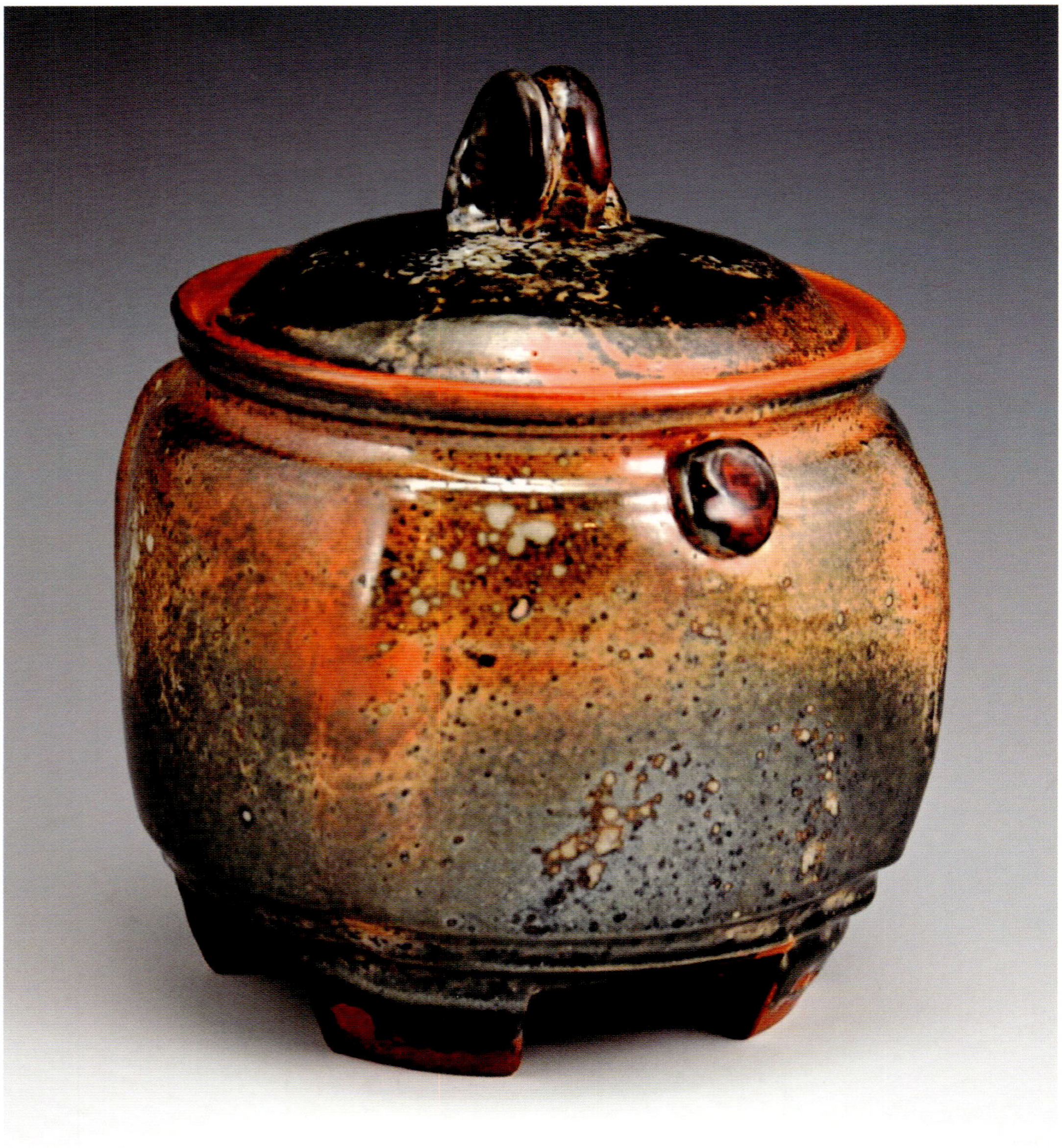

Haddie Hadachek, *Shino Covered Jar*, 2011. Wood-fired B-Clay, wheel-thrown, altered, 7" x 5" x 5". *Photo credit: Haddie Hadachek.*

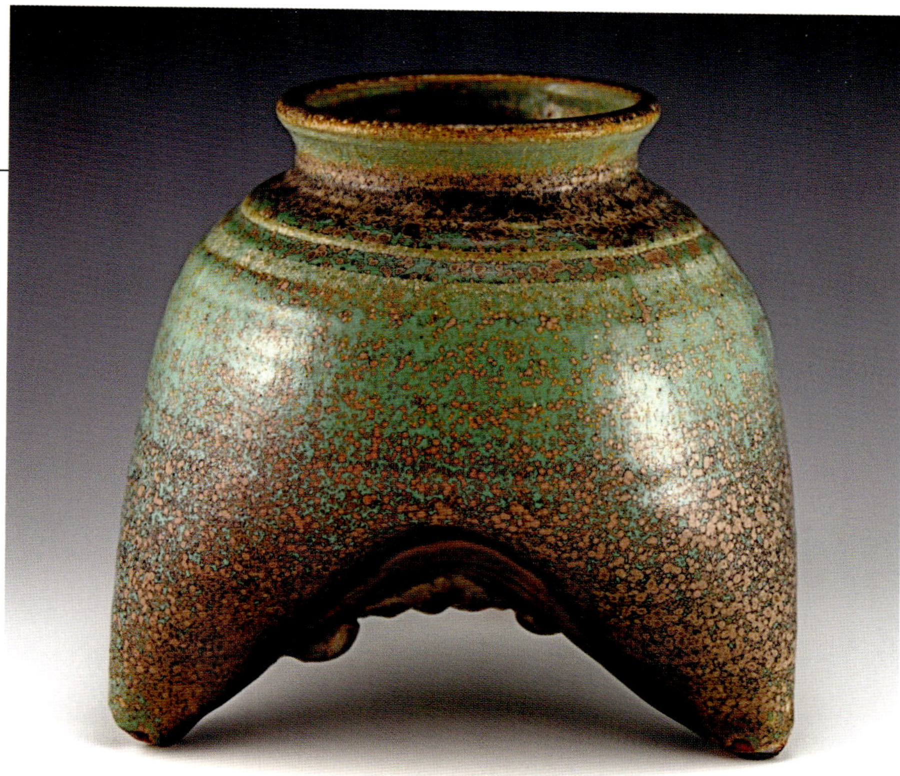

Haddie Hadachek, *Laura's Button Tripod*, 2010. Wood-fired B-clay, wheel-thrown, altered, 7" x 7" x 7". *Photo credit: Haddie Hadachek.*

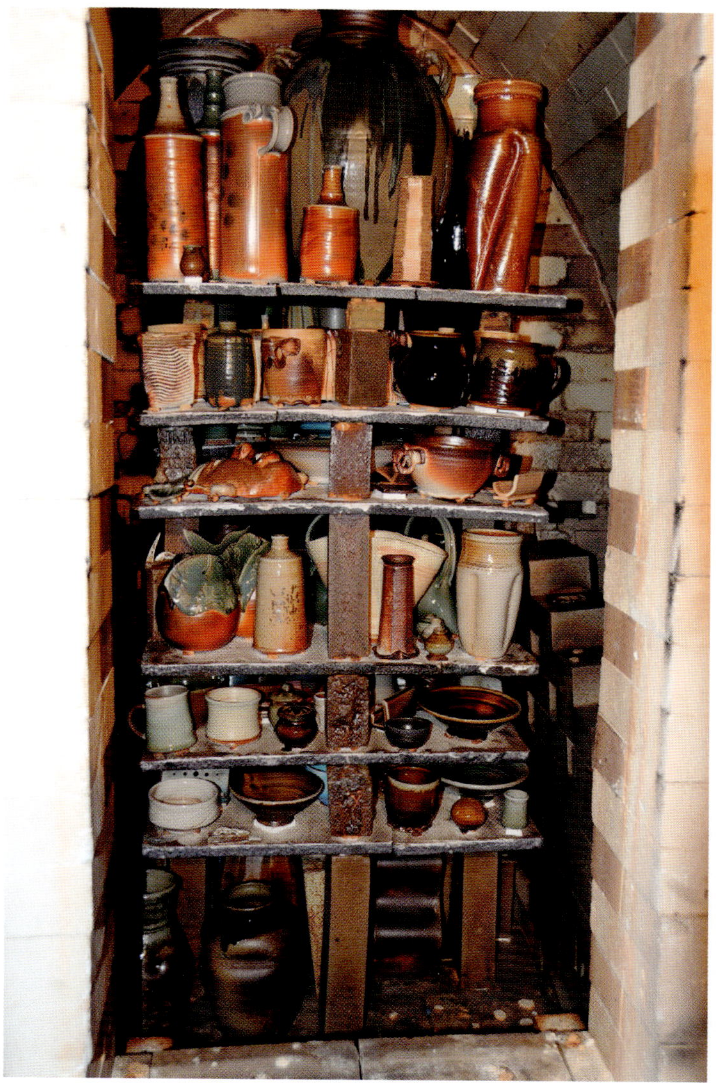

Haddie Hadachek, Middle stack of finished firing, catenary arch kiln. *Photo credit: Haddie Hadachek.*

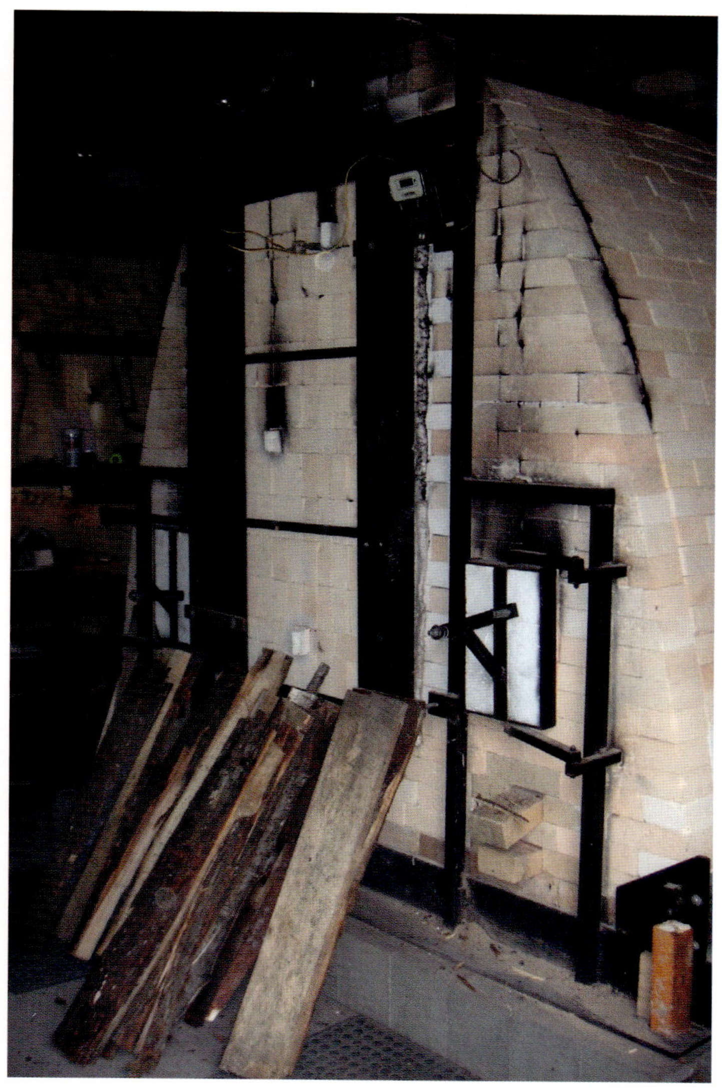

Haddie Hadachek, wood-fired kiln, catenary arch, downdraft, at Black Bear Pottery, finished 2005. *Photo credit: Haddie Hadachek.*

RENA HAMILTON

Flagstaff, Arizona
rhelementalartworks@gmail.com, www.elementalartworks.org

Artist's Statement

My pottery, like my life, is deeply rooted in the philosophy of balance in opposites: masculine-feminine, work-play, whimsical-austere. I fire my wares in a centenary style, cross-draft kiln with wood and soda. As a child of avid antique collectors, I've always had a strong appreciation for things handmade as well as an affinity for the sense of nostalgia and the idealist notion of easier carefree living in decades, or centuries past. Utilizing the atmospheric technique of spraying soda into the kiln enhances a sense of rustic elegance in my wares. Firing with wood satisfies my intense need for physical labor and takes some of the control of the outcome out of my hands, providing a natural, loose, serendipitous surface that compliments my more tightly thrown forms: balance in opposites. I take immense satisfaction from being able to interact and connect with my work throughout the entire making/firing cycle.

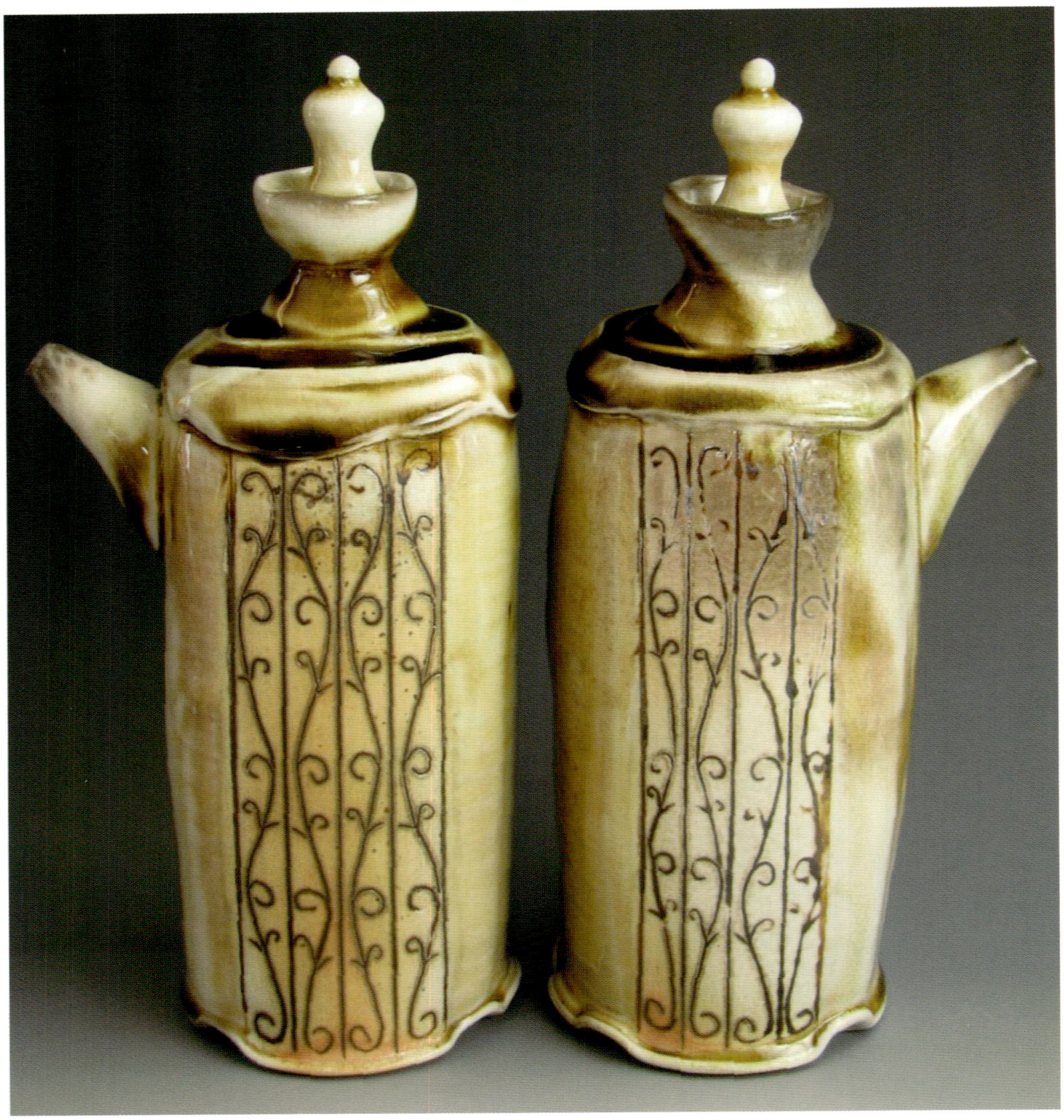

Rena Hamilton, *Two Ewers*, 2010. Wood and soda fired porcelain with inlay scrolls, 9" x 4.5" x 3". *Photo credit: Rena Hamilton.*

Most of my work starts on the wheel and then is altered and assembled. I am increasingly integrating more complex and decorative surfaces such as inlay, underglaze transfers, and slip trailing. My cross-draft style kiln allows me to more strategically predict the path of the flame on my wares providing some protection from heavy wood ash or soda that would mask my laborious efforts in surface decoration. I am delighted to create pieces that are used in moments of celebration and daily rituals; whether it's a gathering of closest friends and family sharing good food and drink or a quiet, contemplative moment alone with a morning cup of coffee.

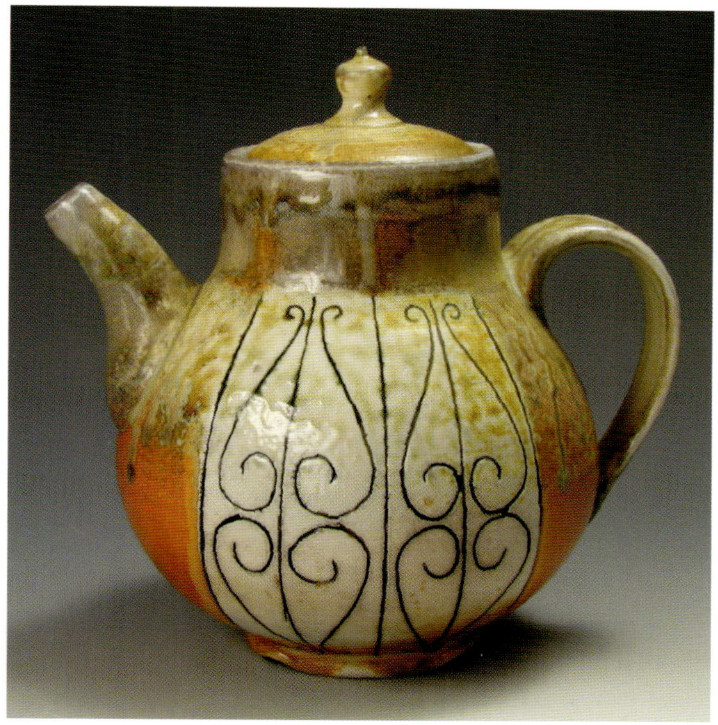

Rena Hamilton, *Teapot*, 2010. Wood-fired porcelain with inlay scrolls, 8.5" x 8.5" x 6". *Photo credit: Rena Hamilton.*

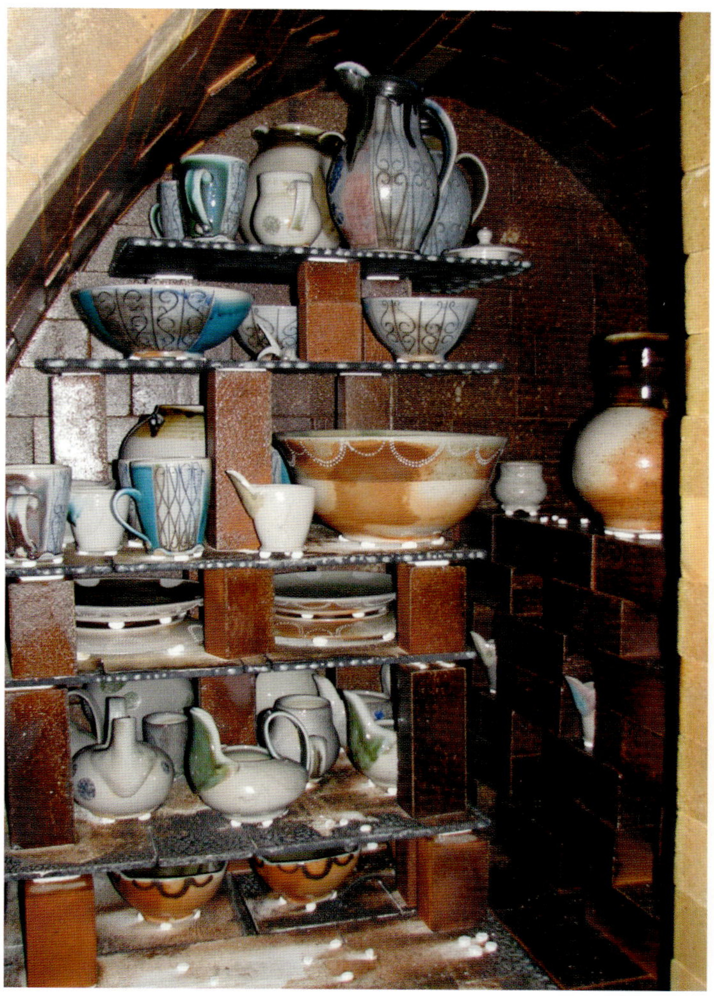

Rena Hamilton, Kiln opening, fall, 2010. *Photo credit: Rena Hamilton.*

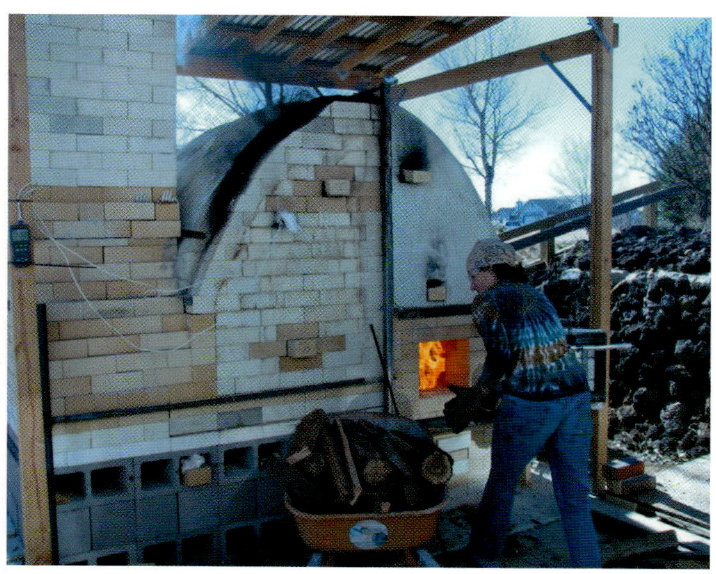

Rena Hamilton, Stoking the kiln, 2010. *Photo credit: Robert Hamilton.*

KJELSTY HANSON

Sylva, North Carolina
kjelsty@gmail.com, www.kjelstyhanson.blogspot.com

Artist's Statement

My work is motivated by people and our relation to the natural world. Wood-firing is directly connected to nature by the pure fact wood is the fuel, and the firings can be done in an environmentally conscious way. I was introduced to wood-firing while attending Linfield College and studying under Nils Lou. The student firings at his East Creek anagama kiln left a mark. The community that is needed and formed around wood-fires is what keeps drawing me back to the process. The many hands it takes to build, load, and fire a large wood kiln enriches the work with purpose and meaning.

After receiving my MFA, I moved to New Zealand and became active in the clay community. I served as president of the Taradale pottery club in the *Waiohiki* art village and was able to help the club raise money and collect bricks to build a wood kiln. The beauty of this kiln and the kiln building process is how it brought clubs/potters from around the island together. The kiln was built completely out of recycled bricks collected from potters around the island. Potters Hooly Thin, John Gisborne, and I designed the *Waiohiki* wood-fired kiln, with each of us having strong opinions on what features should be included. After some compromising, the kiln was built as a catenary arch kiln with a chamber that was large enough to stand up in. At the first firing we had at least thirty potters participate.

A wood-fired kiln is special because the wood fuels the kiln and colors the fired pottery with flame and wood ash. The fire strokes the sculptures'/pots' surfaces, adding natural highlights and shadows, leaving the piece blushing. After the firing, the potters who have pieces in the kiln line up like kids around a Christmas tree to see what presents they got. It is the element of unpredictability that makes wood-firing so enchanting to me, with each piece providing its own unique qualities.

Kokako—your secret is safe with me

The Kokako is an endangered bird of New Zealand and we were foreigners to this country, its nature, and wildlife; with our fresh eyes we looked at it with wonder. I was pregnant when building this sculpture and maybe it was a motherly instinct that compelled me to want to guard and protect. I used the pinch and coil method to make this sculpture and it was fired in the first firing of *Waiohiki* wood-fire kiln.

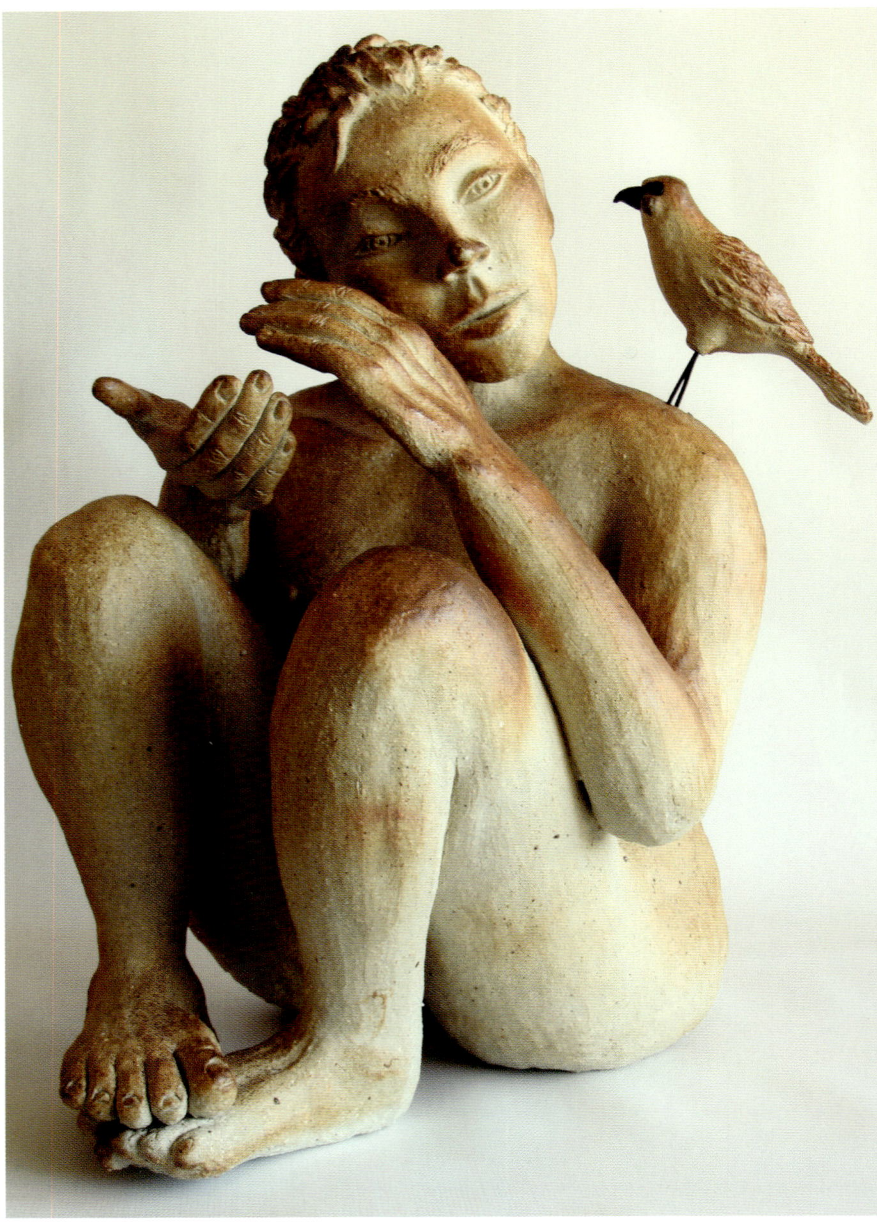

Kjelsty Hanson, *Kokako—your secret is safe with me*, 2008. Wood-fired stoneware, handbuilt pinch and coil techniques, 60 x 42 x 48 cm [23" x 16" x 18"]. Photo *credit: Stephen Burnell.*

Passing Gaze

This piece was built using the pinch pot technique. *Passing Gaze* is a piece that I made to fire at East Creek anagama, a trip back to visit my mentor. The sculptural form emerged from dealing with my father's death and speaks to the ephemeral versus the permanent. In many ways having this piece fired at East Creek anagama was important to me; it was a way of revisiting the past and letting go.

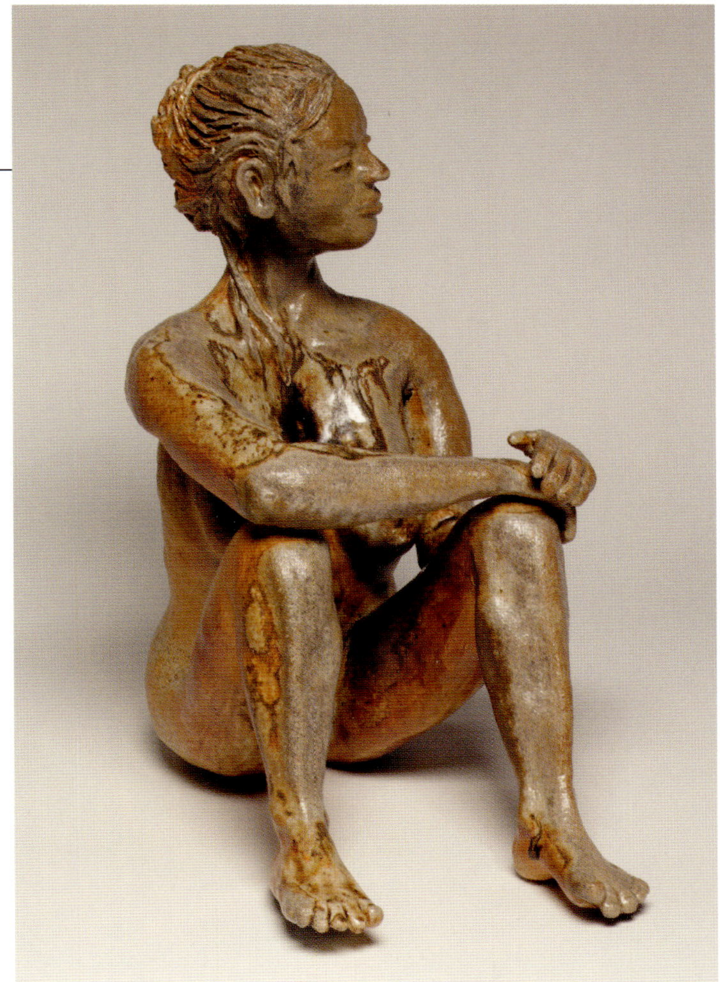

Kjelsty Hanson, *Passing Gaze,* 2011. Wood-fired white stoneware, handbuilt pinch technique, 30 x 28 x 16 cm [12" x 11" x 6"]. *Photo credit: Terri Clark.*

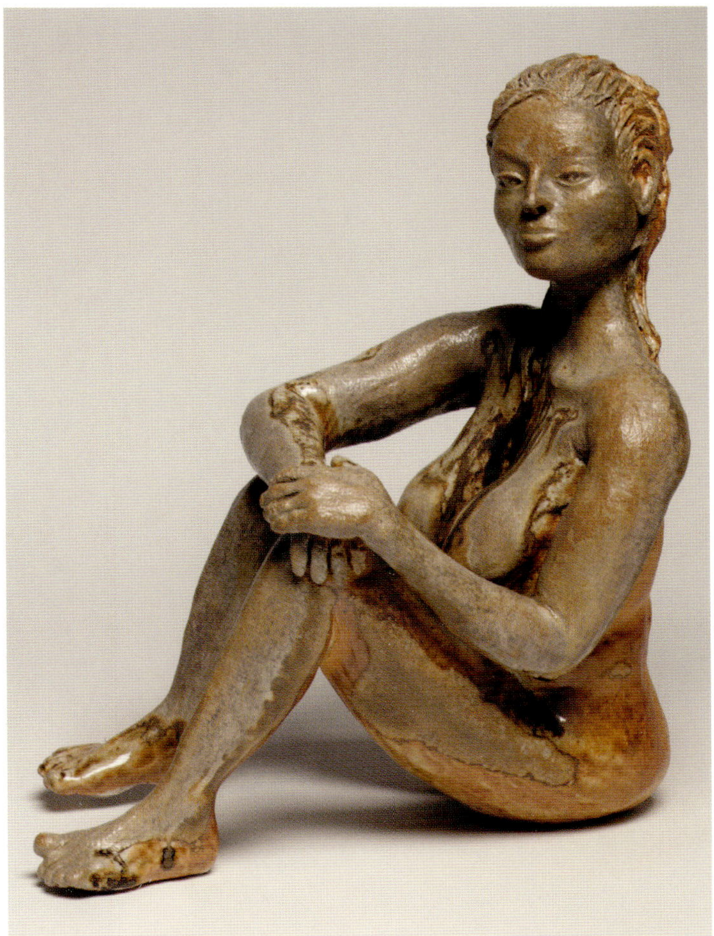

Kjelsty Hanson, *Passing Gaze*, 2011. Wood-fired white stoneware, handbuilt pinch technique, 30 x 28 x 16 cm [12" x 11" x 6"]. *Photo credit: Terri Clark.*

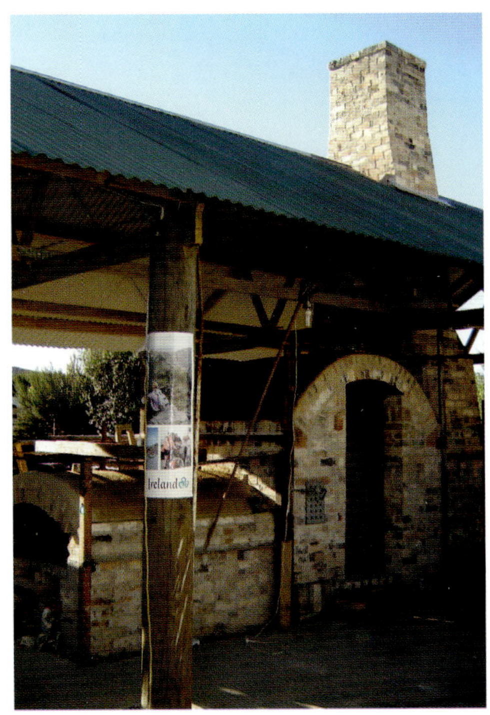

Full image of the Waiohiki wood kiln. *Photo credit: Kjelsty Hanson.*

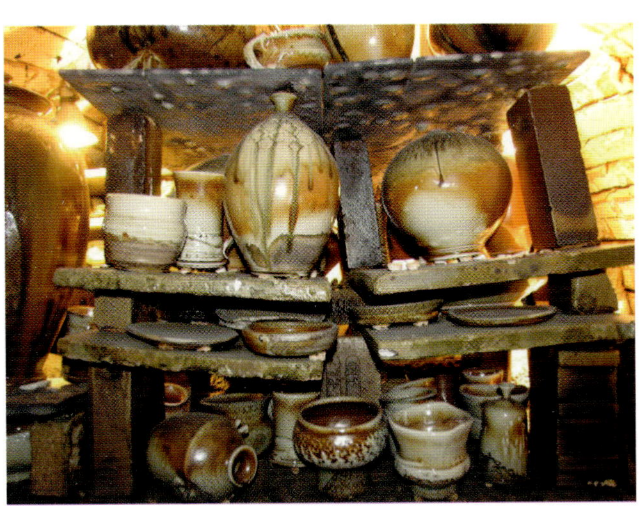

A peek inside the East Creek anagama kiln. *Photo credit: Cindy Hoskisson.*

MAKOTO HATORI

Ibaraki-ken, Japan
makotohatori@gmail.com, www2r.biglobe.ne.jp/~makoto-h/

Artist's Statement

"A Traditional View on Japanese Pottery"
~Makoto Hatori

It is generally considered that there are three important factors in the creation process of Japanese pottery: kiln, material (clay), and form (potter). Only when all these three factors are highly achieved and well balanced is a resultant piece considered to have high artistic value. Sometimes kiln is considered to be the most important of the three, which means that the baking process determines the overall artistic value of the piece. The extreme example of this emphasis on kiln can be observed in non-glazed (naturally glazed) pottery, a type of Japanese traditional ceramics most typically exampled in Bizen. In these cases the "natural spirituality" of the pots is so emphasized that the intentions of potters are excluded as much as possible.

This "natural spirituality" can be traced back to "Jomon" era, around 12,000 BC. The earthenware in this era (Jomon Earthenware), which is now considered to be one of the oldest earthenware in the world, is used in religious ceremonies and thus regarded as something sacred. In Shinto, a traditional Japanese religion, Jomon style earthenware is used in religious activities. Because of this relation between this sort of earthenware and the religion, this type of earthenware is considered to have "natural spirituality," rejecting the artificial "form" imported by human beings (especially by the eyes of human beings). The same interpretation can in fact be recognized in the structure of a Shinto-shrine.

In many Japanese traditional crafts or constructions, it is said that spirituality is put above the artificial form or structure. It is not difficult to explain the simple structure and gray-based coloring of many shrines and temples from this "spirituality." Rejecting clear forms and colorings are interpreted as the manifestation of this "spirituality."

The same mentality is observed in the people who appreciate the beauty of naturally glazed ceramics. In the traditional firing with firewood, natural glaze comes from the firing point. As a result, pieces far from the firing point in the kiln, as well as the far sides of pieces do not have natural glaze. As a result, the symmetry of the wheeling is destroyed and negated. This also means the exclusion of the intention of the potters.

At the end of sixteenth century, this natural spirituality was highly praised in *Wabicha* (tea ceremony) advocated by Rikyu. Rikyu claimed that there is special beauty in ordinal ceramics in Korea and Japan, which is simple but extraordinary. He explains this beauty with the "natural spirituality," which exists outside human intention. Since then simple, naturally glazed pots have been used for tea ceremony as well as fully colored pots originated in China. And in the tradition of tea ceremony, these two are uniformly appreciated from the point of view of natural spirituality; the former as a result of the negation of human intention and the latter at the extreme of human intention.

This mentality observed in tea ceremony, i.e., trying to regard everything from the point of view of spirituality, in fact is omnipresent in Japanese who appreciate traditional Japanese ceramic of any type. In this sense, what many of the Japanese try to find out in traditional Japanese ceramics is a sense of catalysis supported by this tradition.

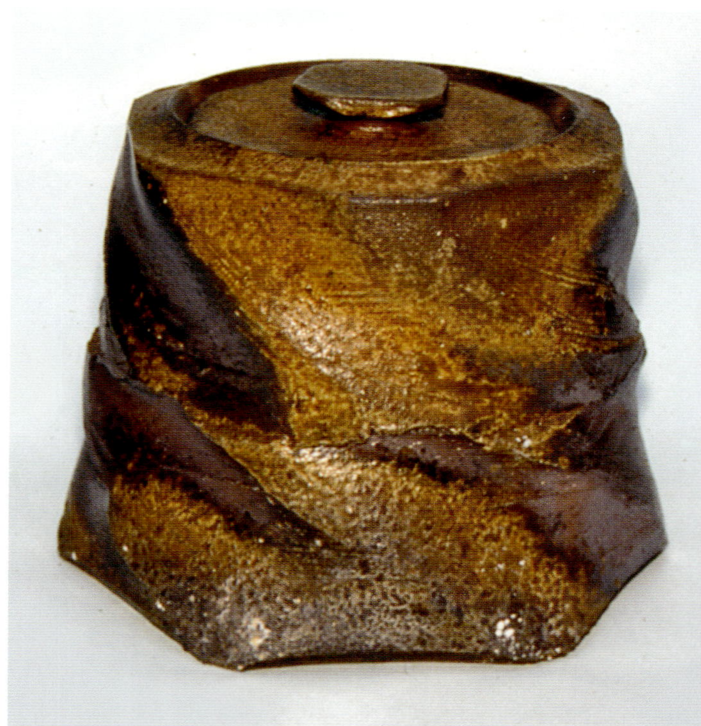

Makoto Hatori, *Bizen-style Cylindrical Lidded Pottery Mizu-sashi,* 1995. Thrown and altered stoneware with natural wood ash, sesame seed fired decoration, Bizen traditional way fired to 1300 degrees C. oxidation, eight-day firing, 6.7" x 7.5" x 7.5" Collection of The British Museum (U.K.), 1996. *Photo credit: Makoto Hatori.*

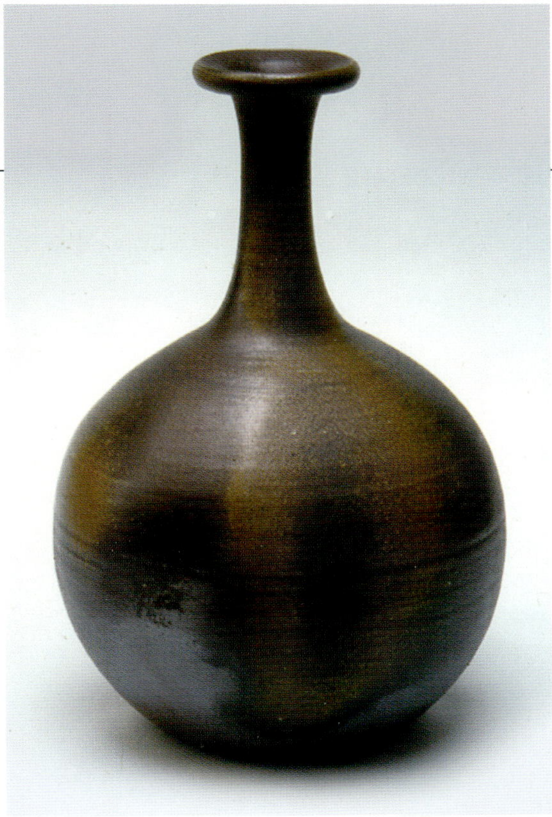

Makoto Hatori, *Bizen-style Wide-mouthed Bulbous Vase,* 1995. Wood-fired stoneware with natural wood ash, Bizen traditional way fired to 1300 degrees C., eight day firing, 12" x 8" x 8", Collection of The British Museum (U.K.), 1996. *Photo credit: Makoto Hatori.*

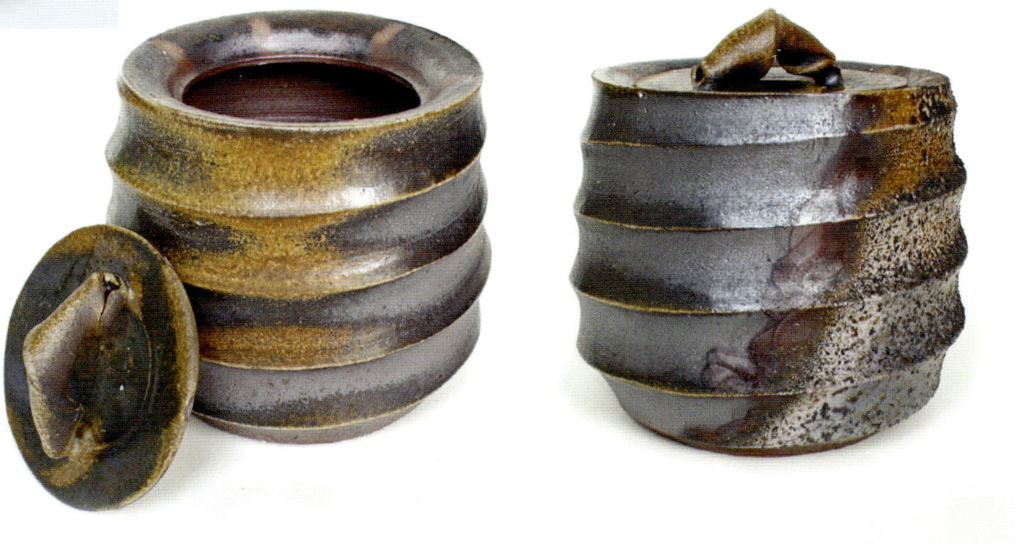

Makoto Hatori. *Fire Change Spherical Lidded Pottery Mizu-sashi,* 2009. Thrown and altered stoneware with natural glaze, fire change, traditional way fired to 1250 degrees C. reduction, eight day firing, 8" x 7" x 7". *Photo credit: Makoto Hatori.*

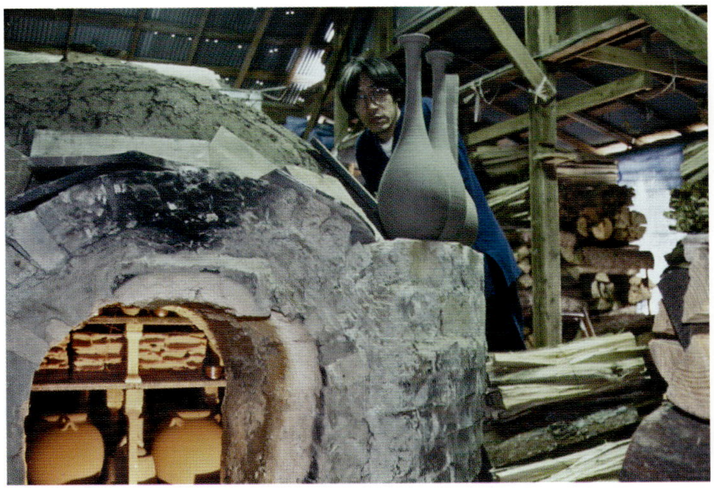

Makoto Hatori beside his climbing wood kiln, stacked firewood and pots ready to be loaded into the kiln, 2006. *Photo credit: Makoto Hatori.*

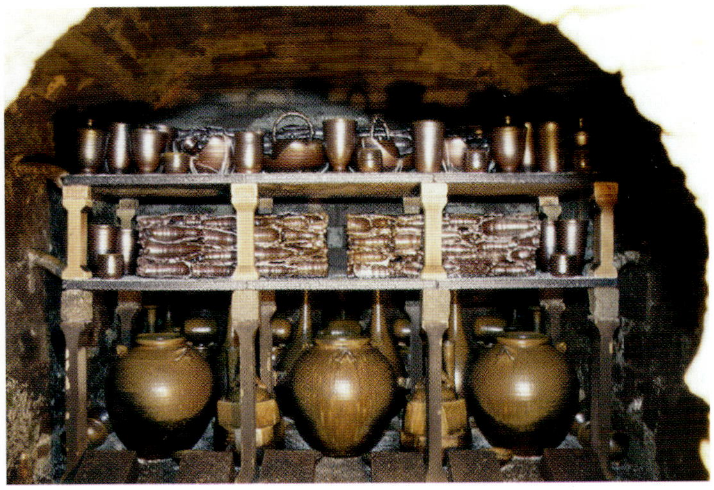

Results of the firing, showing shelf configuration and pots, the first chamber of Makoto Hatori's climbing wood kiln, 2006. *Photo credit: Makoto Hatori.*

FRED HERBST

Painted Post, New York
fred.threestones@gmail.com, www.fredherbst.com

Artist's Statement

My current work references the organic shapes found in the gardens I have tended throughout my life. Gourds, squash, peppers, and a variety of plant forms have become inspiration for the jars, bottles, teapots, and other utilitarian forms I create. Another aspect of my interest in these shapes stems from my concern for the huge loss in genetic diversity in our food sources. Most of the types of food plants developed over hundreds of years by small growers and family gardeners have disappeared under the overwhelming weight of agribusiness. The recent resurgence of heirloom plants, those based on older varieties not suited for large scale commercial production, is a hopeful sign that many are concerned about the loss of this vital diversity and an obvious reaction to industrial food production. The growth of the "slow food" movement also honors the unique flavors developed by generations of gardeners. As an artist, I work to make permanent these same plant forms that are vital to sustaining our civilization. In turn, my utilitarian forms become the host to the foods that are their inspiration.

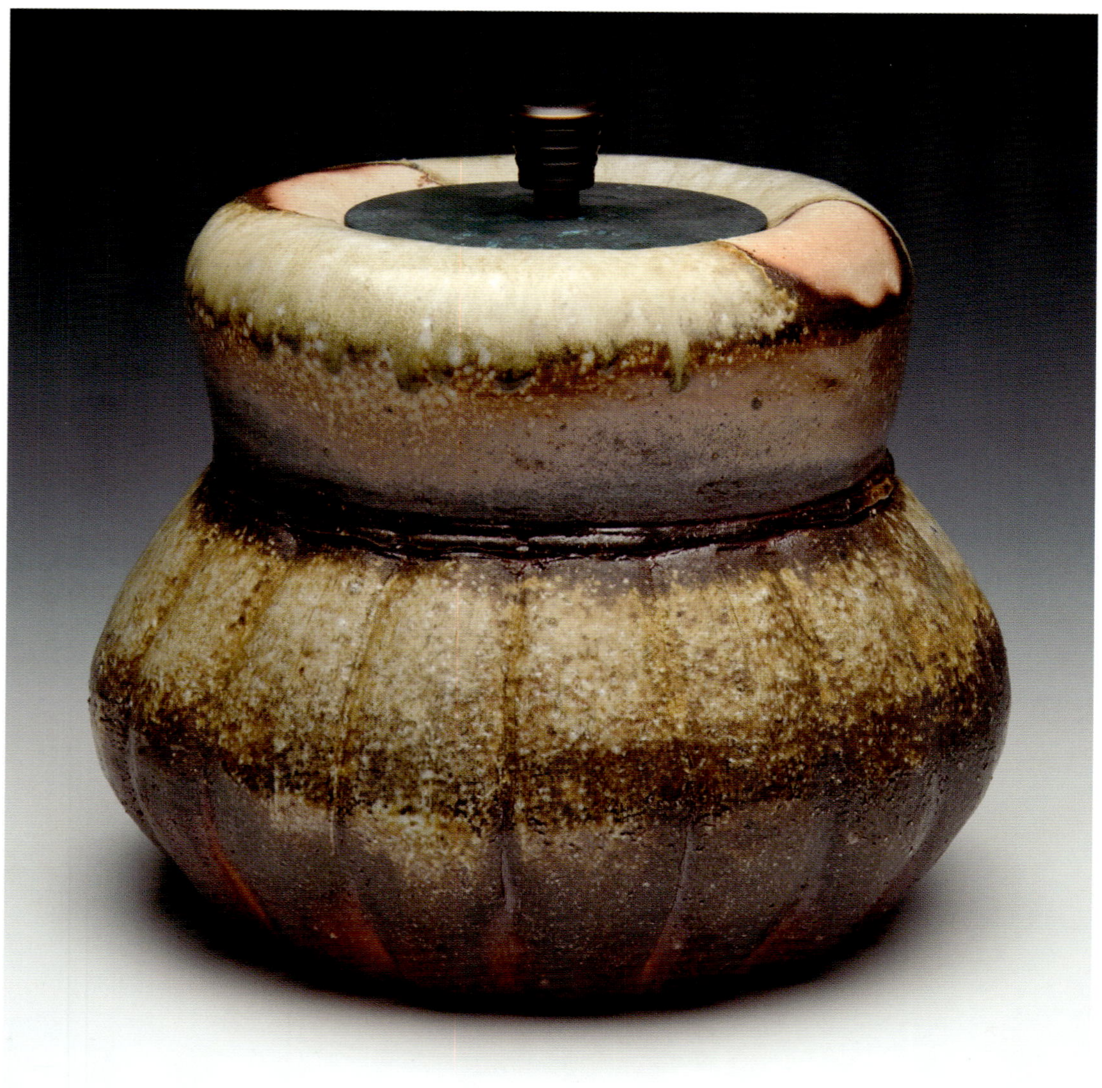

Fred Herbst, *Gourd Jar*, 2010. Wood-fired stoneware with copper and bronze, wheel-thrown and altered, 9" h x 9" x 9". *Photo credit: Ann Cady.*

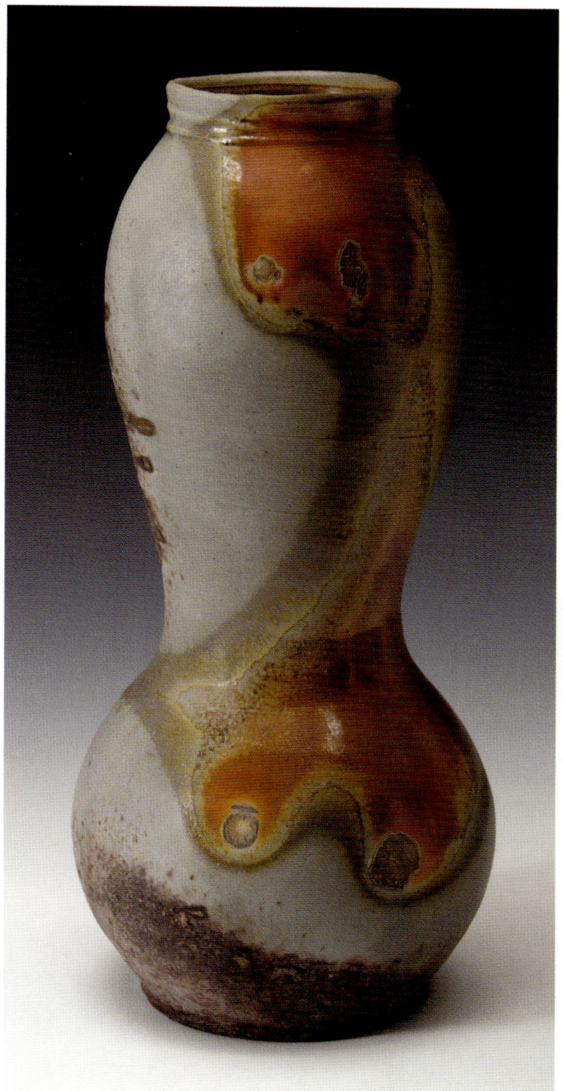

Fred Herbst, *Gourd Bottle*, 2011. Wood-fired stoneware, wheel-thrown, 11" x 5" x 5". *Photo credit: Ann Cady.*

The wood-firing process I use results in a dynamic surface that adds another layer of interest to these small-scale functional sculptures. The anagama and "*glassagama*" (a new type of kiln I designed for wood-firing ceramics and blowing glass simultaneously) kilns I use were designed to allow for the most direct effects on the work. Clay bodies and specific shapes are selected for particular zones of each kiln to maximize the potential for interesting surface patterns and coloration. Each piece is loaded in the kiln with concern for how the flame and ash produced during the firing will complete its long journey from idea to object.

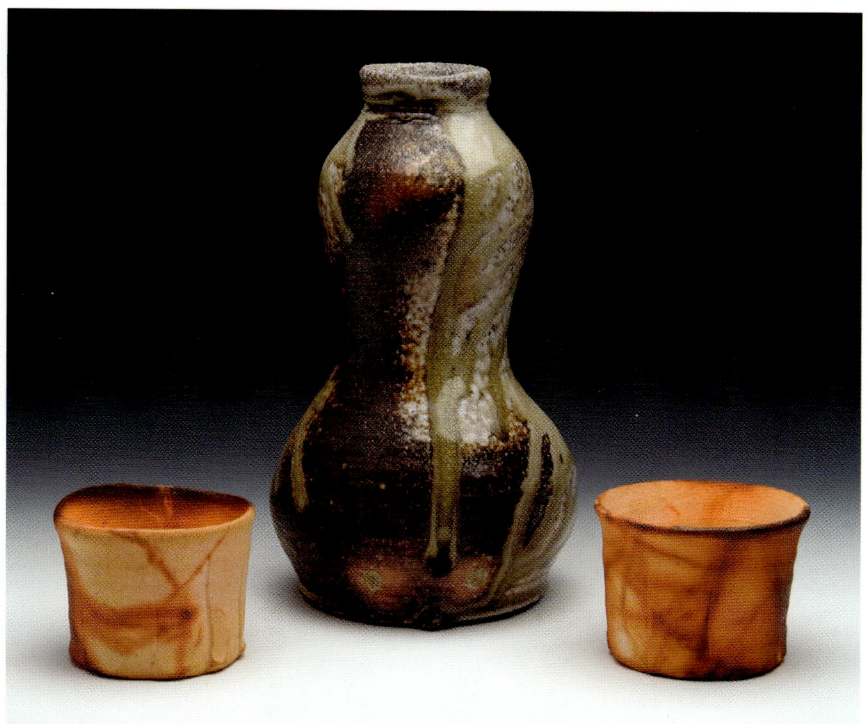

Fred Herbst, *Antidote Set*, 2010. Wood-fired stoneware, wheel-thrown, cups with straw flashing, bottle 9" x 5" x 5", cups 3" x 3" x 3". *Photo credit: Ann Cady.*

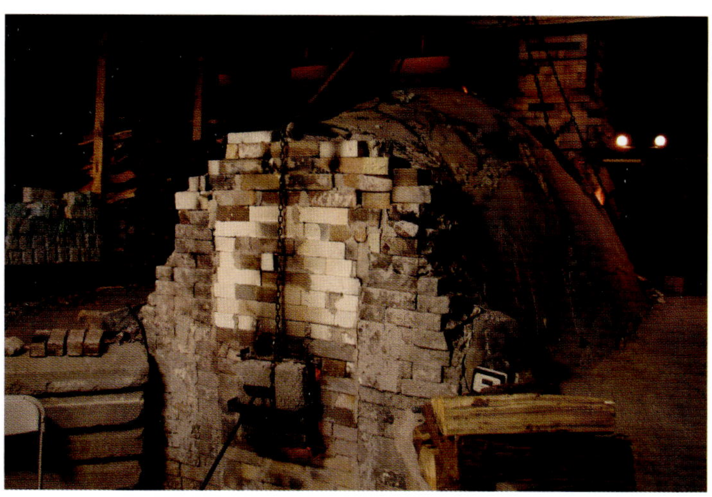

Front of Corning Community College anagama during firing. *Photo credit: Fred Herbst.*

JASON HESS

Flagstaff, Arizona
Jason.Hess@NAU.EDU

Artist's Statement

I have always enjoyed the infinite variety of surface and texture that are attainable through the wood-fire process. I am drawn to the way in which volatile alkalies from certain varieties of wood or introduced into the kiln atmosphere as a solution, react with silica and iron in clay. These marks from ash and vapor dramatically enhance the forms that I make and I have been unable to achieve comparable results with other firing processes.

The infinite and elusive variety of texture and color attainable through the various making and firing techniques that I use has generated an interest in presentation. I enjoy presenting my work so that a viewer might notice subtle differences in form and surface. The vast majority of what I make is wheel-thrown from either an iron rich stoneware or porcelain. The iron rich stoneware's are reduction cooled in a Train Kiln while the porcelain is typically fired in wood/soda.

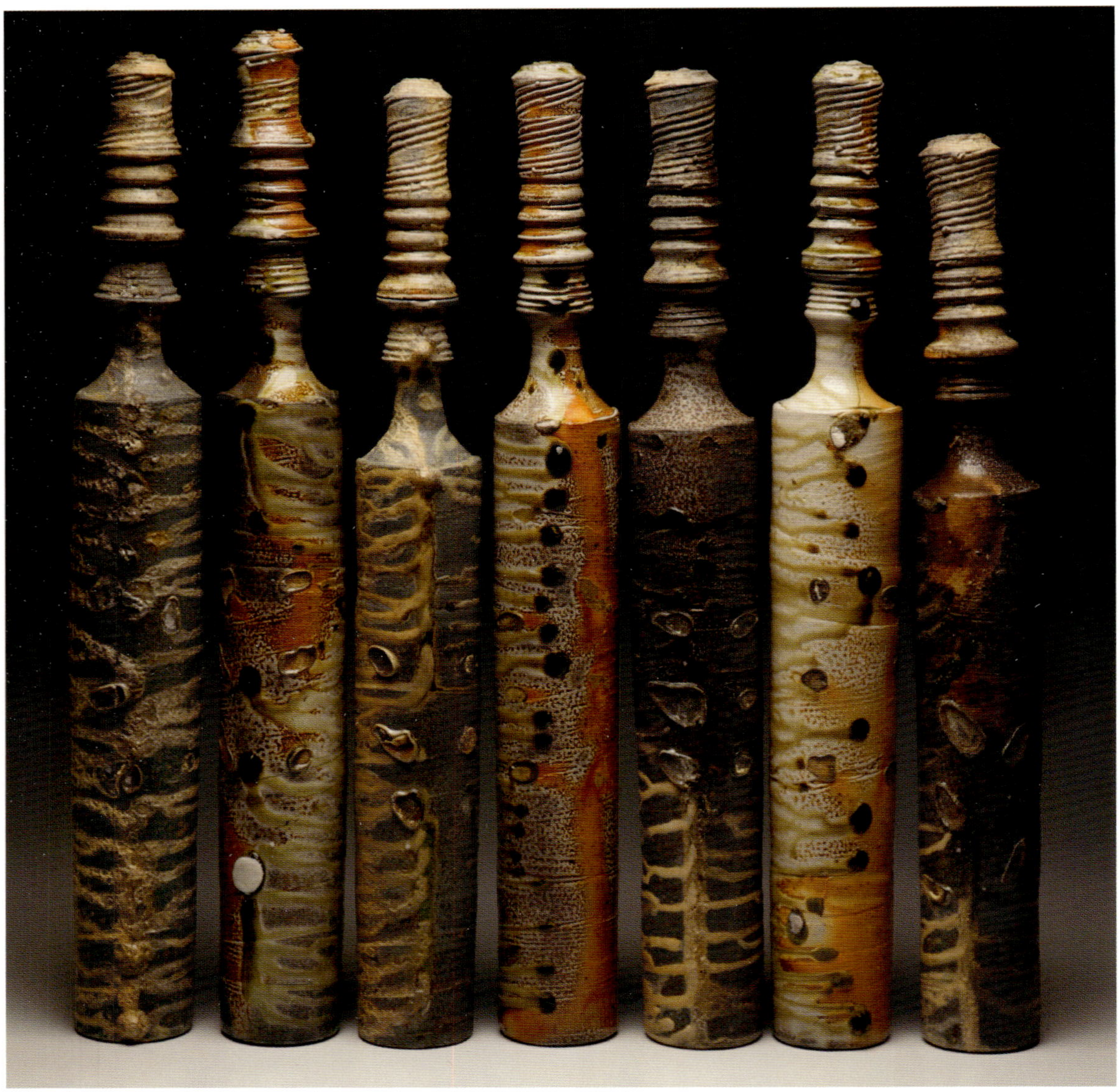

Jason Hess, *Seven Tall Bottles*, 2011. Combination of reduction cooled white stoneware and wood/soda fired porcelain. These bottles were side fired in the front of the train kiln or along the bag wall in the wood soda kiln. 24" x 3.5" x 3.5" (average bottle). *Photo credit: Jason Hess.*

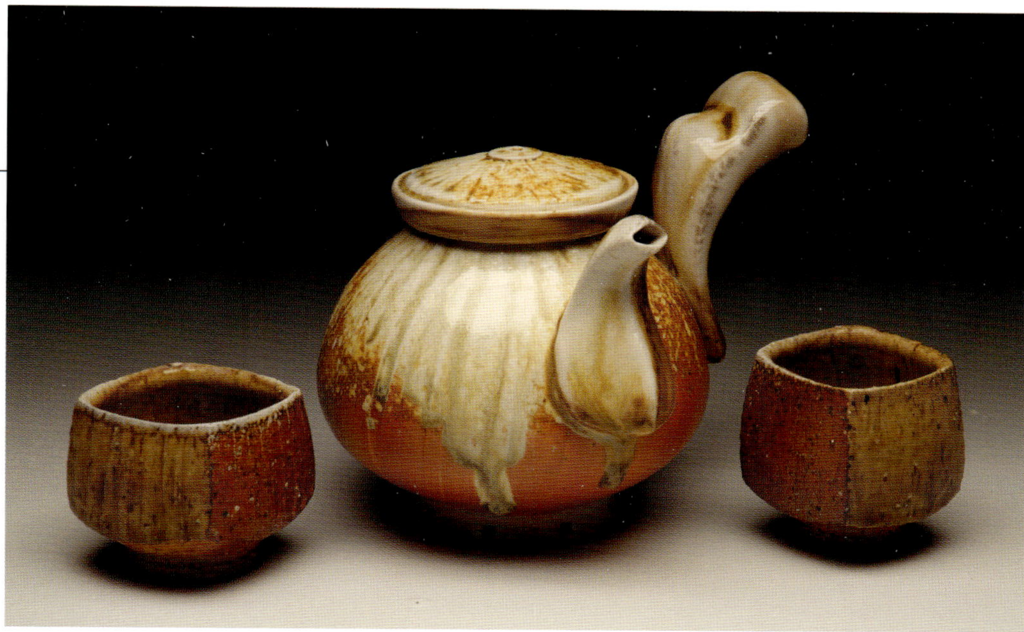

Jason Hess, *Teapot with Cups*, 2011. Porcelain, kaolin flashing slip, press molded handle and spout. The cups are standard stoneware with glaze. Fired for 30 hours in wood/soda with 4 lbs of soda introduced at the end, 5.5" x 5.5" x 6.5" (teapot) 2" x 2" x 2" (cups). *Photo credit: Jason Hess.*

Jason Hess, *Bottle with Cups*, 2011. Porcelain with titanium flashing slip (the yellow). The cups have glaze accents, fired for 30 hours in wood/soda with 4 lbs of soda introduced at the end, 8" x 4" x 4" (bottle) 2" x 2" x 2" (cups). *Photo credit: Jason Hess.*

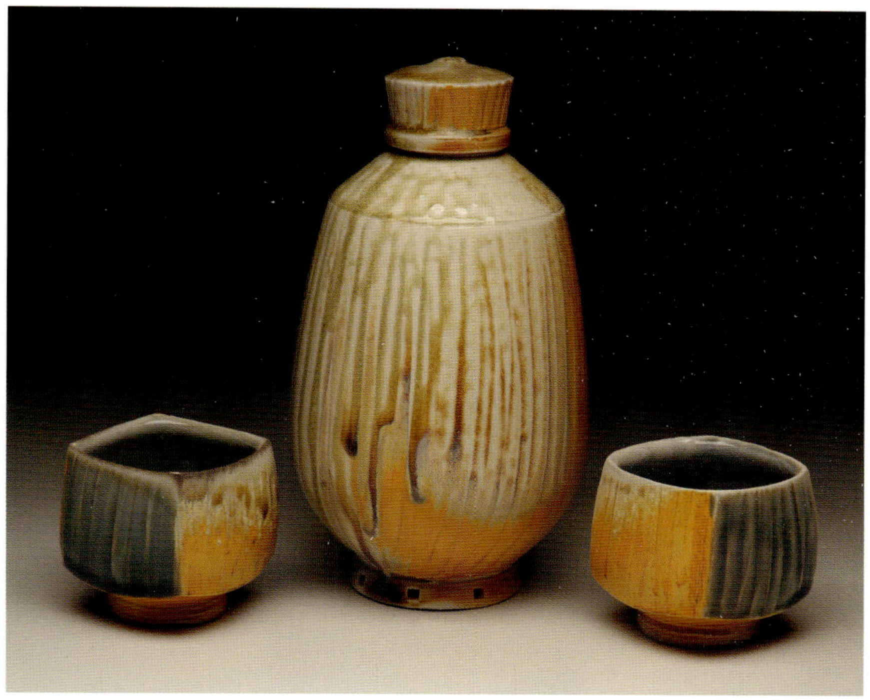

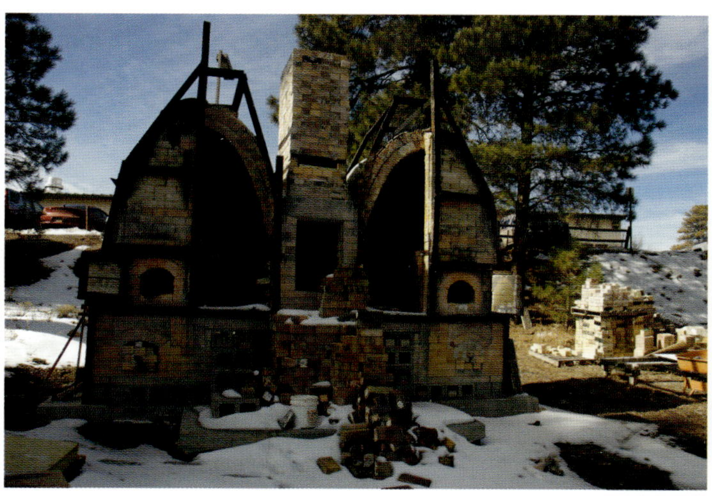

Double Catinary Arch Wood / Wood Soda Kiln. The right side is wood/soda and the left side is just wood. These chambers can be fired together or independently. The door at the base of the chimney is an extra stacking area that makes use of both chambers. *Photo credit: Jason Hess.*

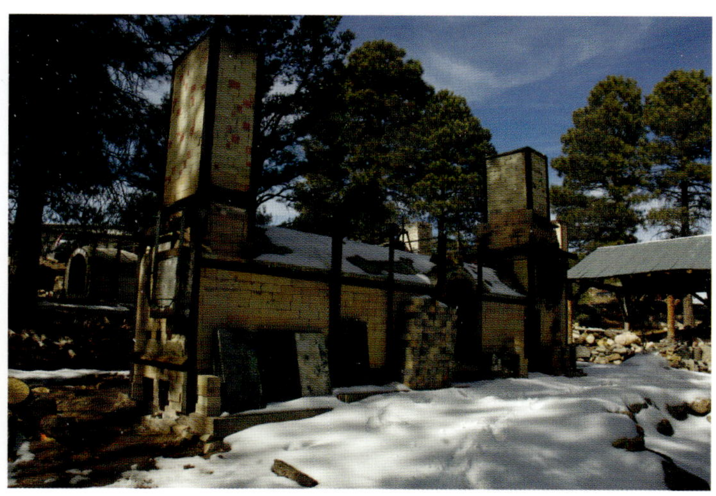

Opposing firebox end of Switch Back Train. This firebox has removable grates in the floor and is a cross draft design. Loading of ware takes place in the middle of the kiln. *Photo credit: Jason Hess.*

GARY HOOTMAN

Swisher, Iowa
gkhootman@aol.com, www.garyhootman.com

Artist's Statement

Through successful experimentation with different clay bodies and handbuilding techniques, I have allowed myself greater freedom in how I express myself through the manipulation of clay and further allowing the clay to keep its integrity. For the last few years, I've been working on both sculptural and functional directions for my work, bringing them together as a whole. I work with the intention of creating an interaction of negative and positive space. The object itself is broken into a collage of layers and a textured line created by the way the clay is formed. In some cases, I used found objects such as a paddle or a stamp to texture the clay. Hand manipulation is used to soften some lines or to take them out completely. My objective is to avoid allowing the viewers an easy passage around the piece and make them deal with the segments that make up the whole. I am searching for objects that are universal in nature. This universal quality may be illustrated by observing a particular phenomenon. When a two or three-year-old child is given a tool to mark with and something to record this mark on, they will always appear the same. It makes little difference what walk of life or what culture they come from, the children will show these marks with a sense of pride and accomplishment. I think they assume we see the same thing they do. It is from this point we try to make ourselves understood. This is not to say that I wish to achieve what some would call primitive art, but rather to make objects that span the history of art forms in a contemporary sense. I do not want my work to be understood only by people of academia, but also by people who can find a connection to my work, which I term universal. The universal is what I continue to strive toward. Wood-firing, in my opinion, is the one thing that ties all of my work together into a successful cohesiveness. Wood-firing and my work cannot be separated.

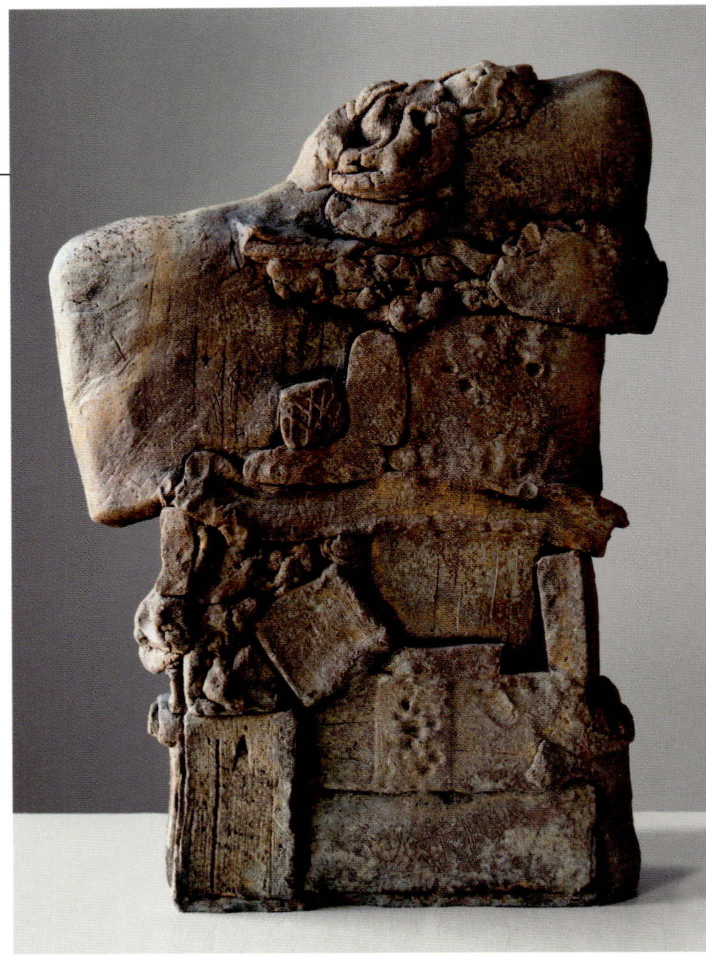

Gary Hootman, *Construct #3*, 1996. Wood-fired stoneware, handbuilt, stamped, cut, 33.5" x 25". Photo *credit: Mark Tade.*

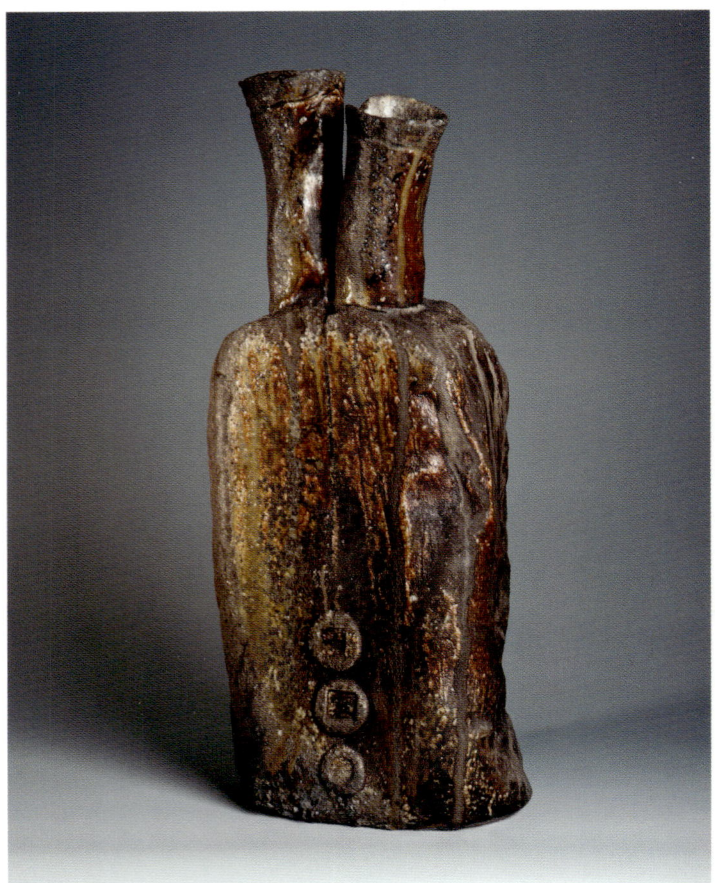

Gary Hootman, *Double Necked Vase*, 2007. Wood-fired stoneware, applied stamps and applied copper leaf, 18" x 7.5". *Photo credit: Mark Tade.*

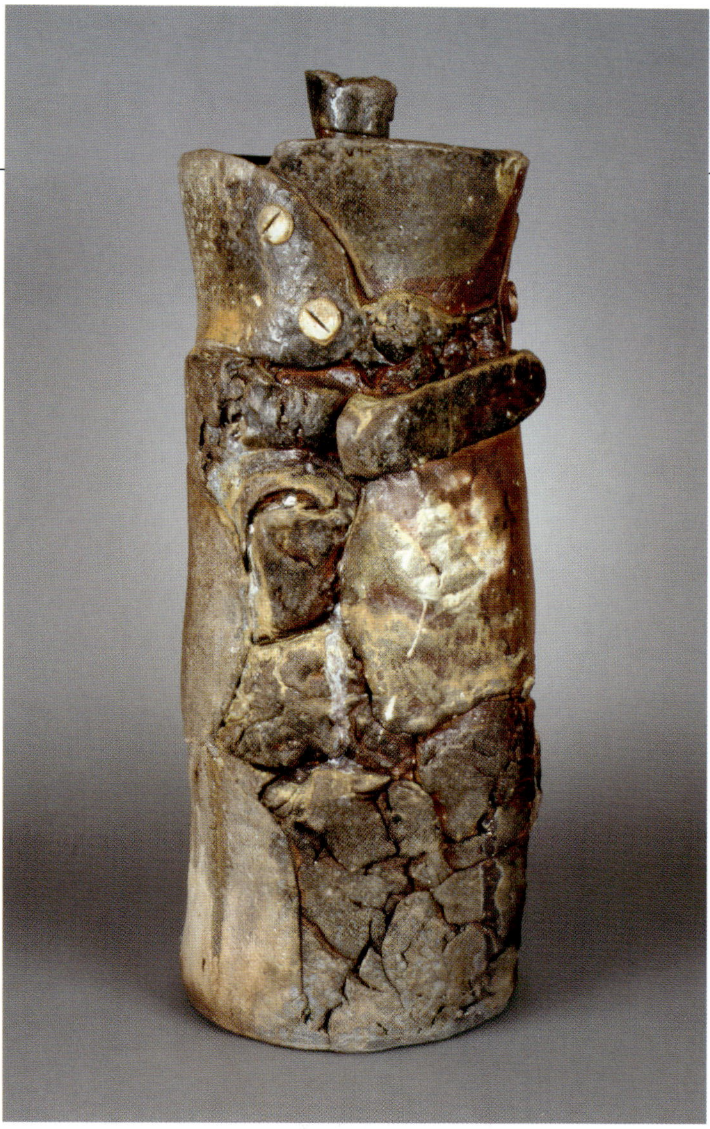

Gary Hootman, *Lidded Construct*, 2003. Wood-fired stoneware, slab built with hand formed pieces, stamped and cut with splashes of white slip, 25" x 13.5". Photo credit: *Mark Tade.*

Over the past thirty years, wood fueled kilns have influenced my firing process, product, and outcome. The anagama (literally, "cave kiln") has been my guide in a continuing journey of artistic discovery. For me, there is a dialogue between the anagama and myself in which I make statements in the form of clay. The anagama responds in ways of its own in dialogue with the forms I give it. I learn from each firing and try to restate and add to the continuing conversation in the next one. It has progressed to the point where the vocabulary of artist, anagama, and clay are inseparable. This relationship has even found expression in names for my kilns—currently *Hootagama 2*.

Many of my pieces, or as I call them "constructs," begin as an additive process that resembles drawing in three dimensions. Over the years I have evolved from functional pottery to sculptural forms and now do a combination of the two, although a sculptural aesthetic is never far from my mind. Some have called this work Jungian and even guttural. I don't know about that. All I know is the conversation continues.

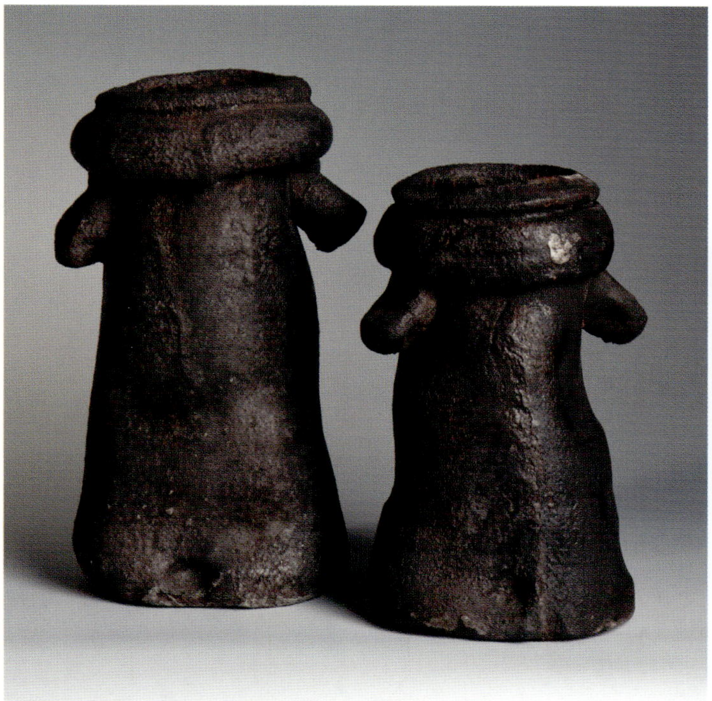

Gary Hootman, *Black Pair*, 2003. Wood-fired black stoneware, wheel-thrown and added handles fired in *Hootagama 1*, 9" x 5.5" and 7.5" x 5". *Photo credit: Mark Tade.*

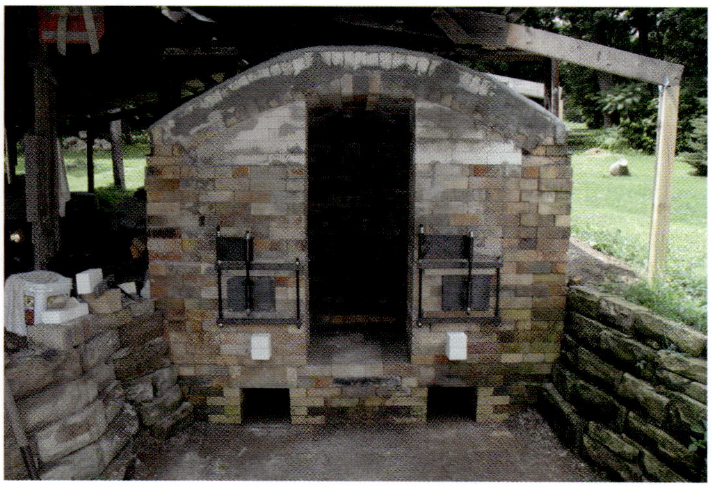

Hootagama 2: construction finished, detailed image showing double hinged swing doors for both fireboxes, two lower fireboxes and back wall inside, 2011. *Photo credit: Gary Hootman.*

DALE HUFFMAN

Pittsburgh, Pennsylvania
dhuffman@carlow.edu, www.dalehuffman.net

Artist's Statement

I designed and built my kiln in 1998. It is a tunnel kiln, a modified anagama, based upon those I fired in Tokoname, Japan, in 1996.

Each time I make pots, each time we load the kiln, each firing, each area of the kiln during that firing, all are unique. These pots freeze moments in time. They are snapshots of the ever-changing flow of my work and my life. On good days I am able to play and dance with the clay. The flow can be magical. Close to the edge of what I understand, sometimes days, weeks, or even months pass before I am able to grasp what we have done together. Then the fly ash produced by the fire, burning in some cases inches away from the pots, will clothe them in glaze. There will be differences between windward and leeward sides of the pots, and there will be scars from pots sticking together when they warp and shift from days in the white heat. These will provide decoration to the shapes that the clay and I have created.

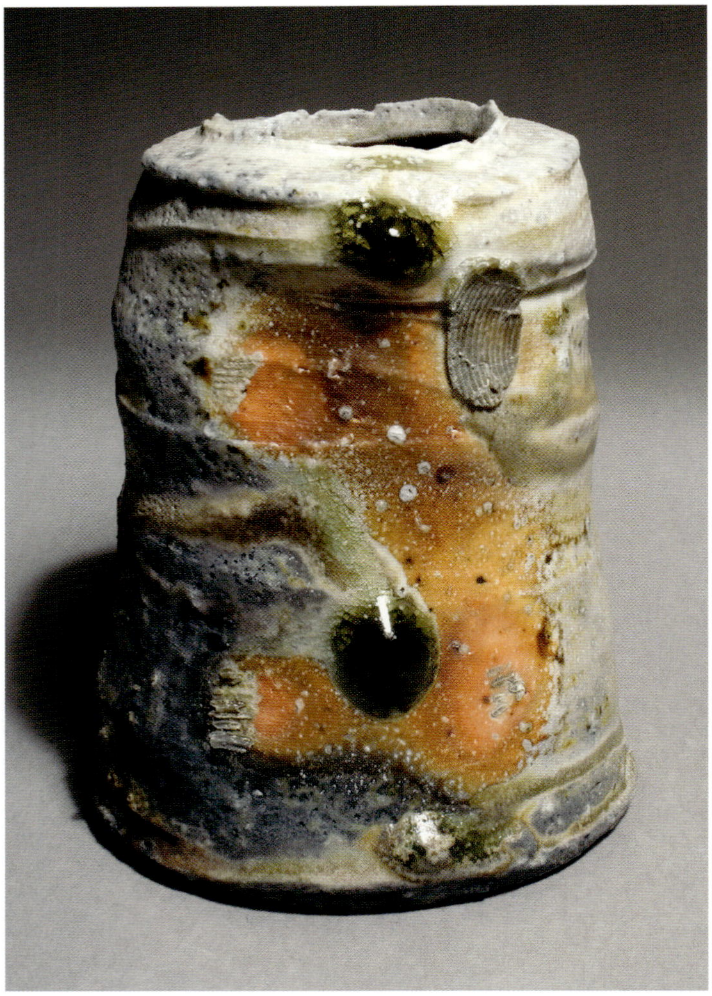

Dale Huffman, *Vase*, 2006. Wheel-thrown stoneware, from side stoke area just off the firebox, natural ash glaze, 4" x 3.5" x 3.5". *Photo Credit: Dale Huffman*.

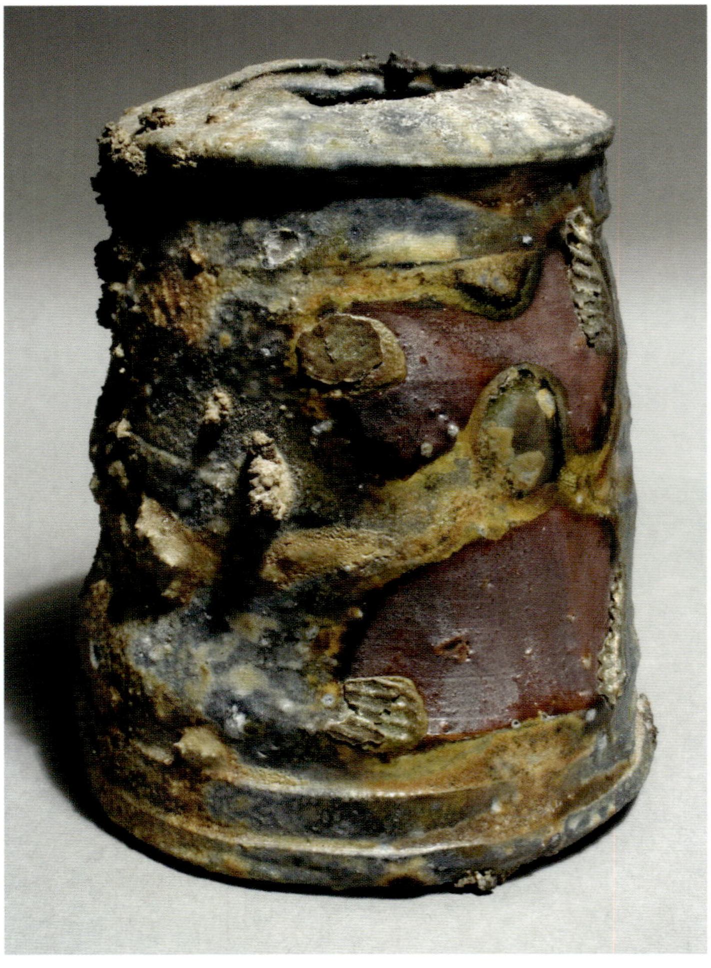

Dale Huffman, *Vase*, 2006. Wheel-thrown stoneware from side stoke area just off the firebox, natural ash glaze, 4" x 3.5" x 3.5". *Photo Credit: Dale Huffman*.

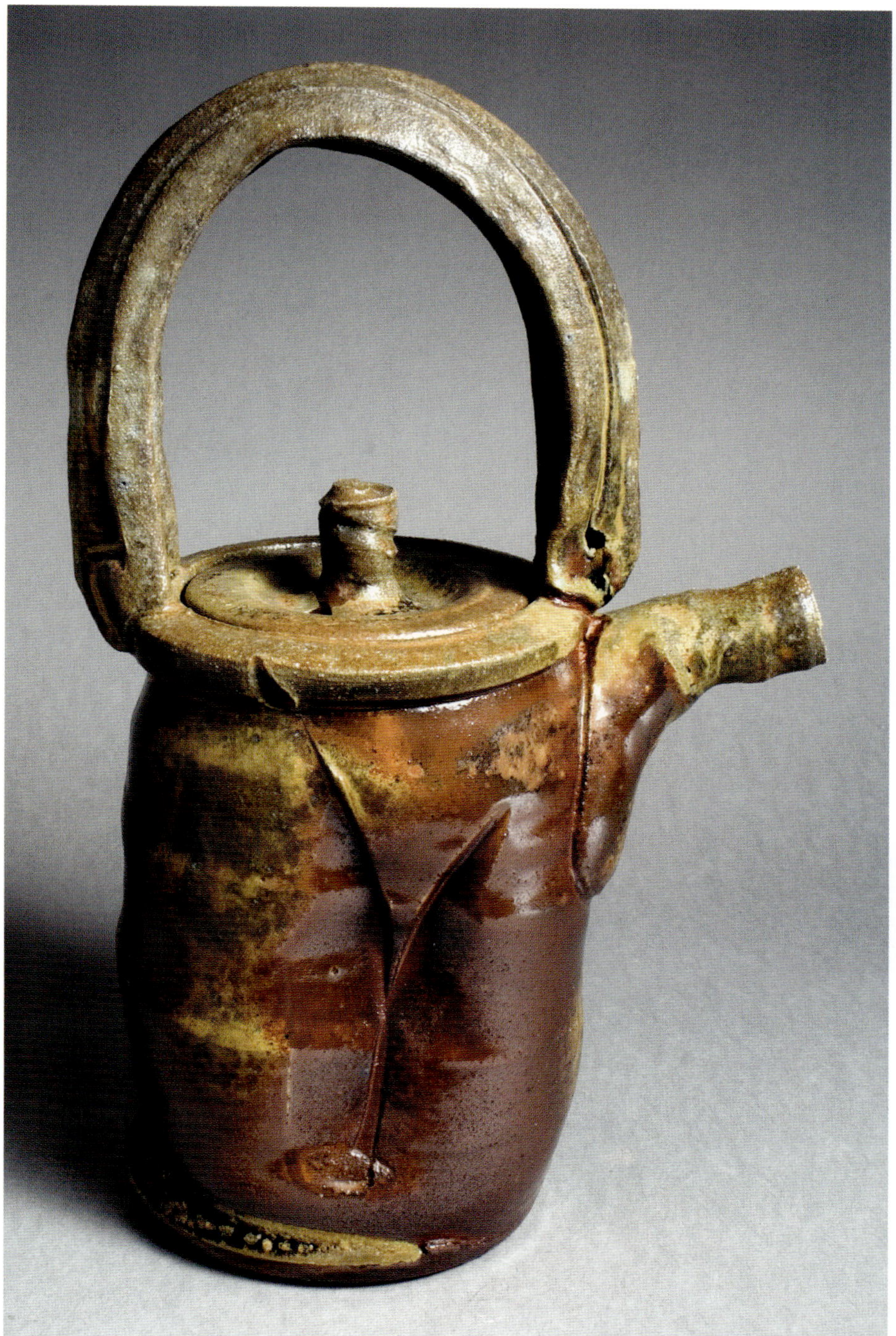

Dale Huffman, *Teapot*, 2011. Wheel-thrown stoneware, from the middle area of the kiln, natural ash glaze, 11" x 5" x 3.5". *Photo Credit: Dale Huffman.*

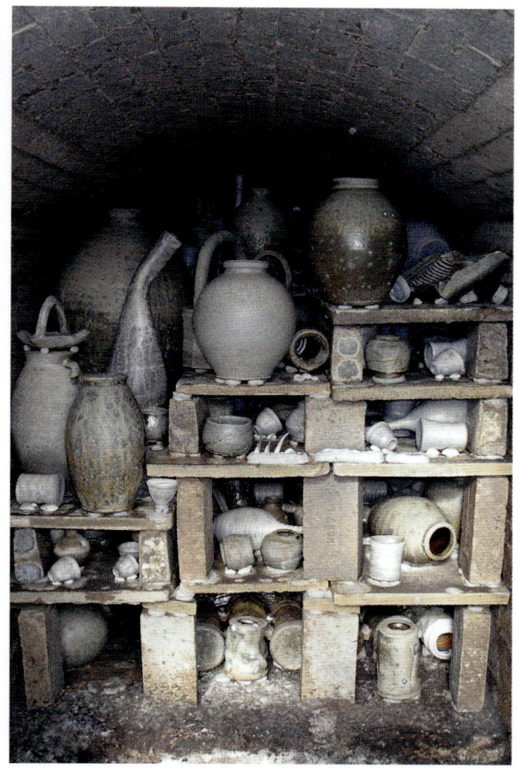

Kiln Stack After Firing. Photo Credit: Dale Huffman.

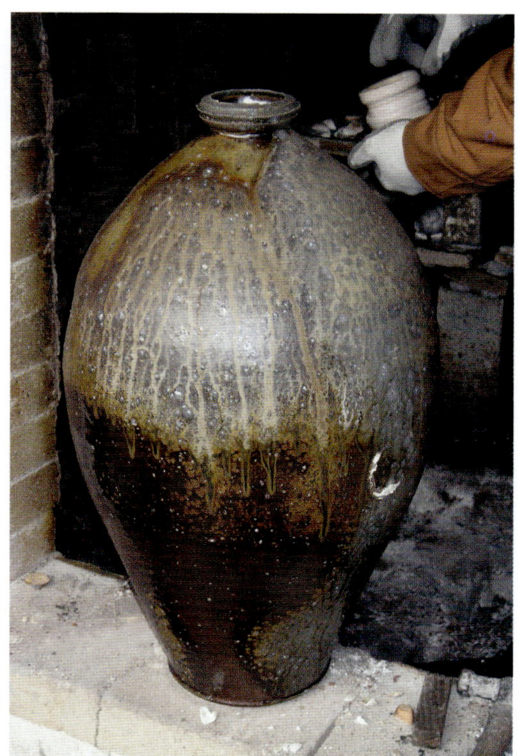

Kiln Unloading. Photo Credit: Dale Huffman.

RICHARD JAMES

Bloomington, Indiana
claymeddler@gmail.com, www.richardwjames.com

Artist's Statement

My work primarily focuses upon dualities, in particular, the interacting forces of attraction/repulsion, freedom/confinement, and seen/unseen. I am currently using the visual language of a molecular level to explore the relationship between individuals and their macro environments. Each piece has two very distinct sides that reflect a dichotomy that I find endlessly fascinating.

Firing with wood offers me a process that embodies my interest in that duality. It is a craft to be refined through experience and at the same time an event that facilitates something that can only be described as magic. In a world where a wealth of information is at my fingertips, I still feel the need to seek a process that is not fully understandable, a way for beautiful occurrences to develop internally. For me, this particular balance between knowledge and chance happens most fruitfully in wood-firing. Allowing something elemental, something wild, to have its hand in the final product offers a shot at creating something larger than the sum of its parts. It is a chance to make something birthed from the inside out (a concept mirrored in my forms.) Good work from a wood kiln doesn't have surfaces, only outside edges of the form.

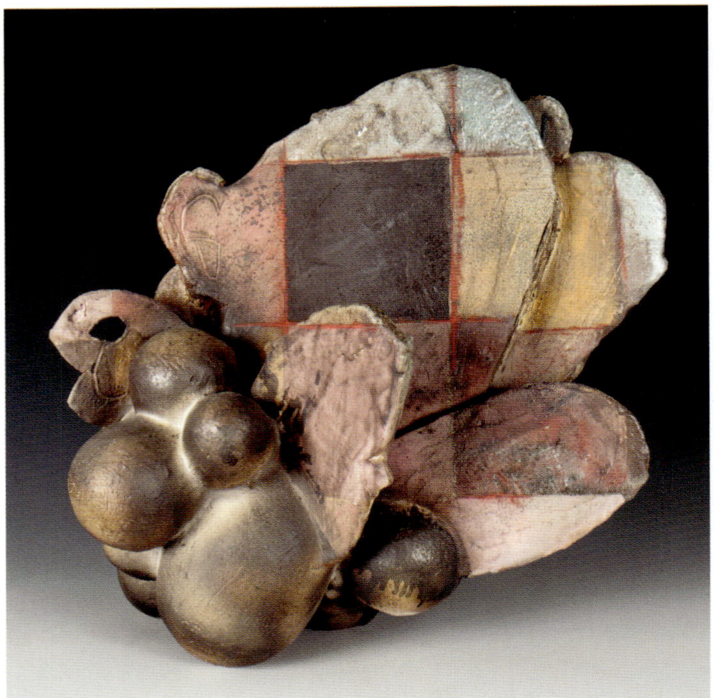

Richard James, *18-bit Mitosis*, 2012. Wood-fired earthenware (paperclay), wire, cone 01,thrown and altered, slab, and coil built. slips, underglazes, terra sigillata, raw stain, and soda ash. 13" x 13" x 10". *Photo credit: Rachel Ankney.*

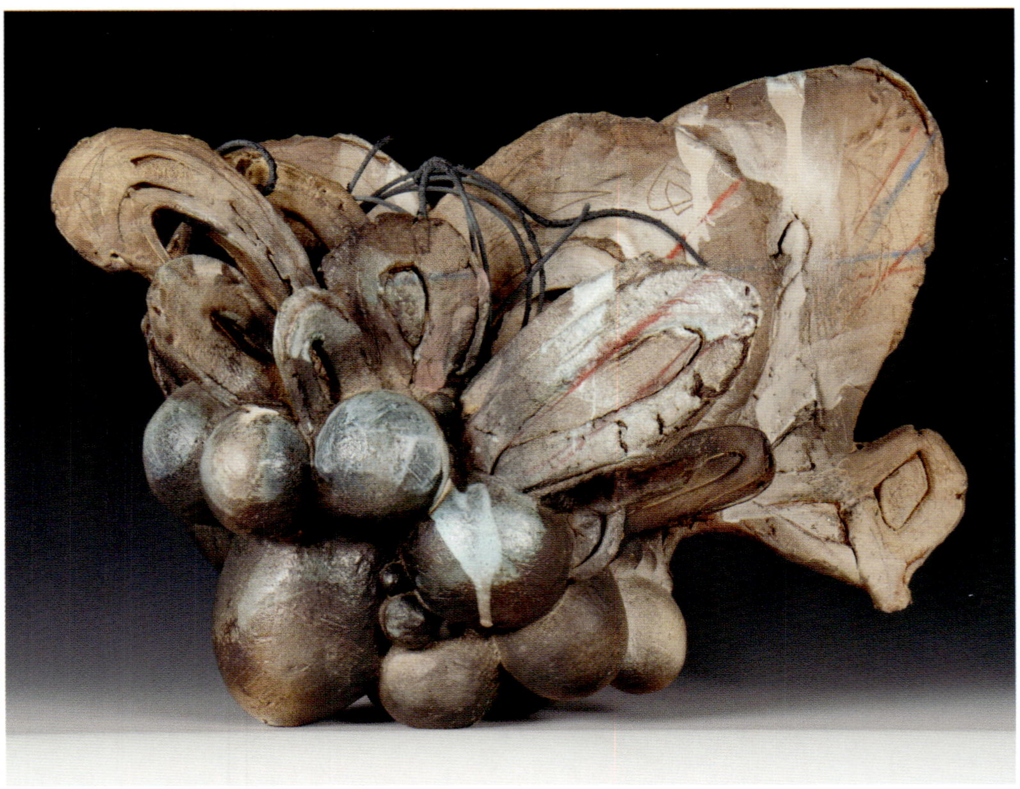

Richard James, *Lorentz*, 2012. Wood-fired earthenware (paperclay), wire, cone 01, thrown and altered, slab, and coil built, slips, underglazes, terra sigillata, raw stain, and soda ash. 14" x 18" x 10". *Photo credit: Rachel Ankney.*

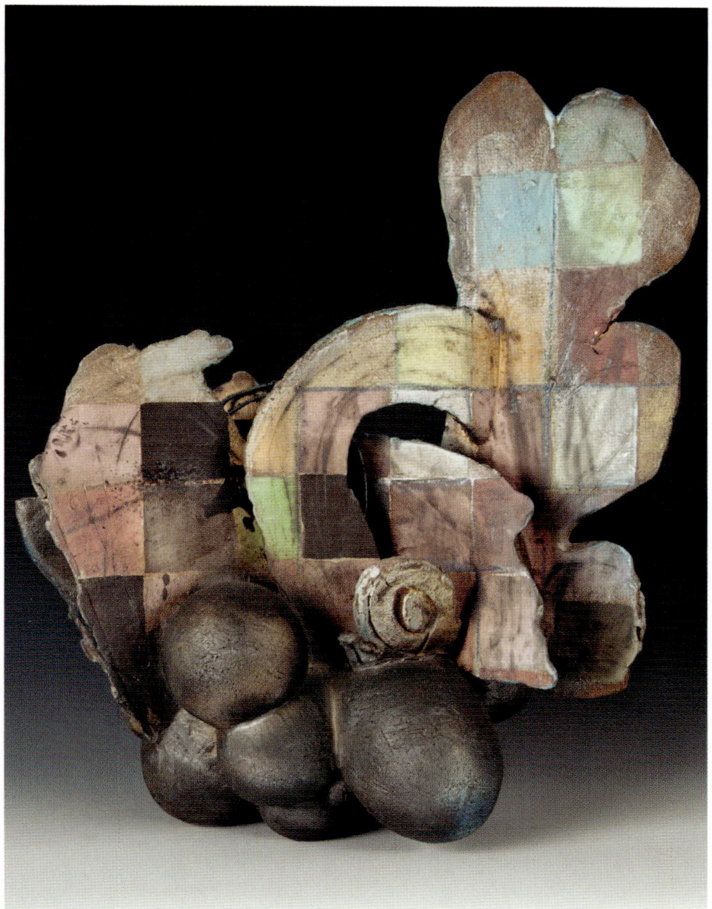

Richard James, *Cloud Chamber,* 2012. Wood-fired earthenware (paperclay), wire, cone 01, thrown and altered, slab, and coil built, slips, underglazes, terra sigillata, raw stain, and soda ash,18" x 15" x 7". *Photo credit: Rachel Ankney.*

In 2010, I designed and built a small catenary arch cross draft (bourry box) for Indiana University, Bloomington, with specific academic goals in mind. The kiln needed to be small enough to have quick feedback on firings, but still hold enough to make firing the kiln worth the effort. It was also constructed to offer a wide possibility of atmospheres and amounts of fly ash. Since its construction, I have been exploring different ways in which to use a wood kiln. In late 2010, I began investigating its use in firing earthenware. This method of firing retains the spontaneity of wood-firing while opening up the opportunity for color, along with the added benefit of saving time and resources. Because there is very little in the way of documentation of this aesthetic, its pursuit has been both exciting and challenging to me.

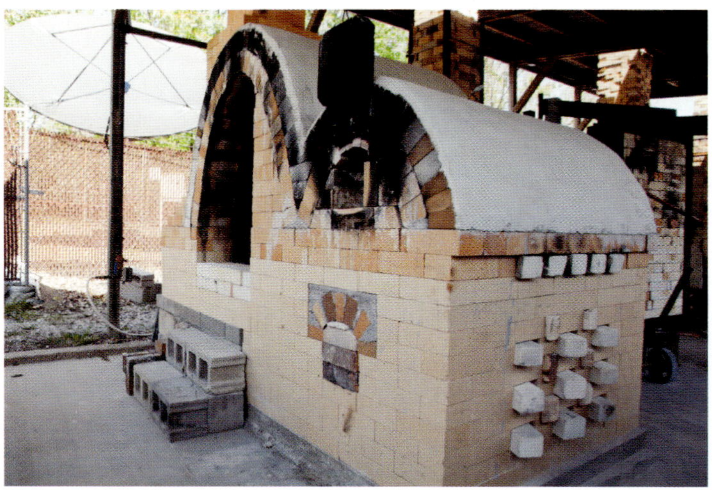

Richard James, Kiln side view, completed 2010. Catenary bourry box, 10' x 6' x 4.5'. *Photo credit: Rachel Ankney.*

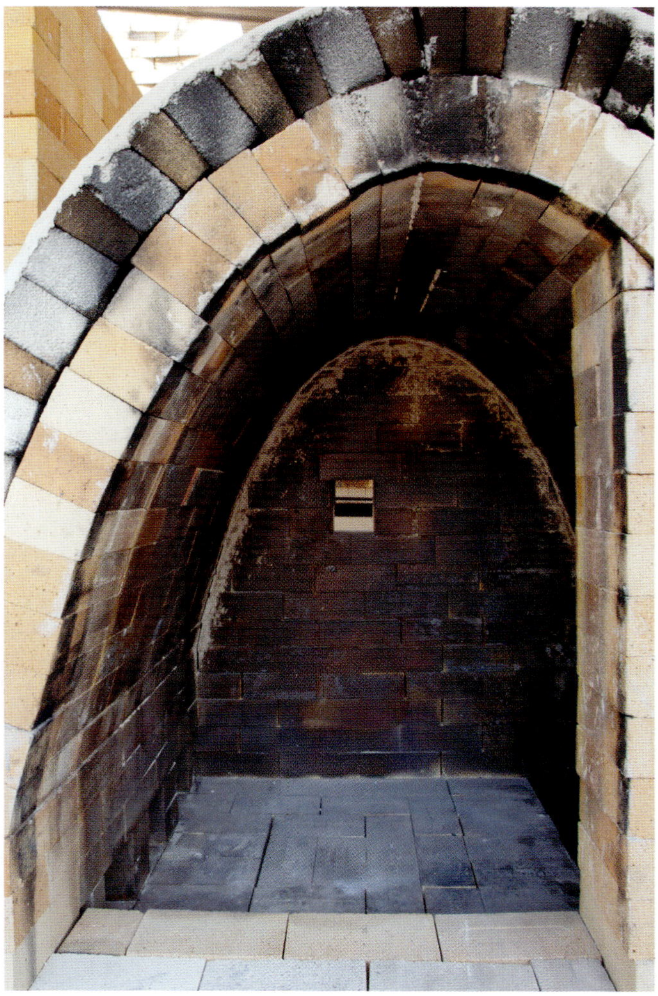

Richard James, Kiln interior, completed 2010. Catenary bourry box, 10' x 6' x 4.5'. *Photo credit: Rachel Ankney.*

DOUG JEPPESEN

Sugar Grove, Illinois
doug@dougjeppesen.com, www.dougjeppesen.com

Artist's Statement

Today we live in a world of convenience rather than one of necessity, and to make functional wood-fired pottery is a reaction against this convenience. I'm interested in the personal relationships between handmade pottery and the industrialized objects used daily in our society. From a solitary morning cup of coffee, a noisy family dinner, or sipping a cup of whiskey with a friend, pots are used every day by millions of people as they record and discuss their ideas, activities, and experiences.

I have made a conscious decision to make pots; it is not one of necessity, but one of passion. It is a decision that speaks to who I am as a person, I enjoy work, and I enjoy making objects from ceramic material. I like the physicality and complete engagement that the material requires, from processing clay dug from a riverbank to prepping and firing wood kilns. I enjoy working with my hands to produce objects and to discover new ways of looking at this ancient material. I am also intrigued by both the physical and mental challenges that the material brings along the way.

Even though we live in a technology-driven world, I choose to fire with wood because of the way time is naturally recorded by flame and ash on each individual piece. I enjoy the decision-making involved in creating work for this particular firing process. Additionally, I like the challenge of loading the kiln to create specific situations as to where the flame will travel during the firing, leaving a permanent record of the object's experience in the kiln. Firing with wood is a laborious process, but I truly enjoy gathering, splitting, and stacking wood in order to fire for three to four days. Finally, once the kiln has cooled for a week, it is a tremendous feeling to pull those pots out and see all the hard work completed and then to get back to work to start developing future ideas.

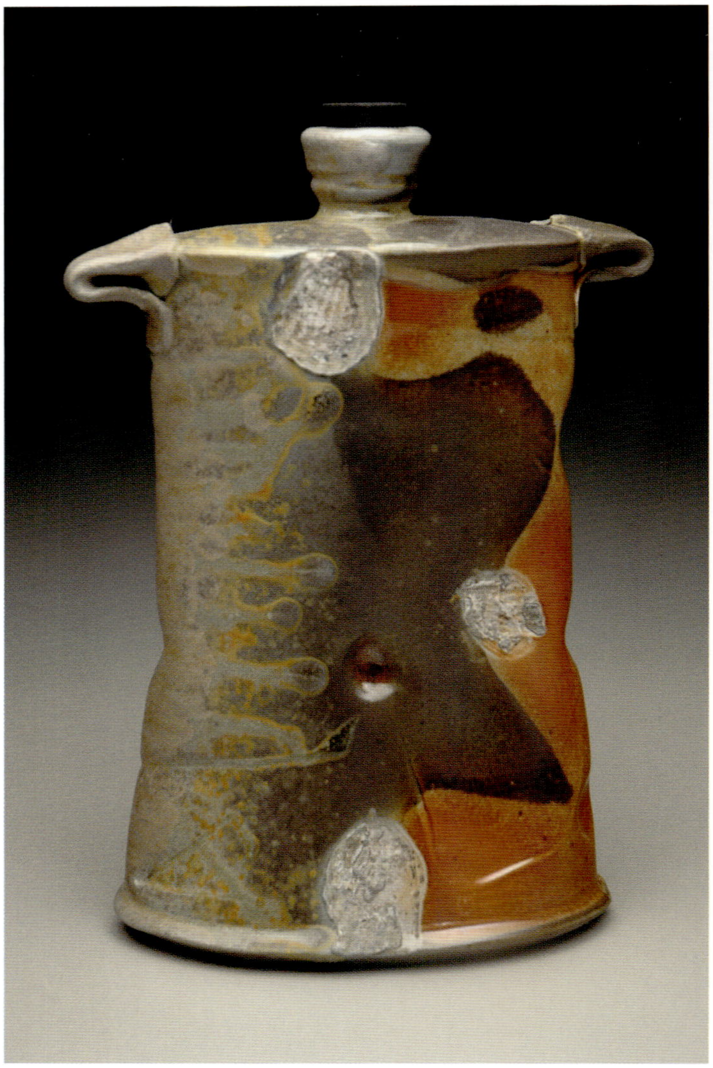

Doug Jeppesen, *Bourbon Bottle*, 2008. Wood-fired B-Mix, wheel-thrown, altered and assembled, 10" x 5" x 3", Collection of Corning Community College, Corning, New York. *Photo Credit: Doug Jeppesen*.

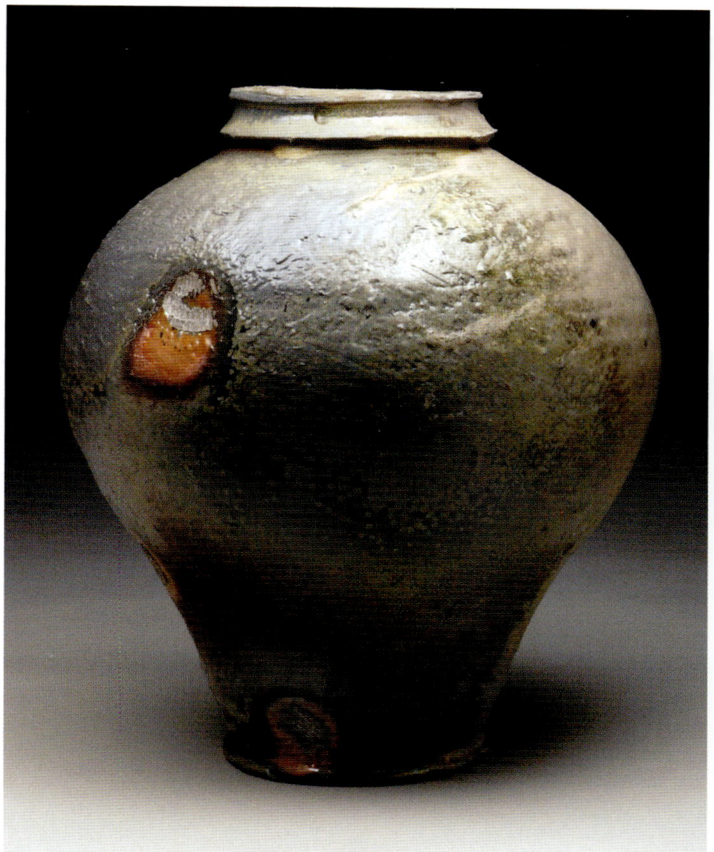

Doug Jeppesen, *Jar*, 2011. Wood-fired B-Mix, wheel-thrown and coil built, 14" x 10" x 10". *Photo Credit: Doug Jeppesen.*

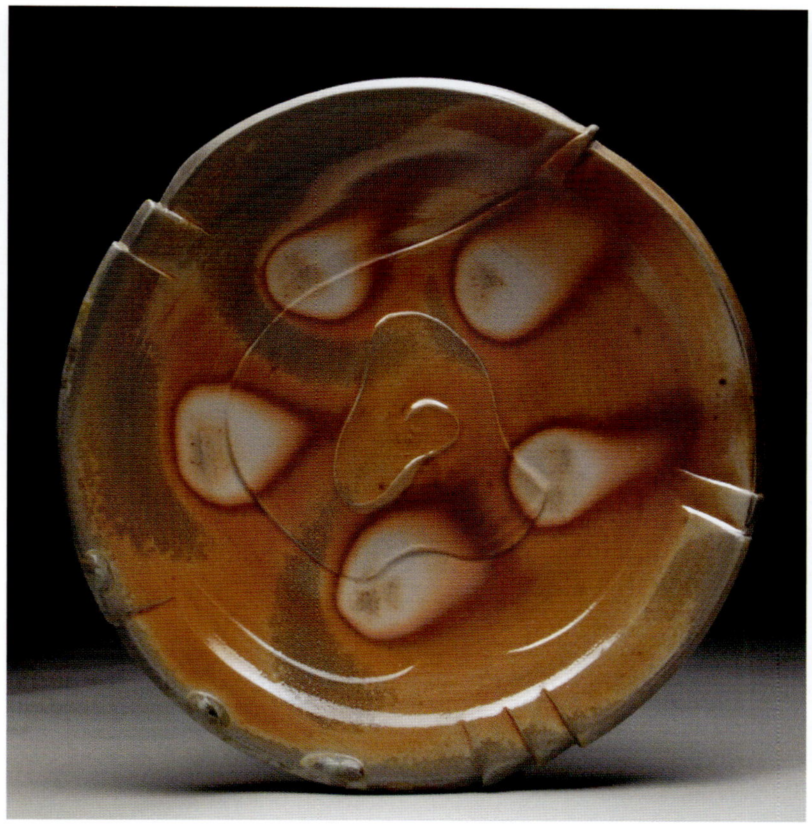

Doug Jeppesen, *Platter*, 2010. Wood-fired B-Mix, wheel-thrown and altered, 14" diameter. *Photo Credit: Doug Jeppesen.*

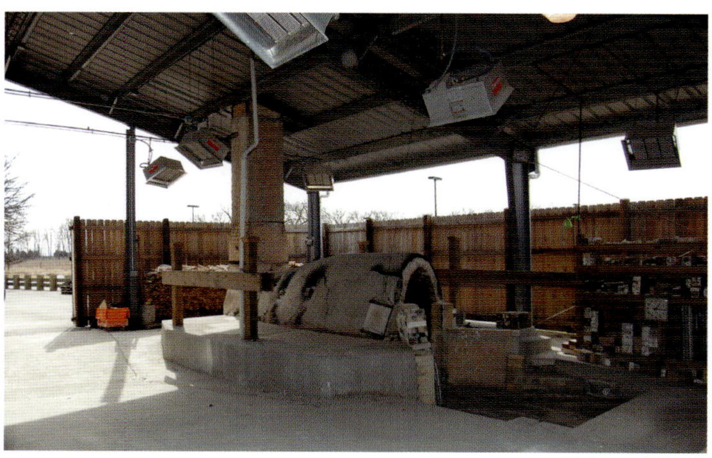

Doug Jeppesen, "*Honey*" Waubonsee Community College anagama, Sugar Grove, Illinois, built fall 2006. *Photo Credit: Doug Jeppesen.*

LISA JETONNE

Fairfield, California
lisajetonne@gmail.com, www.lisajetonne.com

Artist's Statement

Typically I fire stoneware, bisque and bare, without any stain or glaze. Usually the firing lasts for seven days, often including several reductions, then the kiln is sealed tightly and allowed to cool slowly over seven days. Because I fire in a large (150 cubic feet) kiln, I have many options for locations of pieces in the kiln, and the effects particular to each area within the kiln differ remarkably from one another. At each firing I take advantage of the range of effects available. Because each specific area tends to reliably produce the same effect firing after firing, and because I know this kiln quite well, I can consider this information when I design a new piece, planning where it will go in the kiln as I'm building it.

I fire in a community kiln. It was the beast, the furnace, the kiln itself that drew me to this technique. I happened to be modeling for a figure sculpture course in a classroom next to the wood-fire kiln, at the time a lighting ceremony was scheduled; we all took a break to participate. At that point, I'd never taken an art class. For ten years I'd been a professional model in all media, acquiring a solid foundation of knowledge along the way, but I had no studio experience. Participating in that lighting ceremony led me to enroll in my first ceramics lab. Within a few months I was making sculptures *just so I could fire them in a wood kiln.*

Now, many firings and years later, I know that it's the camaraderie of the labor that I love most: the loading and hauling of huge stumps and logs, the hours of splitting, the meditative absorption of stacking eight cords of wood, the late nights loading the kiln are all shared activities. For several weeks we work together preparing the kiln, for seven days we work together firing it. Even though we divide tasks based on physical ability, our shared sense of non-ownership of the kiln means we each value everyone else's contribution to the fullest. Regardless of individual strength, endurance, or amount of time each person contributes, each of us during the preparation and firing influences the final result of everyone's pieces. Our work emerges from the firing having been touched, albeit invisibly, by the hands of all the artists.

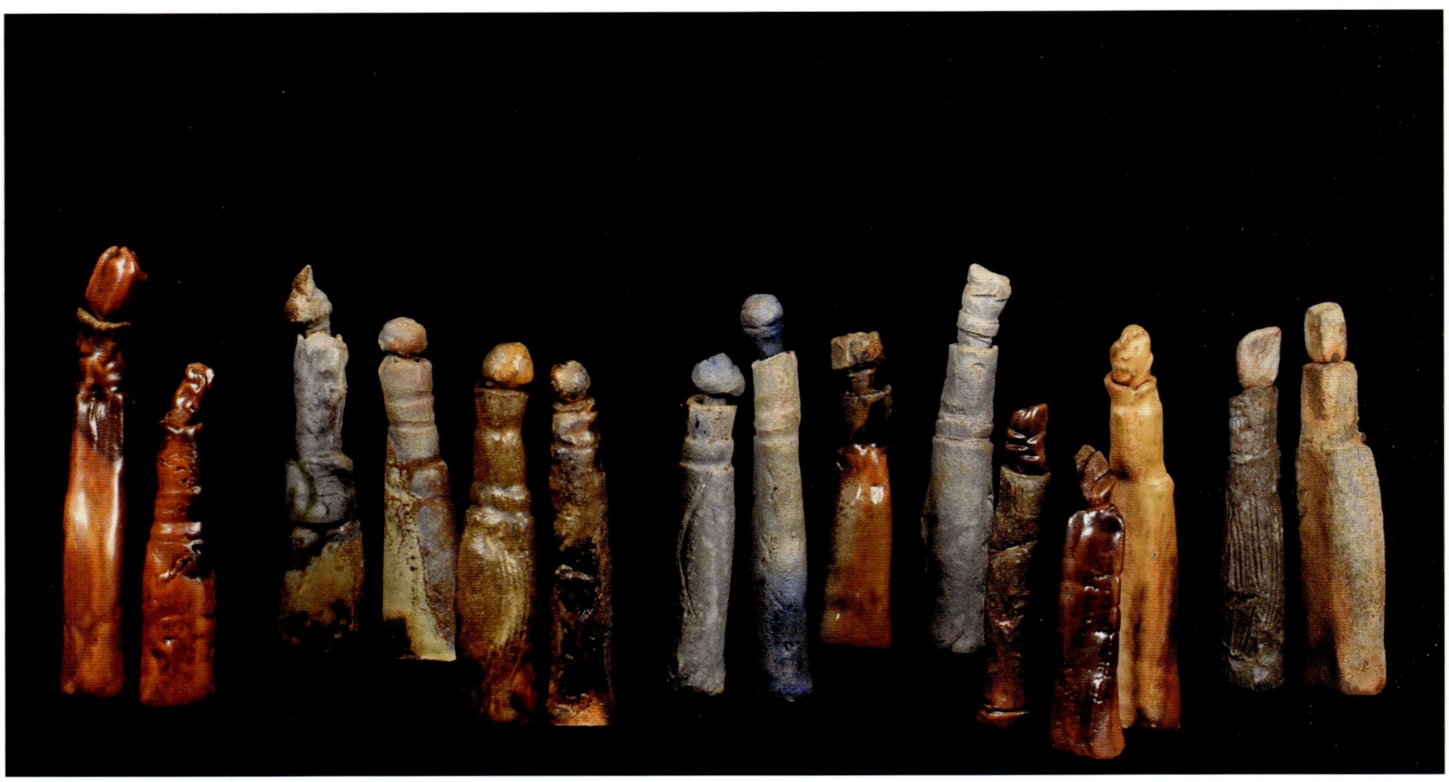

Lisa Jetonne, *Therapeutic Conversations,* 2010. Stoneware, abstracted couples (and one group of three), all fired in the same kiln during the same firing, but scattered throughout the kiln; each figure, couple, or group was fired next to or very near the other. Variable between 5" to 7" tall. *Photo credit: Lisa Jetonne.*

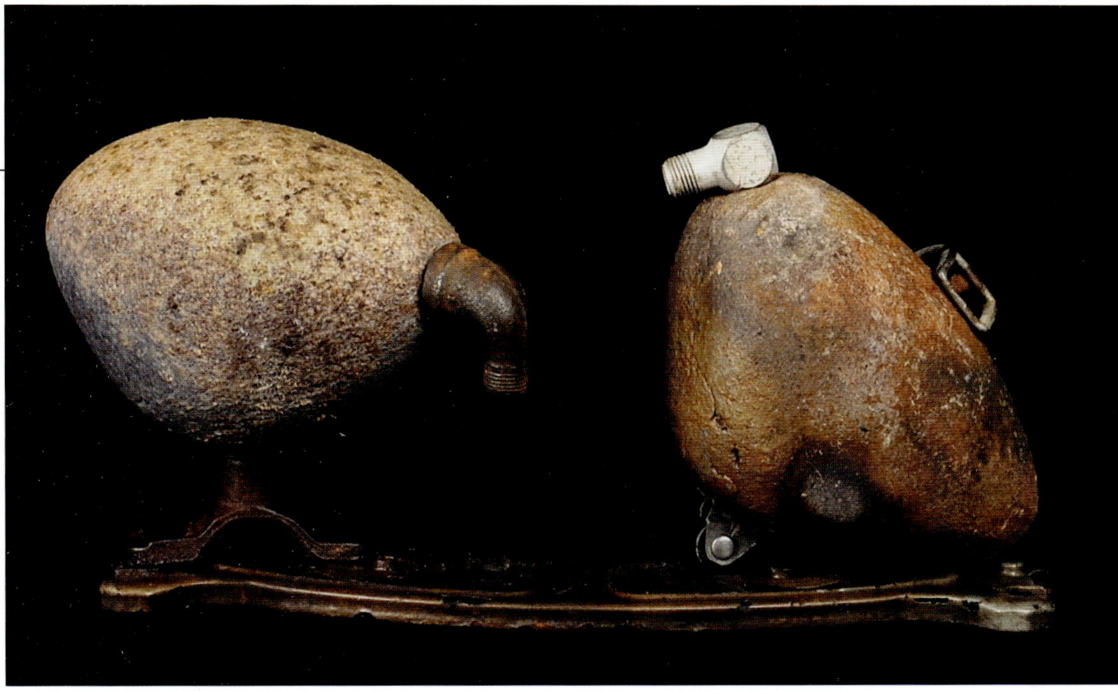

Lisa Jetonne, *Along The Swash*, 2011. Stoneware with salvaged metals, 7" x 12" x 3". *Photo credit: Lisa Jetonne.*

Lisa Jetonne, *Writer's Toolbox*, 2009. Eight stoneware kiln gods in a vintage galvanized metal-lined carpenter's nail box; tie wire and a handmade pencil. (Only 4 kiln gods visible), 11" x 17" x 9". *Photo credit: Lisa Jetonne.*

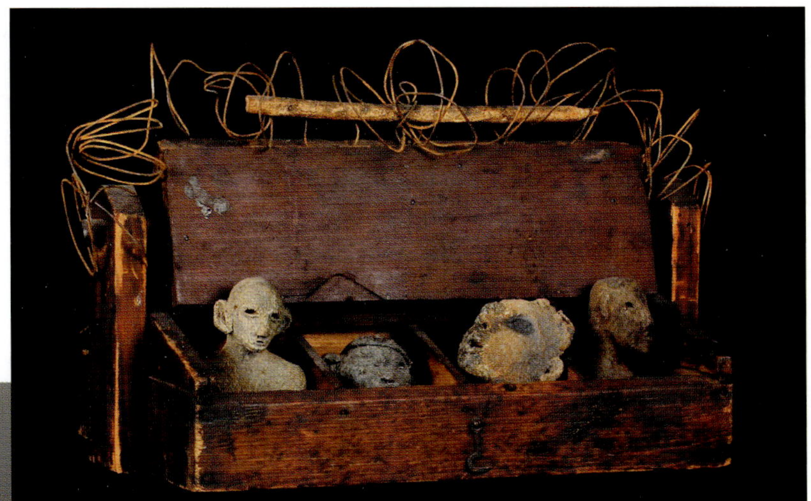

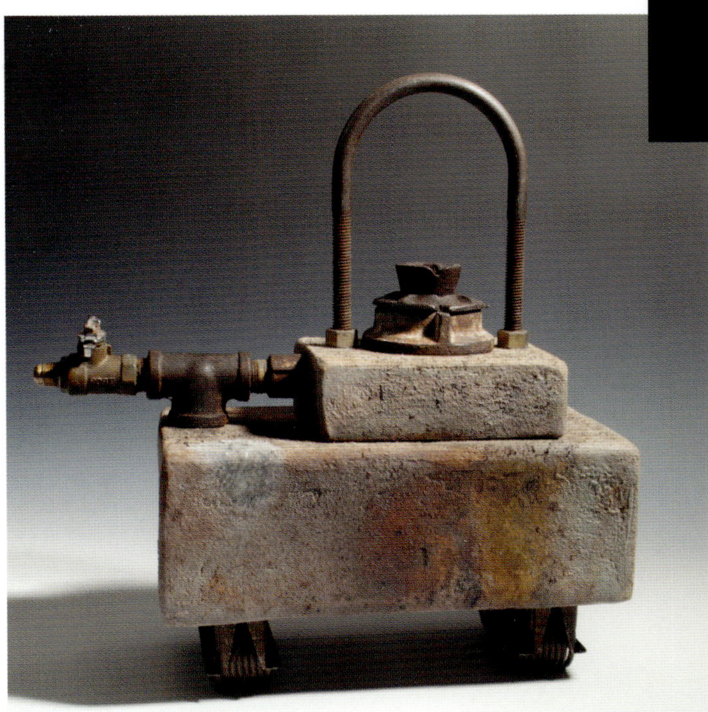

Lisa Jetonne, *Made in USA*, 2010. Slab-built stoneware with salvaged metals: brass vintage platform rocker springs, chrome, steel, iron, and brass fittings. Fired in side stoke area for seven days; heavy reduction on the seventh day; banked with hardwood coals. *Made in USA* is from my series interpreting tea-related vessels, called *The Mechanic Dreamed He Was A Tea Master*, 16" x 16" x 6". *Photo credit: Kathy Kearns.*

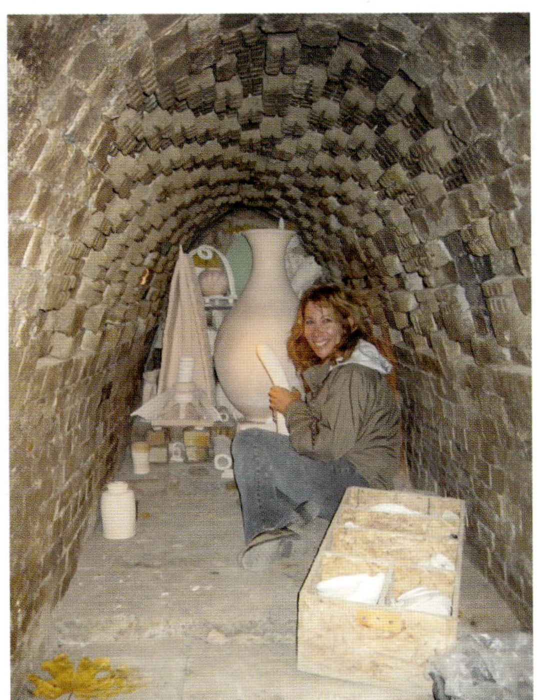

Lisa Jetonne loading other artists' work into the front chamber of Dancing Fire Wood Kiln. *Photo credit: Nora Roush.*

DANIEL JOHNSTON

Seagrove, North Carolina
daniel@danieljohnstonpottery.com,
www.danieljohnstonpottery.com

Artist's Statement

I have been trained in the Leach and Cardew school of making pots. For eighteen years I have devoted my time to understanding pottery traditions that embody this school of thought. Studying the historical traditions of North Carolina, England, and Thailand has taught me to use local materials and fire large wood-burning kilns. I mine local clays and materials for making and glazing pots. These raw materials are far superior to commercially refined products if one invests the time to understand their unique properties. The effects of my large tube wood kilns enhance the beauty of the local clays and glazes, allowing their range of properties to be experienced. Understanding these traditions makes my pots relevant in my own tradition and culture.

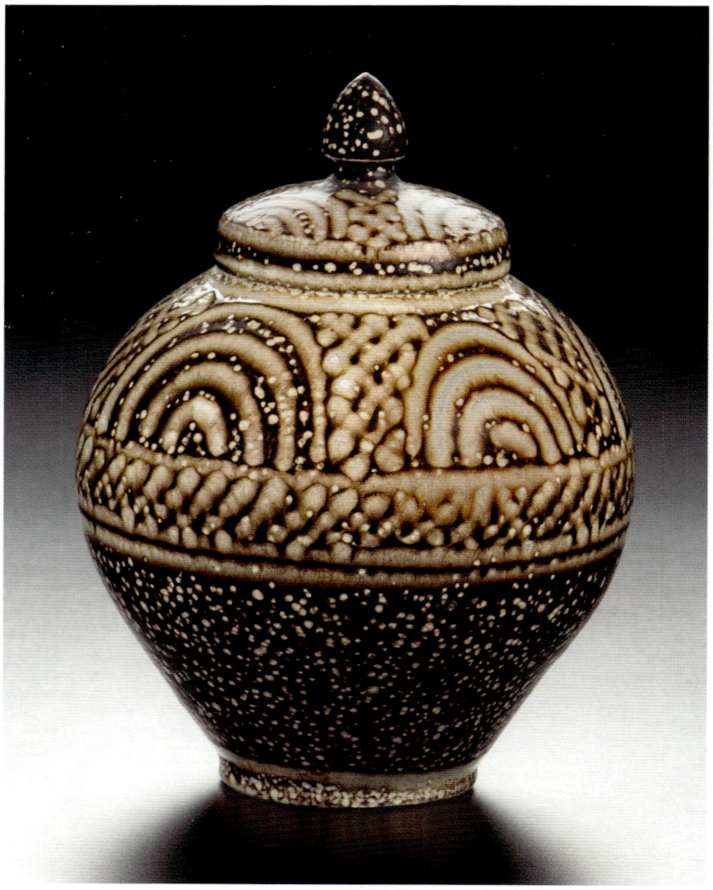

Daniel Johnston, *Jar,* 2008, local wood-fired salt glazed stoneware, 10" x 8". *Photo credit: Juan Villa.*

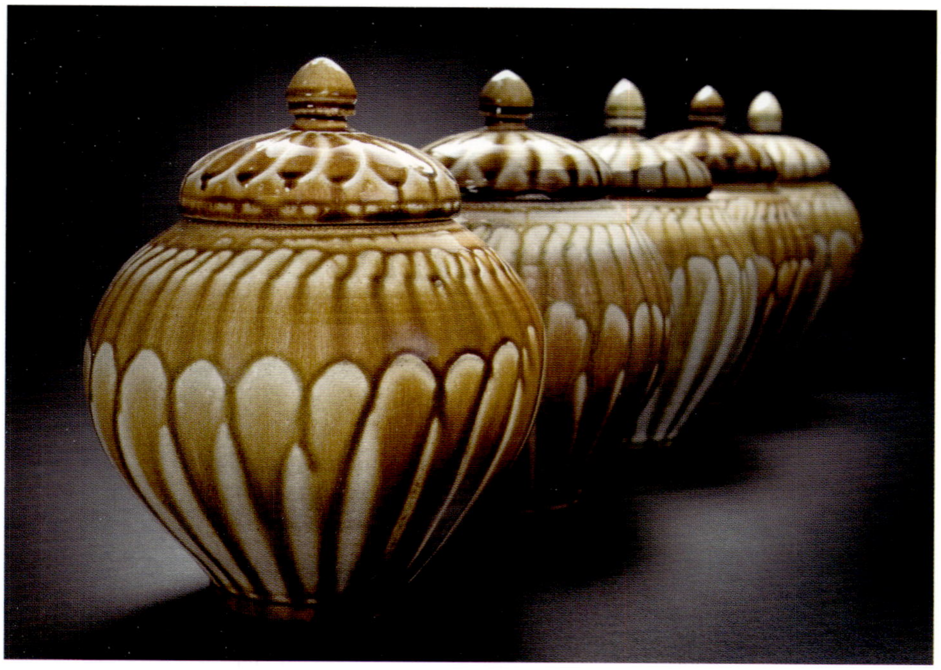

Daniel Johnston, *Jar Lineup,* 2009, local wood-fired salt glazed stoneware, 14" x 12". *Photo credit: Juan Villa.*

I dig most of the materials I use to make and glaze my pots. My pots are fired in a large 900 cubic ft. wood kiln. I enjoy the hard work and it leaves no part of the process separated from me. I do not try to control my materials; rather I try to understand them. From digging the clay to firing the kiln, I put all my effort into creating pots that have a powerful presence. It is important to me to create pots that are timeless but reflect the culture and times in which I live.

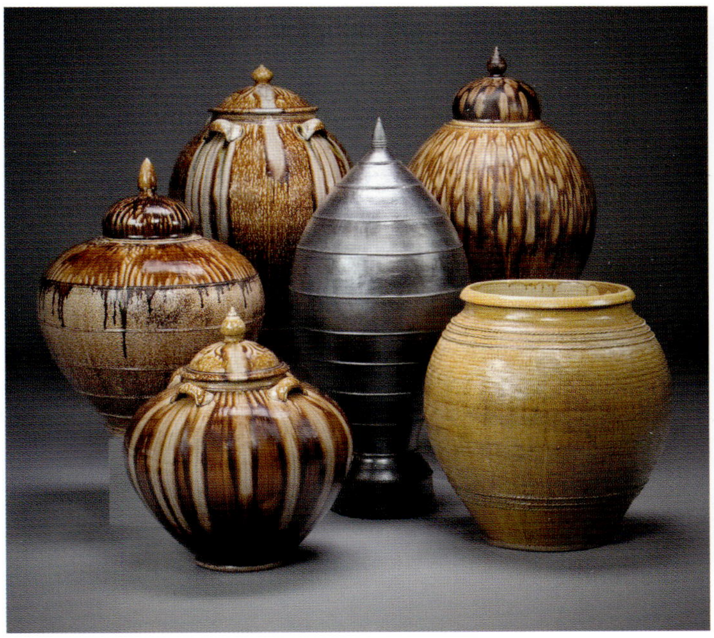

Daniel Johnston, *Large Vessel Grouping*, 2010. Local wood-fired salt glazed stoneware, 60" x 30" (largest piece). *Photo credit: Juan Villa.*

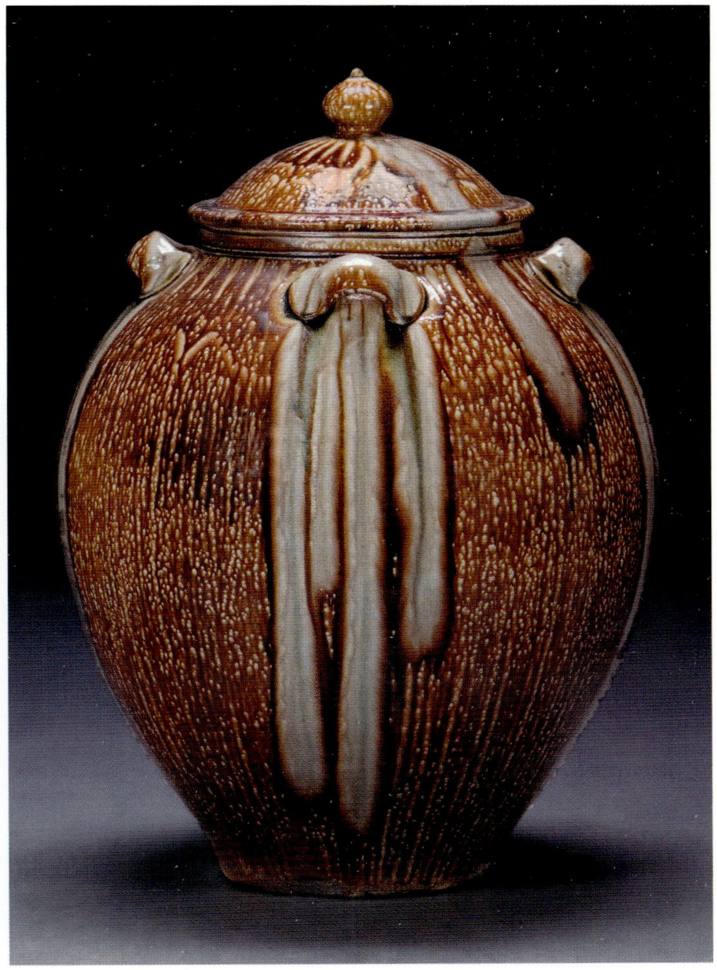

Daniel Johnston, *Large Jar*, 2009. Local wood-fired salt glazed stoneware, 36" x 26". *Photo credit: Juan Villa.*

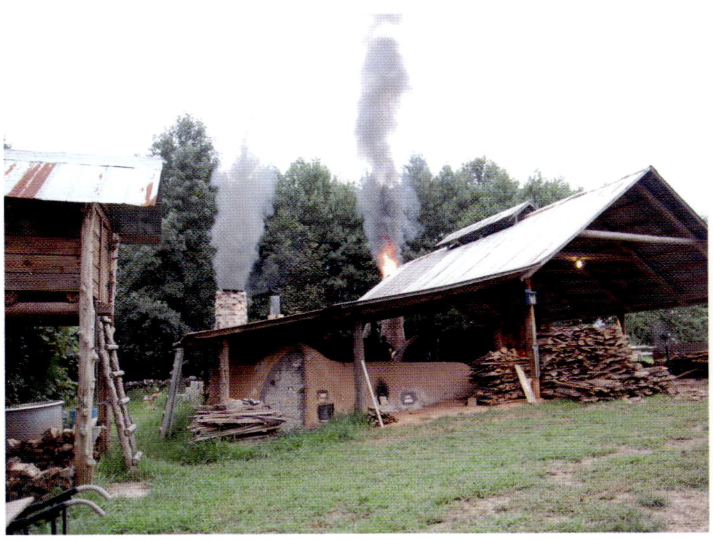

Two Kilns Firing, 2010. Photo *Credit: David McClelland.*

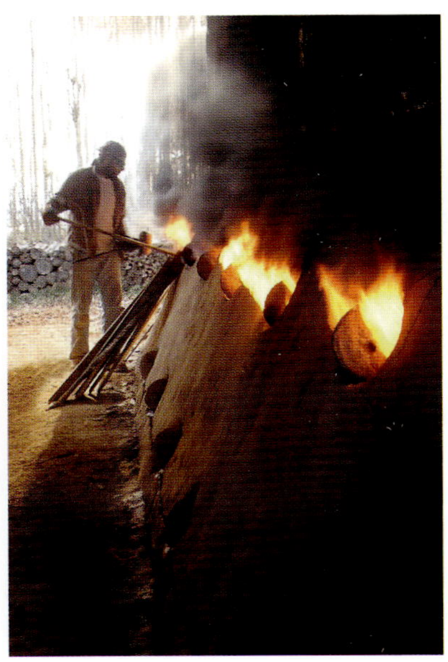

Chad Brown stoking Daniel Johnston's kiln, 2010. Photo *credit: David McClelland.*

RANDY JOHNSTON

River Falls, Wisconsin
randy.johnston@uwrf.edu, www.randyjohnstonceramics.com

Artist's Statement

As I work in clay, the reality that is the starting point is the choice to investigate the formal range of the vessel (pot) structure in clay, and the belief in the potential that the pieces must entertain, suggest a narrative, and allude to things outside of themselves. The largest question is how to invest my art with life, force, dignity, and a sensibility to the process and material. I am interested in this process as a means to manifest ideas and form. Categories are not important. The ongoing pursuit to enlarge the boundaries of conventional perceptions is essential.

The architecture of wood kilns makes them wonderful objects—creaking, belching, and breathing. It is an architecture completed by fire and containing the violence out of which living pots emerge. Firing for days with wood creates a complex of layers and structures on the work and in one's mind. Awareness of this layering brings me to a sacred place, a memory of a perfect place where we can stop and play with our minds and voices. This space also exists often within the architectural structures of our work as we attempt to define space, design structure, and deal with symbolic interpretations. The complex of social interaction and the place of our humanity within this structure is in large part a continuing expression of my conversation with the material and process of clay.

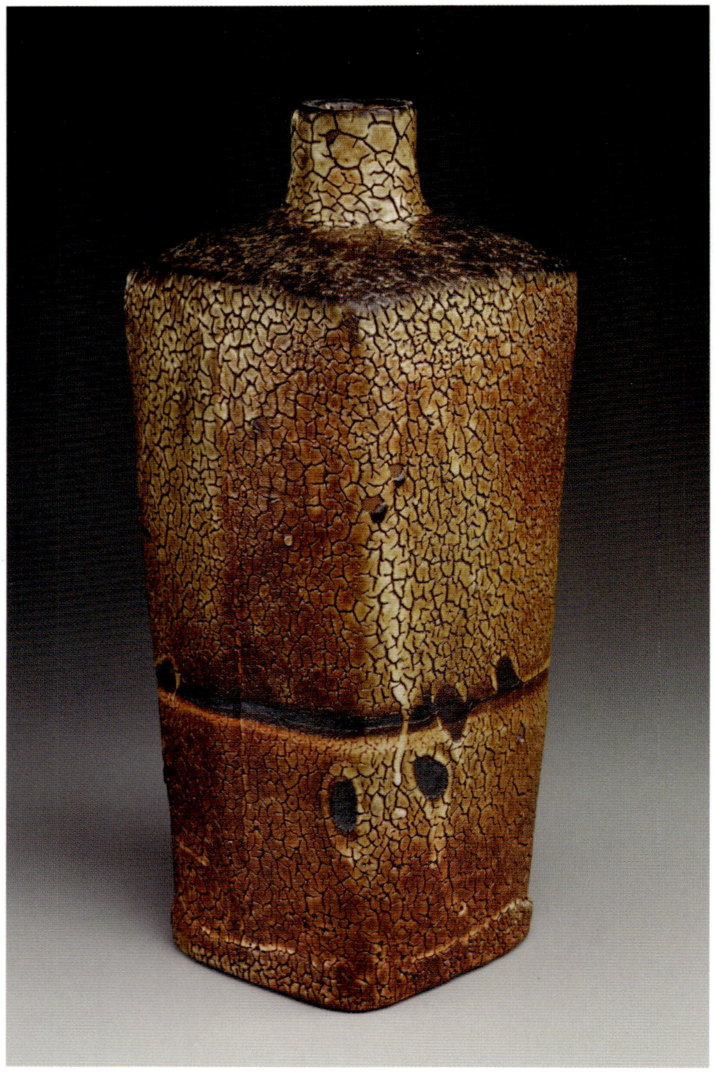

Randy Johnston, *Squared Vase Form*, 2010. Natural ash glaze with overglaze decoration, this vase is hand formed from a slab and paper pattern, Fired in the Noborigama kiln 3.5 days and redecorated with overglaze, 16.5" x 7.5" x 7.5". *Photo Credit: Peter Lee.*

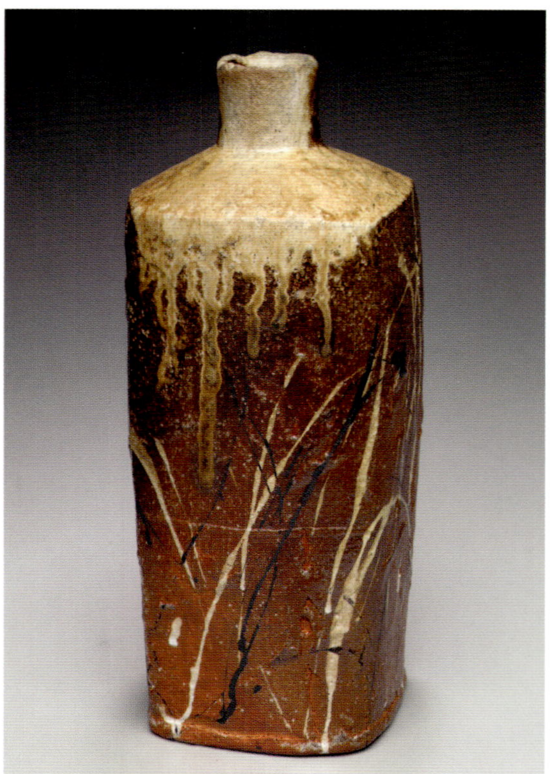

Randy Johnston, *Squared Vase Form Shino Glazed with Iron Slip*, 2008. Hand formed from a slab and paper pattern. Shino Glaze is applied over an iron slip. Fired in the Noborigama 3.5 days, size 16" x 7.75" x 7.75". *Photo Credit: Peter Lee.*

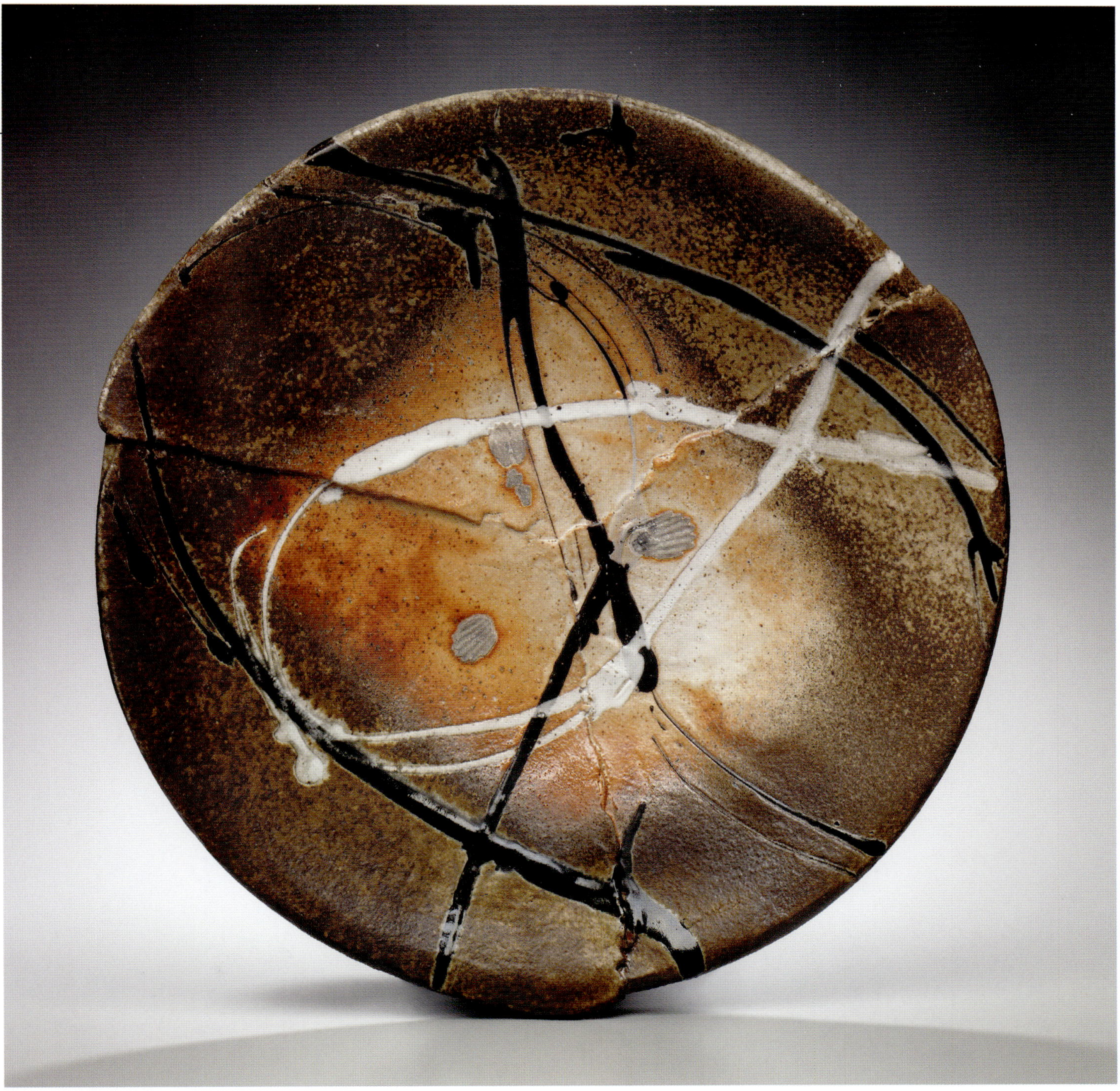

Randy Johnston, *Large Triangulated Platter*, 2008. Shell marks with white and black trailed decoration, Anagama, fired 5.5 days, square shadow from piece stacked on top, 4" x 23" x 23". Photo *Credit: Peter Lee.*

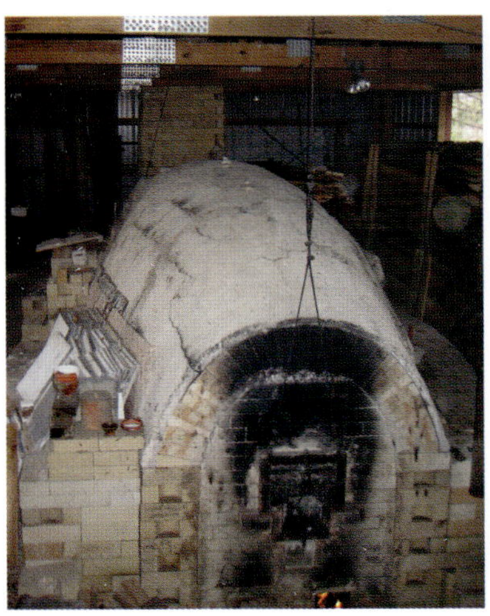

McKeachie Johnston anagama Kiln front view, 6' x 6' x 24' inside, 2010. *Photo Credit: Randy Johnston.*

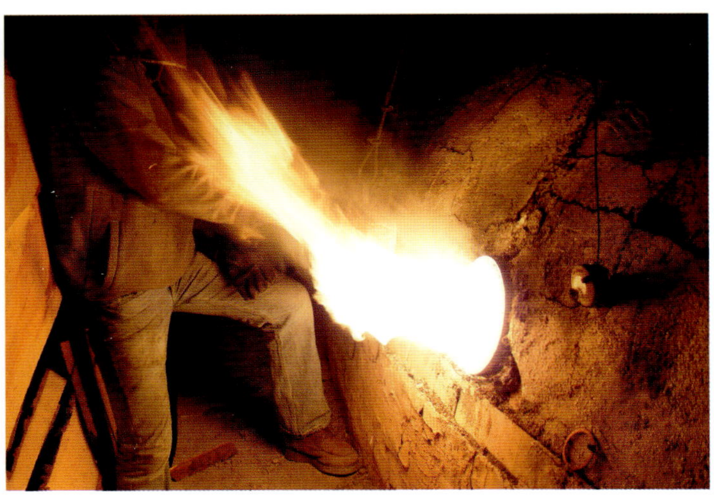

Side Stoking Randy Johnston and Jan McKeachie Johnston's anagama. *Photo Credit: Randy Johnston.*

JODY JOHNSTONE

Swanville, Maine
jody@jodyjohnstonepottery.com,
www.jodyjohnstonepottery.com

Artist's Statement

I built my anagama kiln in 1997 after returning from a two-year apprenticeship in Bizen, Japan, with National Living Treasure, Jun Isezaki. My kiln is twenty-four feet long and holds about 800 pots. I do two, eight day firings a year. A crew of four stokes the kiln. Each firing requires about six cords of wood that includes a mixture of local Maine hardwoods: ash, beech, maple, oak, a little birch, scrap cedar, and poplar from local sawmills. Most of my pots go into the kiln with at least the outer surface unglazed and rely on clay body, natural ash deposits, loading, flame patterns, and placement in the kiln to achieve a wide range of effects. Flatware is generally stacked with rice straw between the pieces to achieve further decorative effects.

I fell in love with pottery and wood-firing together, and to me they are inseparable. I've never done anything else. I want my pots to be both strong and soft, to embody my sense of beauty and rightness. The long firing adds richness and depth, and introduces just the right element of chance and mystery to keep me from trying to be too precise with my work.

Wood-firing is a lifestyle too and the anagama fills the needs I have to make lots of pots, to stack wood along with doing other physical, mind-clearing work, and to sit and enjoy a quiet shift with the kiln. I relish the community of potters and firing friends that gather around the kiln twice a year. I enjoy the rhythms of the process and the scale that I work on as it makes me feel connected to the seasons, strong and competent.

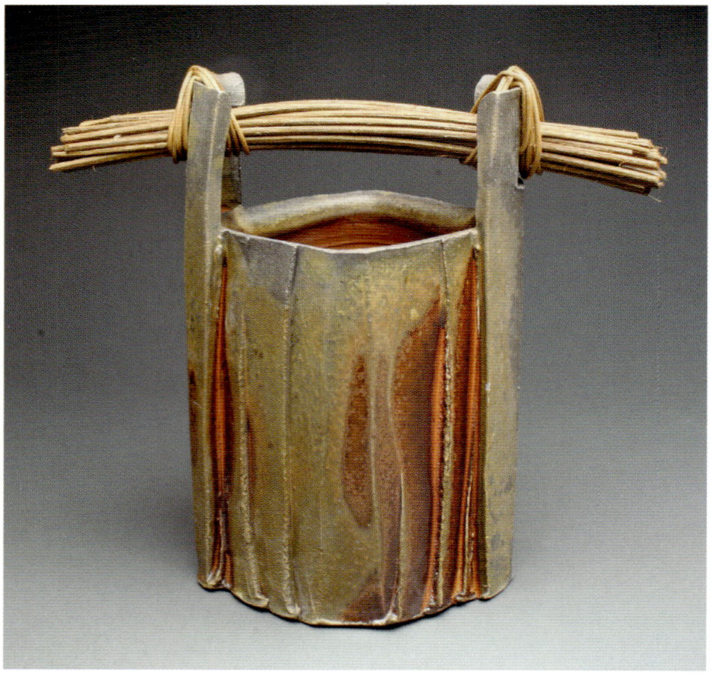

Jody Johnstone, *Basket Vase,* 2008. Anagama fired stoneware, wheel-thrown, assembled and altered, woven rattan handle, 7" x 9.5" x 5". *Photo credit: Jody Johnstone.*

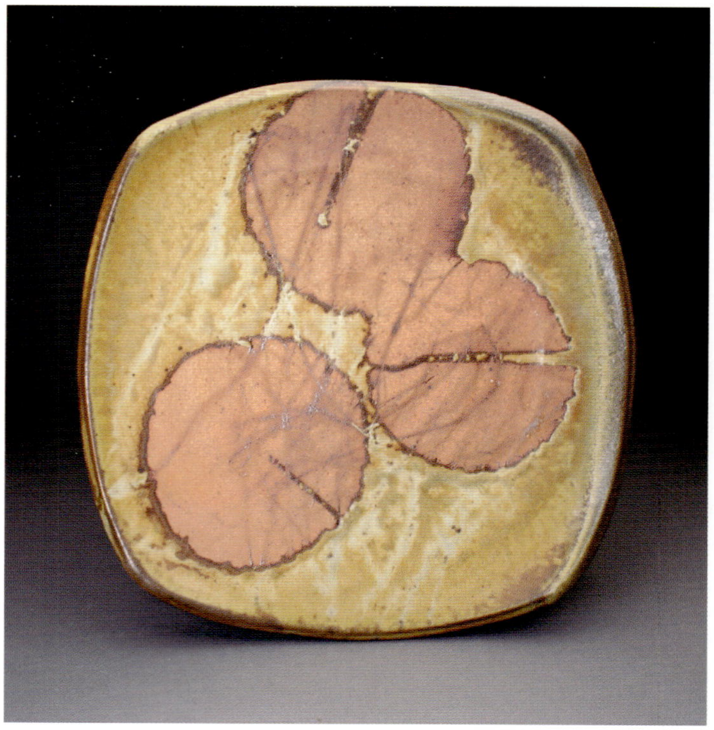

Jody Johnstone, *Lily Pad Plate,* 2010. Anagama fired stoneware, wheel-thrown, altered, rice straw and wadding "cookies" to achieve pattern, 1.5" x 11.5" x 12". *Photo credit: Jody Johnstone.*

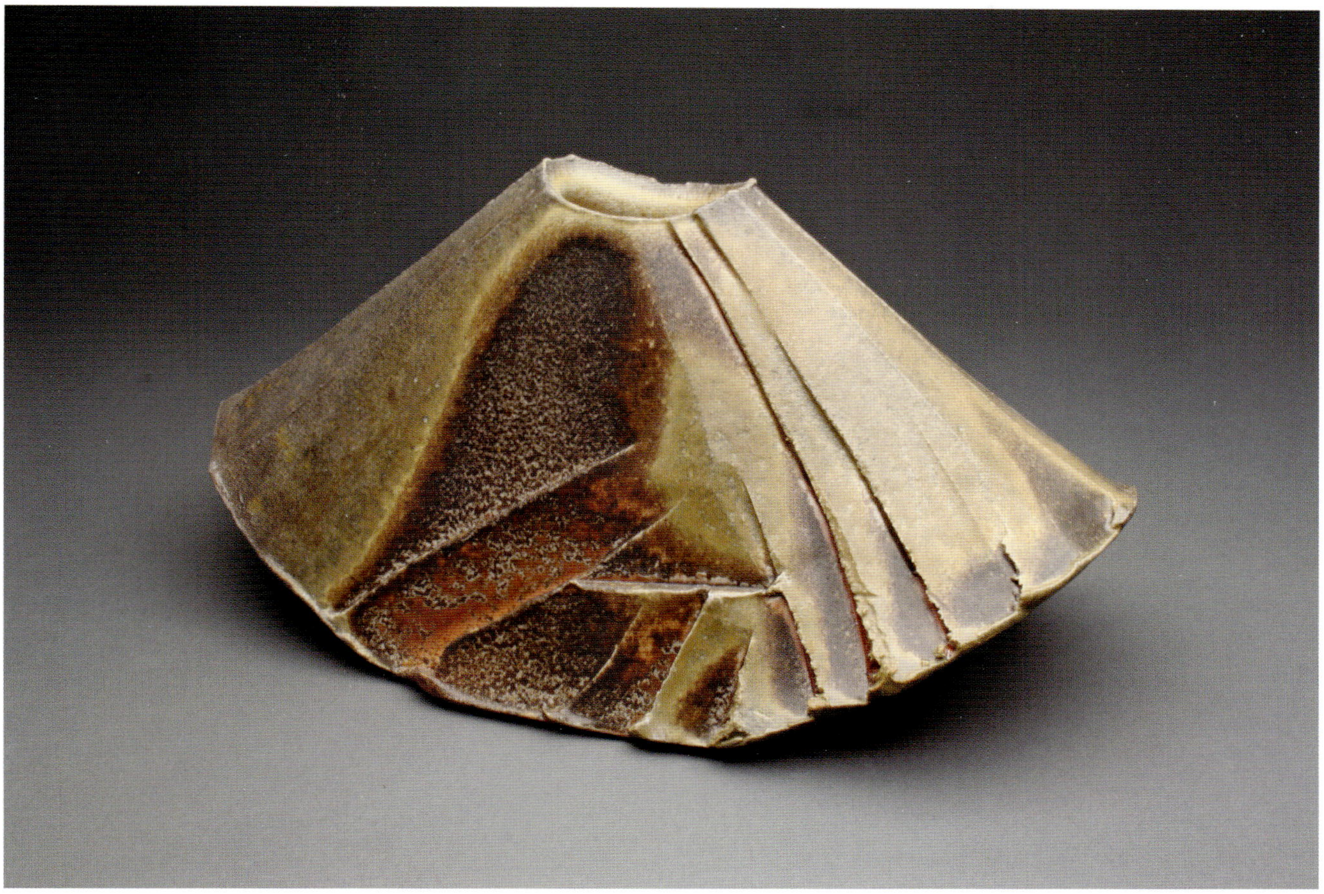

Jody Johnstone, *Fan Vase*, 2008. Anagama fired stoneware, wheel-thrown, assembled and altered, 6" x 10" x 5". *Photo credit: Jody Johnstone.*

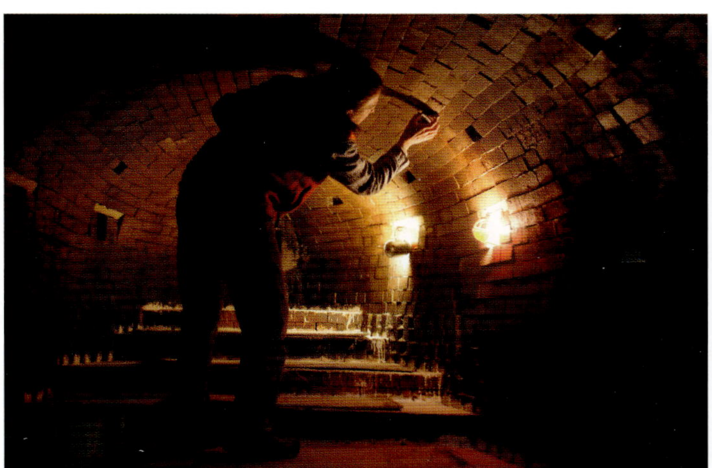

Jody inside the kiln, Spring 2010. *Photo credit: Marti Stone Photography.*

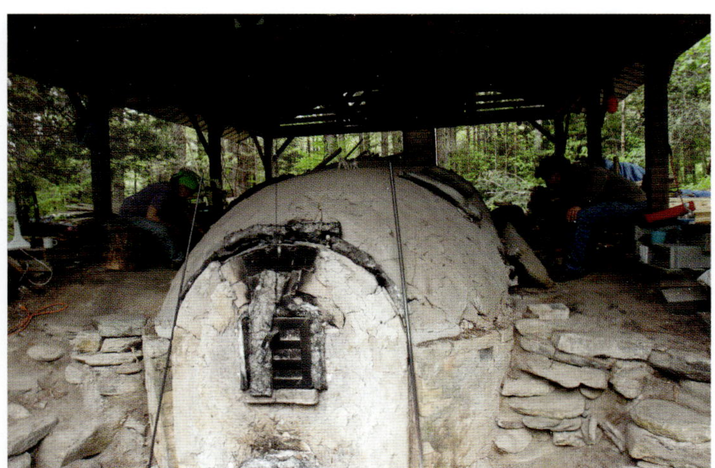

Jody and Betsy Levine side stoking, Spring 2010. *Photo credit: Marti Stone Photography.*

MASAKAZU KUSAKABE

Fukushima, Japan
Miharu1015jp@yahoo.co.jp, www.miharuarts.com

Artist's Statement

I designed my smokeless kiln myself. It is one of the many smokeless kilns that I have been designing since my first Ombu kiln, built in Canada, Burnaby city, in August 2001. This kiln is very ecological, being wood efficient, easy to reach high temperatures, and when fired correctly, is perfectly smokeless. The firing effects give strong results in less than two days. Later in 2001 I also designed my Sasukenei kiln, which has now been built by many others in over seven countries worldwide. It has proven to be a wonderful solution to the problems of smoke and wood-firing in urban areas for people who would otherwise be forced to stop wood-firing.

Wood-fired ceramics comes from and creates an echo of nature. Working with nature, earth, and fire reflecting the human heart, mind, and body. Wood-firing brings people together, sharing, exchanging, and, metaphorically, fires both ceramics and human relationships. My ceramic works are based around this theme of encounter.

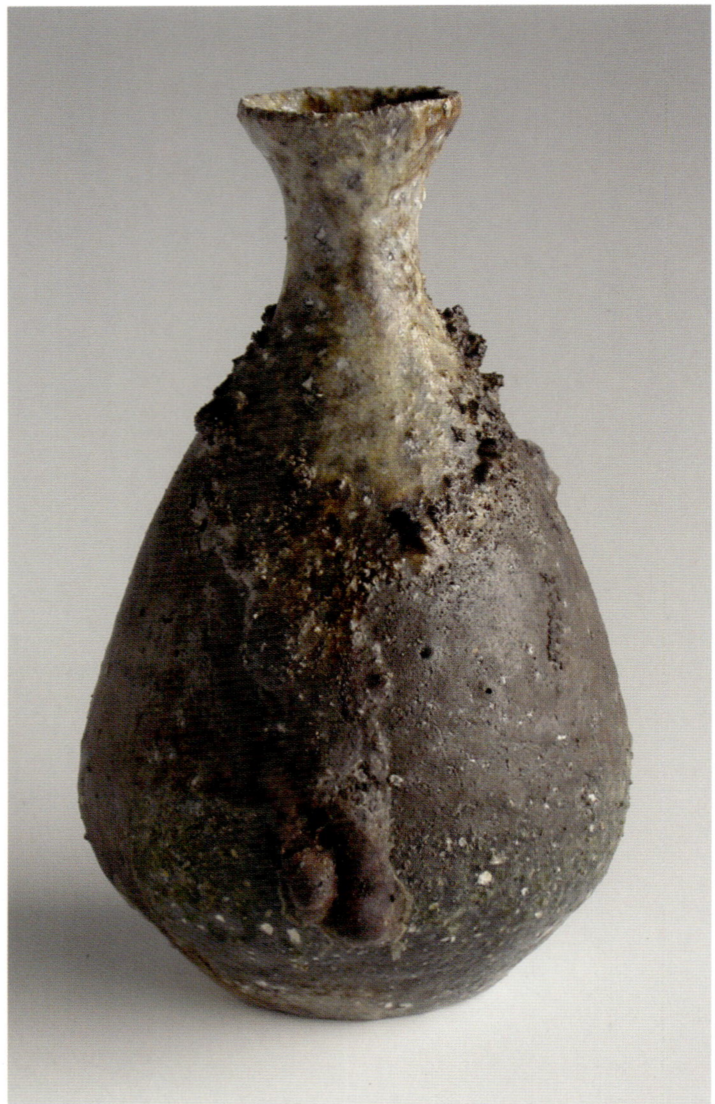

Masakazu Kusakabe, *Unglazed Sake Bottle*, 2008. Earthenware, wheel-thrown fired in the flue between the firebox and the chamber for two days in a smokeless kiln, 6" x 4" x 4". *Photo credit: Masakazu Kusakabe.*

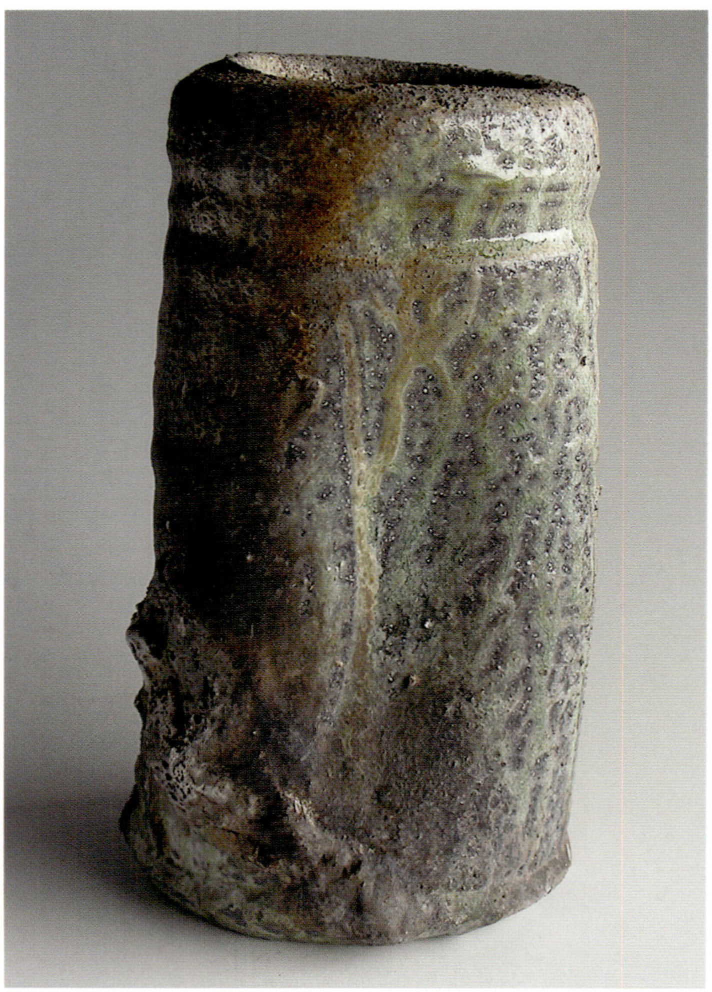

Masakazu Kusakabe, *Unglazed Hanaire (vase)*, 2008. Fired on the chamber floor under the side stoke position for two days in a smokeless kiln, 9" x 4" x 4". *Photo credit: Masakazu Kusakabe.*

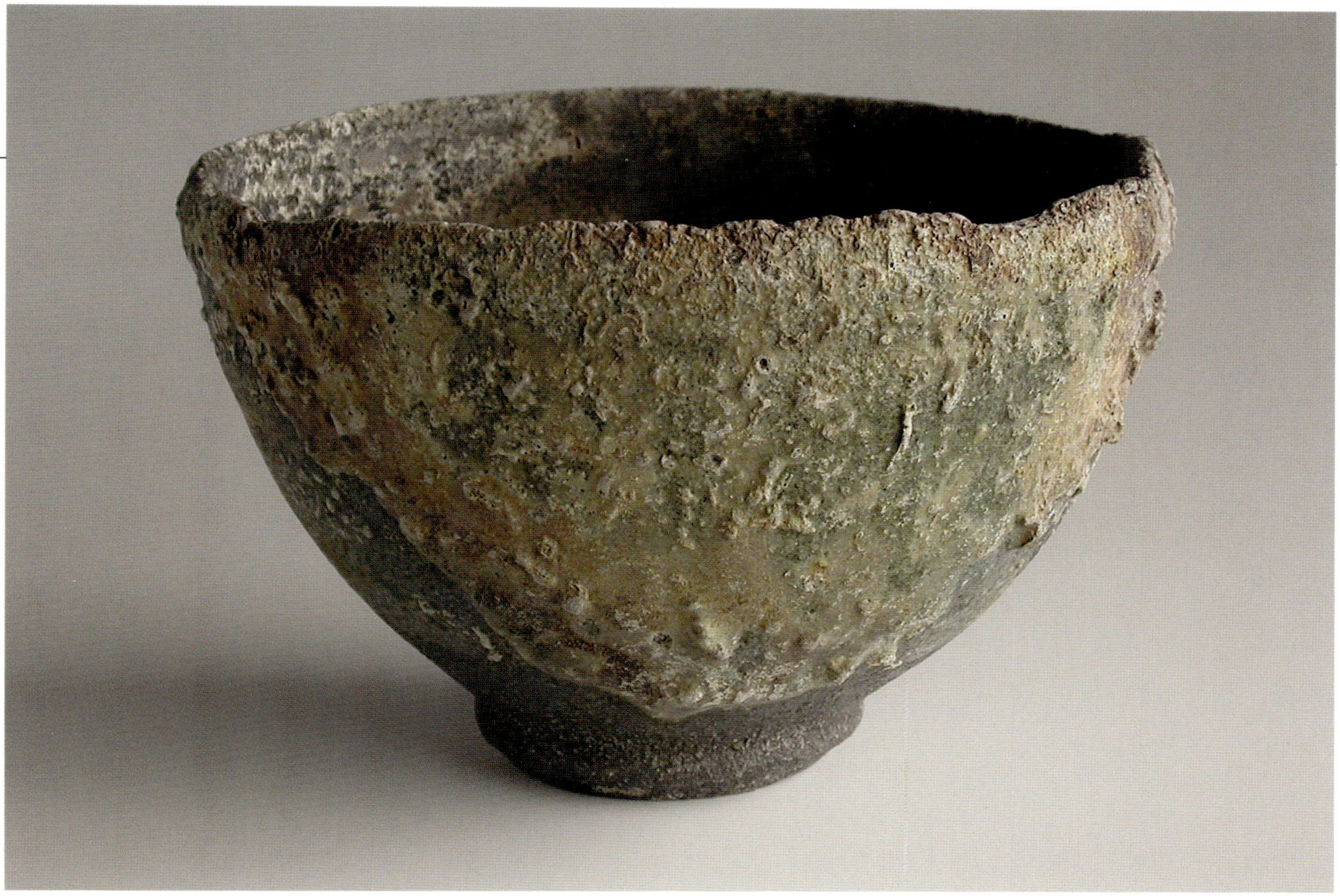

Masakazu Kusakabe, *Ash Glazed Tea Bowl*, 2005. Earthenware, wheel-thrown, re-fired under the firebox in the ember section for two days in a smokeless kiln, 4" x 5" x 5". *Photo credit: Masakazu Kusakabe.*

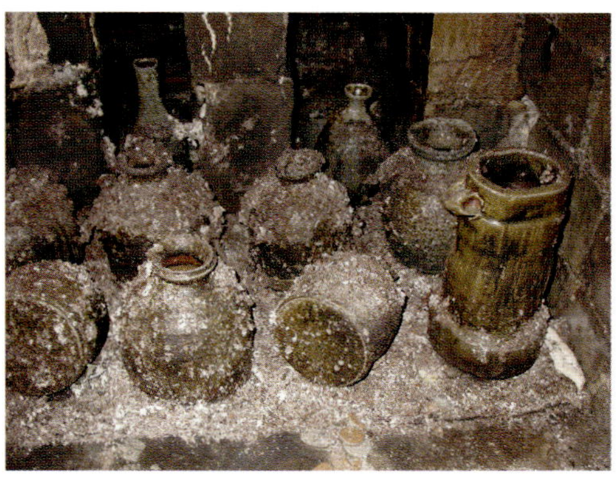

Unloading near firebox flue.

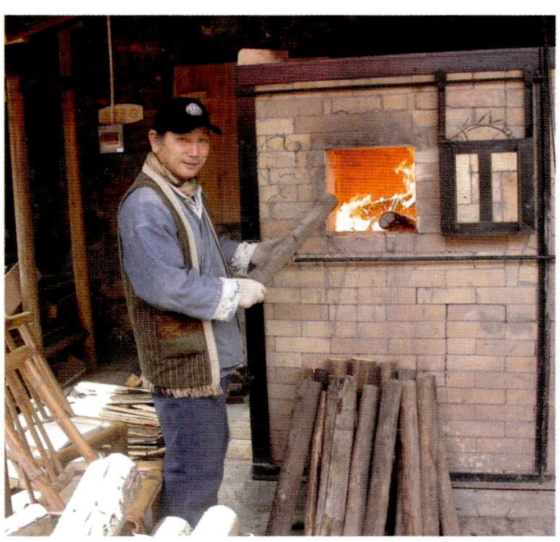

Sasukenei kiln firing.

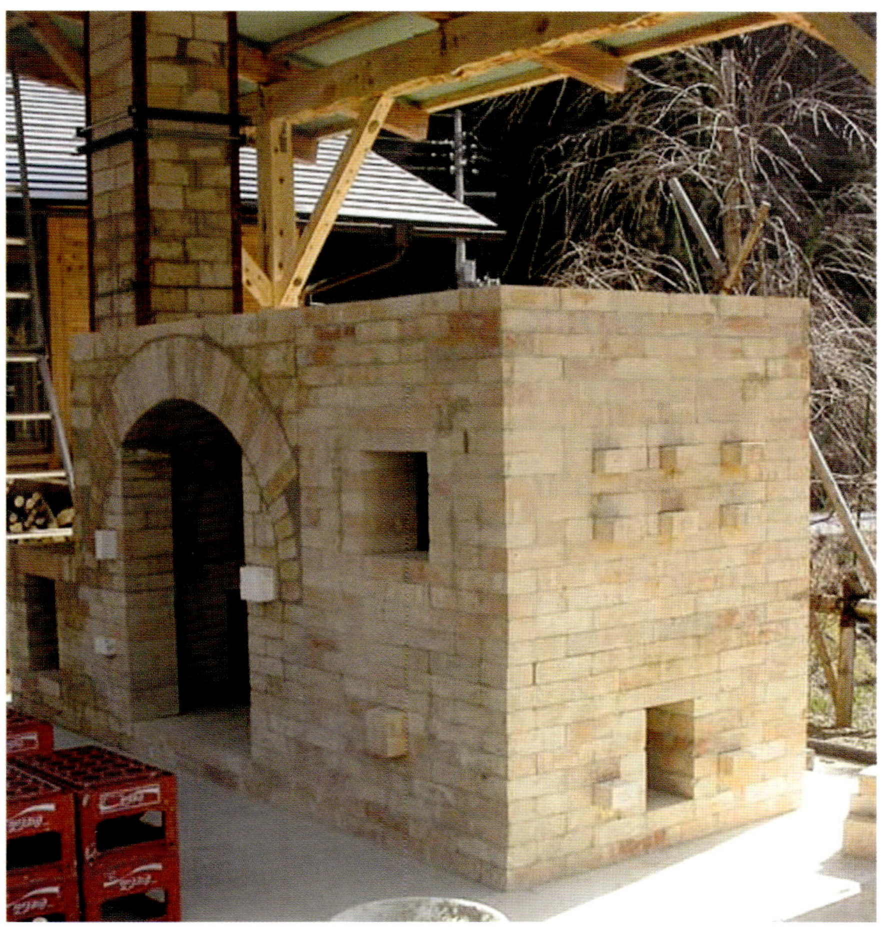

Smokeless Sasukenei kiln.

LUCIEN M. KOONCE

Haydenville, Massachusetts
lmkoonce@mindspring.com,
www.lucienkoonce.com

Artist's Statement

I approach the composition of my work with spontaneity and immediacy, discovering the form during the process of making it. My alteration and manipulation of solid clay emphasizes the plasticity and gestural qualities of that material while achieving asymmetry. I seek to push the disorganization and the subsequent reorganization of the vessel from being a member of the 'pot' realm to becoming an abstract object in the sculptural realm while retaining its function. Beginning with a solid mass of clay, I use either a metal blade or cutting wire, and sometimes both, to shape the exterior surface.

While I may have preconceived notions of the form, it is through spontaneous manipulation that the exterior of an object is derived. Once the clay stiffens, I hollow out the interior by hand with a carving tool, a process known as '*kurinuki*'. This technique, in which a form is dug or carved out of solid clay instead of being shaped on a potter's wheel or made from coils or slabs, is a process that allows me to intently focus on the inner shape. I strive to keep a balance between the outer and inner movements of an open form.

Wood-firing in an anagama adds another dimension to the composition, ultimately helping to define ones interpretation of the whole. The inherent nature of continuous flame, intense heat, and ash upon the clay, whether glazed or unglazed, adds color and textural effects that are congruent to each piece. The unpredictability of the firing, juxtaposed to the implementation of as many controlled variables as one can, creates random visual beauty, or landscape, which harmonizes with the physical form.

Subsequently, the form, whether functional or non-functional, has become like a diary, recording the thoughts and process of the maker and the kiln's fire.

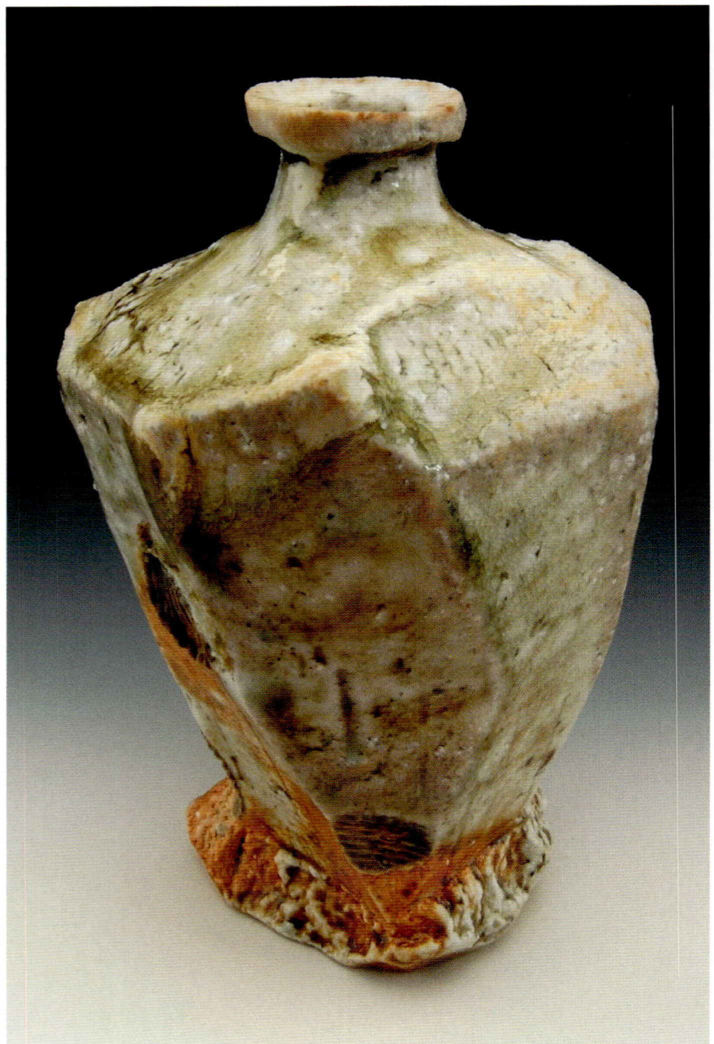

Lucien M. Koonce, *Tokkuri (Sake Flask)*, 2011. Hand carved (kurinuki technique) stoneware clay, feldspar inclusions, fired on sea shells, natural ash glaze, wood-fired (Gustin anagama) to cone 12+ for three days, 5.75" h x 3.5" w x 3.5" d. *Photo credit: Lucien Koonce.*

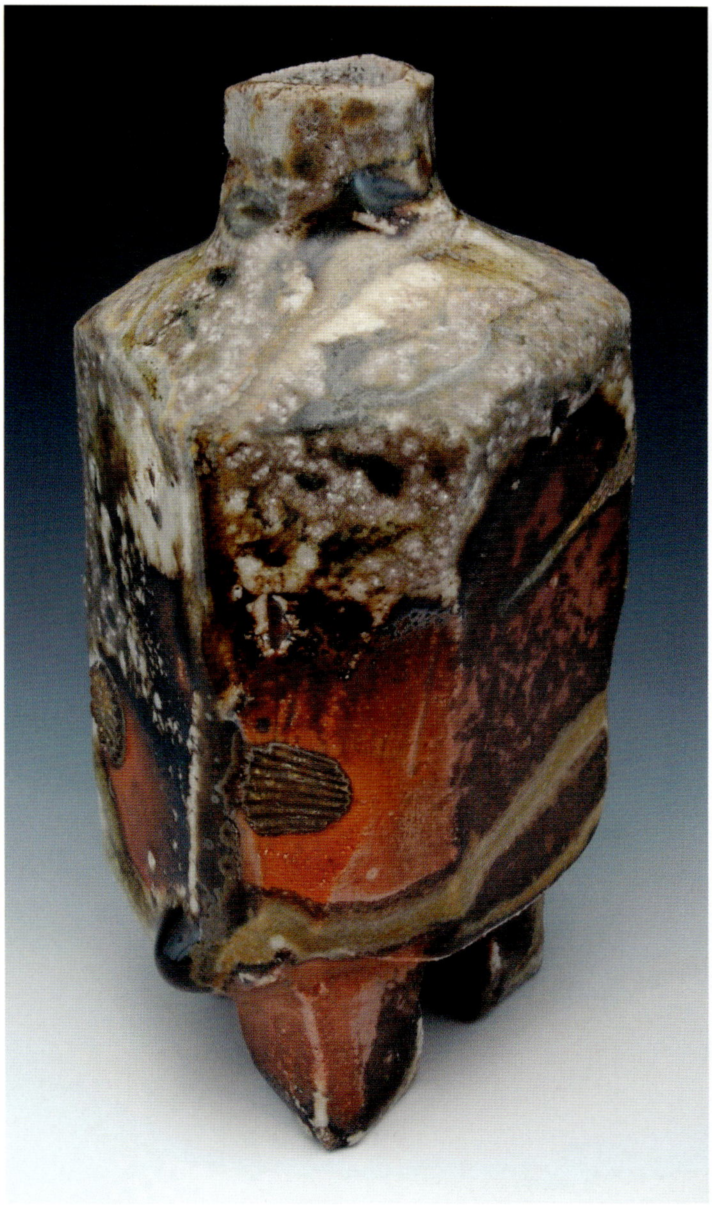

Lucien M. Koonce, *Tokkuri (Sake Flask)*, 2010. Hand carved (kurinuki technique) stoneware clay, feldspar inclusions, iron-rich slip, fired on sea shells, natural ash glaze, wood-fired (Gustin anagama) to cone 12+ for three days, 5.5" h x 3" w x 3" d. *Photo credit: Lucien Koonce.*

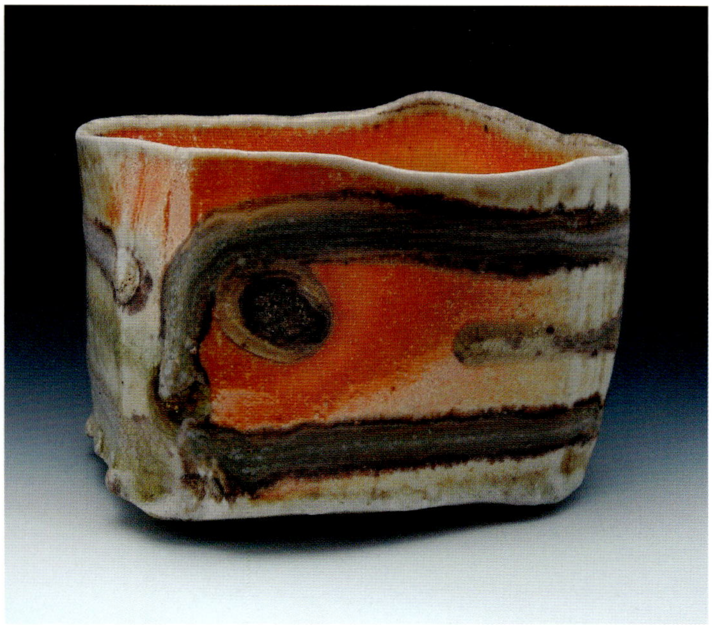

Lucien M. Koonce, *Chawan (Tea Bowl)*, 2010. Hand carved (kurinuki technique) stoneware clay, feldspar inclusions, fired on sea shells, natural ash glaze, wood-fired (Gustin anagama) to cone 12+ for three days, 3.25" h x 4.5" w x 4.5" d. *Photo credit: Lucien Koonce.*

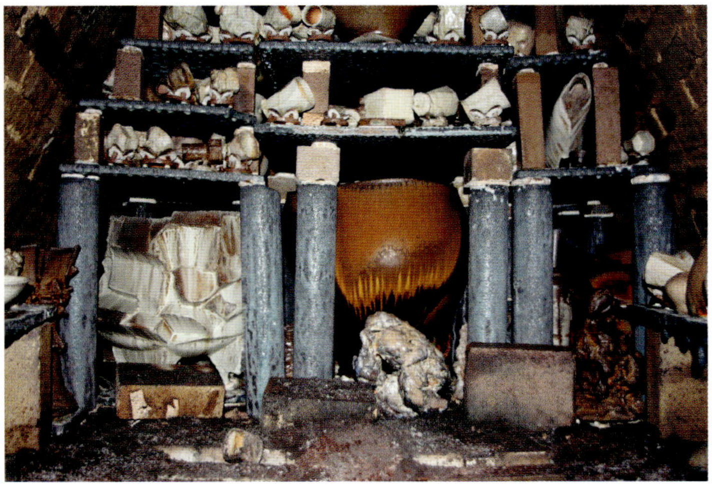

Front Stack, Post Firing, 2011. *Photo credit: Chris Gustin.*

HEIDI KREITCHET

Pomona, California
hkreitchet@gmail.com, www.hkwoodfire.com

Artist's Statement

I chose wood-firing for my work because the time involvement, camaraderie, process, and atmospheric finish complete my pieces in a manner that no other kiln has been able to do. The effort it takes to wood-fire is incomparable to any other kiln. One must love being outdoors, doing wood preparation, loading the kiln sometimes in the cold winter months. There is a sense of accomplishment once the firing is complete and the artist has seen their pieces go from a pile of clay to finished wood-fired ceramic.

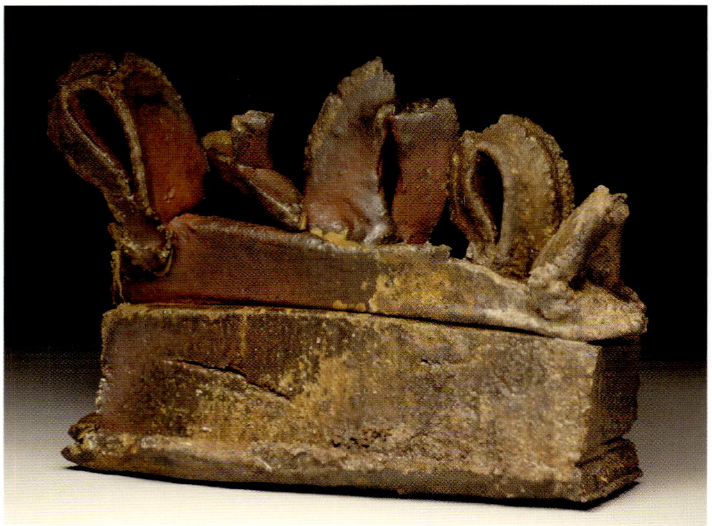

Heidi Kreitchet, *Loaf*, 2009. Wood-fired stoneware, handbuilt, 8" x 11" x 5". *Photo credit: Heidi Kreitchet.*

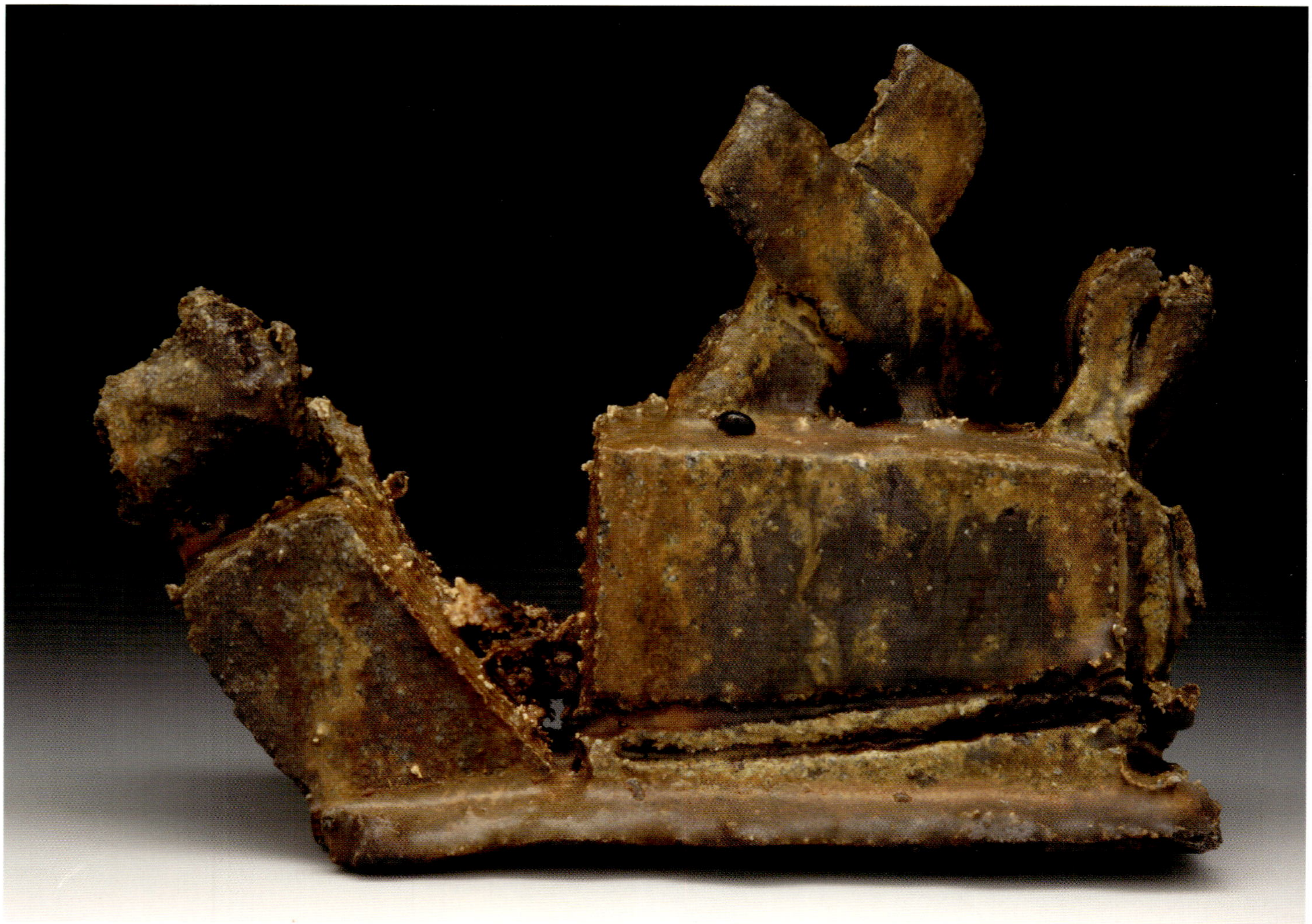

Heidi Kreitchet, *Burnt Toast*, 2009. Wood-fired stoneware, handbuilt, 6" x 8" x 4". *Photo credit: Heidi Kreitchet.*

The train kiln is the most reliable, efficient, and manageable kiln for me and a friend or small group to fire. While the anagama is my favorite kiln because of the firing experience, it is not the most convenient type of kiln for my current lifestyle. The train kiln is much smaller and does not involve as much effort, wood, and time as the larger types of wood kilns.

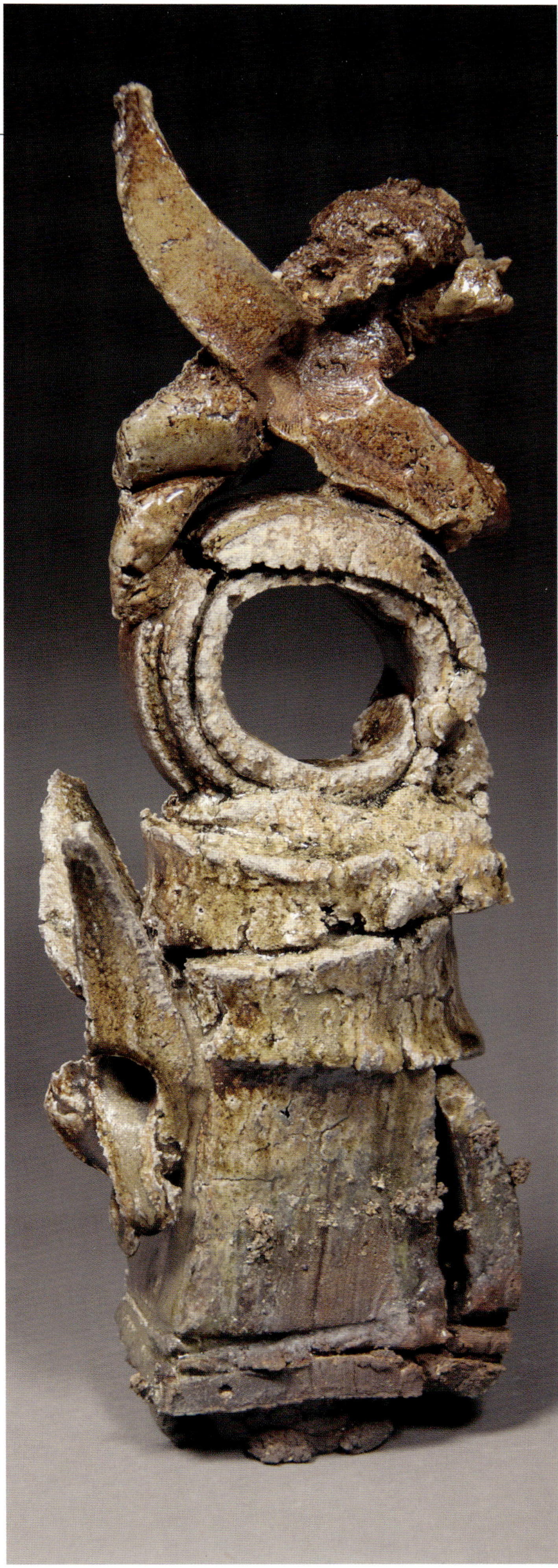

Heidi Kreitchet, *Do you like my hat?* 2011. Wood-fired stoneware, handbuilt, 17" x 6" x 5". *Photo credit: Dave Tarullo.*

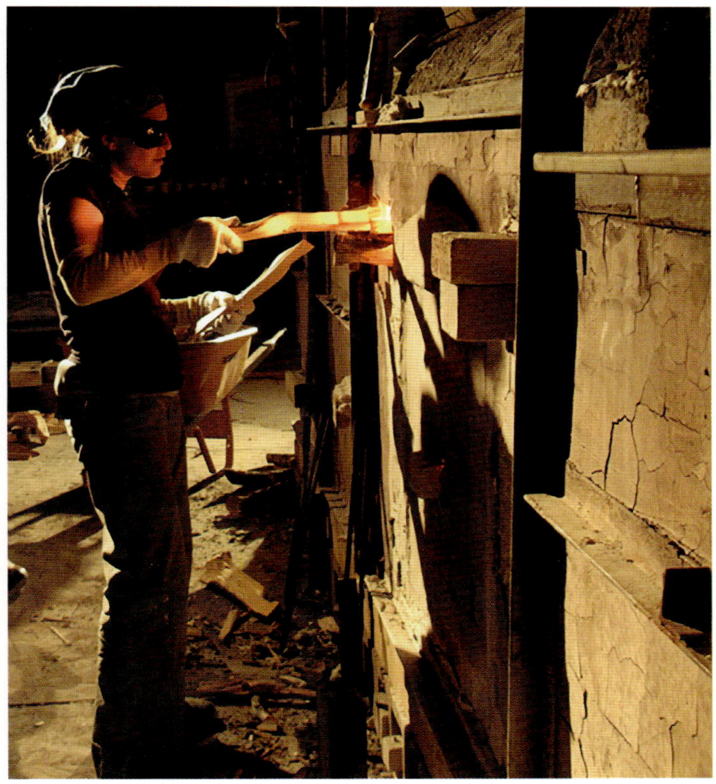

Heidi Kreitchet, firing train kiln. Photo *credit: Kelly McLendon.*

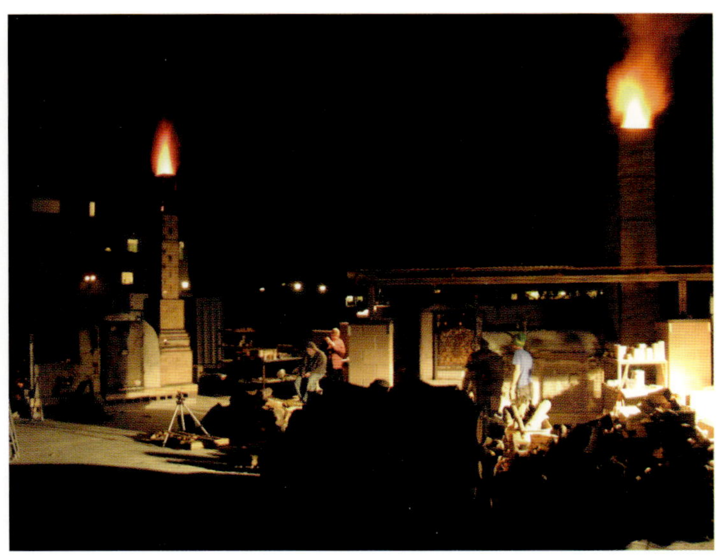

Heidi Kreitchet, firing train kiln and wood soda kiln. *Photo credit: Kelly McLendon.*

EVA KWONG

Kent, Ohio
3kwong3@gmail.com,
www.personal.kent.edu/~ekwong/gallery.htm

Artist's Statement

Firing the large seven chambered wood kiln in 1974 changed the way I looked at the firing process and ceramics forever. The huge kiln huffed and puffed, rose and fell with each stroke of wood. It was a breathing, belching, live dragon/chamber kiln at the "Earth, Air, Fire, Water" workshop at Grass Valley, California. It was a revelation for me; this experience made me feel and understand the intimate connections between the clay, the fire and the flame, the heat and the atmosphere. I was hooked on wood-firing.

The next transforming experience was studying with Katsu Sakazume at Peters Valley Craft Center in 1980. This workshop involved working with local, natural clays, culminating in the ten day firing of the large anagama kiln. I learned about the Japanese ways of prospecting for local clays, making, stacking, and firing.

Both of these intensive experiences led me to build my own wood-kiln in 1981–83 with Kirk Mangus in Pennsylvania.

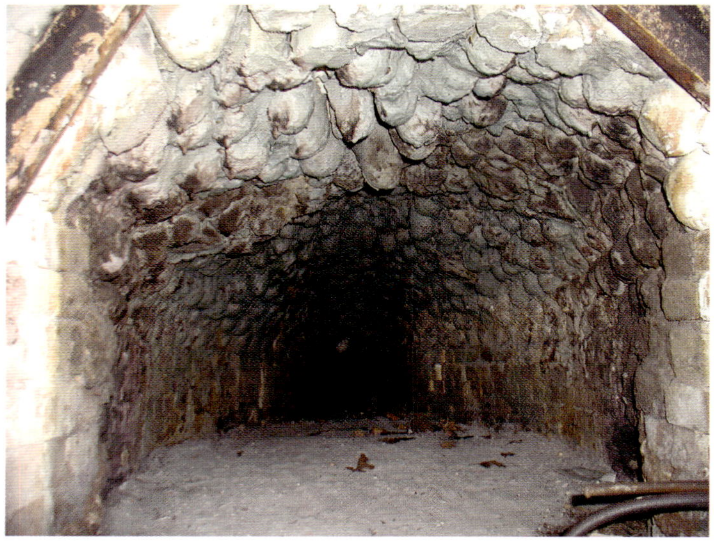

Interior view of kiln. *Photo credit: Kirk Mangus.*

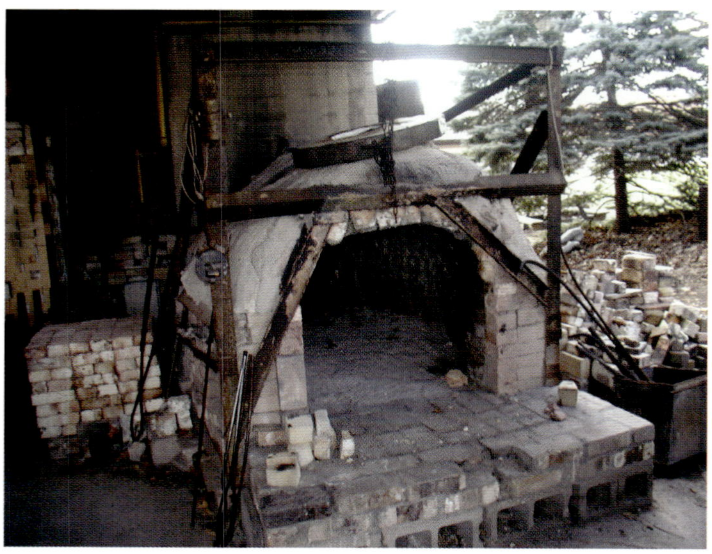

Anagama kiln that I built with Kirk Mangus in 1980–1981, made of handmade clay bricks. Photo *credit: Kirk Mangus.*

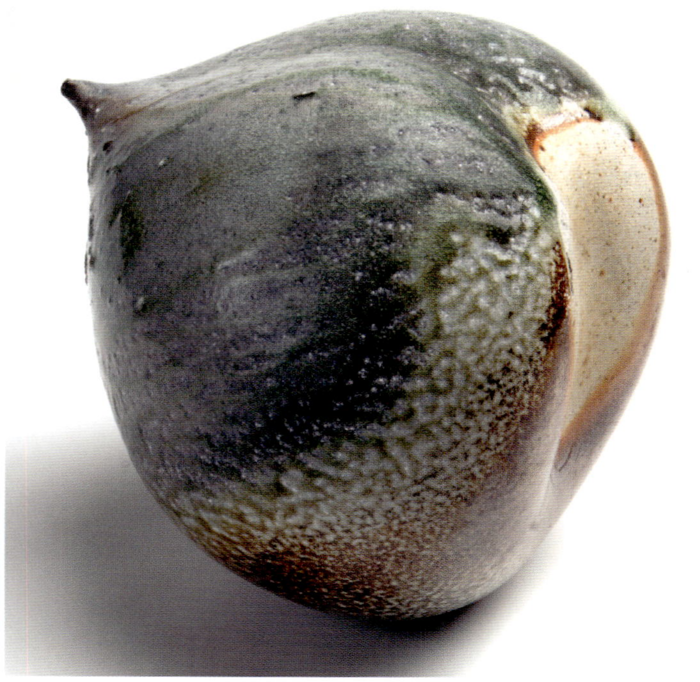

Eva Kwong, *The Immortal Peach,* 2002. Porcelain, wheel- thrown, wood-fired. 8" x 8" x 9.5". *Photo credit: Kevin Olds.*

I choose to wood-fire for the process, for the interaction of the flame to the surface, what I call the "Radiant Blush" effect. Like the game of mah-jongg, wood-firing involves both skill and chance. The flame and ash create that "Radiant Blush," recording the intensive burning on the surface of my work much like the blush that the sun leaves on the side of an apple.

The recent "Immortal Peach" series are wheel-thrown and altered forms based on the Immortal Peaches from the Chinese folk tales. The peaches are symbolic of fertility and immortality/mortality.

I am interested in the sensuous and feminine qualities of the peach form. The molten ash and flame patterns of the Radiant Blush enhance the soft and voluptuous form of the Immortal Peaches.

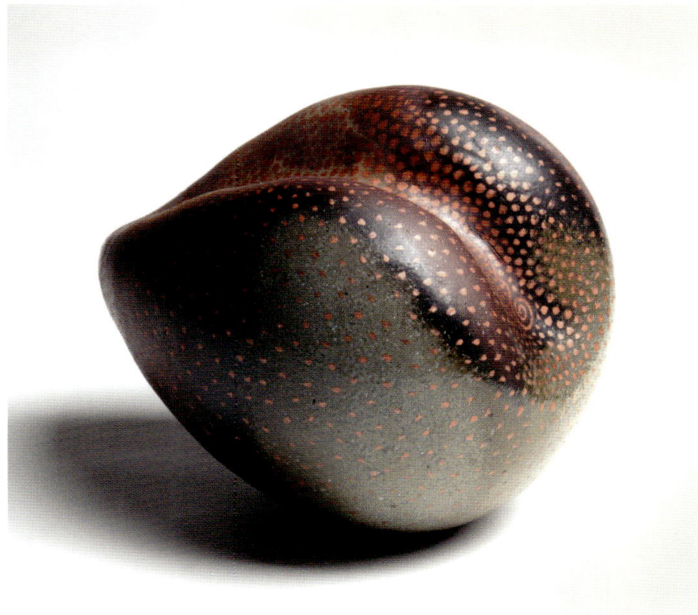

Eva Kwong, *Ruby Spotted Peach*, 2005. Stoneware, wheel-thrown, wood-fired, ruby luster, 6" x 6" x 6". *Photo credit: Kevin Olds.*

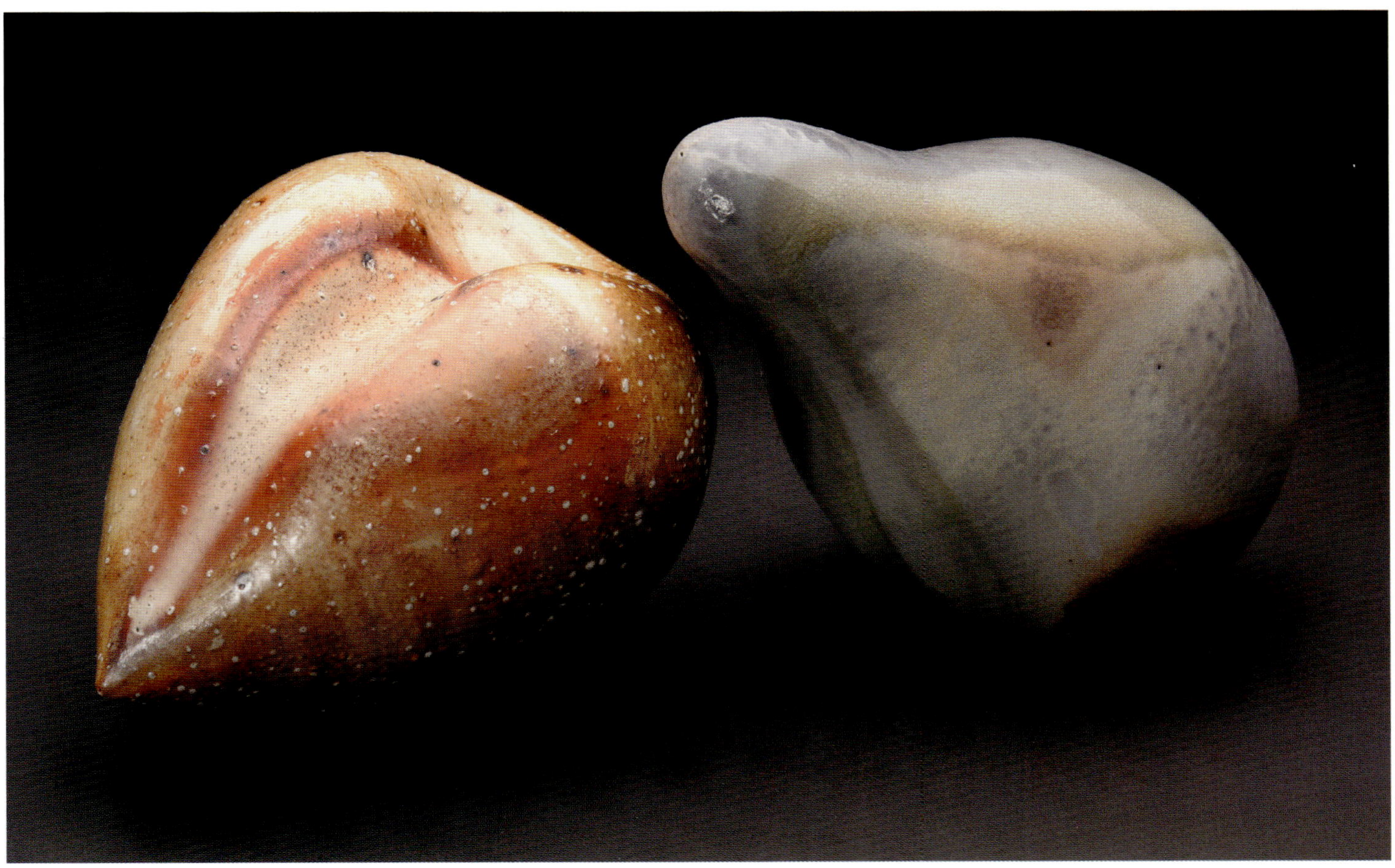

Eva Kwong, *Orange Blushing Peach and Fossil*. Stoneware, wheel-thrown, wood-fired. Orange Blushing Peach: 8" x 8" x 9", Fossil: 5.5" x 8" x 5.5". *Photo credit: Kevin Olds.*

JUSTIN LAMBERT

Jupiter, Florida
justin@liveoakpottery.com, www.liveoakpottery.com

Artist's Statement

An interest in objects that serve a specific purpose motivates me to make functional pottery. I am also interested in how pottery can inherently initiate a certain situation with a single user and companion. It is through the grouping of particular pots that I am able to suggest a special moment to occur.

It is the interaction of my pots that led to certain scenarios alluding to the ideas of companionship and solitude. Groupings of bottles or cups are about inviting myself, and the viewer, to slow down and take notice of the subtle diversities in form and the infinite variety of surface texture and color attainable through wood and soda firing. The scale of my work brings the viewer in close to examine the subtleties of form and surface, and creates a more intimate experience through its utilitarian qualities.

The firing process I choose provides a direct interaction between the clay and the user. My work is not covered with any glaze; rather the firing itself glazes the work, enriches the surface and brings out intrinsic color from the clay. My investigation into high alumina clay bodies in both wood and soda firing leads my research. I reduction cool these kilns to achieve deeper colors, and to explore a palette somewhat unknown. Frosty, dry, movement rich glazed surfaces provide information for future work, and my careful analysis of surface to form integration provide insight to new formula's and firing schedules.

My work is influenced by pottery from all cultures, but more specifically Southeast Asia, including Oceania and Africa. I feel these cultures' primary concern when making pottery was the pottery's function. Form simply followed function and some of the most amazing pottery in the world was made. Their honesty, simplicity, necessity, and beauty are the qualities I strive for in my own work.

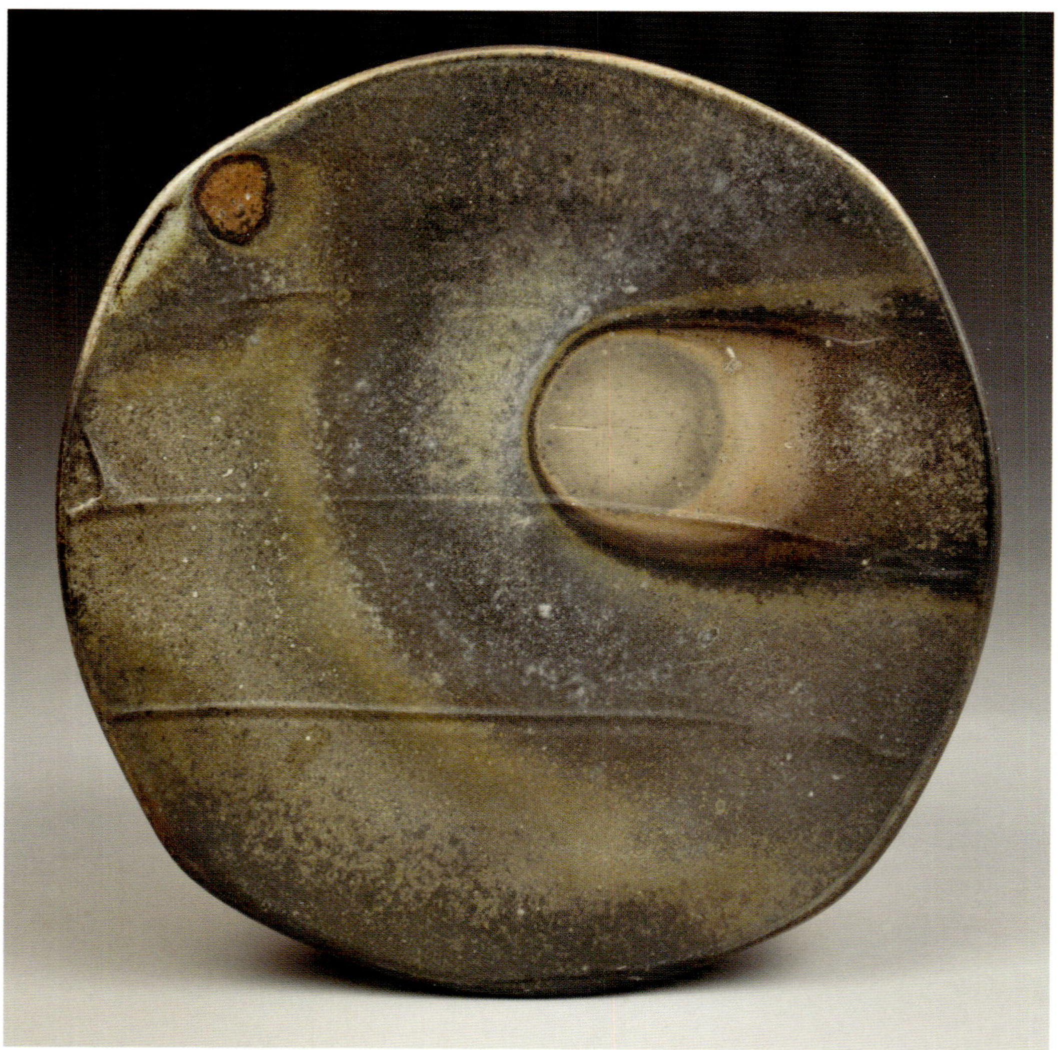

Justin Lambert, *Dinner Plate*, 2011. Anagama fired stoneware, cone 10, cooled in reduction, 10" wide. *Photo credit: Justin Lambert.*

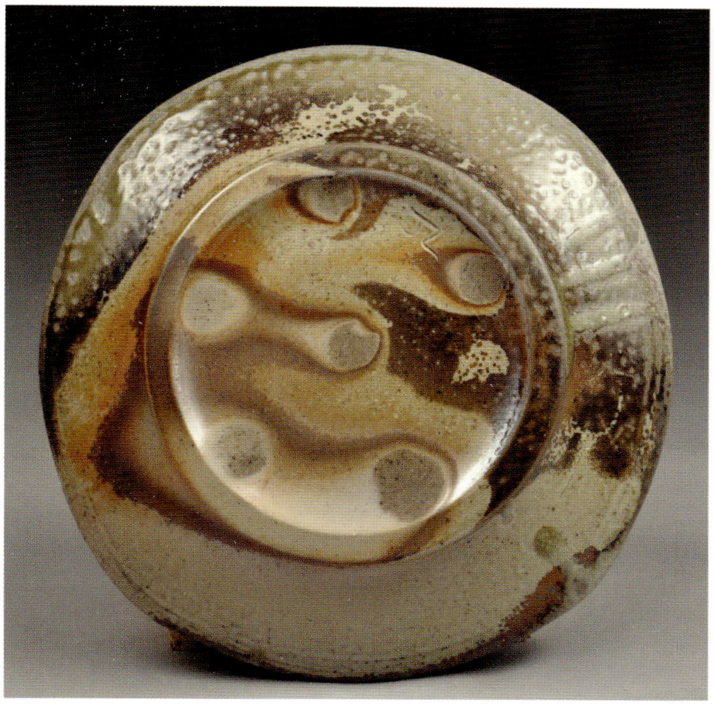

Justin Lambert, *Dinner Plate* (Bottom Detail from #4), 2011. Anagama fired stoneware, cone 10, cooled in reduction, 10" wide. *Photo credit: Justin Lambert.*

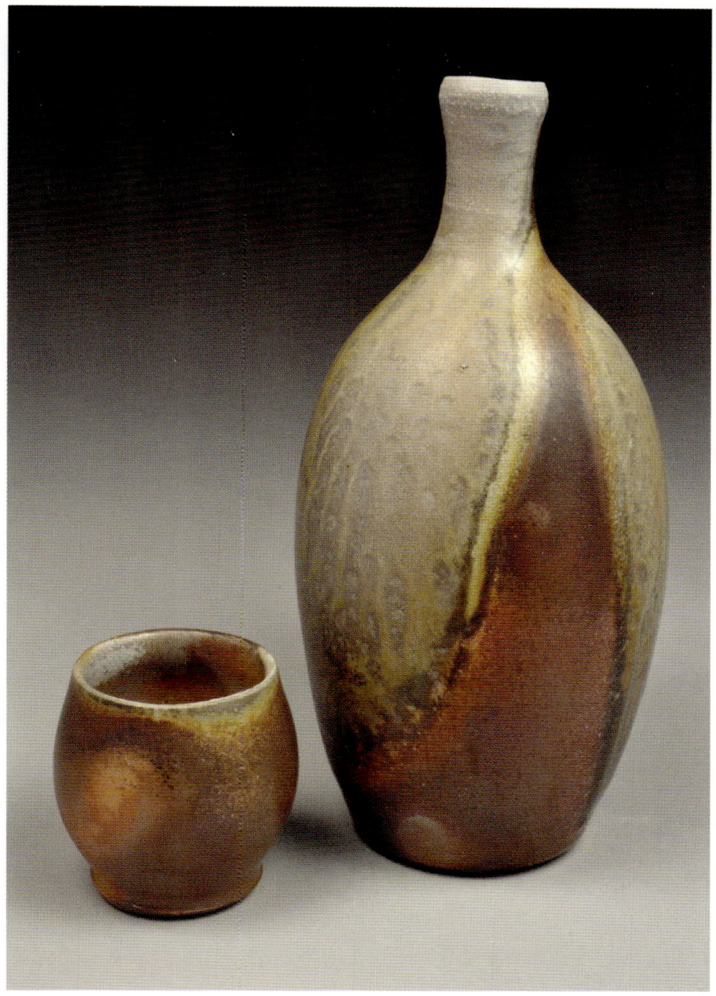

Justin Lambert, *Liquor Bottle and Cup*, 2011. Anagama fired stoneware, cone 10, cooled in reduction, 9" tall. *Photo credit: Justin Lambert.*

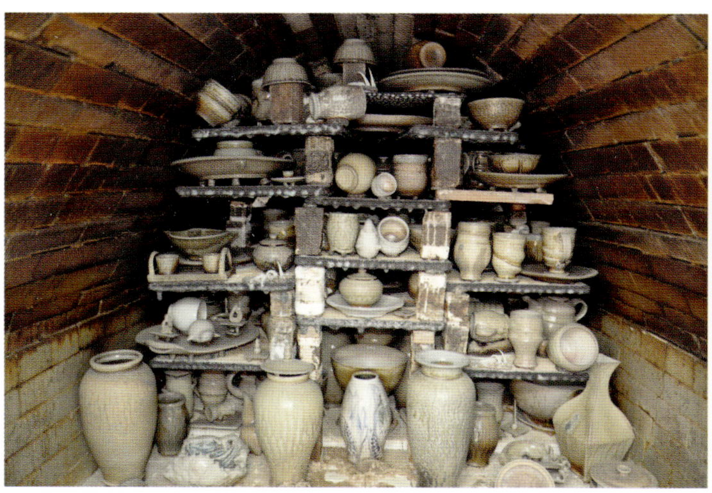

Justin Lambert, anagama front view, cone 10 down, cone 11 standing, cooled in reduction, 78 hour firing December 2011. *Photo credit: Justin Lambert.*

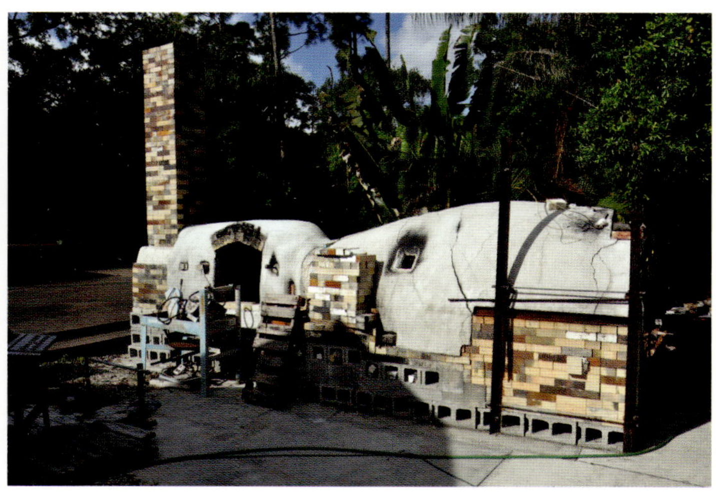

Justin Lambert, anagama kiln with soda/post combustion chamber, 2012. *Photo credit: Justin Lambert.*

DEREK LARSEN

Kyoto,Japan
dezzalarsen@yahoo.com, www.dereklarsenceramics.com

Artist's Statement

My ceramic work is an exploration in my relationship with the natural world and a personal reaction to folk craft traditions. I create rugged, irregular vessels for ritual domestic purpose that provide visual, emotional, and psychological stimulus in one's daily life. Influenced by the Arts and Crafts Movement and the Japanese *wabi* aesthetic, I view my work as a contemporary extension of these movements.

Since the industrial revolution, most domestic objects have transformed into mundane, stereotypic utensils, and finding beauty in these ordinary objects is no longer cultural commonplace. Serving as a counterpoint to these manufactured objects, my work challenges contemporary culture to examine its own relationship with objects and nature.

I possess a deep appreciation for artifacts and for all things affected and patinized by nature over long expanses of time, and utilize glazes and kiln firing techniques that emulate these time-laden surfaces. Often fired multiple times or with multiple glazes, I seek out a layering of surface information to create visual surface depth. Born in the intense heat of anagama wood kilns, my pots are subjected to a harsh, yet enchanting natural phenomenon yielding natural colors, textures, marks, and scars. These surfaces are deposited only from the power of the flame, my sensibilities and selection of natural materials, and my control of the kiln environment. The anagama process allows direct interaction with the creation of these fired surfaces. Continually stoked in rhythm for days, kiln firing is a demanding, labor-intensive test of spirit. Through this process and in these vessels I record the beauty of my materials, and my life as a craftsman in a growing tradition.

The small jars were coil built and then quickly thrown using a variety of local clays found in the Shigaraki area. All the jars were fired for one week in an anagama kiln to 2550 degrees Fahrenheit [1400 degrees Celsius] stoking red pine wood.

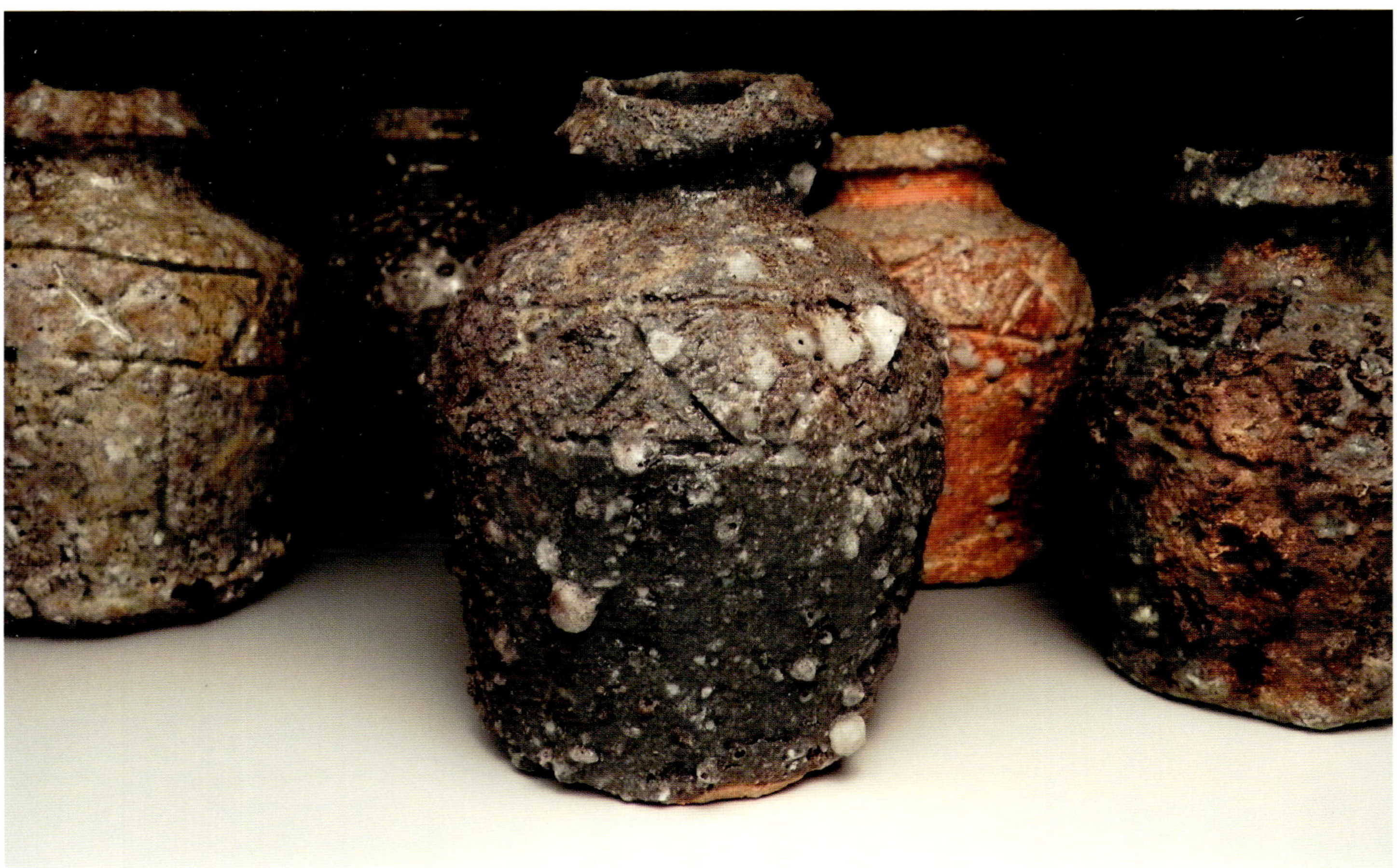

Derek Larsen, *Small Jars*, 2011. Anagama fired mountain clay with stones, wheel-thrown and handbuilt, to 6" x 5" x 5" [15 x 13 x 13 cm]. *Photo credit: Derek Larsen.*
The small jars were coil built and then quickly thrown using a variety of local clays found in the Shigaraki area. All the jars were fired for one week in an anagama kiln to 2550 degrees Fahrenheit [1400 degrees Celsius] stoking red pine wood.

The jar was coil built then wheel-thrown using a local clay found in the Shigaraki area, then fired for four days in an anagama kiln to 2460 degrees Fahrenheit [1350 degrees Celsius], stoking pine and hardwoods. This jar demonstrates unique natural glaze found in the first meter or firebox of the kiln. The work was side fired to create rivulets of natural ash glaze.

The jug was coil built, then wheel-thrown using local clay found in the Shigaraki area, then fired for four days in an anagama kiln to 2460 degrees Fahrenheit [1350 degrees Celsius], stoking pine and hardwoods. This jug demonstrates unique natural glaze found in the first meter or firebox of the kiln. The interaction with the charcoal and extreme temperature have produced a unique textural ash glaze.

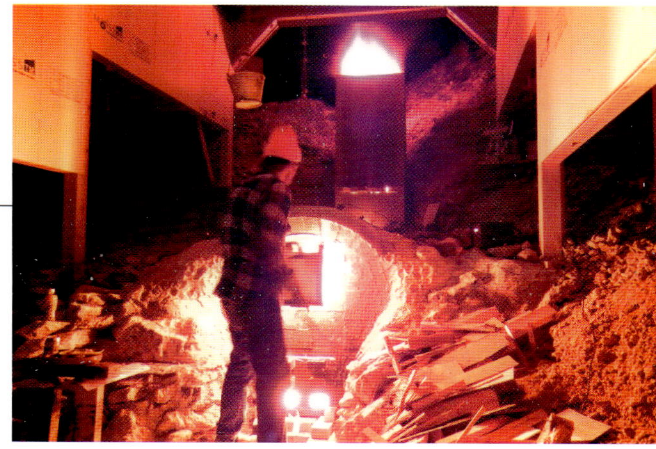

Kakidaira Kiln, 2011. Stoking the kiln. *Photo credit: Derek Larsen*

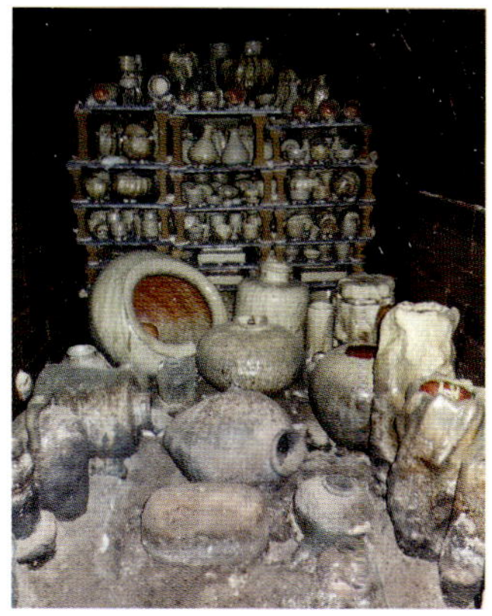

Derek Larsen, *Kakidaira Kiln*, 2011. View of firebox after four day firing. *Photo credit: Derek Larsen.*

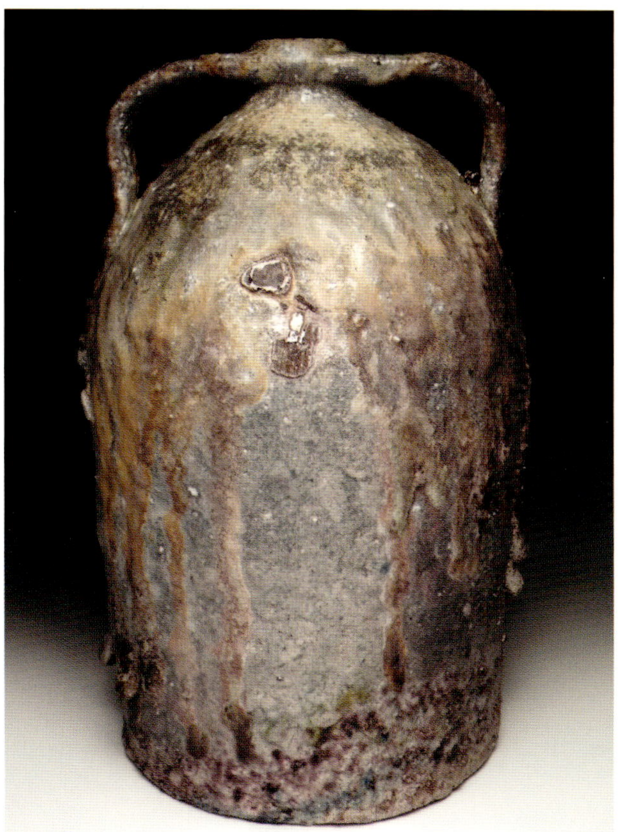

Derek Larsen, *Jug*, 2011. Anagama fired mountain clay with stones wheel-thrown and handbuilt, 11.5" x 8" x 8" [29 x 21 x 21 cm]. *Photo credit: Derek Larsen.*
The jug was coil built, then wheel-thrown using local clay found in the Shigaraki area, then fired for four days in an anagama kiln to 2460 degrees Fahrenheit [1350 degrees Celsius], stoking pine and hardwoods. This jug demonstrates unique natural glaze found in the first meter or firebox of the kiln. The interaction with the charcoal and extreme temperature have produced a unique textural ash glaze.

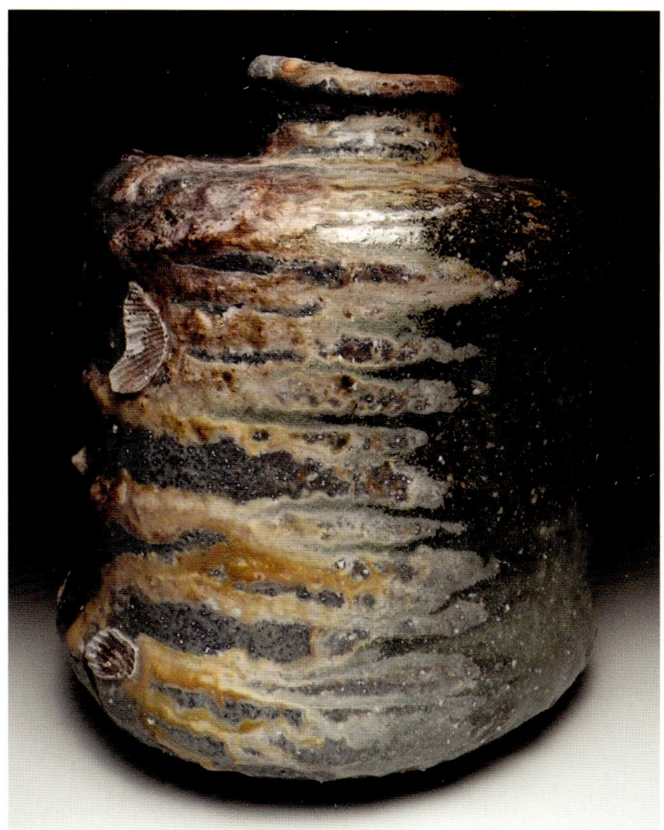

Derek Larsen, *Jar*, 2011. Anagama fired mountain clay with stones, wheel-thrown and handbuilt, 13" x 11" x 11.5" [33 x 28 x 29 cm]. *Photo credit: Derek Larsen.*
The jar was coil built then wheel-thrown using local clay found in the Shigaraki area, then fired for four days in an anagama kiln to 2460 degrees Fahrenheit [1350 degrees Celsius], stoking pine and hardwoods. This jar demonstrates a unique natural glaze found in the first meter or firebox of the kiln. The work was side fired to create rivulets of natural ash glaze.

JOHN LEACH

Langport, Somerset, England
john@johnleachpottery.co.uk, www.johnleachpottery.co.uk

Artist's Statement

In this very immediate, push button, plastic age we live in, it is a real privilege to be able to make functional and beautiful designs by hand, on the potter's wheel, using the humble raw material of clay.

I choose to wood-fire my pots because I love the variegated, toasted finish. I particularly enjoy the effects from the volatile wood ash flying through the kiln impinging on the surface of the pots in a natural and haphazard way. This goes back to 1960 during my apprenticeship with my grandfather, Bernard Leach, where there was a Japanese climbing kiln (noborigama) built in 1920, the first one to be built in the West. The Muchelney Pottery kilns are derived from the same design that I grew up with. The current one is a wood-fired three-chambered climbing kiln.

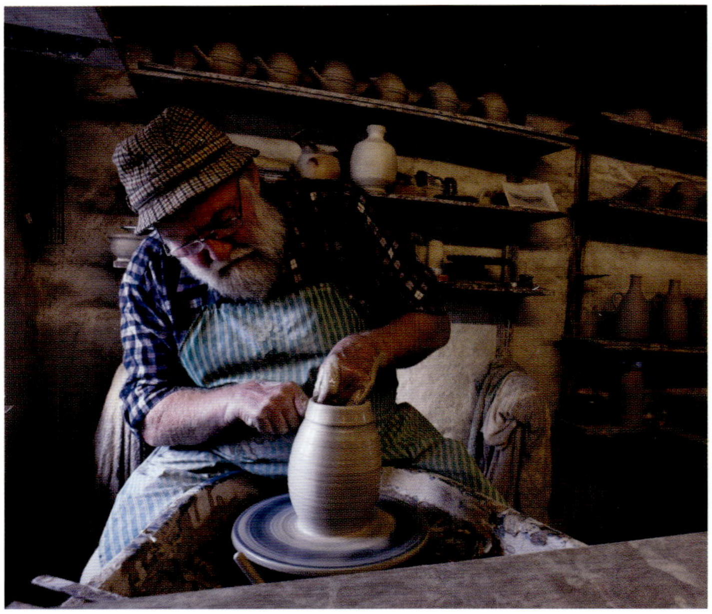

John Leach, Wheel throwing, 2011. *Photo Credit: Nick Fleming.*

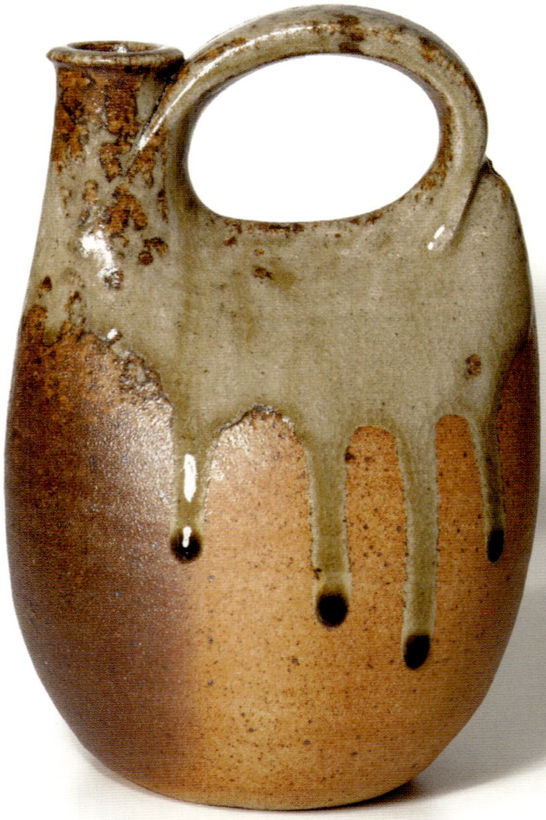

John Leach, *Cross-handled bottle with ash dribble glaze*, 2012. Wood-fired stoneware, wheel-thrown, 12" x 6". *Photo credit: Carole Melbourne.*

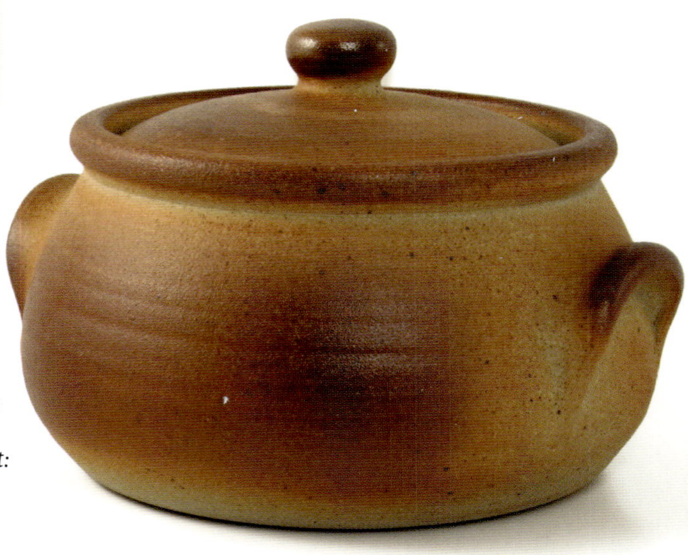

John Leach, *Medium squat casserole dish*, 2012. Wood-fired stoneware, wheel-thrown, 4.5" x 6". Photo *credit: Carole Melbourne.*

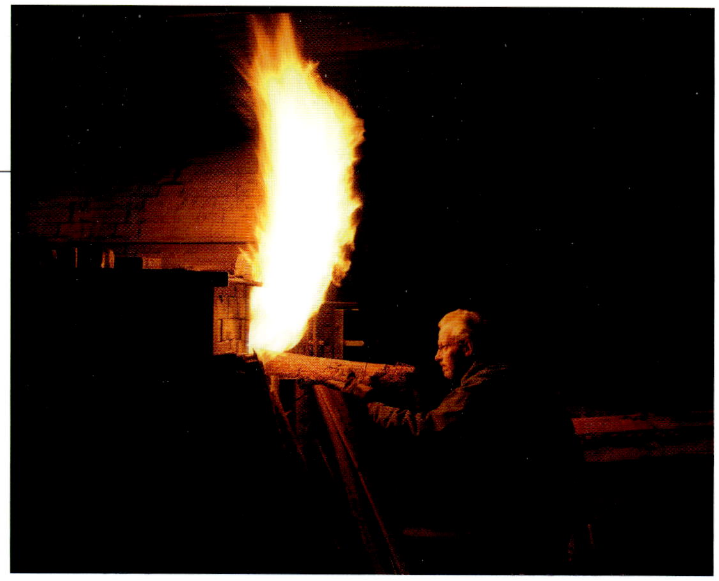

Nick Rees, stoking the *Muchelney Pottery* three-chambered climbing kiln, 2012. *Photo credit: Carole Melbourne.*

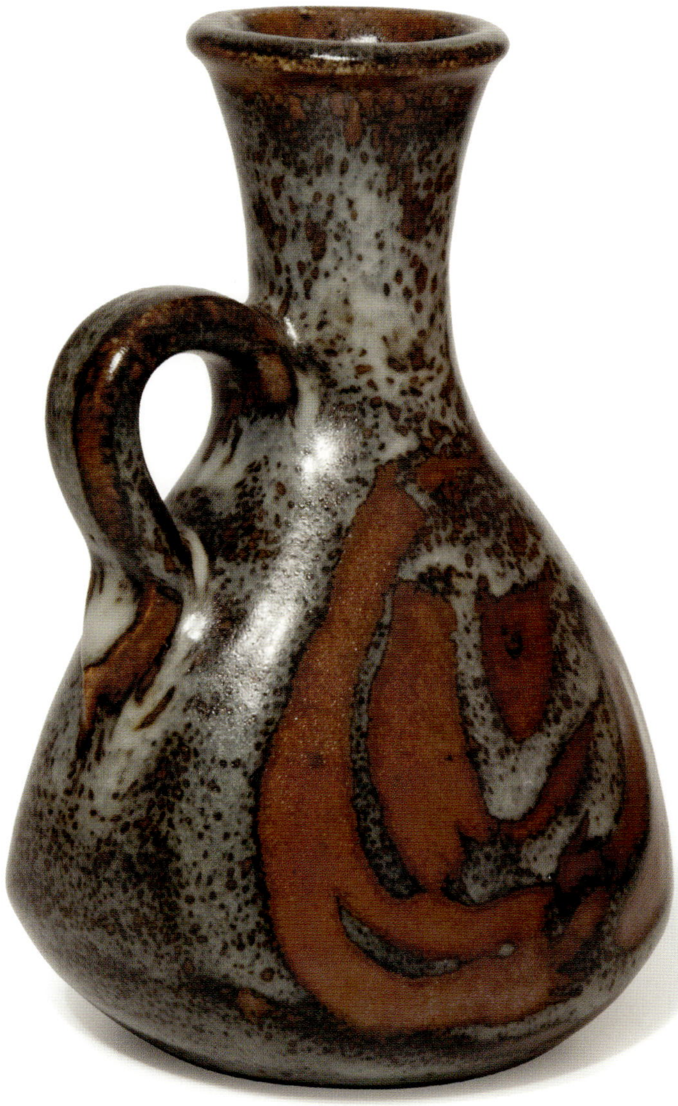

John Leach, *Bottle with wax resist decoration*, 2012. Wood-fired stoneware, wheel-thrown, 8" x 4.75". *Photo credit: Carole Melbourne.*

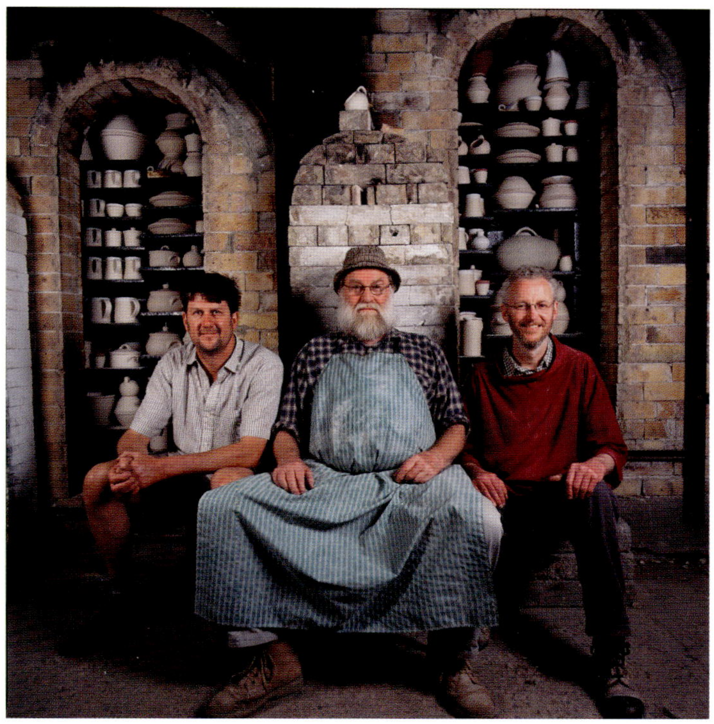

The Muchelney Pottery Crew, (l-r) Mark Melbourne, John Leach, Nick Rees, 2010. *Photo credit: David Barnes.*

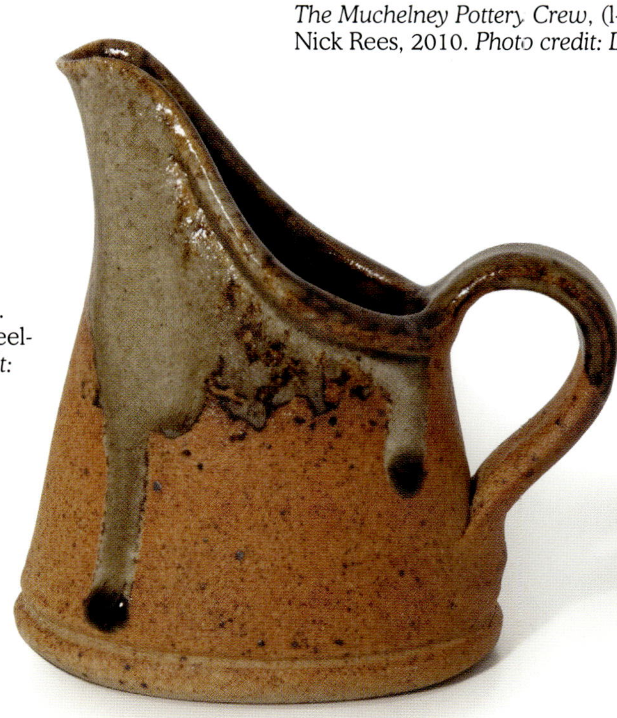

John Leach, *Hod jug*, 2012. Wood-fired stoneware, wheel-thrown, 4" x 3". *Photo credit: Carole Melbourne.*

DICK LEHMAN

Elkhart, Indiana
dicklehman@aol.com, www.dicklehman.com

Artist's Statement

I experience a kind of reverence in wood-firing that I find in no other ceramic pursuit: I stoke the kiln with wood that has grown over the last thirty or fifty, or perhaps seventy years. During those years the tree, being true to its own biological requirements (and subject to the particularities of the roughly two hundred cubic yards of soil atop of which it sits, and through which its roots traverse), has quietly but steadily stored away in its bark and cambium layer a peculiar set of soluble minerals and metallic salts.

When the wood, which I thrust into the blazing twenty-five hundred degree firebox fairly explodes into combustion, minuscule trace amounts of these minerals and salts (which are not combustible) hitch a ride on the fly-ash and start a journey through the kiln. The ash swirls and eddies around pots, gets lifted with the heat of combustion to higher elevations within the kiln, then cooling a bit begins to descend through the pots and shelves, being inexorably pulled by the chimney-draft to a small exit flue hole at the bottom of the kiln. If by some chance of the swirling tides of flame-currents within the kiln the ash has avoided direct contact with the pots, the fly-ash exits the chimney, eventually returning to the earth to fertilize another generation of forests. But should the fly-ash, during its dance among the pots, come in direct contact with the red-hot molten sticky surface of a pot, the ash adheres, leaving the smallest imaginable trace of flux and hitchhiking-glaze-chemistry on the surface of the pot.

After seven to fifteen days of these chance encounters, there begins to collect a formidable swell of these glaze-making traces, and soon a new kind of flow emerges: an oozing sticky mass of improbable collaborators, flowing down the side of the pots, drawn now by gravity's rule to find their way to the lowest point on the pot.

It is this unlikely blend of biology, chemistry, physics, and intentionality that leads me to a sense of reverence concerning these wood-fired forms. The arbitrary quality of the flames and the fuel sources create never-to-be-repeated surfaces that are rich with clues, hints, and information to the inquiring eye.

I believe that one can never really MAKE wood-fired pots. One can only work alongside the trees and the clay and the flames, and RECEIVE the gifts of the kiln with awe and appreciation.

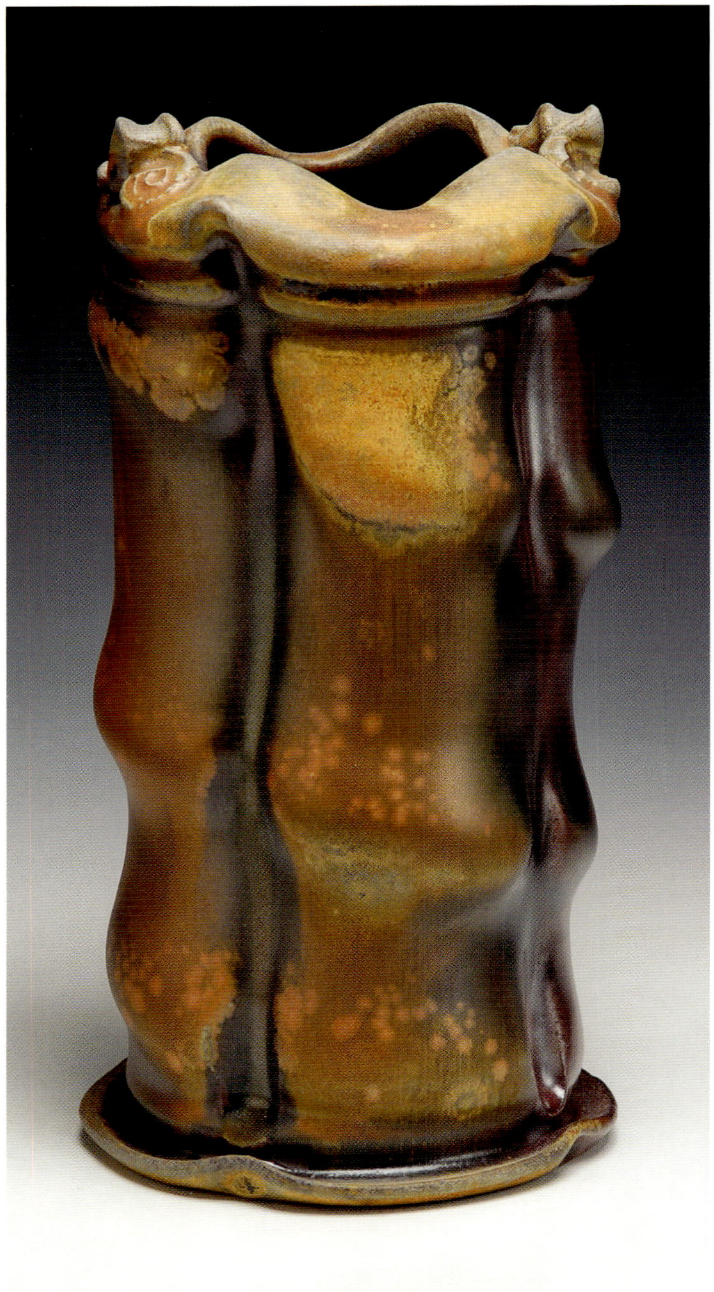

Dick Lehman, *Squared-Spiral Vase*, 2011. Glazed with iron-saturated glaze and wood-fired for three days with Chinese Elm, wheel-thrown and altered, 10" x 5" x 5". *Photo credit: Dick Lehman.*

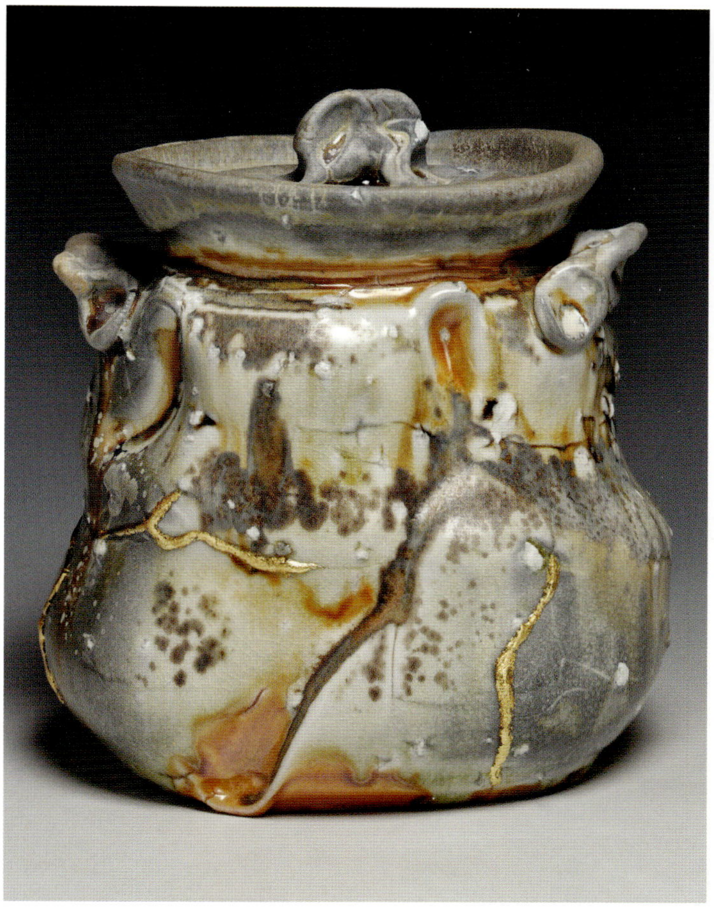

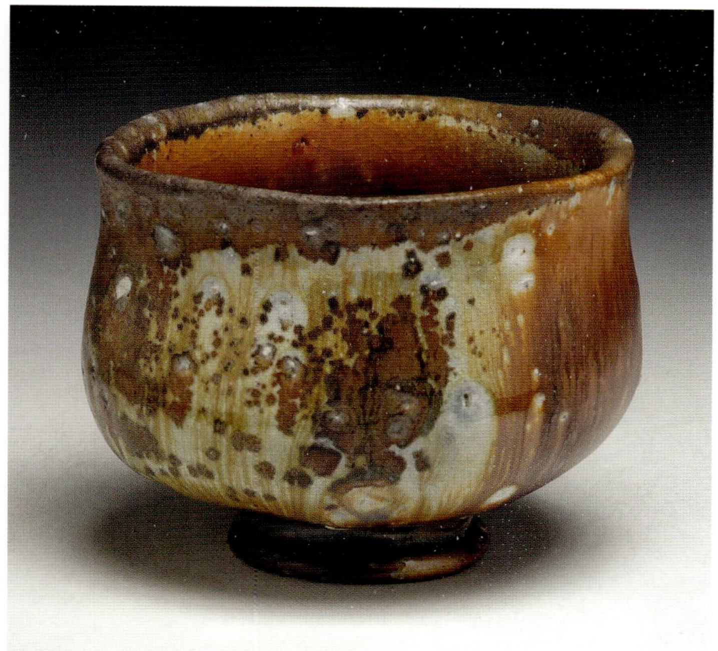

Dick Lehman, *Tea Bowl*, 2008-2009. Twice wood-fired for 10 days each firing, with 20 tons of mixed hardwoods, all natural ash glazing, wheel-thrown, 3.5" x 5" x 5". *Photo credit: Dick Lehman.*

Dick Lehman, *Mizusashi*, 2010. Wood-fired twice for three days with Chinese Elm, wheel-thrown, cracks repaired with epoxy and gold leaf, 7" x 6.5" x 6.25". *Photo Credit: Dick Lehman.*

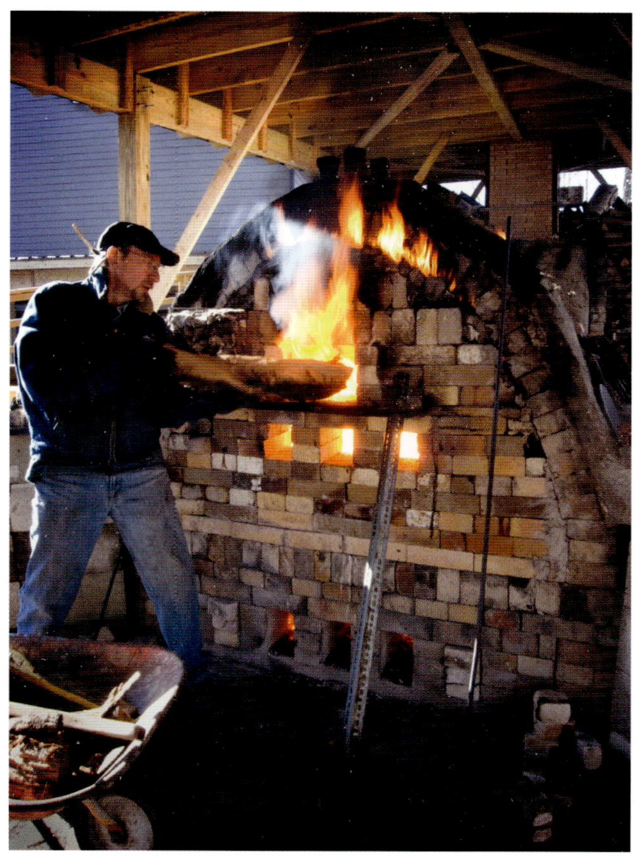

Dick Lehman firing the Rothshank wood-firing kiln, 2011. *Photo credit: Dick Lehman.*

KEVIN LEHMAN

Lancaster, Pennsylvania
kevin@klpottery.com, www.klpottery.com

Artist's Statement

Creativity is my passion. It is spontaneous and limitless. I embrace these qualities as I work with clay and the wood-firing process. The clays I use to create functional, architectural and sculptural forms vary from unrefined, hand-dug clay, full of rocks and other impurities, to highly refined porcelains, providing me with endless possibilities for expression.

Wood-firing provides a finishing process that contains a certain amount of unpredictability that I cherish. Pieces can be placed in the kiln or "set-up" for a certain look but the actual results are not predictable. For instance, pieces that are placed in the back of the kiln have less fly ash and a very different look from those placed in the firebox that receive a heavy dose of ash and a crusty finish. The type of look will be known but the actual amount of ash, flashing patterns, and the colors are unknown until the unloading.

To achieve various looks, my work is fired in several local wood kilns. Most of the kilns are tube type kilns that are fired for various lengths of times, ranging from twelve hours up to twelve days, depending on the size of the kiln and look that is desired. We often flatten cone 13 in the front and cone 11 is flat in the back of the kiln. We also wood-fire a catenary arch kiln several times a year that Amedeo Salamoni and I own. We fire it to cone 10 and add salt or soda toward the end of the firing.

Another aspect of wood-firing that I am interested in is the group effort that is needed for loading and firing the kilns over multiple days. The process goes beyond the individual and includes a group that must work together to achieve quality results, much like communities must work together for common goals.

As an advocate for ceramics in my region, I educate the community about the unique qualities of wood-fired ceramics. Through a community ceramic center that I established in Lancaster, Pennsylvania, Kevin Lehman's Pottery, I provide opportunities for others to participate and learn the wood-firing process. In turn, expanding the local knowledge about the unique aesthetic of wood-fired ceramics.

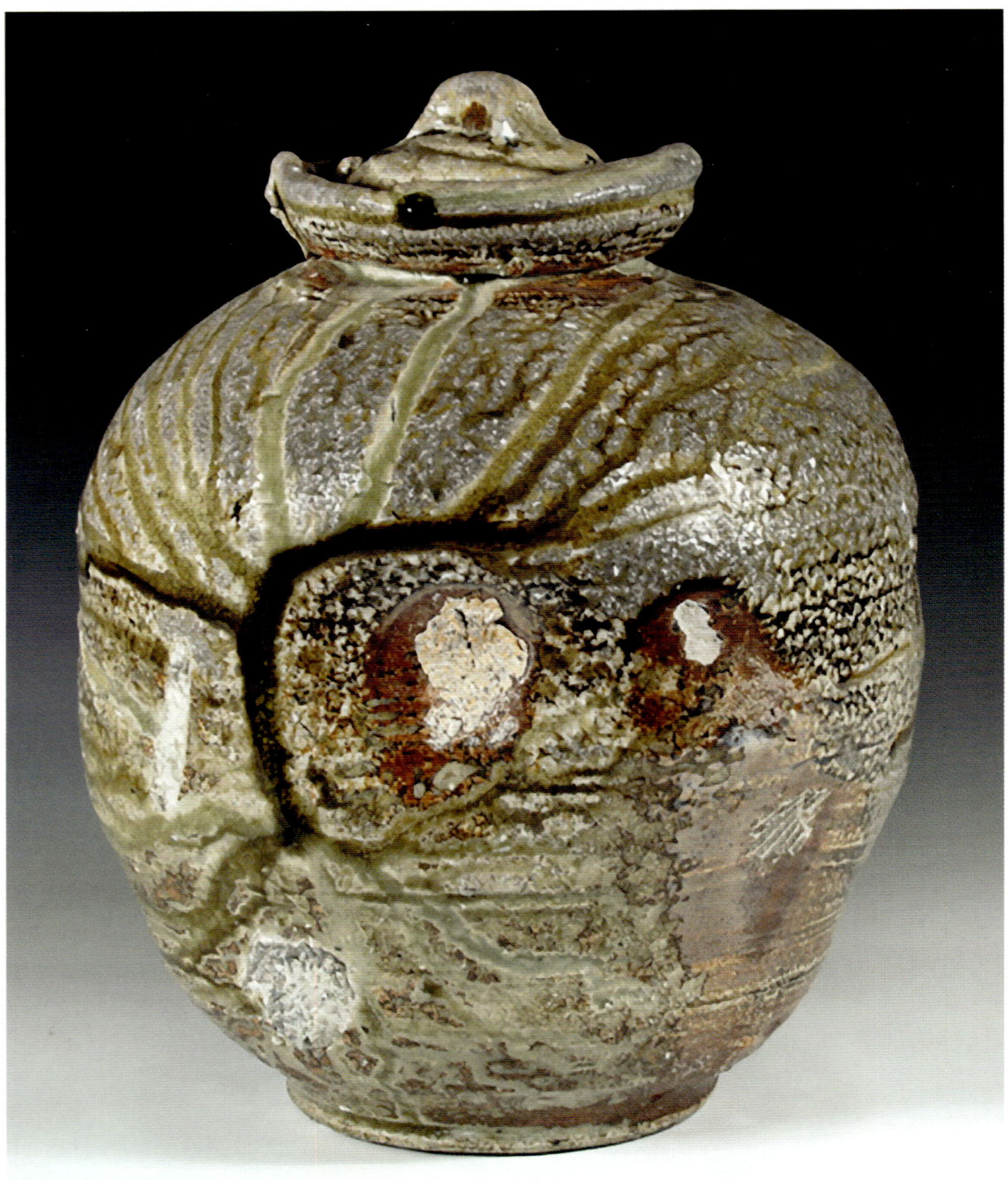

Kevin Lehman, *Painted by Process: Jar*, 2010. Wood-fired ceramic with natural ash glaze, wheel-thrown, 14" x 13" x 13".
This piece was side fired on a shelf at the top of the kiln in the first row of the firebox. Ash was shoveled on the piece multiple times as well as it fell into the firebox and was being re-set on the shelf. Total firing time 36 hours.

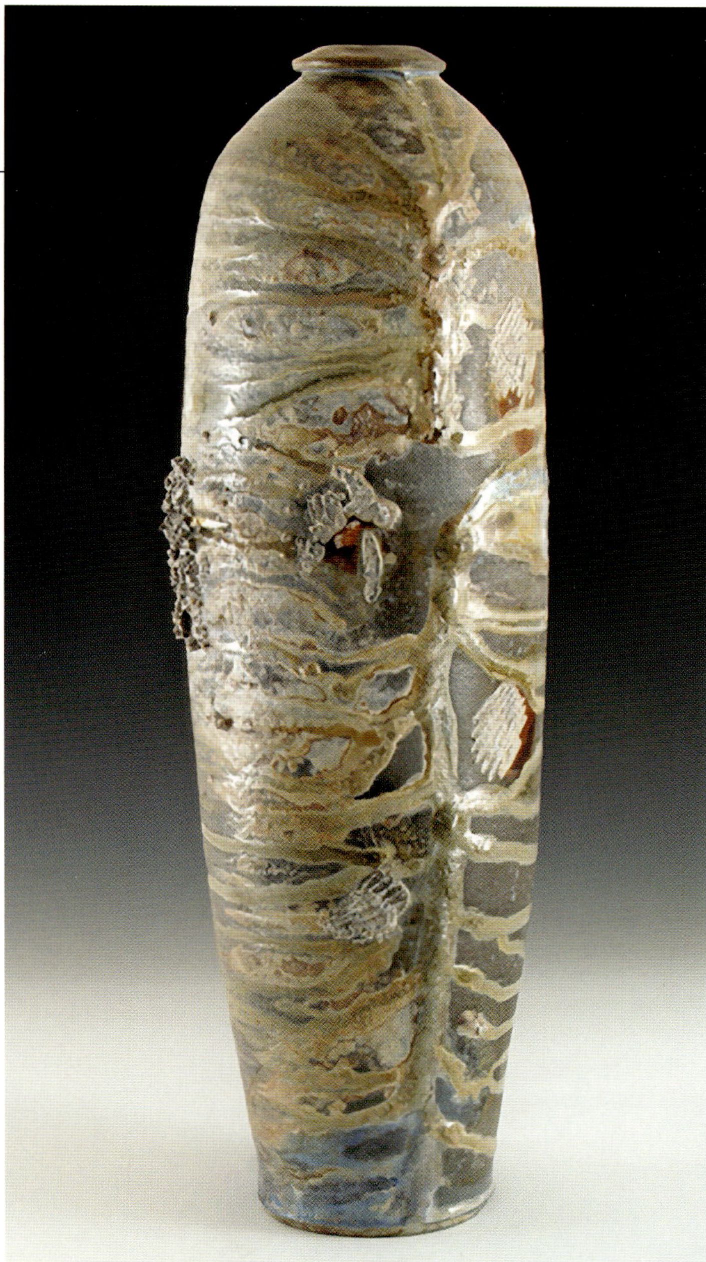

Kevin Lehman, *Painted by Process: Side Fired Vessel*, 2010. Wood-fired ceramic with natural ash glaze, wheel-thrown, 22" x 7" x 7". This piece was side fired in the firebox and ash was shoveled on the piece multiple times throughout the 36 hour firing.

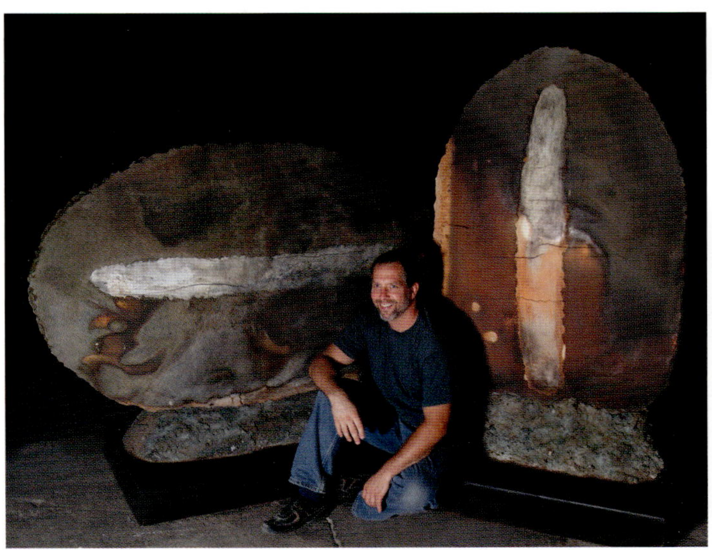

Kevin Lehman, *Monoliths: Stable Growth*, 2007. Wood-fired natural ash, 4' x 6' x 1' and 5' x 4' x 1'. Fired for 14 days.

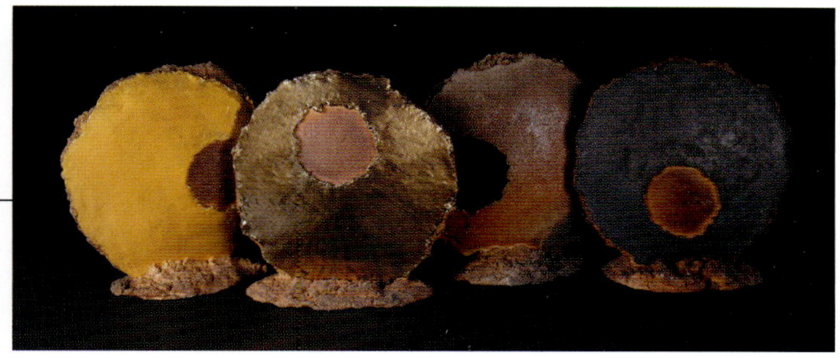

Kevin Lehman, *Monolith Study: The Four Seasons*, 2007. Wood-fired and electric re-fired, slab built, 12" x 12" x 4" each. These pieces were fired in multiple locations in the wood kiln and then re-fired multiple times in an electric kiln. 16-20 hour firing.

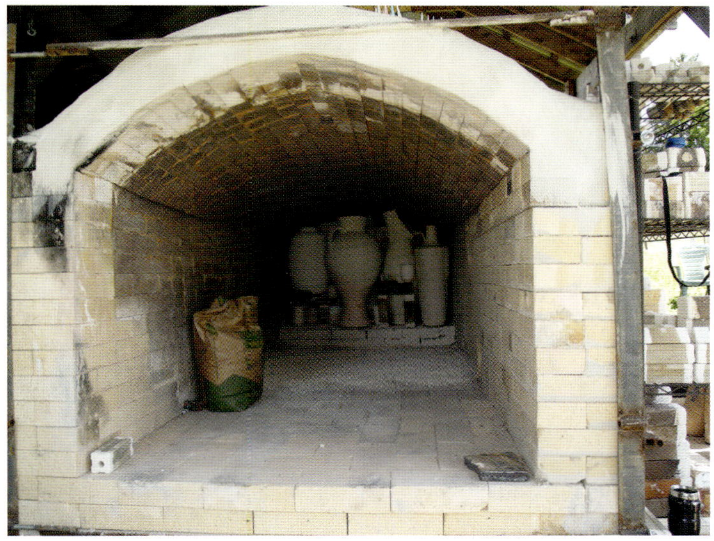

Loading Gill Mallinckrodt's big kiln, Kirkwood, Pennsylvania.

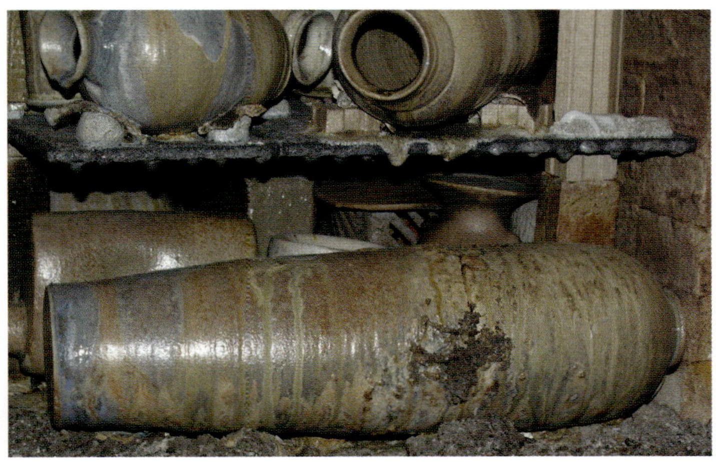

Side fired vessel in front section of kiln near firebox, Gill Mallinckrodt's kiln, Kirkwood, Pennsylvania.

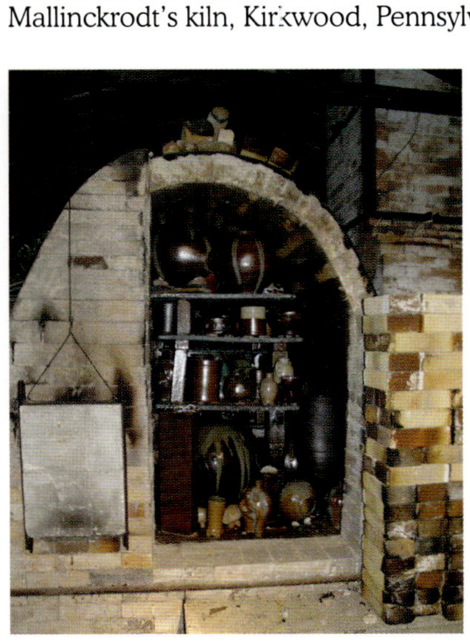

Unloading the kiln, modified catenary arch cross-draft kiln built by Kevin Lehman and Amedeo Salamoni, Holtwood, Pennsylvania.

SIMON LEVIN

Gresham, Wisconsin
Simon@Woodfire.com, www.simonlevin.com

Artist's Statement

Connection to our world and the people around us is the failed promise of technology. Technology alone cannot bring us closer. Touch, interaction, engagement, and conversation are the inroads to intimacy. Functional pottery in the home, in the hand, can begin to forge connection between artist and user, between object and human, between process and people. I see that my work as an artist is to forge deeper connections; to create interactive forms that communicate across the synapses of our compartmentalized world. My work speaks eloquently about my love, exploration, and development of a hugely rich and ancient process of wood-fired ceramics. I use technology as a tool to create opportunities where the intimate connection to clay can occur. My website, Woodfire.com, is a resource for information, relevant links, images of work and kilns, published articles, and, recently, candid photos of pottery in other people's homes.

I believe that our lives are richer for being connected to our everyday objects. Artistic potters are privy to a unique and subversive role in contemporary art. Whereas much of late twentieth century art has been an attempt to merge art and life, pulling art out of galleries and bringing life scenes into the setting of the museum, the functional pot continues to hold a place in everyone's home. A cup is one of the first things we hold in the morning and often one of the last things we touch at night. When we touch our cups to our lips, it is an intimate moment. Honored by that intimate space, I hope to fill it with deep connections between user and maker, clay and fire, in an ageless exploration of emotion.

I believe in making the pots with a revelation of process. I integrate throwing lines and evidence of human touch into surface design. I do not hide attachments of handles or spouts. I seek the inherent qualities in the marks tools make and I approach each individual pot with a similar feel, though the tool might differ. This revelation of process harmonizes with the choice of wood-firing. Ware from a wood kiln speaks loudly about its making. The end result is a narrative of the journey of raw clay to usable pot.

I believe clay accesses timeless forms of communication. I work to explore a universal language of utility and creation. Within my pots I seek to express a cohesive statement of emotion. Each pot holds an exploration of mood. I harmonize the foot of the pot with the rim and the handle with the spout in an attempt to create a unified work. Much of my work reflects a playful humor that pervades my life and feeds my spirit, but whatever idea I choose to explore I seek to reiterate it throughout the pot.

I believe in making pottery in dialogue with my processes. Firing with wood has several inherent qualities. Along with a distinct wind and directionality, high temperature wood-fired pottery has the narrative quality of a surface created through process. Ash from the burning wood melts into glass, creating a glazed surface over days of firing. I have developed clay bodies to be reactive to variations in temperature, atmosphere and ash. As each body of work is fired, questions are answered and new ones are posed in a progressive conversation with the firing technique. Of late I have been working to slow down the flame, achieving much rounder, softer flashing marks. My response to these marks has been to round the forms, pillowing out volume and firing to a slightly lower temperature so the clay is visually softer. In 2003, I designed and built my own 140 cubic foot anagama kiln, looking to evoke the softer narrative flame marks on the pottery through architectural decisions in the design of the kiln. Traditionally, anagama kilns are long tunnels that run up hill. I designed mine to be proportionally taller and wider in order to draw the flame upwards and out, decreasing velocity while maintaining volume. The creation of this kiln was a response to surfaces and qualities in my work that I have come to love. The resulting work from the past fourteen firings of this kiln has informed my work and continued the dialogue. I am in an evolving cycle of conversation with form, technique, kiln, process, and utility.

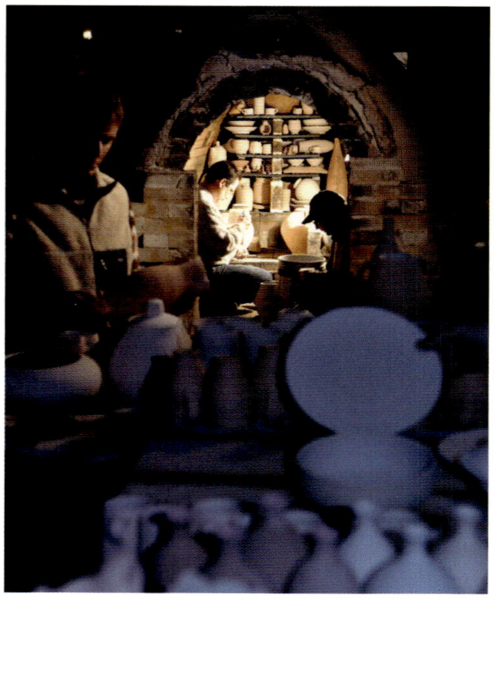

Simon Levin, Loading the anagama. *Photo Credit: Curt Knoke.*

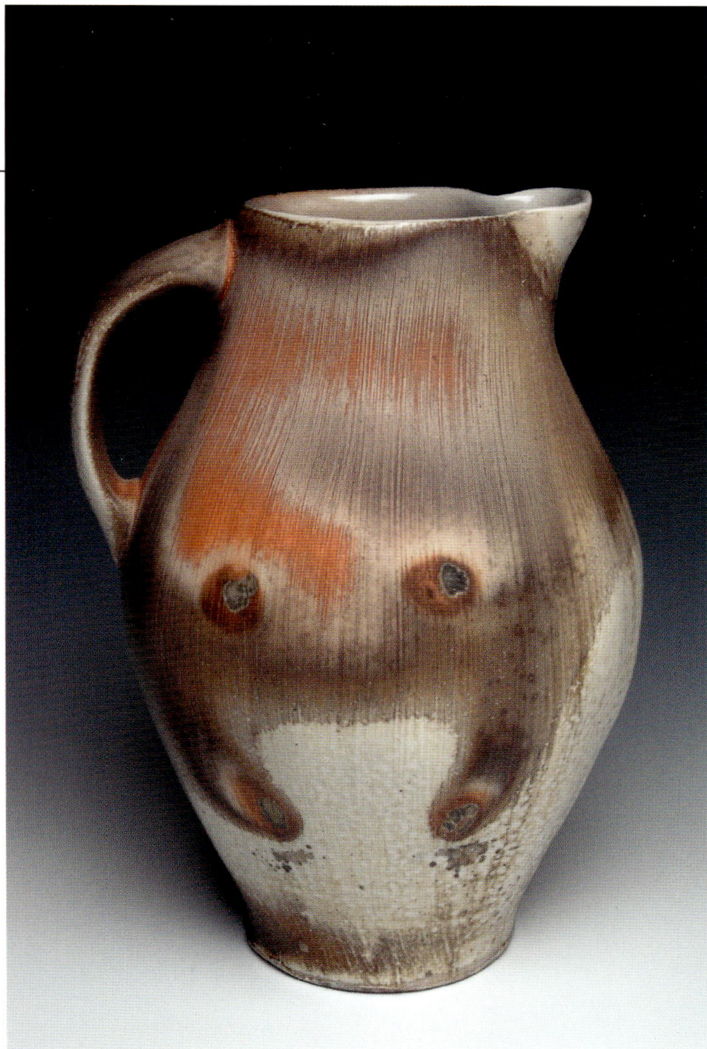

Simon Levin, *Brushed Pitcher*, 2011. Wood-fired porcelain, train kiln, wheel-thrown, 11" x 8" x 7". *Photo Credit: Simon Levin.*

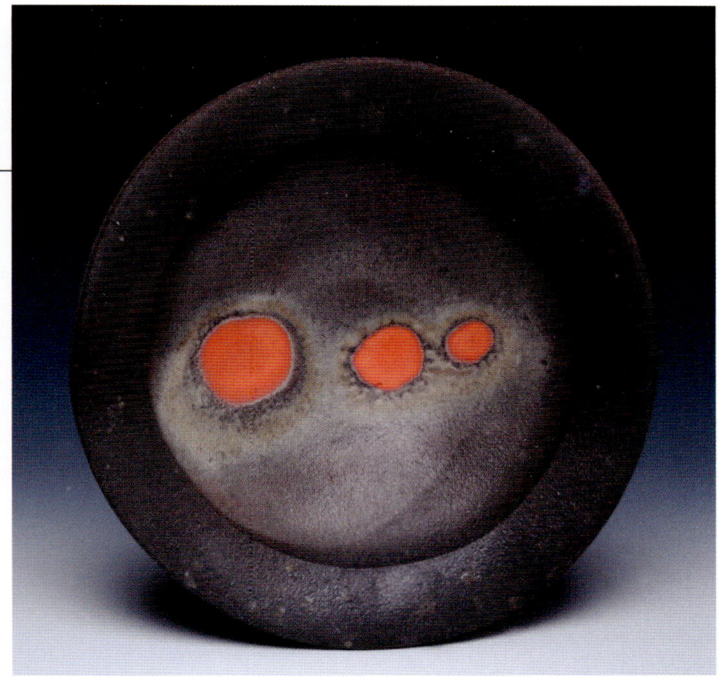

Simon Levin, *Green Halo Bowl*, 2011. Wood-fired stoneware, Anagama, wheel-thrown, 2.5" x 8" x 8". *Photo Credit: Simon Levin.*

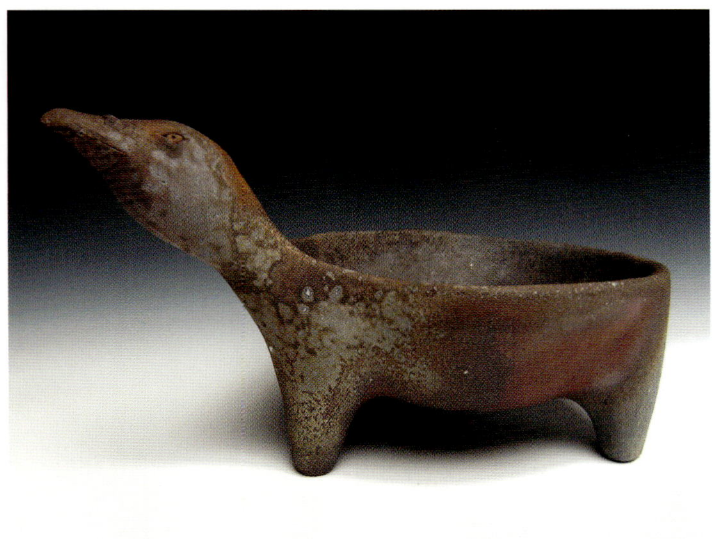

Simon Levin, *Aloof Goose Bowl*, 2011. Wood-fired stoneware, train kiln, handbuilt, 9" x 14" x 9". *Photo Credit: Simon Levin.*

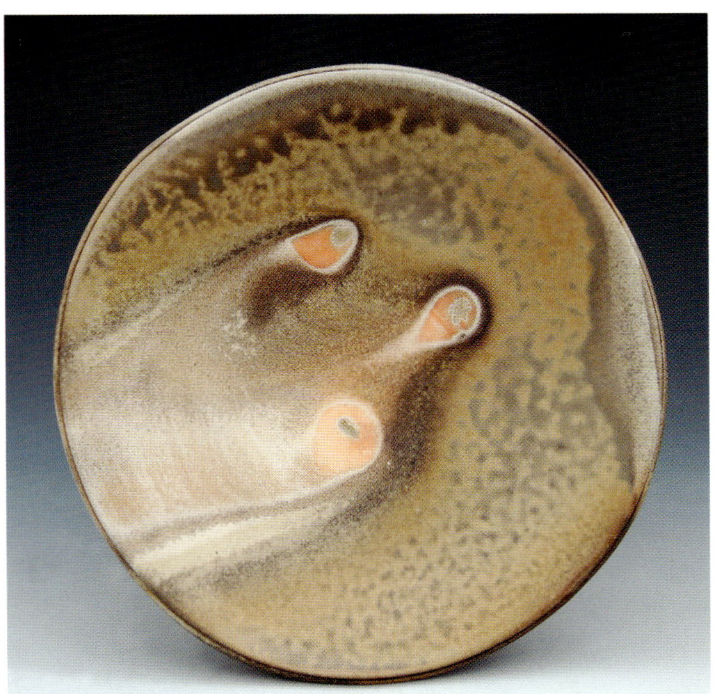

Simon Levin, *Whispy Plate*, 2010. Wood-fired porcelain, Anagama, wheel-thrown, 1.5" x 8" x 8". *Photo Credit: Simon Levin.*

JAN LOVEWELL

British Columbia, Canada
rareearthpottery@gmail.com, www.rareearthpotters.com

Artist's Statement

After several years of making raku pottery, Ron and I wanted to expand our creative capacity. We considered various options and were most interested in firing with wood, both for its aesthetic potential and its use of non-fossil fuel. Wood-firing offers a direct connection with fire, air, earth, and wood. It can build a depth and richness of surface and color that can only be found in nature. As we discovered, it also offers us friendship and camaraderie with like-minded artists in a stunning rural environment.

Ron and I began to gather materials for a wood kiln in 2000. In March 2007, Jinny Whitehead, Pia Sillem, Ron, and I began to build a wood-fired train kiln with sixty-five square feet of stacking space at Pia's home near Earl's Cove. When Jinny, Ron, and I arrived, there was already a beautiful, generous kiln shed, loaded with wood from trees cut from the property to create building sites and roads, and a cement slab for the kiln. For the next two years, the four of us gathered and worked ourselves silly for three days at a time. Our first firing was in January 2009.

Jinny travels north from Vancouver while Ron and I travel south from the coastal village of Lund to the kiln at Earl's Cove. We all have our ware ready to load, which we do on the first day. We share a delicious dinner (always) and next morning we light the kiln. One or two people tend the fire throughout, working in four hour shifts. The sound of the fire is quiet, and we are attentive to its needs. We watch the ash bed, the flame, and the temperature, doing whatever we can to support a good outcome for everyone. We fire for 48–52 hours and complete the firing during daylight hours. Then everyone goes home to rest and wait while the kiln cools.

Unloading takes a few hours, while we observe what happened where in the kiln, remove pots and shelves, and admire the results. Once again we pack up our pots and head home. It takes some time to digest what has happened and to think about what we'd like to do for the next firing.

We have been very pleased with what the kiln has given us. With experience, we are learning more about what to expect from the interaction of our clays, shapes, glazes, and kiln. We are grateful for our partners and to be involved in this creative work.

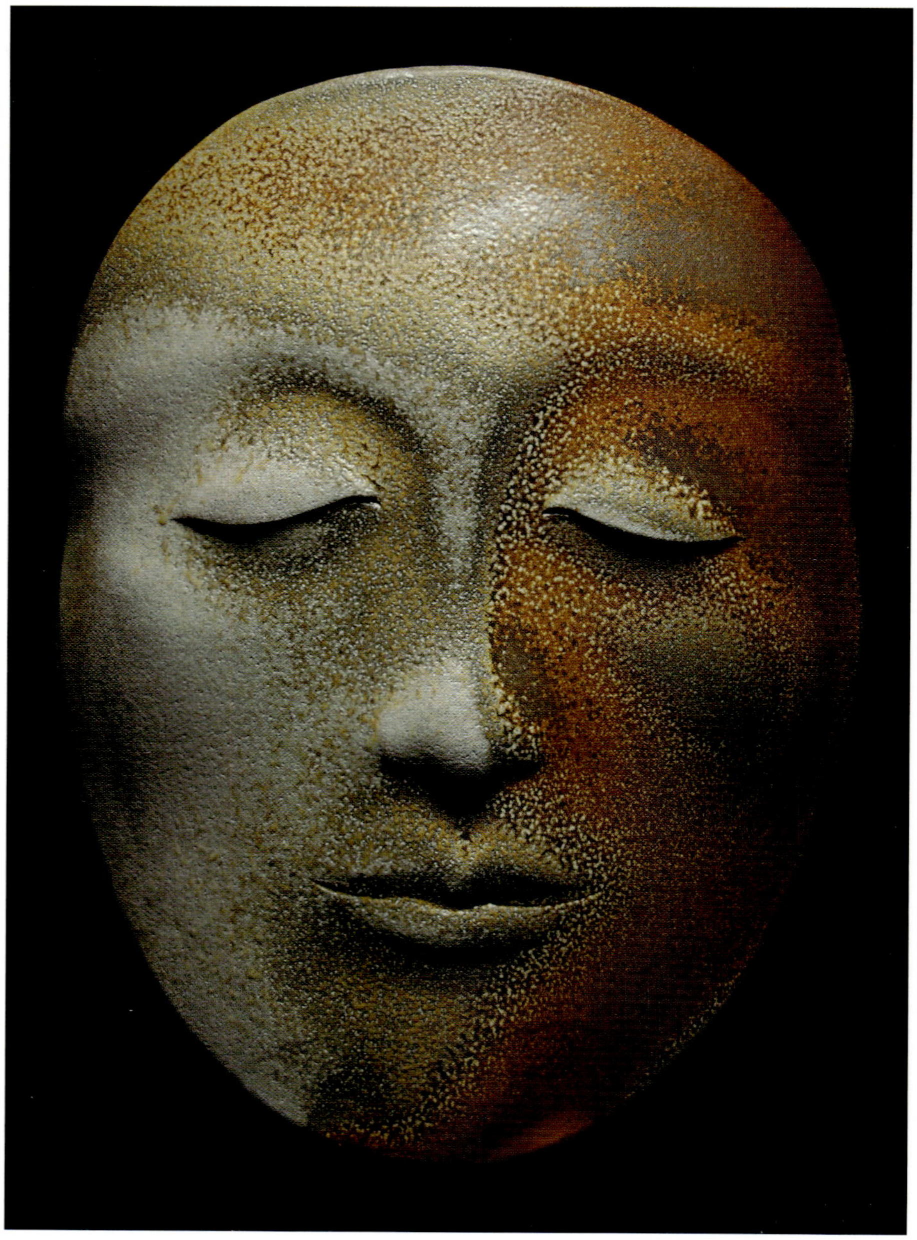

Jan Lovewell, *Mask* 2011. Wood-fired stoneware, altered slab, 16.5" x 12.5" x 2". *Photo credit: Ron Robb.*

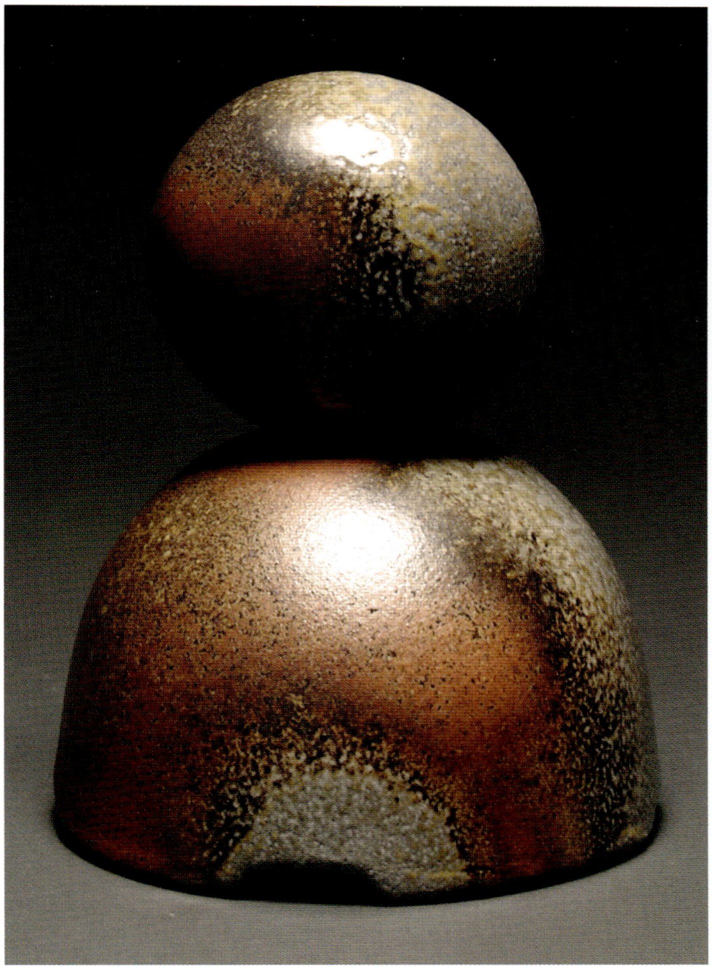

Jan Lovewell, *Form #1*, 2011. Wood-fired stoneware, wheel-thrown & altered, 6" x 4.5" x 4.5". *Photo credit: Ron Robb.*

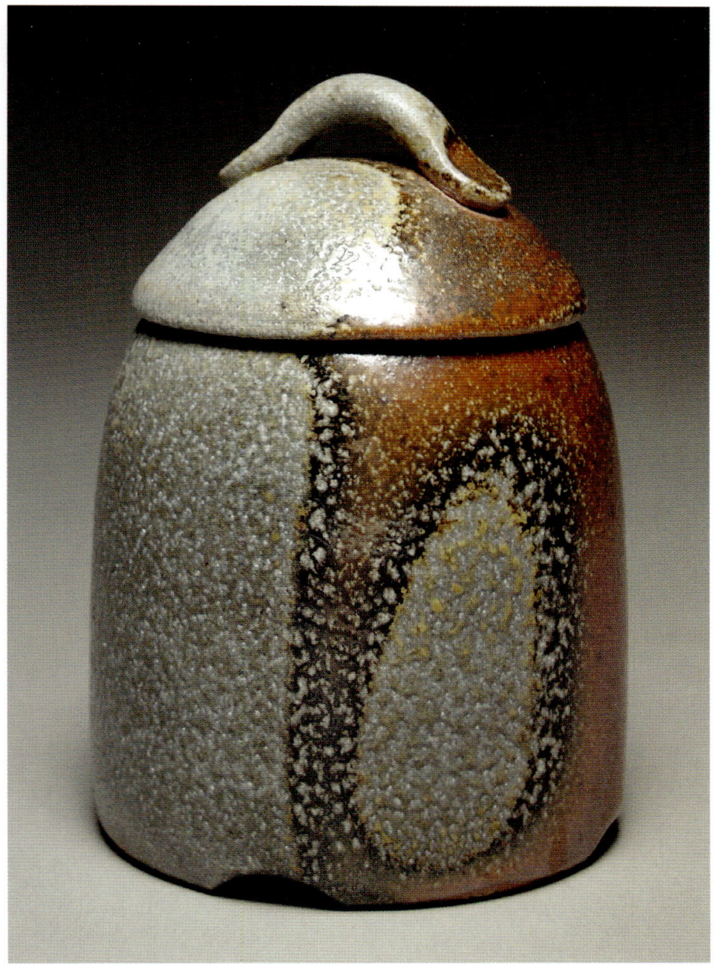

Jan Lovewell, *Jar*, 2011. Wood-fired stoneware, wheel-thrown & altered, 6" x 4" x 4". *Photo credit: Ron Robb.*

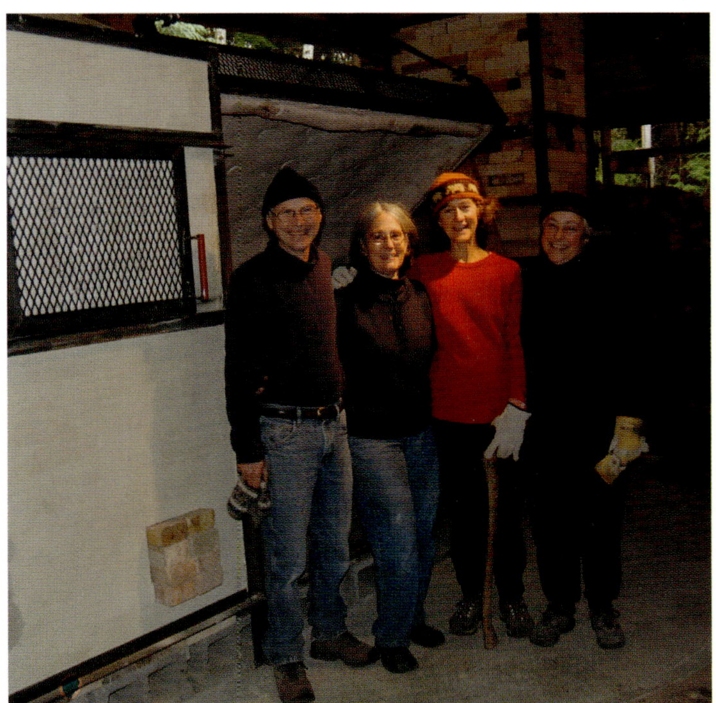

Ron Robb, Jan Lovewell, Pia Sillem, and Jinny Whitehead had just finished building the Earl's Cove Kiln, October 24, 2008. *Photo credit: Jan Lovewell.*

KIRK MANGUS

Kent, Ohio
kmangus@kent.edu,
www.art.kent.edu/faculty/fulltime/Mangus.html

Artist's Statement

I am not a zealot. I have fired with wood, gas, oil, butane, propane, and nuclear powered electricity. Luckily I have been spared firing with tires and coal.

Though in true fairness to coal, it is just very old trees.

I have fired many types of clays at many temperatures. I have been firing with wood for forty years and I had teachers who worked with wood-firing for a lifetime before that. The combination of history and international influences are a liberal strength. I am not stuck with tradition and borrow shapes and concepts freely.

The wood-firing process still has the charm, excitement, and camaraderie it always had. It is still not a solution for bad habits. It is difficult to produce sculpture and pots using local materials and a sense of artistry that combines knowledge, experimentation, and physical involvement. This is a quest for the singular, odd, spectacular object, as well as a host of related vessels. Many of my wood-fire pieces are simply functional, and I like to think beautiful. I cannot reproduce them in any exactitude. My sense of Yankee economy allows for this myriad of related works to be elegant, rustic, vigorous, and intimate as well. This suits my concept of American individual sophistication.

My first experience with a wood kiln was at Penland in 1971. Robert Turner fired a small cross draft pile of bricks that brought out two fabulous pots. I can still see them in my mind's eye. I have fired with Rimas Visgirda and Dick Hotchkiss in Grass Valley, California, in 1972–76. They held month long workshops called Earth, Air, Fire, and Water. Their seven chamber kiln was quite a marvel. We dug native clay and processed ash and clays for glazes. Over the four summers I worked there I gained a respect for a can do attitude. I realized that your materials couldn't be bent just to do what you what them to do. You have to find their right character and work with it. The same goes for fuel. In California we used cedar. In Montana we used oak and pine. In Pennsylvania and Ohio we used cherry, oak, and ash. There is a natural bond with local clays and local wood.

In 1981 I worked with Katsuki Sakezumi at Peter's Valley. His large anagama was to be fired for the first time and Eva Kwong and myself became involved with the workshop. The firing lasted ten days and was quite an undertaking. That fall we started our own anagama style kiln in Pennsylvania. We made the bricks by hand, built them into a hillside, and fired them in place. The kiln worked wonderfully. I located clay in the area and started creating work. We fired it nearly every year for twenty years. I also built a kiln at Kent State University. I used rounded cone shape bricks instead of cubes. These work well for almond shape and dome kilns. They also create drips that land on the work.

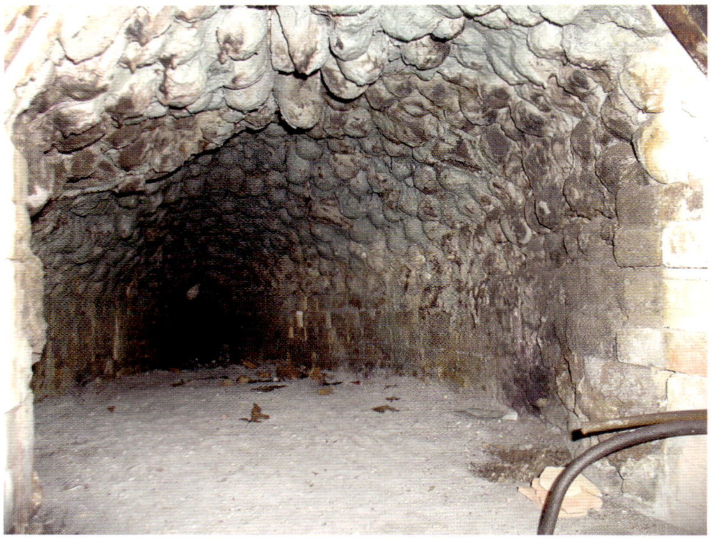

Kent State University anagama, interior, round brick design, with flue, 18' long, 2006. *Photo credit: Kirk Mangus.*

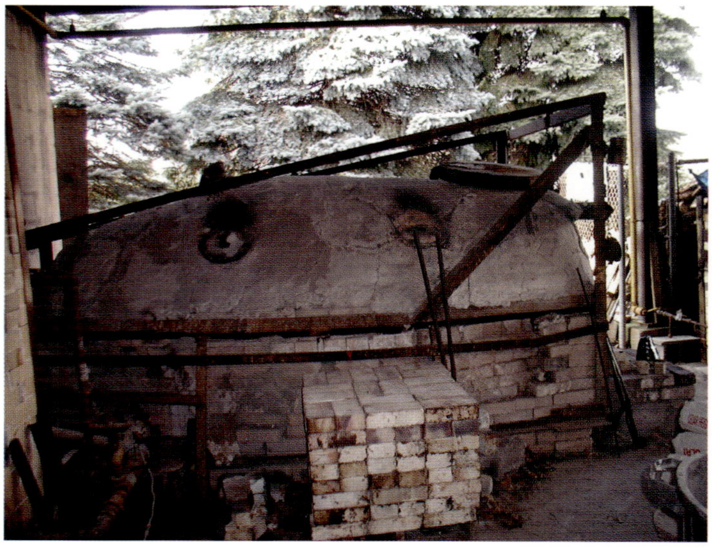

Kent State University anagama, round brick design, 18' long, 2006. *Photo credit: Kirk Mangus.*

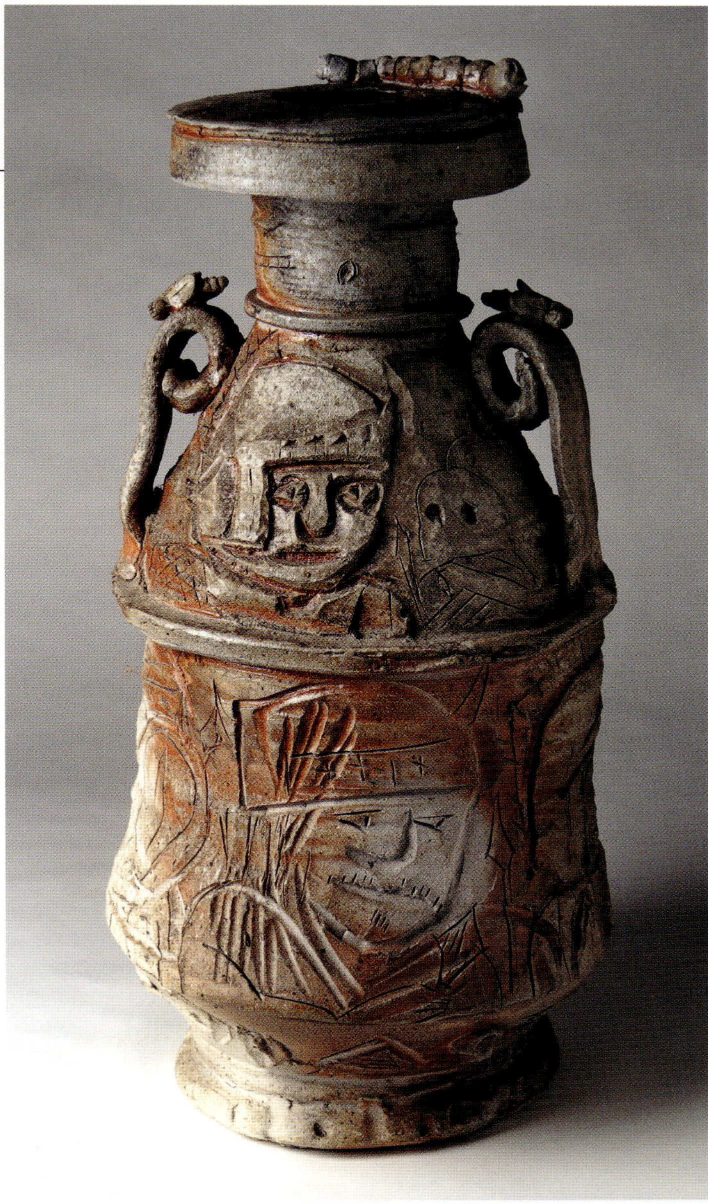

Kirk Mangus, *Barbarian Amphora*. Wood-fired, wheel-thrown, carved. *Photo credit: Kirk Mangus.*

My pottery is either very simple or occasionally paddled. My larger work is heavily carved and sculpted. These pieces are either coil thrown or stacked. My early work tended to be urn shapes or bumps and ziggurats. I was interested in powerful somewhat lumpy pieces. More recently I have tried to use the math of European type pots like ruffled vases or amphorae. Handles dominate or define the forms. I believe the wood ash and flashing illuminate the works.

I have a fondness for Iga pottery. I feel it is too difficult for me to do justice to the forms. I do not have the right attitude or the right clay. I have a mathematical sense about pots and the need for rendering the figure. My pots often have figurative intentions. Iga has a figurative feeling, but I fear my idea for abstraction is not brusque enough to equal Iga's vitality. My work has a tendency for the fluid. I just let Iga inspire me.

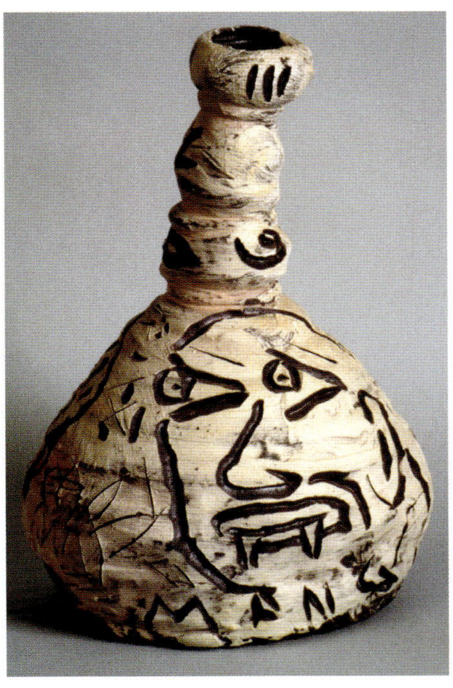

Kirk Mangus, *Sad Guy Jug*. Wood-fired, wheel-thrown, white slip, sgraffito, Iron Clay, 13" x 10" x 8". *Photo credit: Kirk Mangus.*

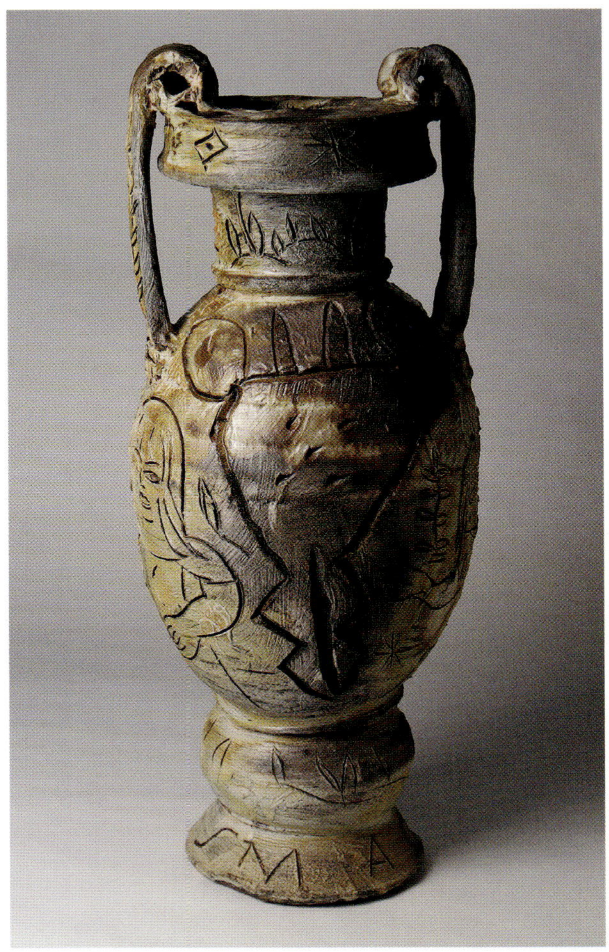

Kirk Mangus, *Sea Life Amphora*. Wood-fired, wheel-thrown, white slip, sgraffito, 33" x 13" x 13". *Photo credit: Kirk Mangus.*

GINNY MARSH

Richardson, Texas
vmarsh@udallas.edu, www.ginnymarsh.com

Artist's Statement

Making pots and firing with wood is, for me, as much about being part of a community as using the pots. We collaborate on our choices of clay, glazes, and the entire process of firing. Of course, we are indebted to and would not have the kiln without the generosity of Louise Rosenfield who owns and built the train kiln we use. She is the consummate hostess and works tirelessly on both the kiln and gathering new participants into our group. We share a vision of work that shows the mark of the flame and ash residue but are not overwhelmed by them so that the work that emerges fits easily into and enhances real lives. We usually fire twice a year, in spring and fall, when the weather in north Texas is fairly comfortable.

Though wood-firing requires intensive labor, the real collaboration and time together make it more than worthwhile. After a group workday to prepare the kiln, three of us usually spend two days loading the kiln. The firing begins the next morning, slowly accelerating so that by evening our two teams take four hour shifts through the night until the middle of the next day. When the kiln reaches temperature, between cones 9 and 11, and the draw trials show that the ash deposit on the clay has melted, we seal the kiln and fire down slowly in reduction to about 1600 degrees. The following week, we get together to unload. Friends come along and neighbors show up to see what surprises are in store for us. Nevertheless, it takes some time after the kiln is unloaded to assess the result and see which pieces really resonate with the balance of form and fire.

I try to make fairly simple, clean forms to give the wood flame a chance to paint its marks on both clay and glazes. To give the flame maximum opportunity, I use porcelain or smooth stoneware. My glazes are usually rather simple, though I have begun using some brushwork on areas, which can be stacked to partially protect them from direct flame. While the forms have to be carefully considered, the whole project is not about a mechanical perfection. It seems to me that they appeal to the users because they are as marked by the flame as we are by our own experiences.

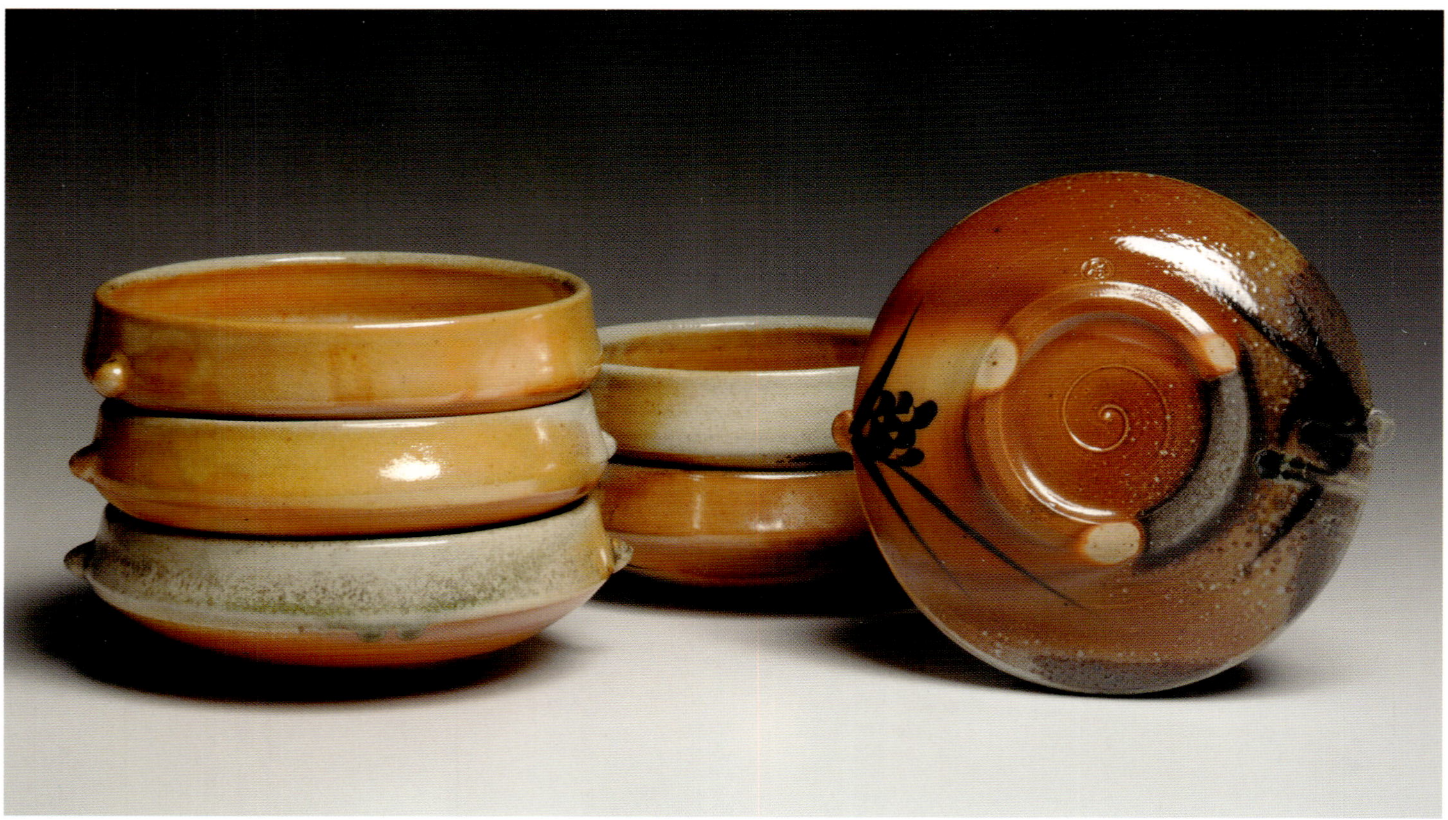

Ginny Marsh, *Set of Six Bowls*, 2011. Wood-fired porcelain, wheel-thrown, 2.25" x 7" x 6.5" each. *Photo credit: Ginny Marsh.*

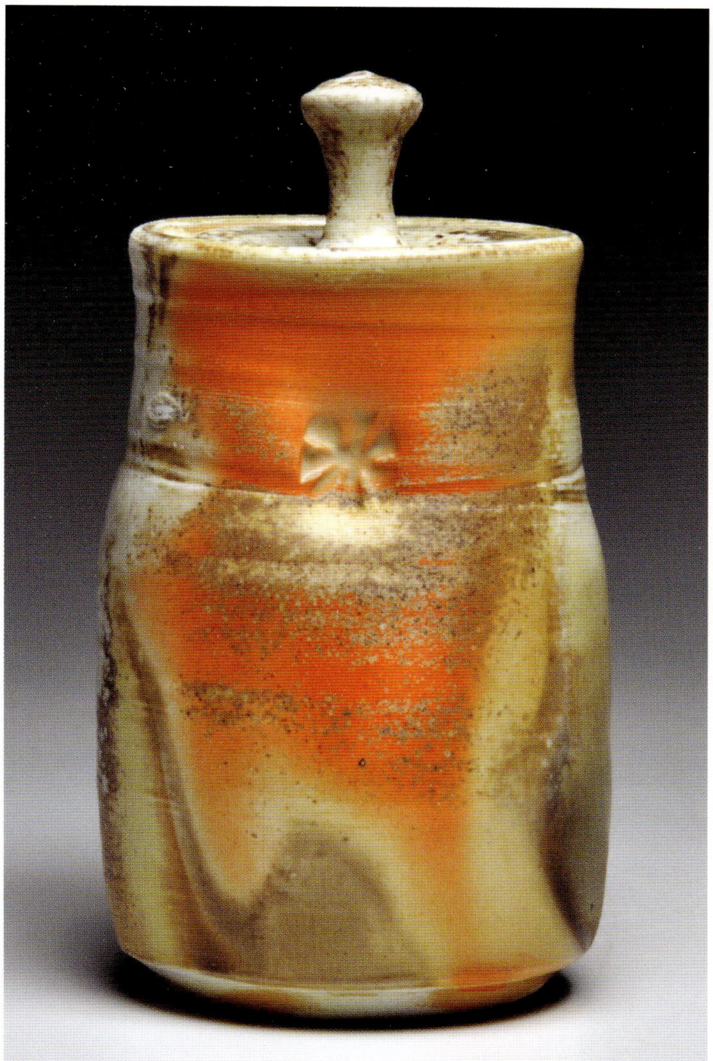

Ginny Marsh, *Lidded Jar*, 2010. Wood-fired porcelain, wheel-thrown, 7" x 4" x 4". Photo *credit: Ginny Marsh.*

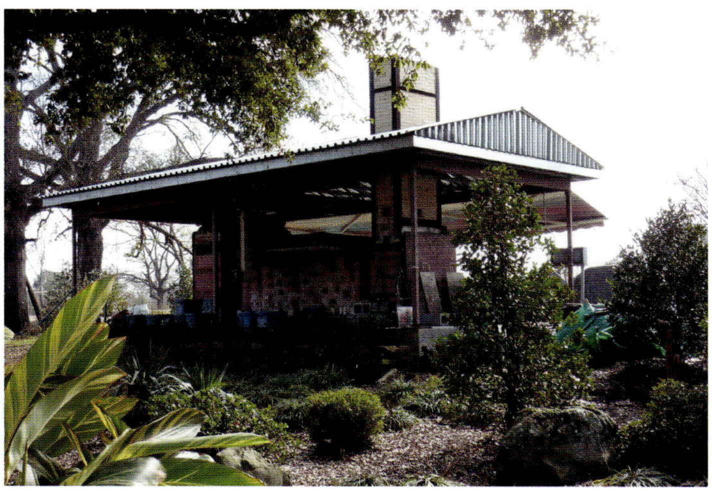

Louise Rosenfield's Train Kiln, Cedar Creek Lake, Texas. *Photo credit: Ginny Marsh.*

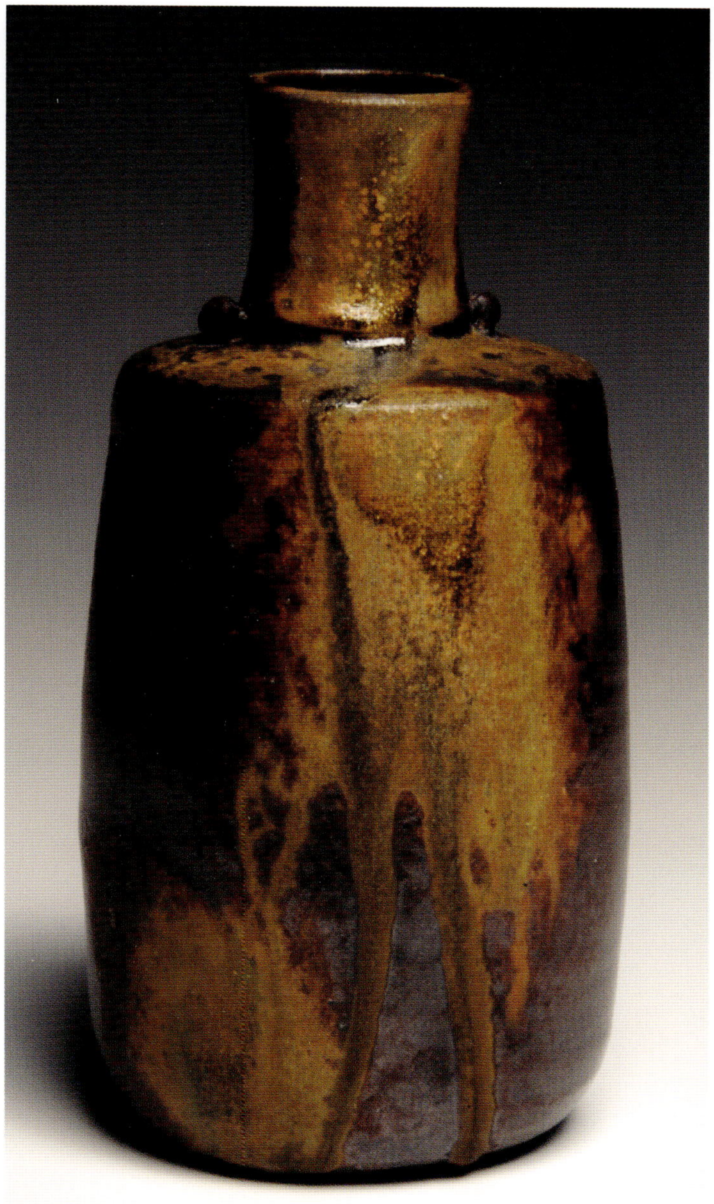

Ginny Marsh, *Vase*, 2011. Wood-fired stoneware, 9" x 4.5" x 4.5". *Photo credit: Ginny Marsh.*

ALEX MATISSE

Marshall, North Carolina
alex@eastforkpottery.com, www.eastforkpottery.com

Artist's Statement

My work is almost all wheel-thrown, varying in size from tea cups to large vases and jars up to four feet in height. About a quarter of the pots are left free of glaze and surface decoration to be fired in the very front of the kiln, where they are exposed to temperatures exceeding cone 14. The rest of the work is slip trailed and then glazed or left bare. I use mostly ash glazes that I find react beautifully to the salt vapors and fly ash within the kiln. I use as many unrefined materials in my glazes as I can; ash from the wood stove or clay from down the road. This lends a subtle depth and quality to the surfaces that I find difficult to produce in any other manner.

My work is made in a fusion of preindustrial country traditions in both process and material. My own clay is mixed predominantly with unrefined North Carolina clays with a small amount of refined materials. It is fired in a large wood-burning kiln for three days and then salted heavily on the final day. The kiln is modeled after many kilns that I have seen and worked with in North Carolina during my three years of apprenticeship. It is an amalgam of different kiln designs, taking elements from all the kilns I have fired and working them into something that suits my own needs. Most of the kilns that I have worked with resembled old kilns either from Thailand, Japan, or North Carolina. My own kiln most resembles a fourteenth century Thai kiln. These labor intensive methods, although time consuming, can yield a pot that has tremendous depth and character.

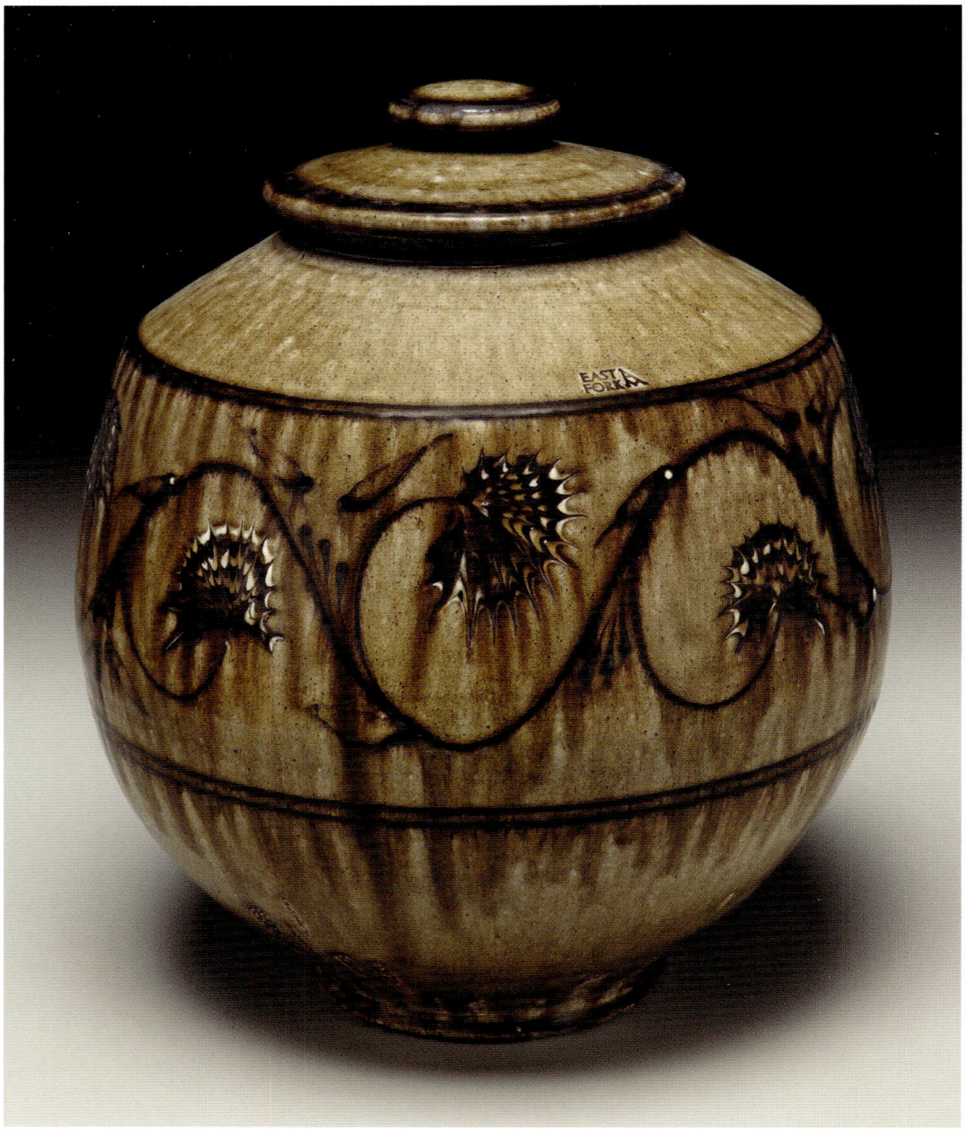

Alex Matisse, *Jar*, 2011. Wood-fired, salt-glazed stoneware with slip trailing and ash glaze, wheel-thrown, 21" x 18" x 18". *Photo credit: Timothy Barnwell.*

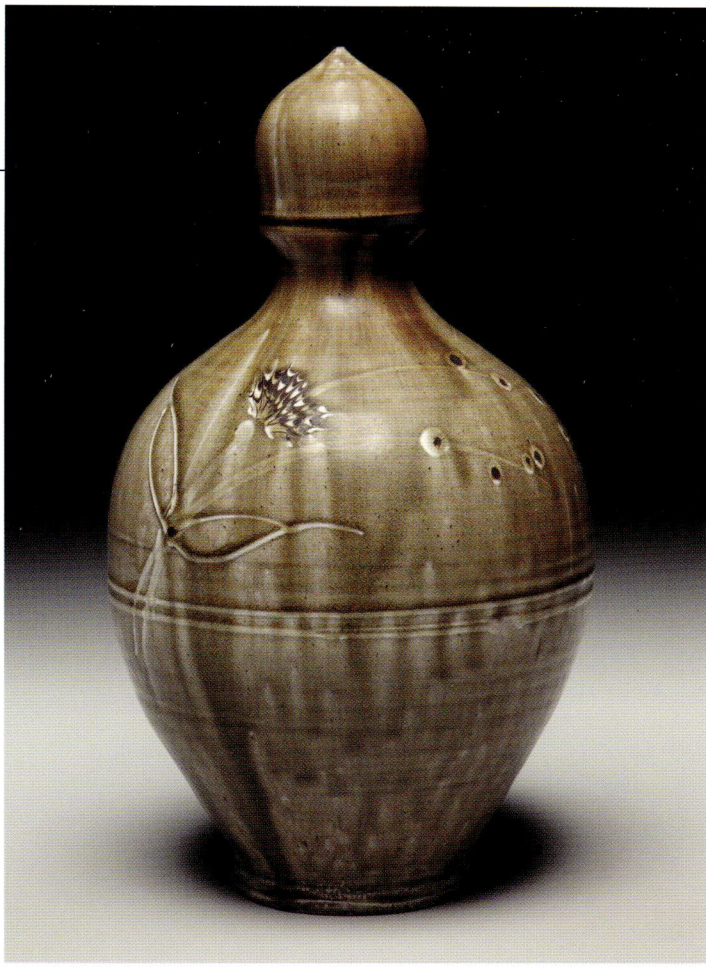

Alex Matisse, *Ometto*, 2011. Wood-fired, salt-glazed stoneware with slip trailing and ash glaze, wheel-thrown, 15" x 8.5" x 8.5". *Photo credit: Timothy Barnwell.*

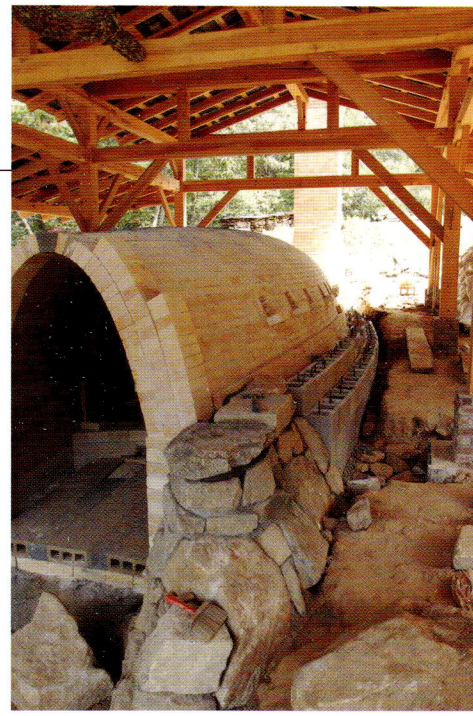

Alex Matisse's kiln nearing completion, Madison County North Carolina. *Photo credit: Connie Coady.*

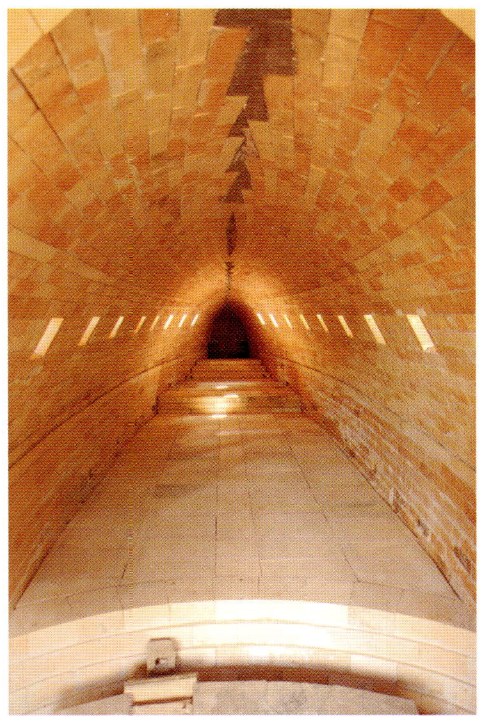

Interior of Alex Matisse's kiln before first firing, East Fork Pottery, Madison County, North Carolina. *Photo credit: Alex Matisse.*

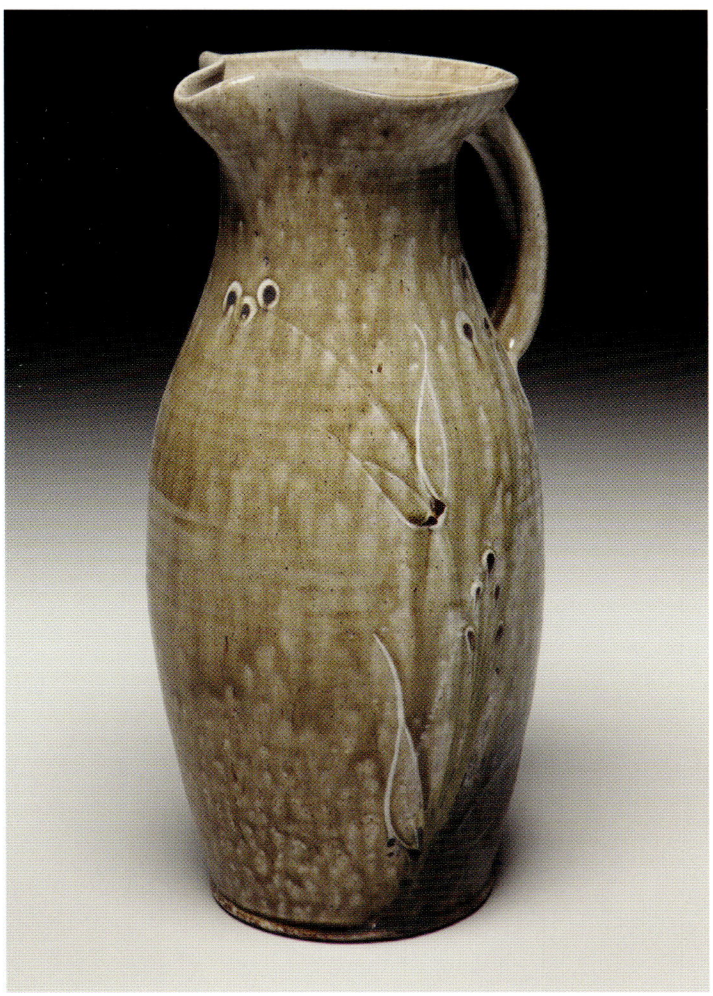

Alex Matisse, *Pitcher*, 2011. Wood-fired, salt-glazed stoneware with slip trailing and ash glaze, wheel-thrown, 16" x 6" x 6". *Photo credit: Timothy Barnwell.*

PETER MEYER

Bend, Oregon
pmeyer@cocc.edu

Artist's Statement

The primary focus my of ceramic work is in refining the shapes and surfaces in preparation for the actions of the wood-fire. I start with a simple shape of pleasing proportion, paying close attention to surface qualities. The proportions of each shape guide the surface and graphic treatments, which include incised lines, inlaid lines, or areas of texture or color. The combination of a variable atmosphere, wood ash, and intense heat create unpredictable effects on the clay. Subjecting the work to such natural forces is a given in the ceramic process, and wood-firing expands the range of possibilities.

I have fired in many different kilns in Oregon but am currently firing in a 125 cubic foot hybrid anagama/ noborigama, located in Madras, Oregon.

A fire typically takes 100 hours and consumes seven to eight cords of wood. The nabori chamber favors pieces glazed with shinos, nukas, celadons, and tenmokus, while the anagama is good for unglazed and slip covered pieces.

We use lodge pole and ponderosa pine, Douglas fir, and species from neighborhood landscapes; elm, birch, maple and apple from local orchards.

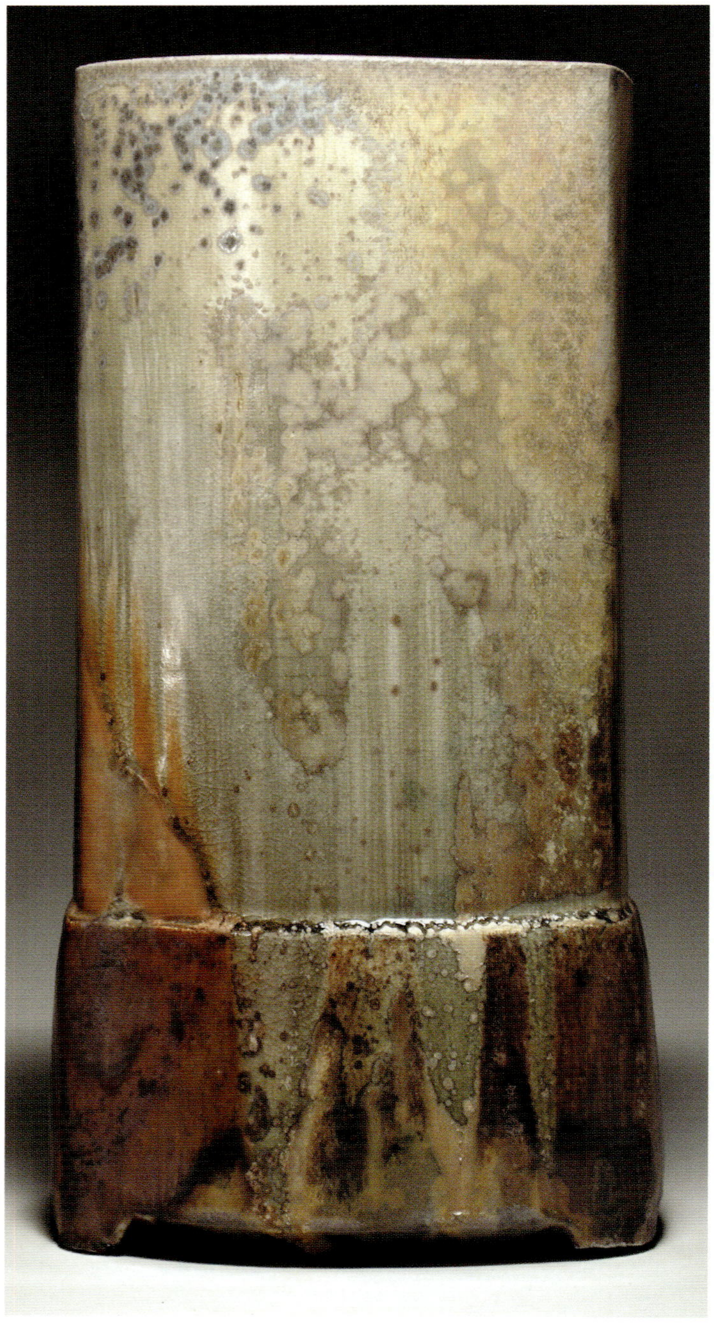

Peter Meyer, *Vase*, 2011. Wood-fired stoneware, wheel-thrown and altered, slips and oxide wash, 24" x 11" x 8". *Photo credit: Peter Meyer.*

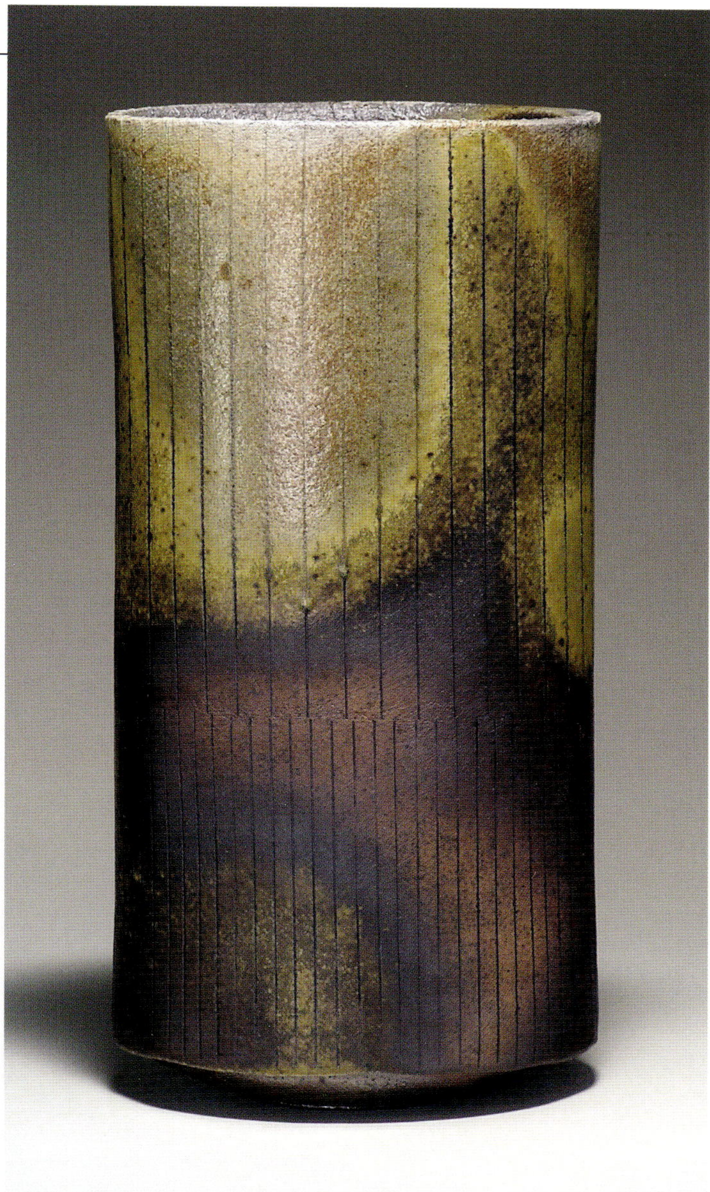

Peter Meyer, *Vase,* 2010. Wood-fired stoneware, wheel-thrown and altered, inlaid slips, 14" x 7" x 5". *Photo credit: Peter Meyer.*

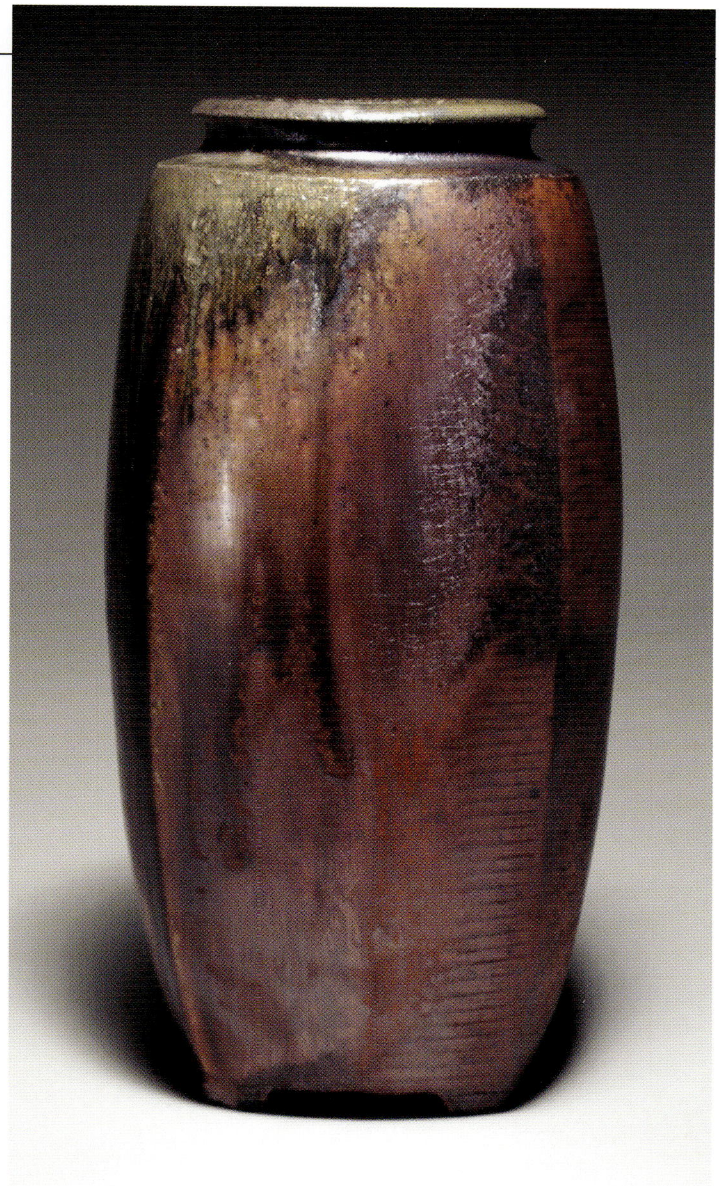

Peter Meyer, *Vase,* 2011. Wood-fired stoneware, wheel-thrown and altered, inlaid slips, 18" x 8" x 8". *Photo credit: Peter Meyer.*

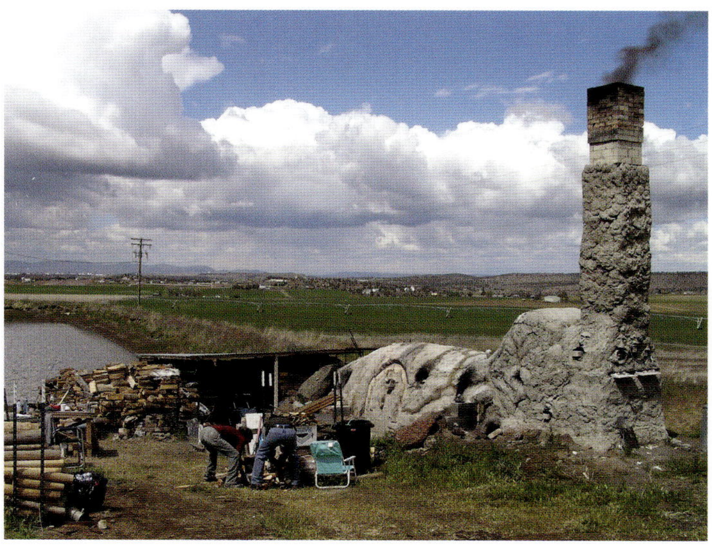

Stoking and stacking wood, 2011. *Photo credit: Peter Meyer.*

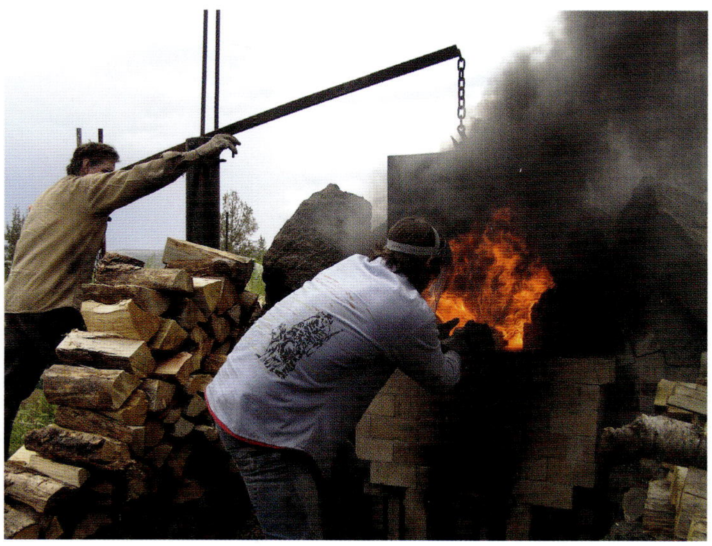

Ryan, front stoking, 2011. *Photo credit: Peter Meyer.*

SCOTT MEYER

Montevallo, Alabama
meyer@montevallo.edu

Artist's Statement

My earliest experiences with ceramics as a creative medium were at Penn State University in the late 1970s where Professor Kenneth Beittel imbued our studio activities with Japanese philosophy, aesthetics, and consequently an approach to materials and process. Ironically, though I was exposed to a couple wood-firing activities then, I felt I needed distance from both the process and its cultural specificity in order to locate my own voice in relation to the tradition. Thus began a personal search that found a kinship with (and was aided tremendously by) the *Primal Spirit* exhibition that toured the US in the early 1990s. It was by no means an exhibition focused on ceramics. But this was a show that made the "voice of materials" palpable and poignant with work that subtly revealed the essence of substances and process. This was a collection of first generation sculptors, often from multi-generational, traditional Japanese craft families. To me, these were heroes who had succeeded in shedding dogma and cultural specificity while becoming closer to the heart of the issues I had learned as a student. It would be another five years before all this came to bear completely in my work, but when it did, both clay and its traditional processes would take on new significance. For me, at the center of this new way of seeing would be wood-fire.

It is essential to my work that I came to this ancient process using this circuitous path. While I have enormous respect for all practitioners of the tradition, I am not interested using the wood kiln as a device with which to replicate objects from the past. Indeed, coming at its practice at age forty, I doubted I could supply any significant technical information that has eluded its 3000 years of collective insight. Rather, I am interested in using my forty-foot anagama as a gathering place for artists who may have no prior knowledge of wood-firing and, perhaps, no experience with ceramic art. I have found the questions generated in this environment to be provocative and invigorating conceptually, aesthetically, and technically. I believe that any solutions respond to the business of contemporary art and the specific charge of the academic climate in which the kiln is located. For me, it is of great importance that the support of the university encourages calculated risk and makes trial and error possible.

For the past five years, my sculpture has employed the crucible form as a unifying element in pieces composed of disparate components. While industry has long provided source material for my work, the crucible brings additional overlays referencing alchemy. The interface between matter, mind and spirit is alchemy and within the thick walls of the crucible, process is elevated to a kind of sacrament. For me, the anagama wood-kiln celebrates what are most timeless and elemental, qualities that are shared by the forms I now fire in it. Hopefully the results offer cohesion between concept and aesthetic not otherwise possible.

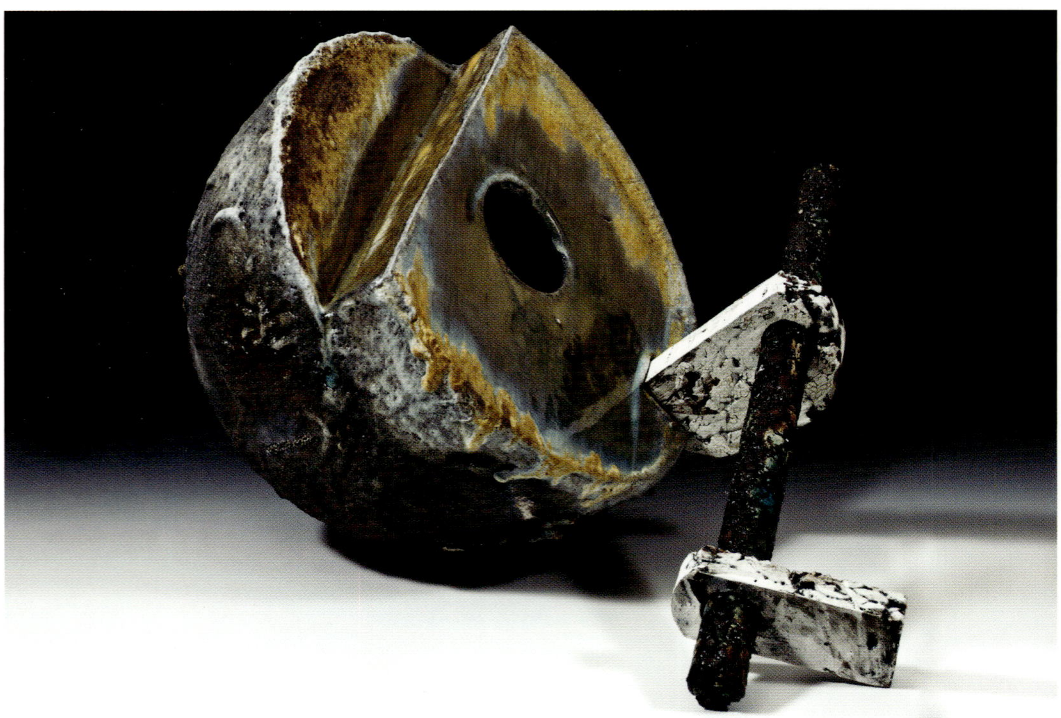

Scott Meyer, *Tipping Crucible 2*, 2012. Anagama-fired stoneware with gas-fired stoneware elements, 25" x 18" x 29". *Photo Credit: Hannah Ronan-Daniell.*

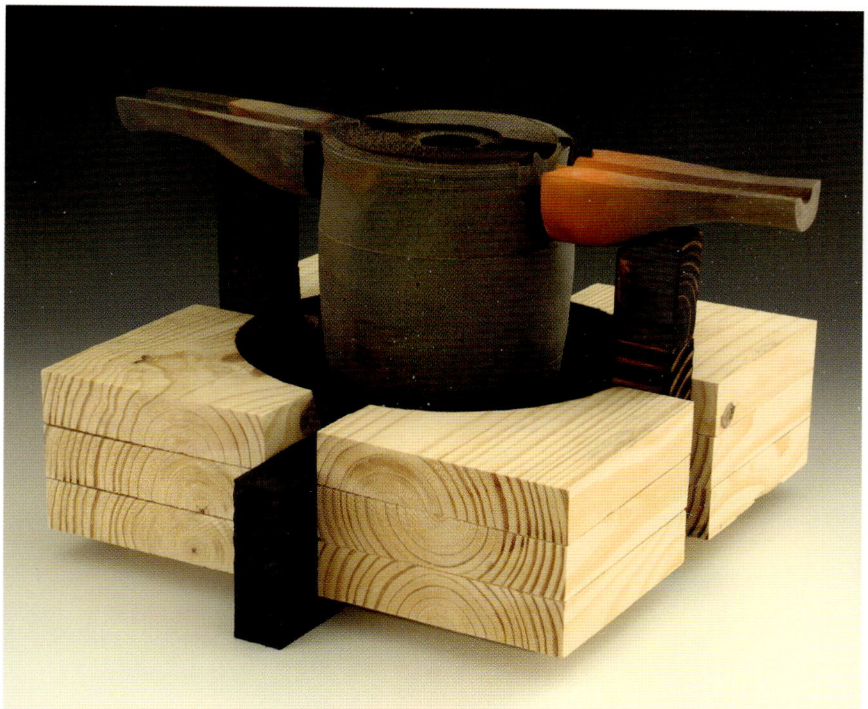

Scott Meyer, *Crucible Assemblage 1*, 2009. Anagama-fired Alabama native clay, charred pine, 10" x 12" x 15". *Photo Credit: Matt Duke.*

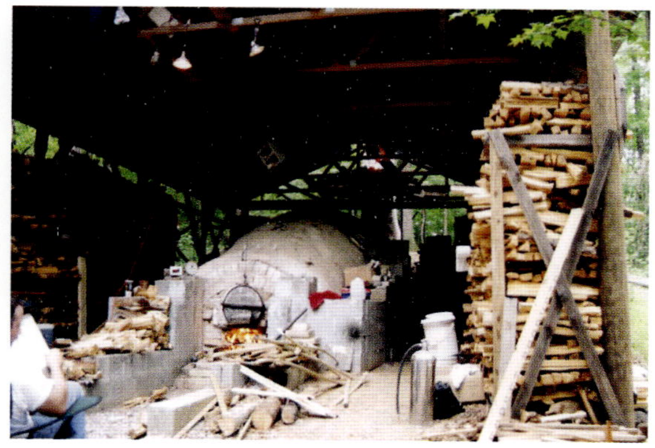

Scott Meyer, students stoke the University of Montevallo anagama firebox, day three of a firing, 2008. *Photo Credit: Scott Meyer.*

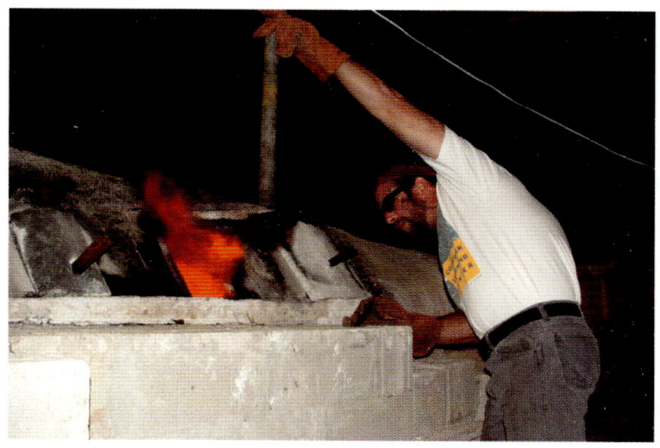

Scott Meyer, University of Montevallo anagama, day one of a firing, 2004. *Photo Credit: Scott Myer.*

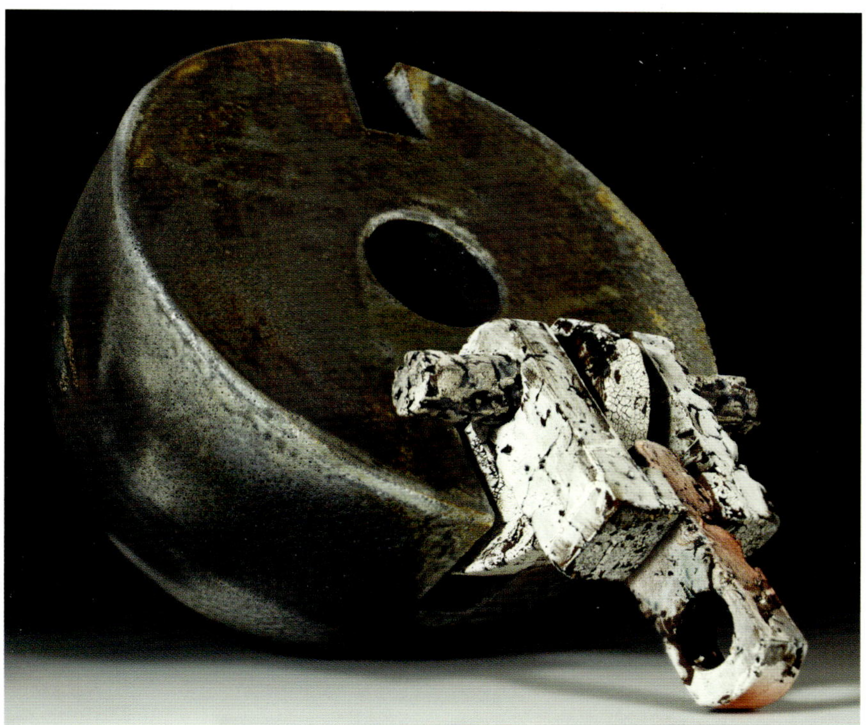

Scott Meyer, *Tipping Crucible 1*, 2012. Anagama-fired stoneware with gas-fired stoneware elements, 24" x 18" x 22". *Photo Credit: Hannah Ronan-Daniell.*

RON MEYERS

Athens, Georgia

Artist's Statement

I built my first wood at the age of seventy in 2004, with the help of my good friend Chuck Hindes. After having some of my work fired in other potters' kilns, "I saw the possibility of what firing with wood could do for my work after seeing the results of many of these types of pots being fired in wood kilns at schools where I have done workshops. The richness to the surface that the firing process can give and it seemed to enhance the marks and imagery." I wanted to start firing with wood because, **"**In contrast to the painted earthenware firing where I'm not asking the kiln to do anything but create enough heat to melt the glaze, I was seeking a process that was more of a risk; a process that I did not quite have control of and was unsure of the end product."

My firings are sixty hours with a reduction cooling segment that adds another six or seven hours to the process. The effects of a long firing and the reduction cooling leave the surfaces of the pots rich in color and texture, having a real sense of mystery to them. As I came to understand how to fire the kiln, the results have varied considerably from firing to firing, but I have always gotten something remarkable. I use clays that are rich in iron that give me a palette of colors that are darker and range from iron reds to purples and oranges. I have also done several earthenware firings in the anagama that have produced some very nice pots as well.

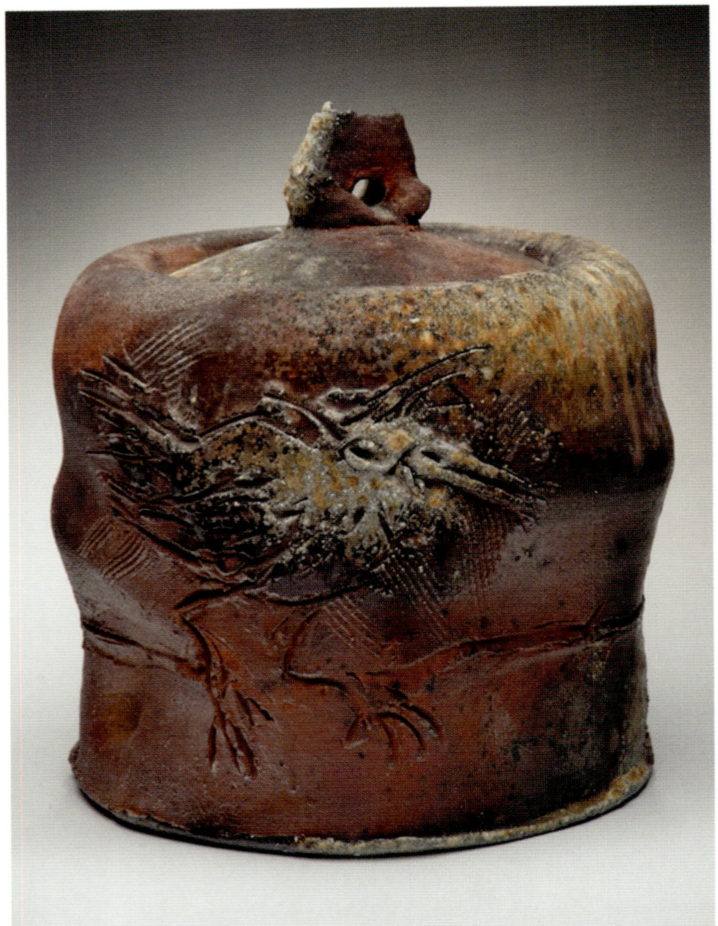

Ron Meyers, *Covered Jar*, 2010. Wood-fire stoneware, wheel-thrown, 10" x 8". *Photo credit: Walker Montgomery.*

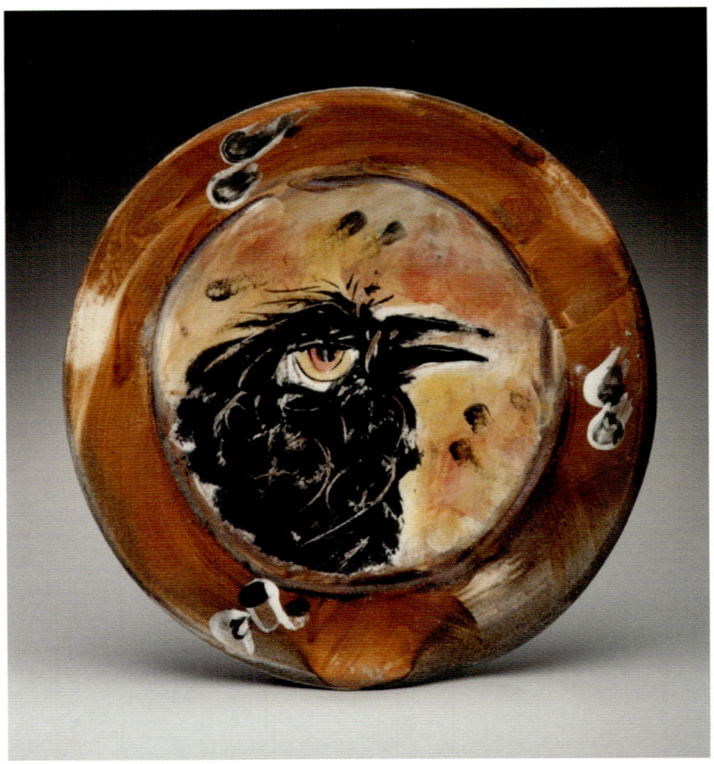

Ron Meyers, *Bowl*, 2009. Wood-fired earthenware, wheel-thrown, 3" h x 9" w. *Photo credit: Walker Montgomery.*

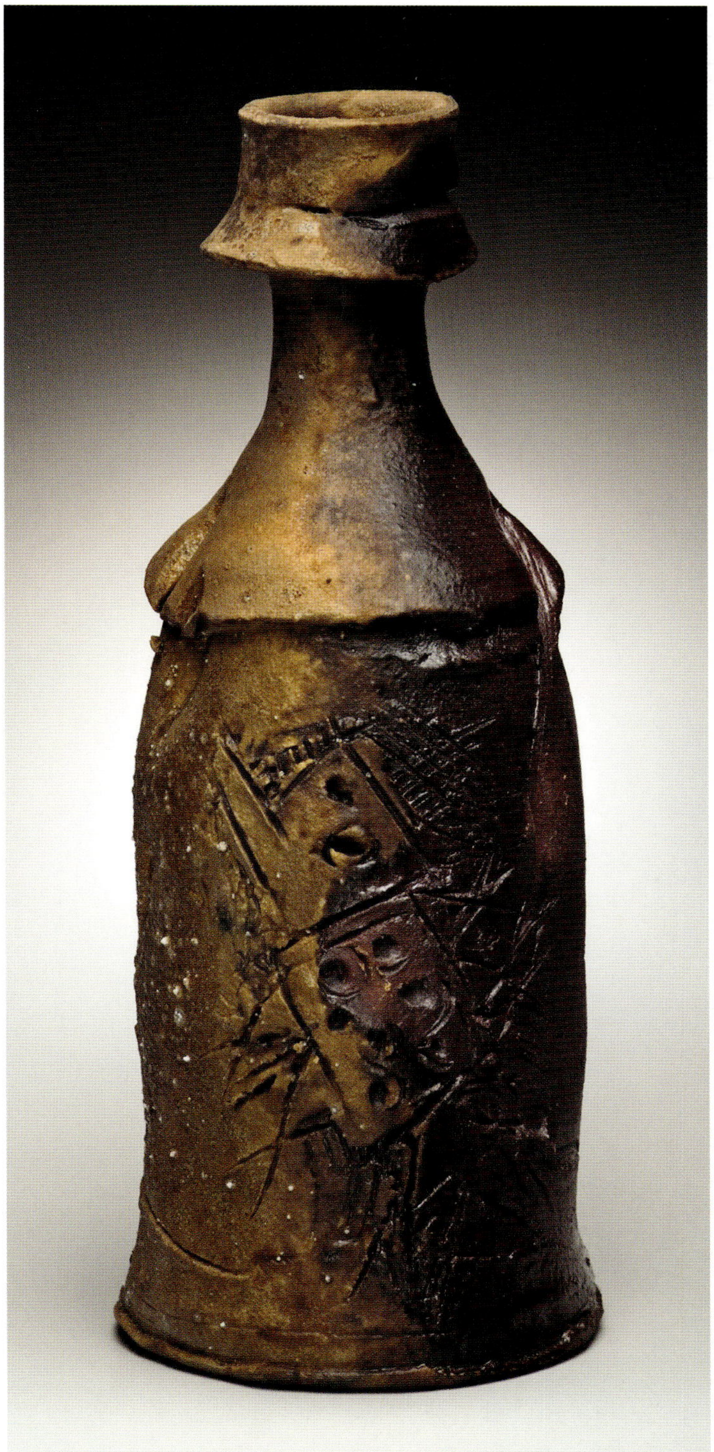

Ron Meyers, *Fish Bottle*, 2010. Wood-fired stoneware, wheel-thrown, 14" x 5". *Photo credit: Walker Montgomery.*

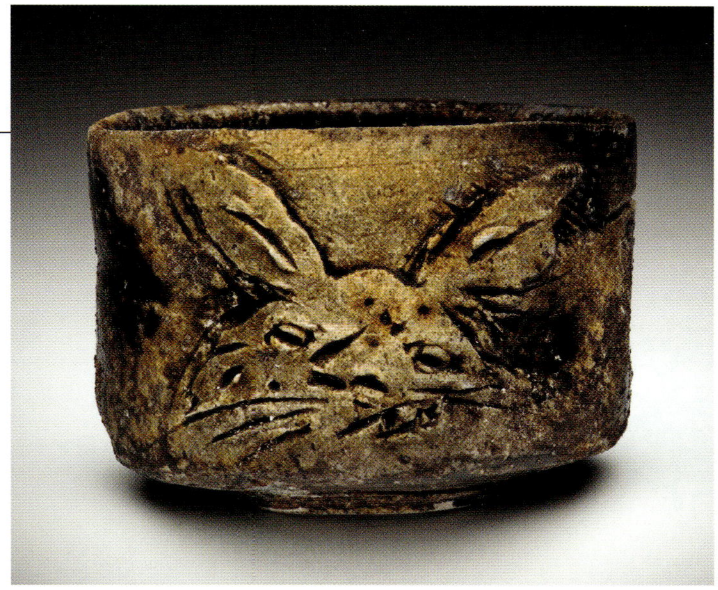

Ron Meyers, *Tea Bowl*, 2010. Wood-fired stoneware, 4" x 4". *Photo credit: Walker Montgomery.*

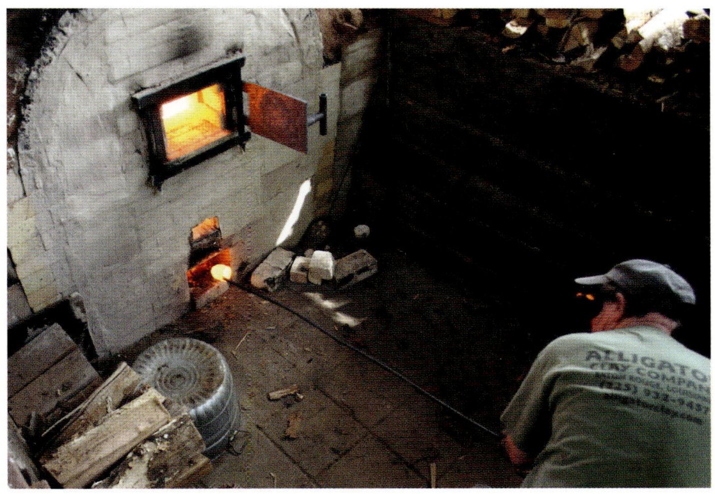

Ron pulling a hikidashi cup out of kiln, 2010. Hikidashi firing technique, Athens, Georgia. *Photo credit: Stephen Driver.*

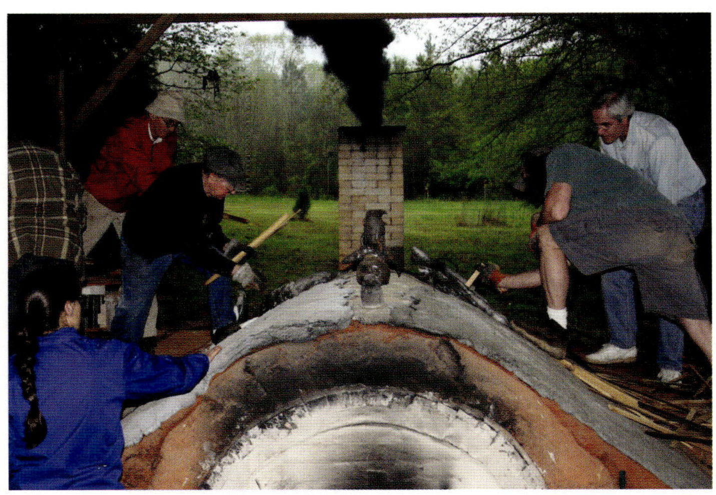

Ron and company side stocking, 2008. Athens, Georgia. *Photo credit: Stephen Drivers.*

TONY MOORE

Cold Spring, New York
TonyMoore@optonline.net, www.TonyMooreArt.com

Artist's Statement

My work is concerned with the relationship of humanity and nature. I conceive of an expanded concept of "Nature" as embodying all existence, both the seen and unseen, socio-political events, daily occurrences, as well as private intuitions that are made concrete through creative action. My objects are places of remembrance where multiplicities of associations take place. Their function is that of contemplation and the transmission of the energy contained within them.

Through additive and subtractive manipulation, the sculptures evolve to where a combination of vision and intent project a final resolution. This process is one of discovery where the known (history) and intuition come into play. One enters into the dialogue of energies brought into being within a specific moment of time. Most recently these have been concerned with issues of the human condition. Often the resultant forms are discovered as though they were pre-existent, and my own actions are to simply reveal their inherent truth.

Releasing the sculpture to the fire of the wood-burning kiln is yet one more step in the transformation of the object. It has taken on the transmission of elemental forces, which radiate to the viewer all that it has become and may further become through time. My concern within this realm of possibilities is where the object's primary function is the expansion of awareness.

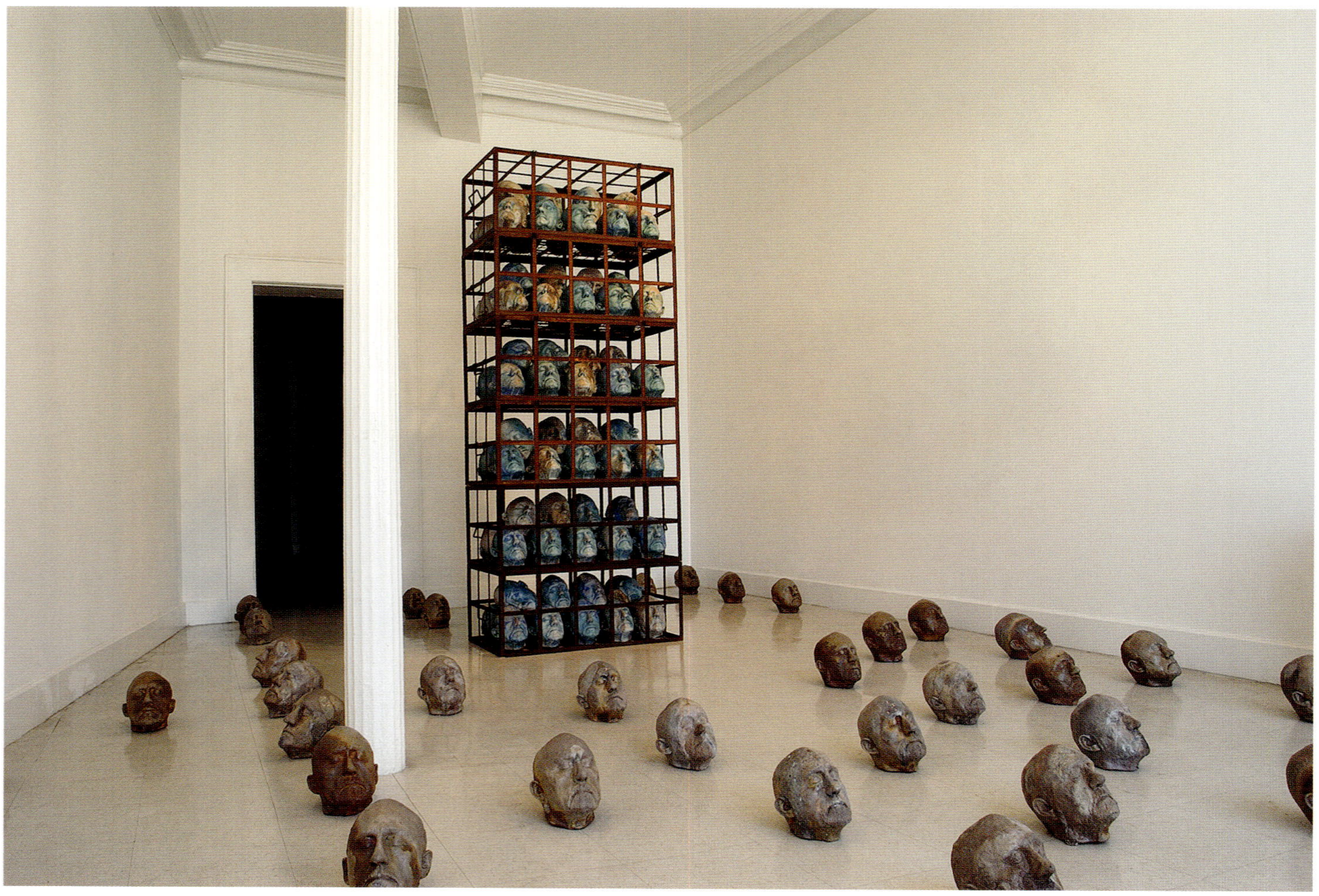

Tony Moore, *Who Knows Why?*, 2006. 150 body-cast human heads, ceramic, steel, barium glaze, wood-fired 5 days, 84" x 300 sq.ft., size variable. *Photo credit: Tony Moore.*
An installation of 150 body-cast human heads, each press molded in coarse sculpture clay. The stacked, vertical, rusted cages are custom made and contain barium glazed heads, fired in the Noborigama, while the heads on the floor are fired in the anagama in a five-day firing to cone 10–12.

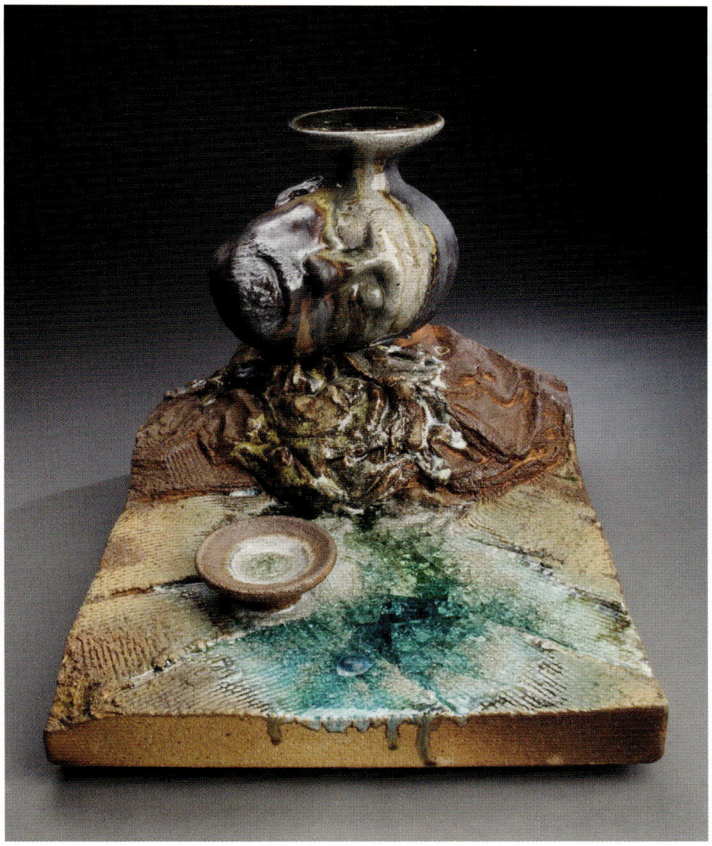

Tony Moore, *Earth and Sky*, 2010. Ceramic, oxide, glass, wood-fired 4 days, 14" x 15.5" x 17.5". *Photo credit: Howard Goodman.* Assembled sculpture using slab, thrown and press molding techniques, copper glaze, iron oxide, and melted glass frit. Fired in the Noborigama for 4 days to cone 10.

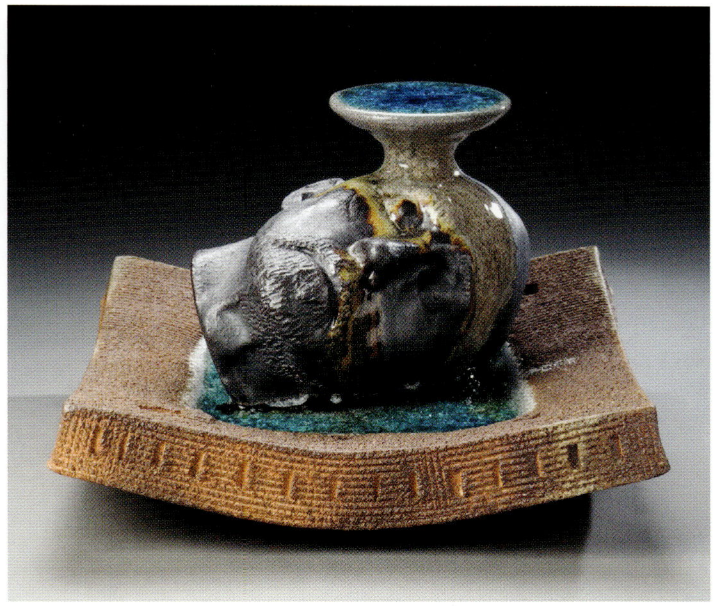

Tony Moore, *Aperture* 2009. Ceramic, glass, wood-fired 4 days, 9.5" x 14.5" x 14.5". *Photo credit: Howard Goodman.* Assembled sculpture using slab, thrown and press molding techniques, iron oxide, and melted glass frit. Fired in the Noborigama for 4 days to cone 10.

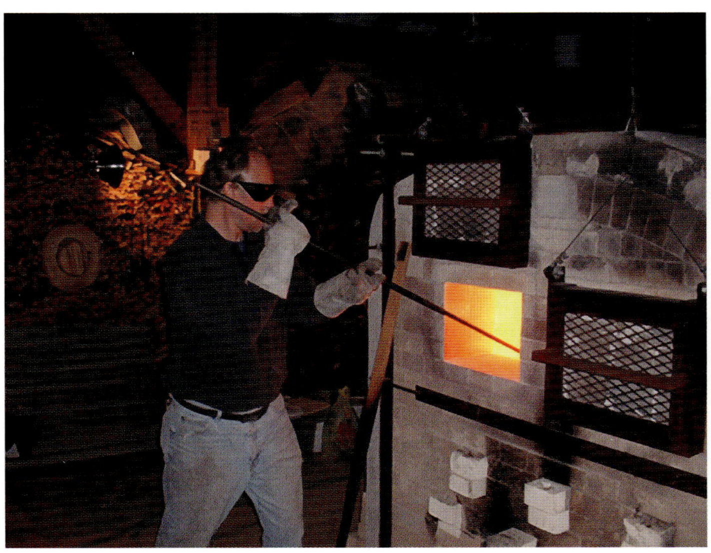

Tony Moore firing his anagama-Noborigama Kiln, using a long steel poker to knock down coals. *Photo credit: Tom Loggia.*

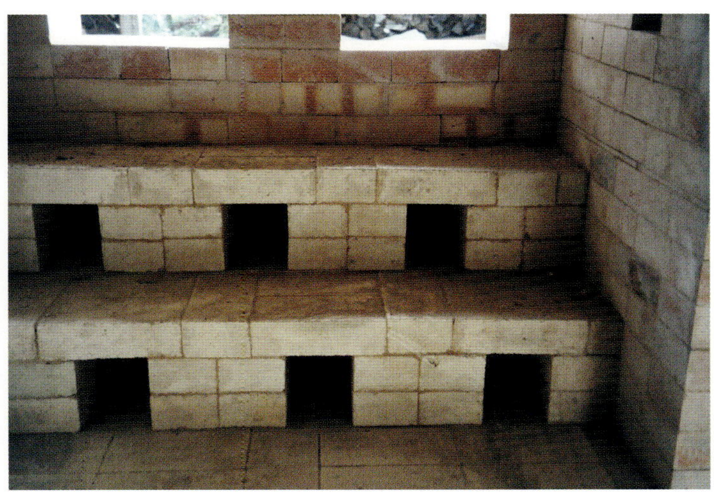

Interior of the anagama firebox,step-grate with air ports under each step. *Photo credit: Tony Moore.*

KRISTIN MÜLLER

Dingmans Ferry, Pennsylvania
mullerclay@sbcglobal.net, www.kristinmuller.net

Artist's Statement

A mysterious creative synergy is at play when I work with clay. While there are some days when every part of the process is deliberate, the most productive are the days in which I wonder, and wander off and lose myself in the process of creating. On these occasions thoughts are channeled through my hands, effortlessly, free from technical constraints. The interaction between self, tools, clay, and environment becomes an expression of a larger intellectual and physical need, beyond my preconceived choices, and beyond the sum of all the parts. What emerges is a whole that brings the tacit into tangible focus.

A similar experience of synergy—of another order—attracts me to wood-firing. The complexities of combustion, the predictable and unpredictable color range and surface qualities are why I choose to wood-fire. For over twenty years I have studied the language of clay and firing to gain sufficient fluency and confidence to communicate complex ideas clearly. The processes required to manipulate clay and wood-fire with an anagama kiln, require a physicality that helps release the intellect. I am passionately drawn to these processes, which interest me as much as the resulting works.

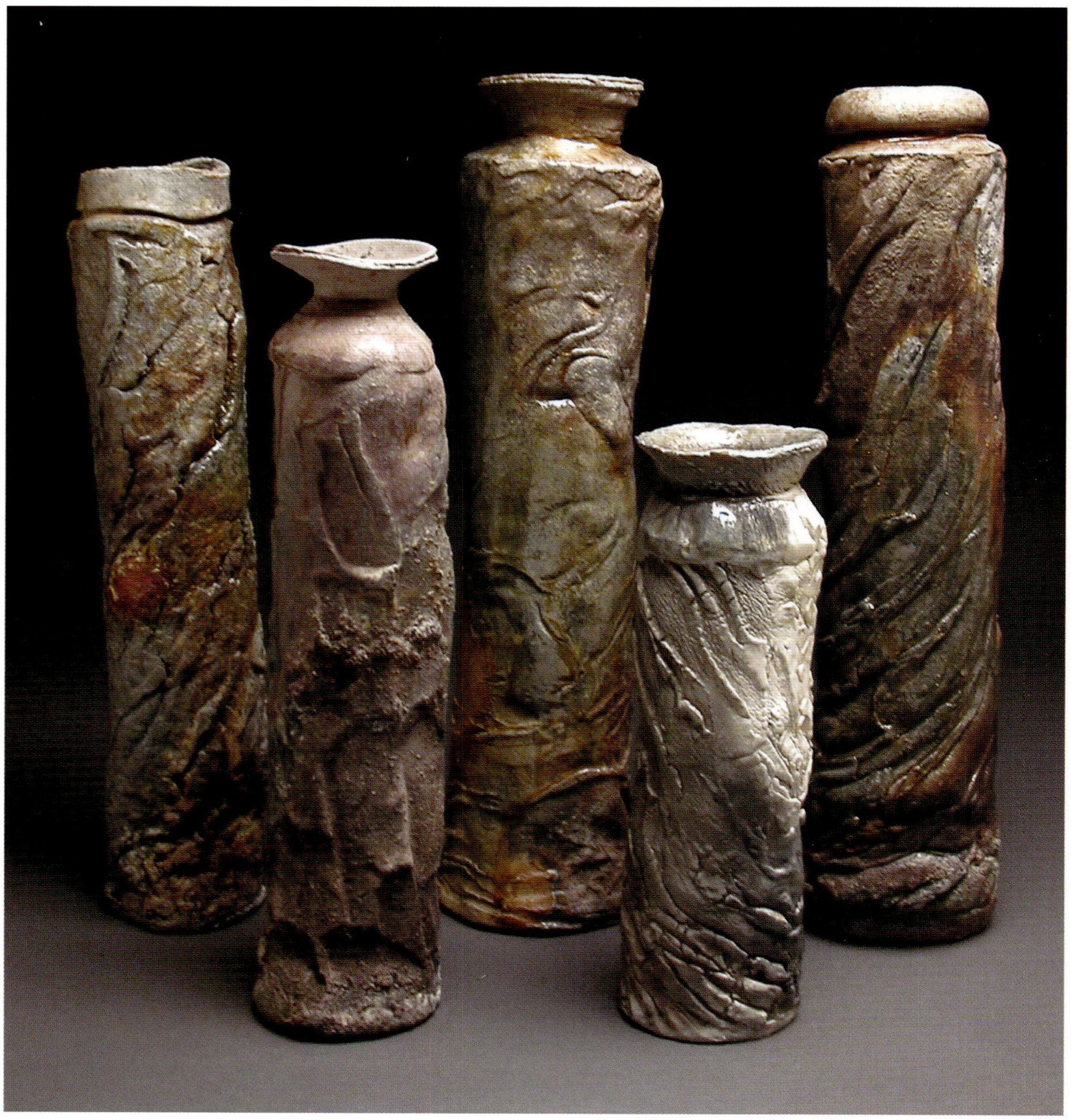

Kristin Müller, *Figurative Landscapes*. Hand-built and wood-fired stoneware clay, pieces vary in sizes. *Photo credit: Kristin Müller.*

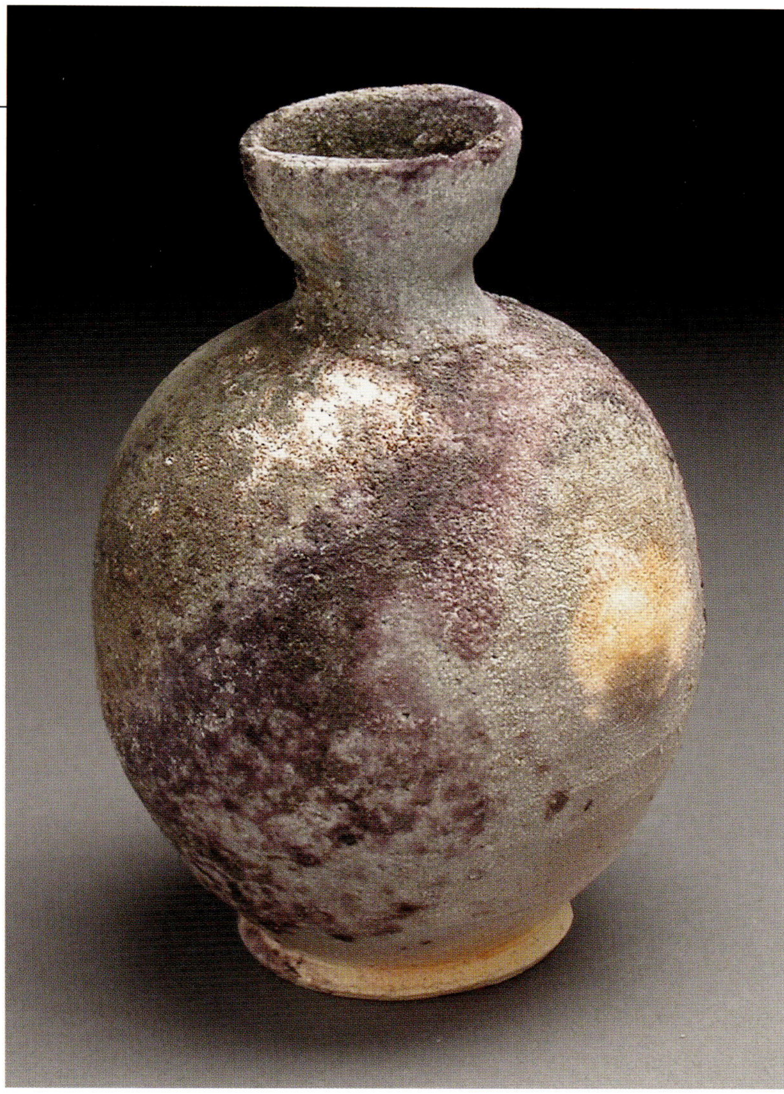

Kristin Müller, Sake *Bottle*, wheel-thrown with natural ash glaze, 5.5" x 3.5" x 1.75". *Photo credit: Kristin Müller.*

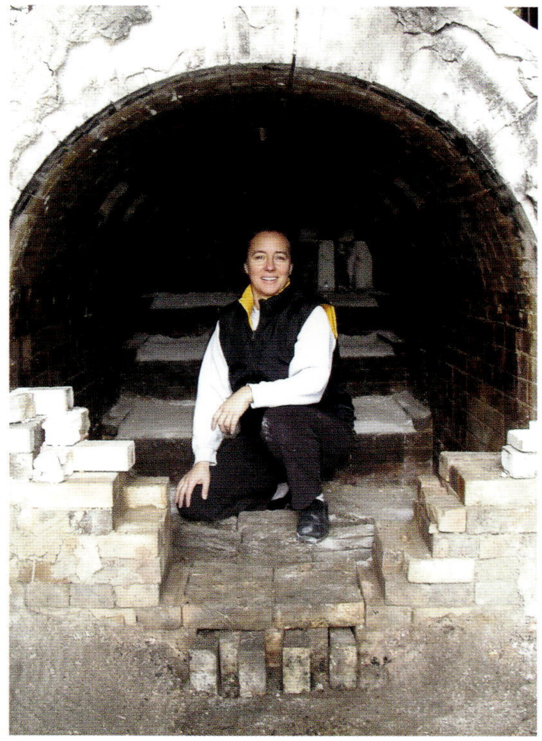

Kristin Müller inside anagama kiln. *Photo credit: Iain Highet.*

Kristin Müller, *Chiyo Hikidashi*, wood-fired stoneware, 18.5" x 6" x 2.5". *Photo credit: Douglas Foulke.*

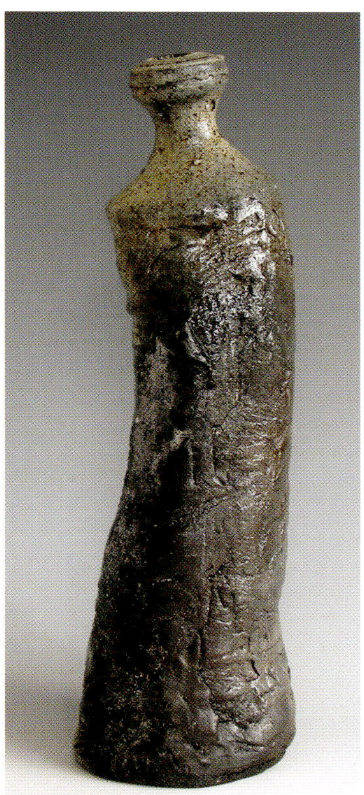

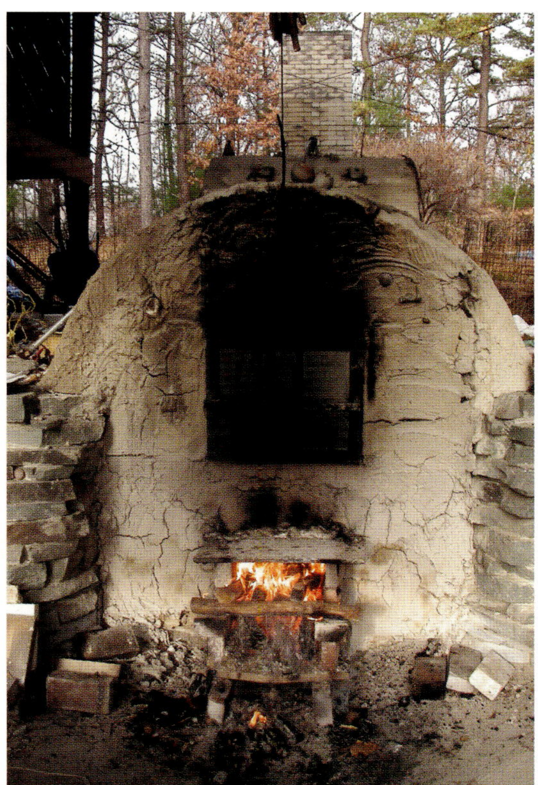

Kiln firing. *Photo credit: Kristin Müller.*

J. DANIEL MURPHY

Logan, Utah
dan.murphy@usu.edu, www.art.usu.edu

Artist's Statement

My ceramic vessels are created swiftly and directly on a slow spinning potter's wheel. I usually work in series, developing one body of work at a time. My goal is to make gestural vessels that reflect my presence in the finished form. This results in families of pots that are inevitably related; yet each piece stands as a unique one-of-a-kind vessel. After the pieces are made most are fired without applied glazes to stoneware temperatures in wood-burning kilns. Colors and textures on the ceramics result from the interaction of wood, fire, and clay. My hope is that each successive generation produces a better piece. I feel a connection to contemporary as well as ancient ceramics, and strive to create objects that will withstand the test of time.

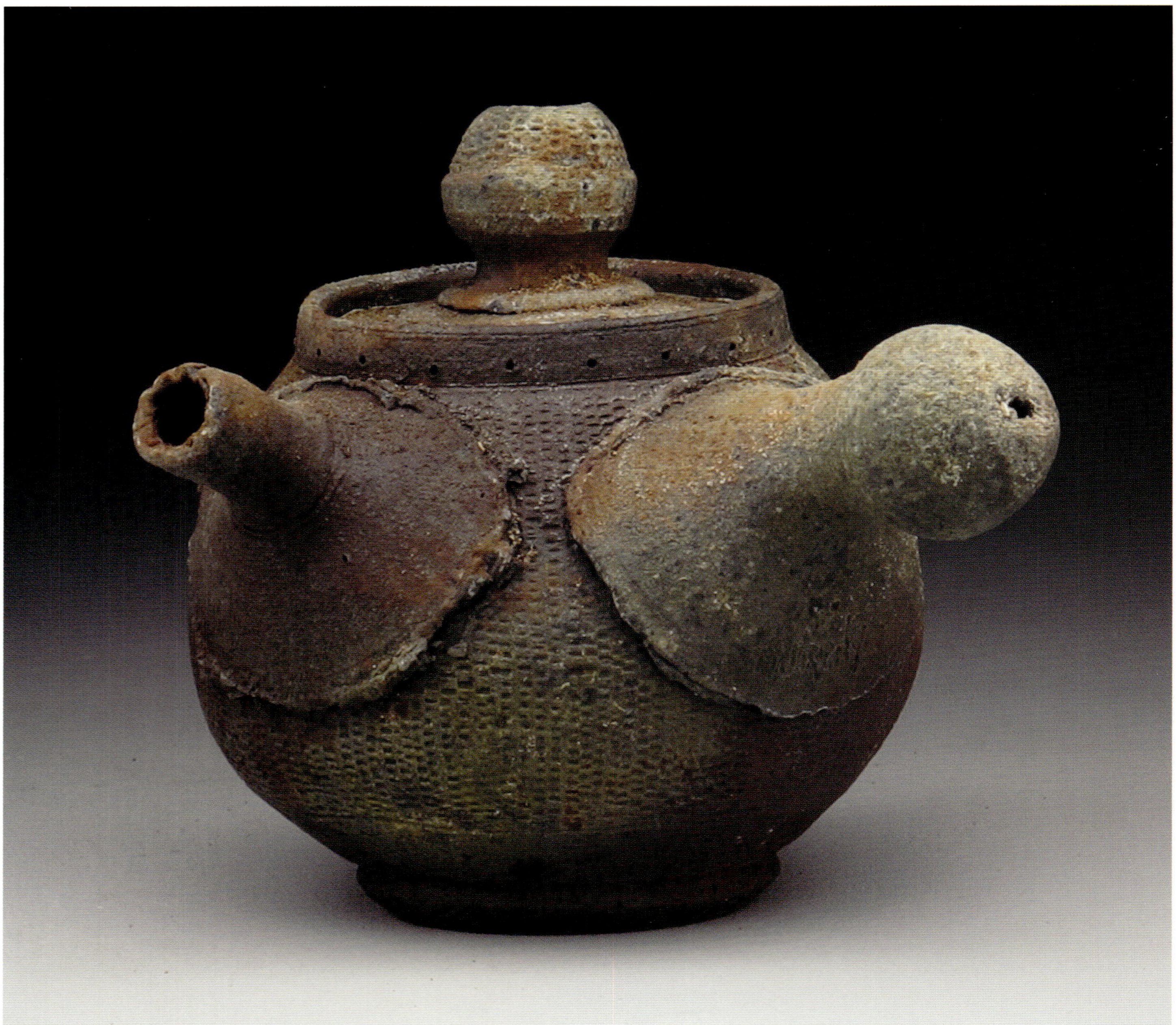

J. Daniel Murphy, *Tea Pot with back handle,* 2012. Wood-fired iron rich clay, natural ash glaze, handbuilt and wheel-thrown parts, 9" x 6" x 6". *Photo Credit: J. Daniel Murphy.*

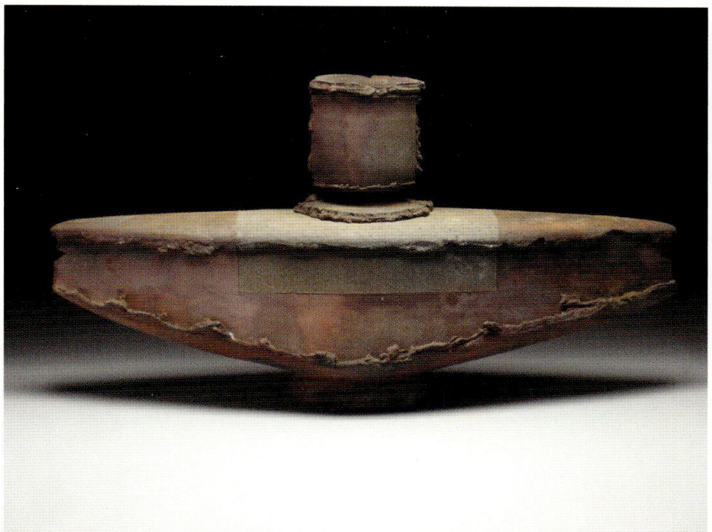

J. Daniel Murphy, *Horizontal Bottle*, 2011. Wood-fired iron rich clay, natural ash glaze, handbuilt and wheel-thrown parts, 36" x 12" x 18". *Photo Credit: J. Daniel Murphy.*

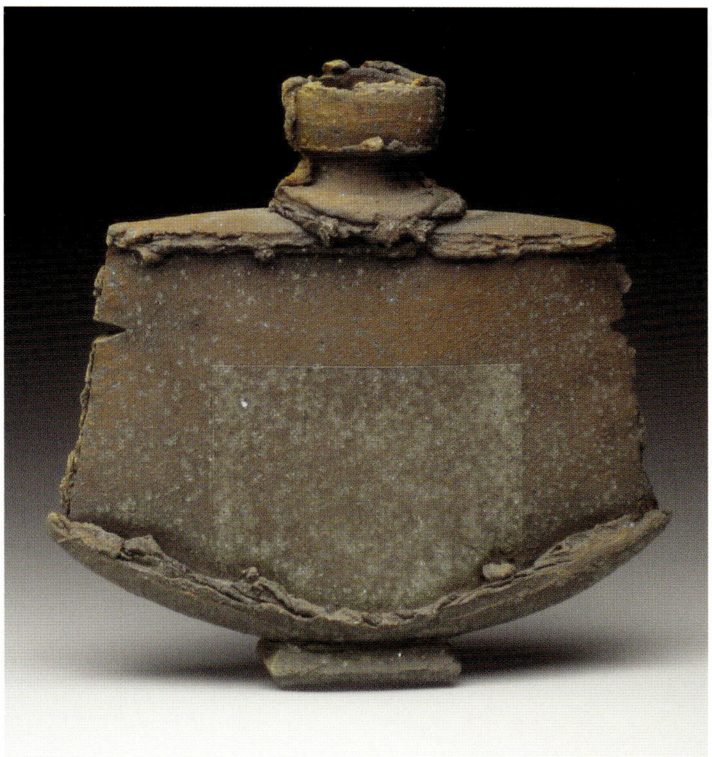

J. Daniel Murphy, *Bottle*, 2012. Wood-fired iron rich clay, natural ash glaze, handbuilt and wheel-thrown parts, 12" x 12" x 3.5". *Photo Credit: J. Daniel Murphy.*

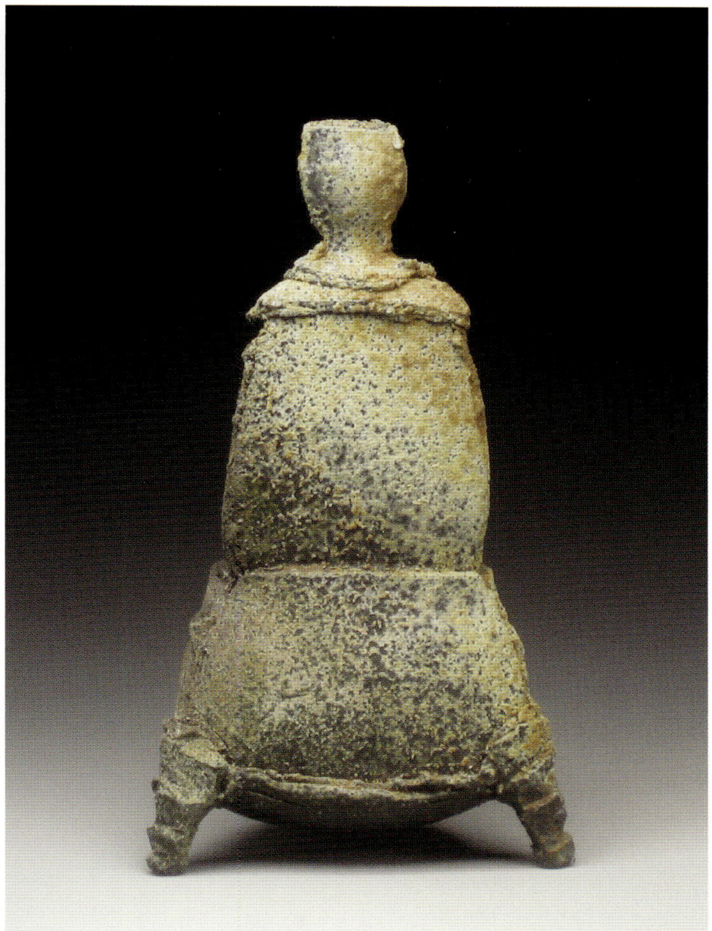

J. Daniel Murphy, *Triangle Bottle*, 2012. Wood-fired iron rich clay, natural ash glaze, handbuilt and wheel-thrown parts, 12" x 6" x 6". *Photo Credit: J. Daniel Murphy.*

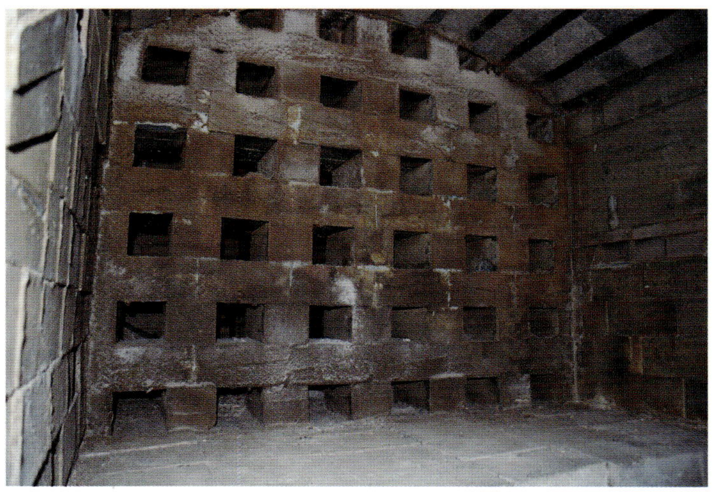

Double-wide kiln, Utah State University, 2009. Interior view: Exit Flue, 54" x 48". *Photo Credit: J. Daniel Murphy.*

HARRY NAKAMOTO

Columbia, California
hnakamoto@sbcglobal.net

Artist's Statement

My work is generally wheel-thrown and of traditional forms. I avoid classifying them as decorative or functional in that the state of being decorative is in itself a function, and, as I have gleaned from the Japanese pottery culture, the decorative aspect of a functional ware is an equally valued characteristic.

My work is fired in a relatively small anagama of my own design. It is constructed of soft insulating bricks backed with ceramic fiber and an outer protective layer of stucco. The kiln is fired for seventeen hours and to cone 13 each of the four days needed to complete a firing cycle.

Why wood-fire? Wood-firing brings to my mind a fanciful notion of kinship with ancient potters and the natural world. Years in growing, a tree comes to me packed with energy to heat my kiln, and minerals drawn through its roots in just the right proportions to glaze and color the pots formed with clay from that same forest. In the kiln, the wood explodes and spews its contents, the energy grows into a long, lazy, and serpentine flame, carrying and depositing the flux and the colorants as it meanders about, and swirls over and around the many shapes as gravity weighs down the cascading sheets of naturally formed glaze. Finally, the finished pots are before me, cooled but richly warm with magnificent colors, patterns, and texture, a scene perhaps not so unlike the earliest stoneware firing. Wood-firing is fun.

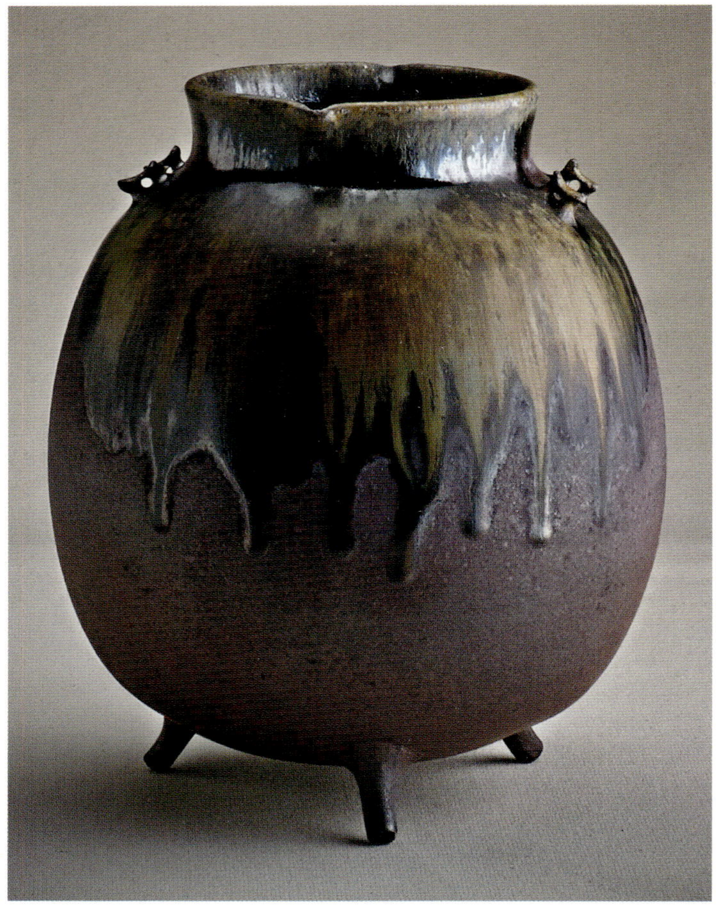

Harry Nakamoto, *Wheel-thrown tripod,* 2010. Wood-fired stoneware, anagama fired to cone 13, 8" x 6.5" x 6.5". *Photo credit: Robert White.*

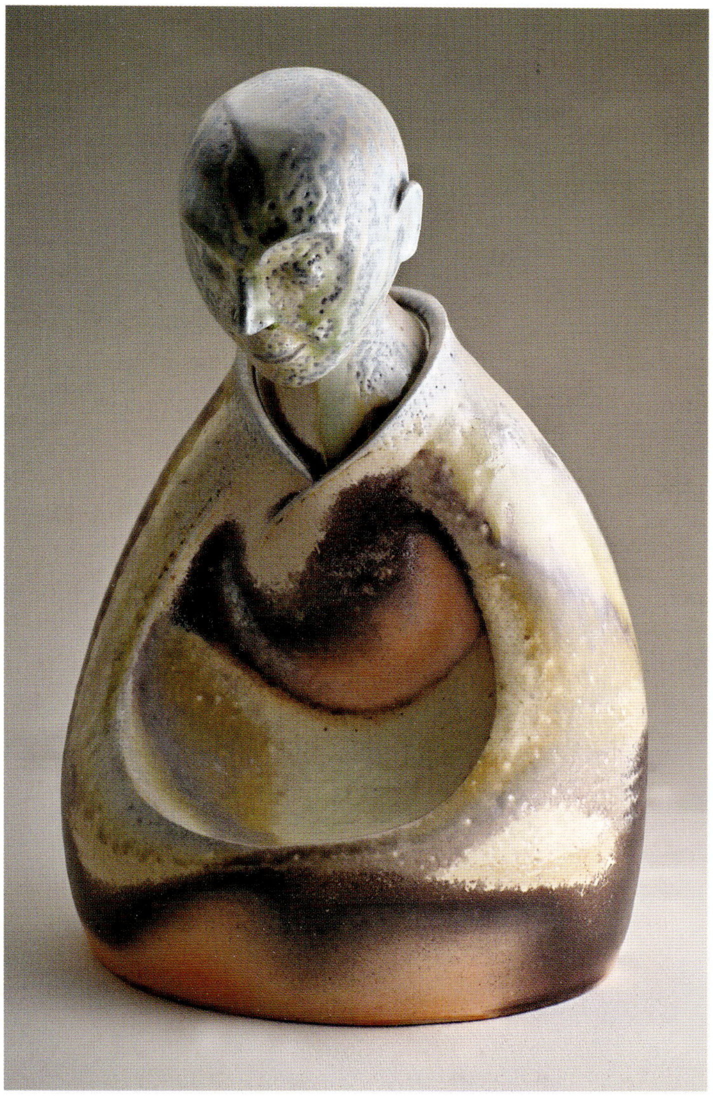

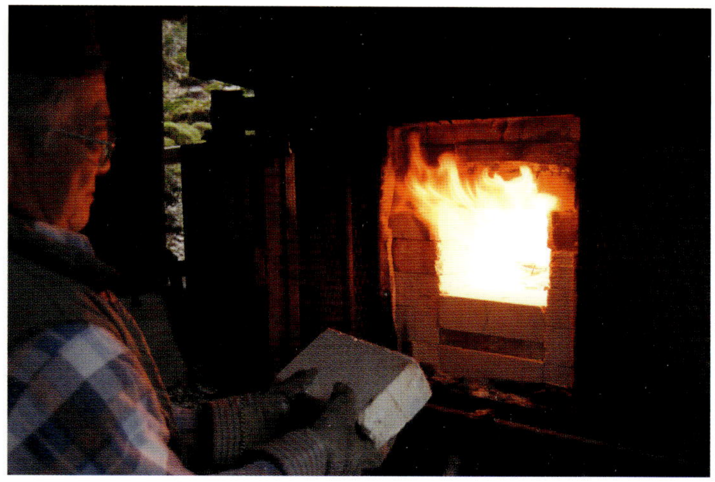

Anagama front stoke port, "pop" bricking front (main) stoke port to begin shutdown procedure. *Photo credit: Marta Magistrali.*

Harry Nakamoto, *Monk*, 2011. Wood-fired stoneware, composite of three wheel-thrown forms, anagama fired to cone 13, 16" x 9.5" x 10". *Photo credit: Robert White.*

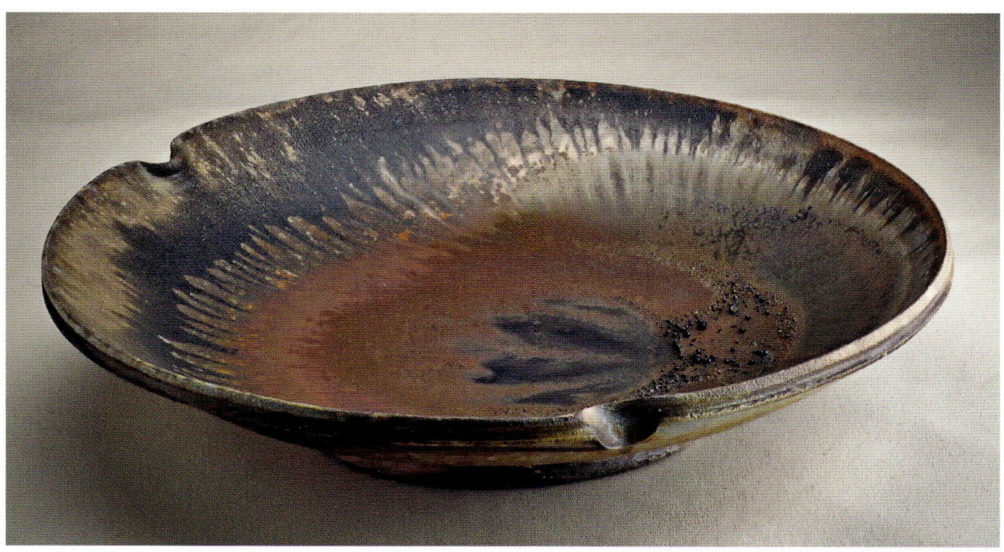

Harry Nakamoto, *Wheel-thrown platter*, 2011. Wood-fired stoneware, anagama fired to cone 13, 31.5" x 17" x 17". *Photo credit: Robert White.*

TED NEAL

Yorktown, Indiana
tneal2@bsu.edu

Artist's Statement

The current sociopolitical and economic environment has led me toward exploring consumption and the use of natural resources as themes in my work. I choose the ceramic vessel as a means to marry industrial form and contemporary iconographic symbols with my thoughts about our consumer culture. I enjoy the interplay of roles between the utilitarian form as an object for use and that of a conceptual vehicle for expression about global consumption. I use the objects as subtle reminders of the true cost of the things that we use.

I fire the majority of my work in a wood-fired train kiln in a reduction cool atmosphere. I use this technique, in combination with a high iron clay body, to mimic the color, texture, and surfaces that industrial objects acquire as they age.

All of the pieces represented are thrown and altered high iron stoneware. Once the peak temperature is reached, the kiln is shut and sealed with the exception of one damper (open a fraction of an inch) and one of the high secondary air holes. Very small pieces of kindling wood are added in the firebox to maintain a small "lick" of flame around the damper as the kiln cools. When the flame retracts from view, another stick is added. This is maintained until the kiln cools to 1450° F. The cooling process adds several hours to the firing. This has the effect of preventing re-oxidation in the cooling cycle until the oxides, primarily iron, can no longer change state, promoting the development of beautiful dark tones and varied surfaces on the clay bodies.

After the work has been fired, I fabricate and then patina the components that will accompany the pieces. In some cases clay is used to make a mold for the bases. I use these molds to cast stained concrete into which I set the steel parts.

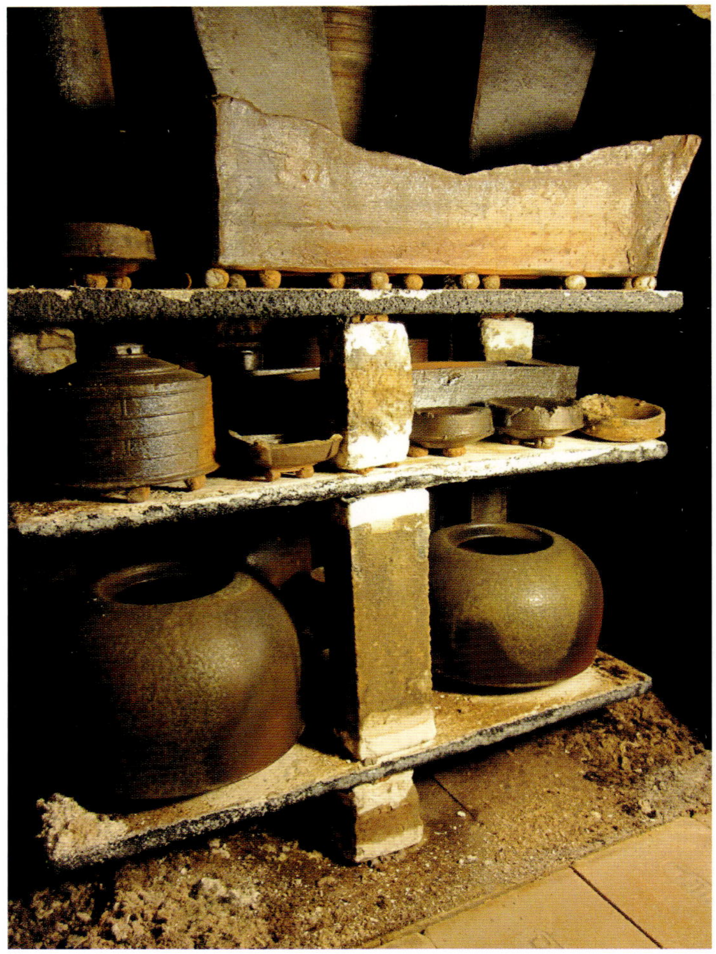

Ball State University Train Kiln, reduction cool surfaces, 2011.
Photo credit: Ted Neal.

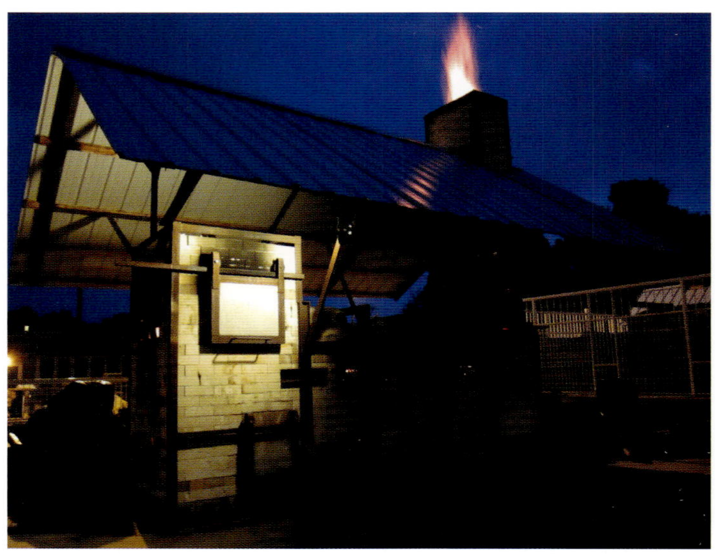

Ball State University Train Kiln, peak of firing in evening, 2011.
Photo credit: Ted Neal.

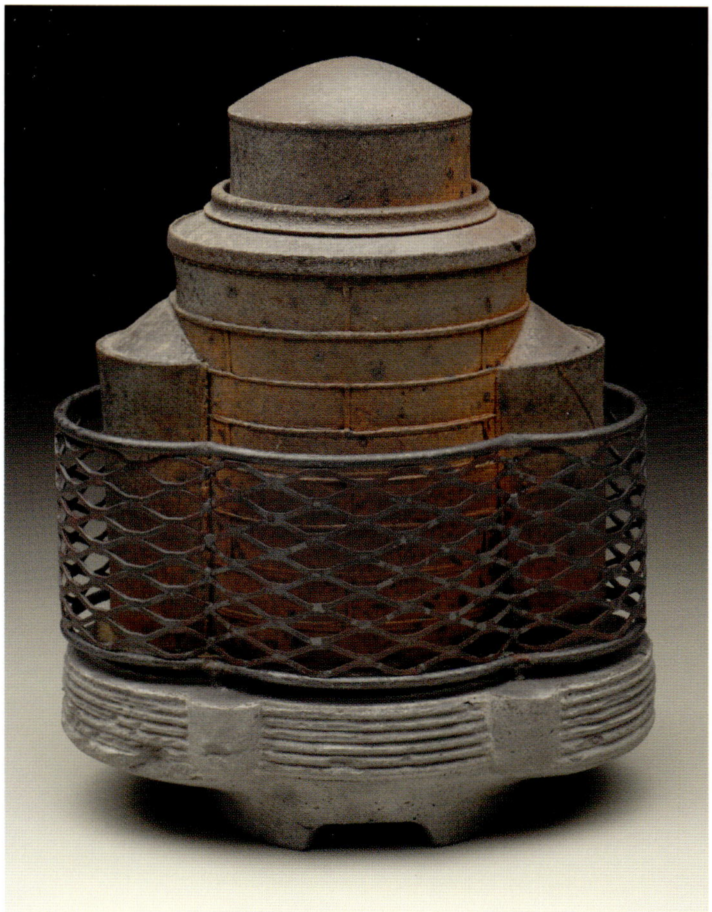

Ted Neal, *Guarded Supply,* 2012. Wheel-thrown and altered, high iron stoneware, fabricated steel and concrete, wood-fired to cone 9 and reduction cooled, 11" x 9" x 7". *Photo Credit: Serena Nancarrow.*

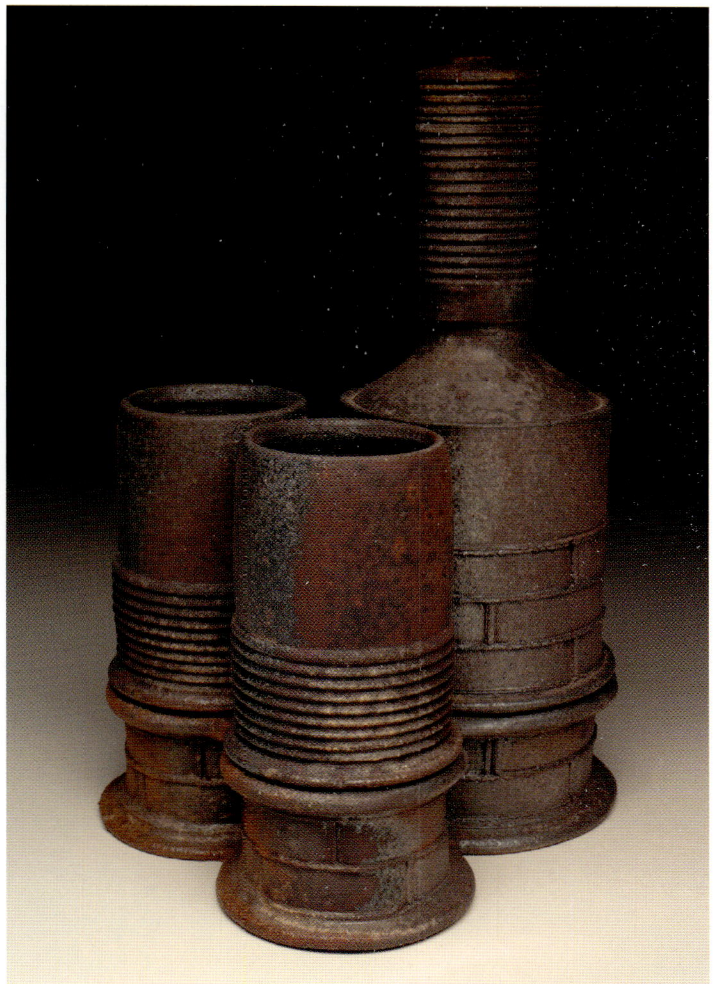

Ted Neal, *Bottle Set,* 2012. Wheel-thrown and altered, high iron stoneware, wood-fired to cone 9 and reduction cooled, 12" x 8" x 7". *Photo Credit: Serena Nancarrow.*

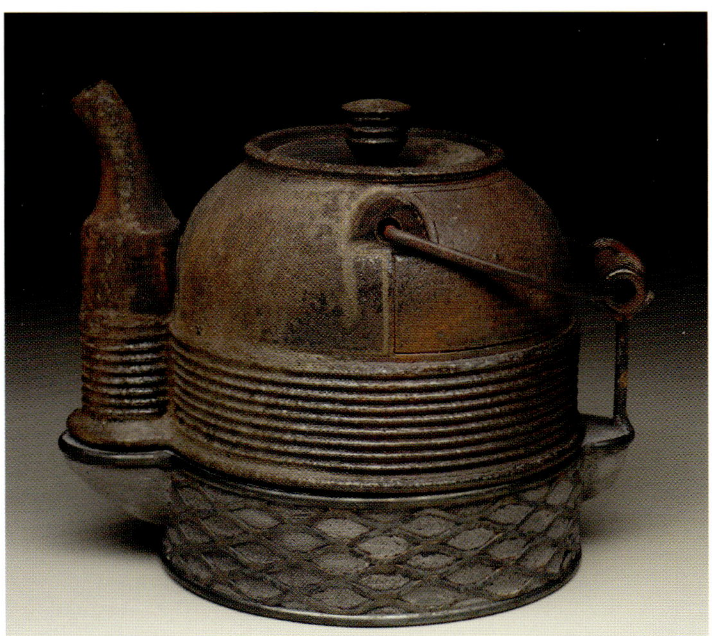

Ted Neal, *Teapot,* 2012. High iron stoneware, fabricated steel and concrete, wood-fired to cone 9 and reduction cooled, 9" x 7" x 8". *Photo Credit: Serena Nancarrow.*

DREW NICKLAS

Seattle, Washington
drewnicklas@gmail.com, www.drewnicklas.com

Artist's Statement

It is difficult to rationalize the wood-firing process, to explain why I do it and why it fascinates me. I feel that most of the easy answers are just that, easy answers. "I like fire, I like to own the firing process, and I like loosely thrown crusty pots." For me, I think the reason I wood-fire my work is because it is hard, it is humbling, I tend to be contrary in nature, and I am not quite sure why I like it. When I describe wood-firing I seem to muddle through personal contradictions. I like the variation in each piece, though I don't particularly like orange flashing. I like the physical work of wood prep and firing, but I am glad I don't have to spend more than two days firing. I think I am really interested in the balance between the process, the surface, and the humbling notion of not knowing all the answers. I think wood-fire potters have to come to terms with an intriguing dialectic, which balances professional practicality with a product whose value is derived by an almost reckless impracticality.

Currently, it is a good time to fire with wood because many of the previous conventions are breaking down and many of the broad generalizations or being overcome. There are new kilns being built and new work is being made. For the same reasons, it is a good time to be a potter. The work of making pottery provides something tangible, something simple, and something artful and delicious. Pottery is a celebration of ideas, work, creativity and community. I think wood-firing pottery is very much in line with this philosophy. It can remind us that it is our duty to own our creativity and invest it in our everyday practice.

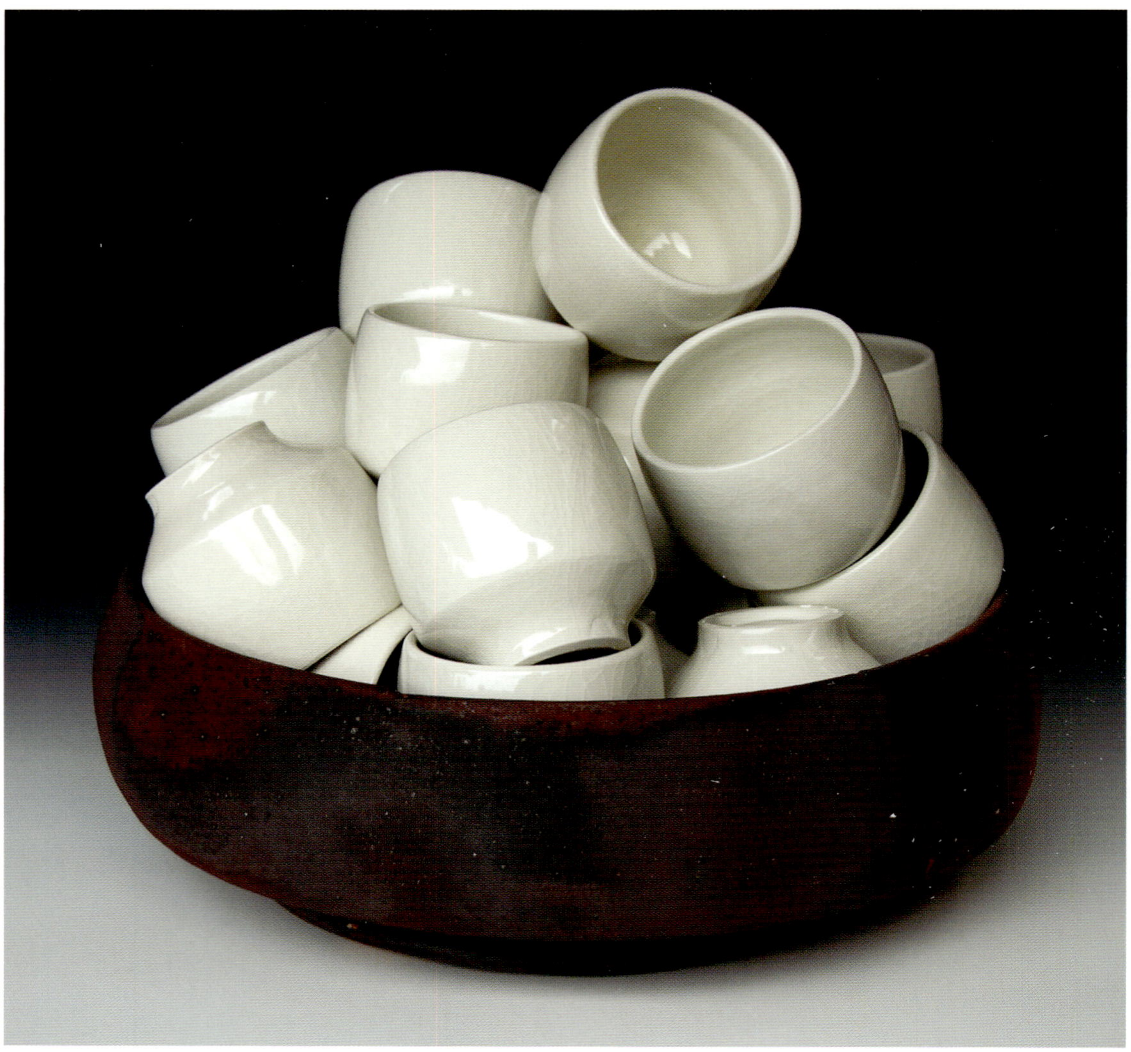

Drew Nicklas, *Platter with Cups*. Wheel-thrown, stoneware, 12" x 17". *Photo Credit: Drew Nicklas.*

I recently built a train style kiln because of the variety of work it can produce, the efficiency of firing, and the ability to teach wood-firing with it. The kiln is wonderfully responsive, easy to load, and can be fired in a couple of days. This makes it quite feasible to fire with normal people who have day jobs and are not interested in spending their hard earned vacation time splitting wood and firing a kiln. Because the train kiln can be fired relatively quickly, the frequent turn around on kiln loads greatly aids experiments, plenty of failures, and subsequent and new ideas. It provides a healthy dose of practical impracticality!

Enjoy the Firing: Enjoy the pottery: Hoorah!

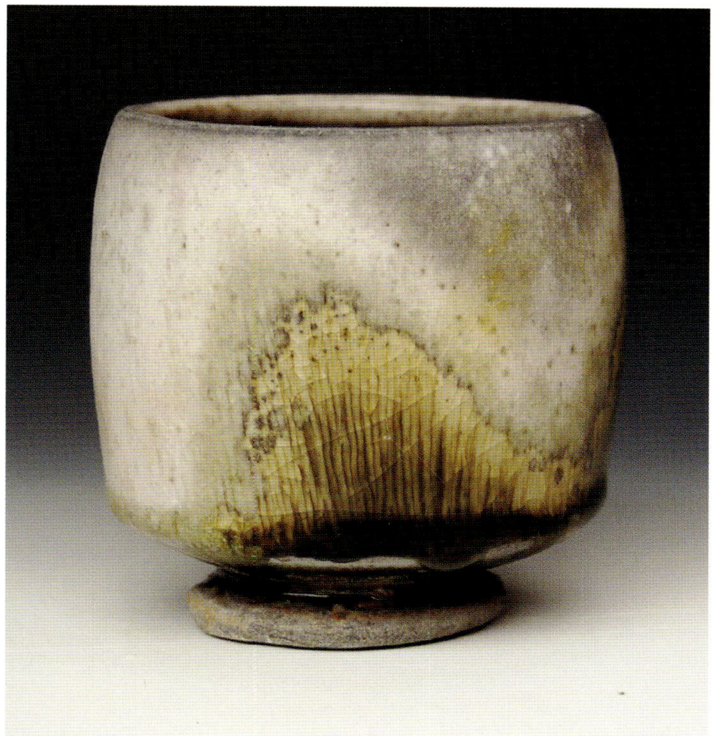

Drew Nicklas, *Cup*. Wheel-thrown, stoneware, 3.5" x 4". *Photo Credit: Drew Nicklas.*

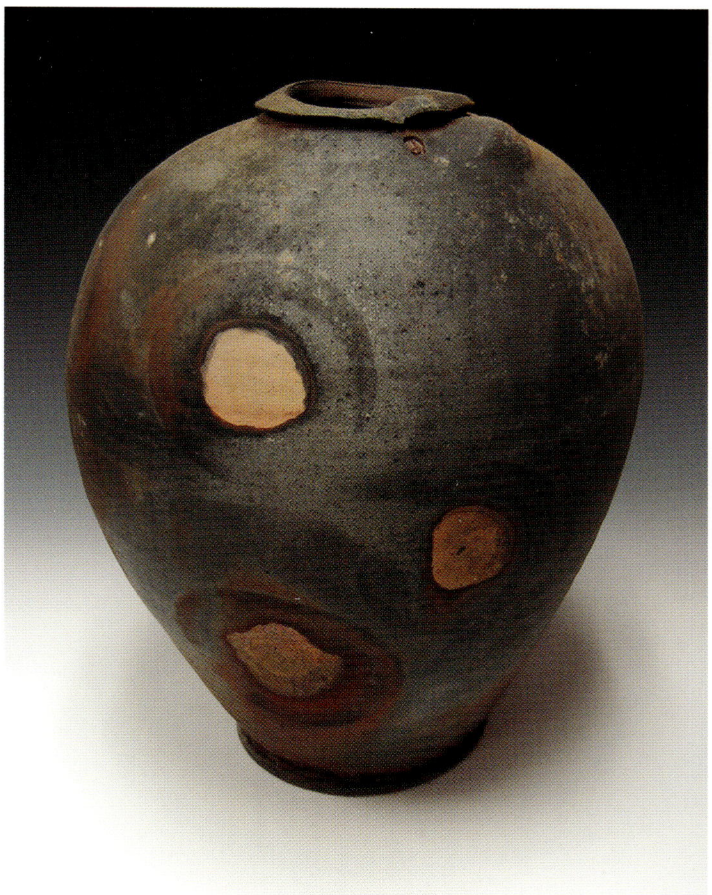

Drew Nicklas, *Jar*, Wheel-thrown, stoneware, 30" x 22". *Photo Credit: Drew Nicklas.*

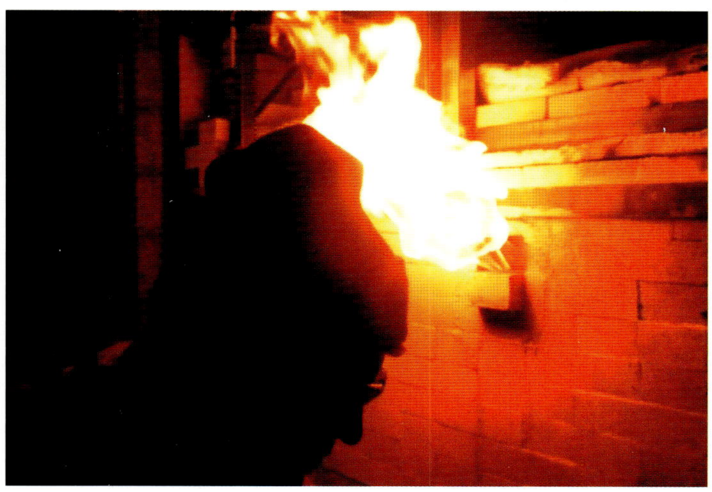

Side stoking at night, Smoke Farm Train Kiln, Arlington, Washington. *Photo Credit: Fionnuala Cormack.*

SHAWN O'CONNOR

Minot, Maine
info@shawnoconnorceramics.com,
www.shawnoconnorceramics.com

Artist's Statement

A home cooked meal is still the essence of good, nourishing food for the body and soul. Society as a whole has forgotten that handmade pots can offer the same thing. The benefit of using handmade pots is that they contain the idea of human endeavor. For embedded within the handmade pot that we use every day resides the memory of our human evolution, an idea that transcends ethnic and racial, economic and class, cultural and national boundaries. The biggest success of handmade pots, unlike other art forms is that it is approachable and accessible. It is an art form that represents conversation between maker and user, and enters the home with the ability to affect and interact with the inhabitants on a daily basis.

My focus on utilitarian objects fills a desire to create useful objects for service in the home. My home growing up was strongly focused around the family. Family dinners were important and rarely missed growing up. My extended family gathered quite often for social events such as birthdays and holidays that always revolved around food. I would like to extend this sense of comfort and warmth through my work to others who use it.

Firing with wood also came with my upbringing. I was raised in a rural Maine home that was heated with a wood stove during the cold winter months. This meant that the fire was constantly being fed in order to heat the house. This required a lot of work and attentiveness to the fire. Preparing a winter's worth of wood required many days of hauling, splitting, and stacking. This process was instilled in my life from an early age. I have always found the physical labor, the rhythm, and the sense of accomplishment that comes with this process, enjoyable.

The work I make is tailored for the process of wood-firing. During the making, I leave the surfaces of the work quiet and relatively unmarked to allow the flame to create the modulated surface that I desire. My desired surface comes from looking at my rural environment. I am interested in worn river rocks, the erosion of land, the old weathered farmhouse, and the rich colors of leaves as they change in the fall. These are all records of time, change, and decay, much like the surface of my work in the wood kiln. The pieces are marked by the flame, colored by the kiln atmosphere, christened by ash deposits, and freckled by erupting impurities. I am fascinated with the rich surface left by this process. No two pieces are exactly the same as the flame's path records distinct marks on each piece. The wood-fired surface is much like the way wind and water erode rock and earth. The flame moves through the kiln, wrapping in and around the work, leaving a mark dependant on what is next to, touching, or above that particular piece. The path of the flame can be controlled when stacking the kiln. Great time and care is spent on each piece as it is loaded, as this will dictate the way the flame moves over and marks the surface of this piece.

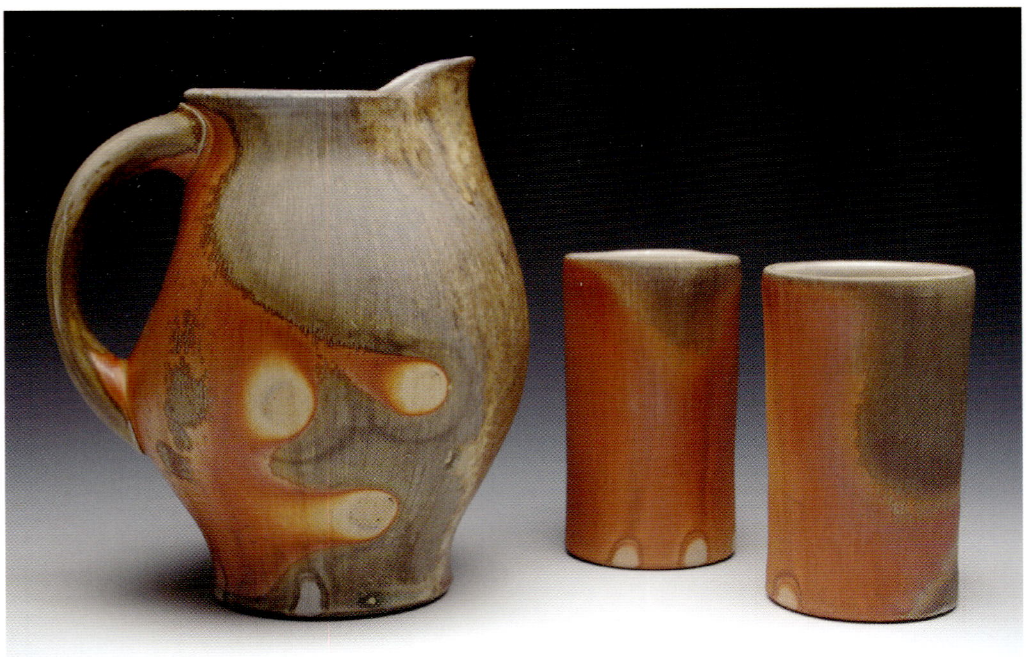

Shawn O'Connor, *Pitcher & Tumblers*, 2010. Wood-fired stoneware, wheel-thrown. Pitcher 10" x 6" x 7"/ Tumblers 5" x 3" x 3" each. *Photo: Shawn O'Connor.*

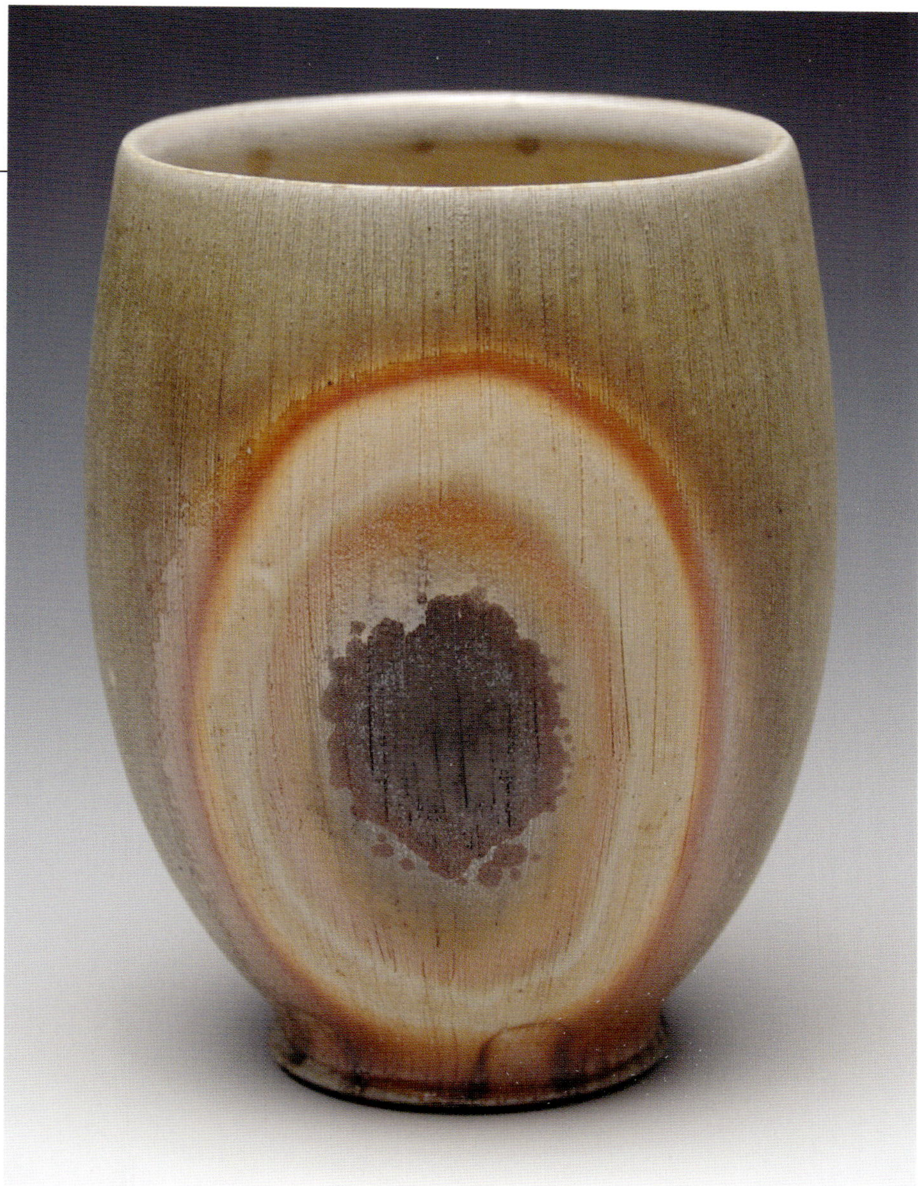

Shawn O'Connor, *Yunomi,* 2010. Wood-fired stoneware, wheel-thrown, 4.5" x 3" x 3". *Photo: Shawn O'Connor.*

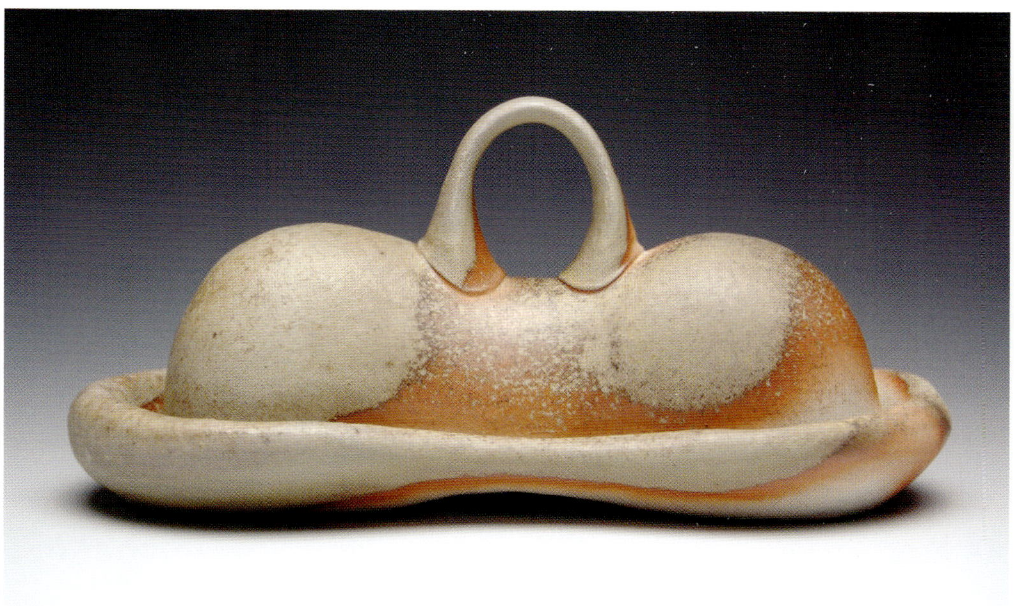

Shawn O'Connor, *Butter Dish,* 2010. Wood-fired stoneware, press molds and handbuilt, 4" x 8.5" x 4". *Photo: Shawn O'Connor.*

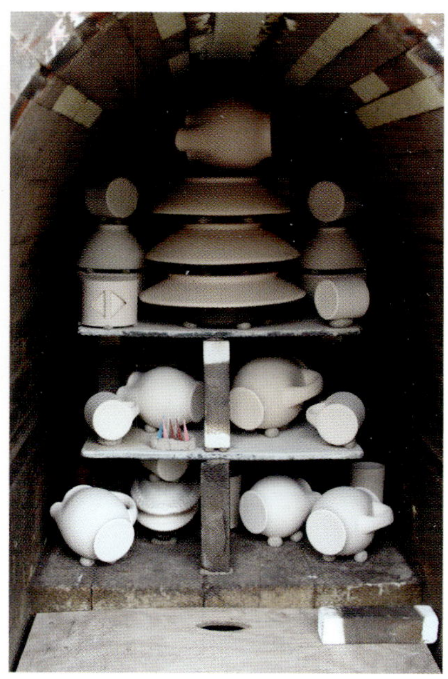

Looking inside the kiln at Arrowmont School of Arts and Crafts, Gatlinburg, Tennessee, after loading. *Photo: Shawn O'Connor.*

LINDSAY OESTERRITTER

Bowling Green, Kentucky
aoesterritter@yahoo.com, lindsay.oesterritter@wku.edu

Artist's Statement

I am intrigued by the slow and natural changes that are constantly yet imperceptibly altering our daily environments; how the seemingly permanent is being perpetually weathered. My current work is punctuated by the concept of the inseparable relationship form and surface develops though subtle and progressive erosion of time.

I find inspiration in commonplace items within my surrounding environment. An ax's blade brand new is shiny and crisp, but through use, its actual shape is altered along with gaining a patina from wear. The form and surface become linked and unique to the singular object. It is my intention to utilize wood kilns in a reduction-cooled atmosphere to highlight form and surface variations, and reference slow and continuous change.

While my wood kiln preference is a train kiln, I also frequently fire the *manabigama,* and occasionally the anagama.

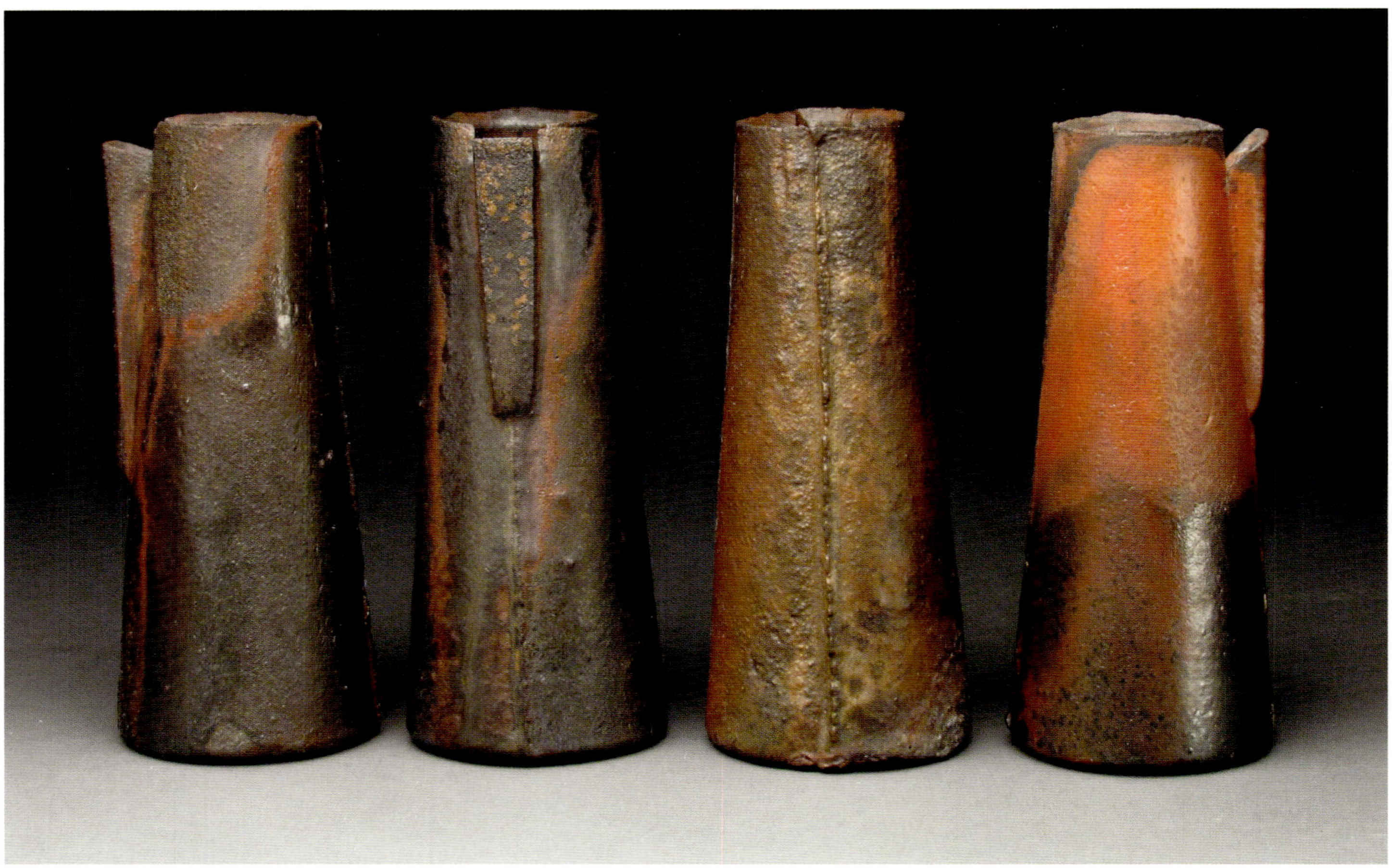

Lindsay Oesterritter, *Pourers*, 2011. Iron rich stoneware, press mold, 7" x 3" x 3" each. *Photo Credit: Lindsay Oesterritter.*
Wood fire reduction cool. These pourers are made with a two-part press mold. I intentionally leave the mold seam, wanting to also reference the process of other materials like metal. Fired in a train kiln cooled in reduction. No glaze.

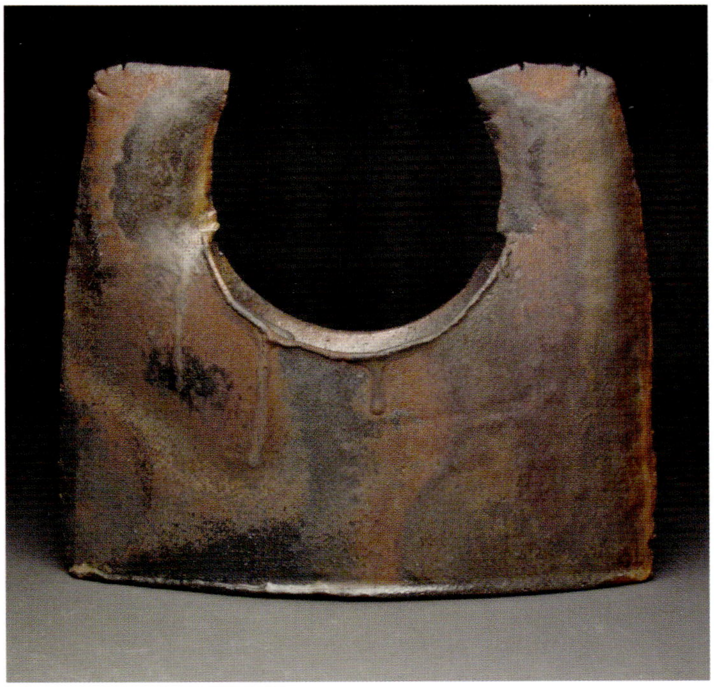

Lindsay Oesterritter, *Blade Vase*, 2011. Iron rich stoneware, slab built, 10" x 12" x 2.5". *Photo Credit: Lindsay Oesterritter.*
Wood fire reduction cool. The blades of both tools and weapons inspired this vase. Built with slabs and fired in a train kiln, cooled in reduction. No glaze.

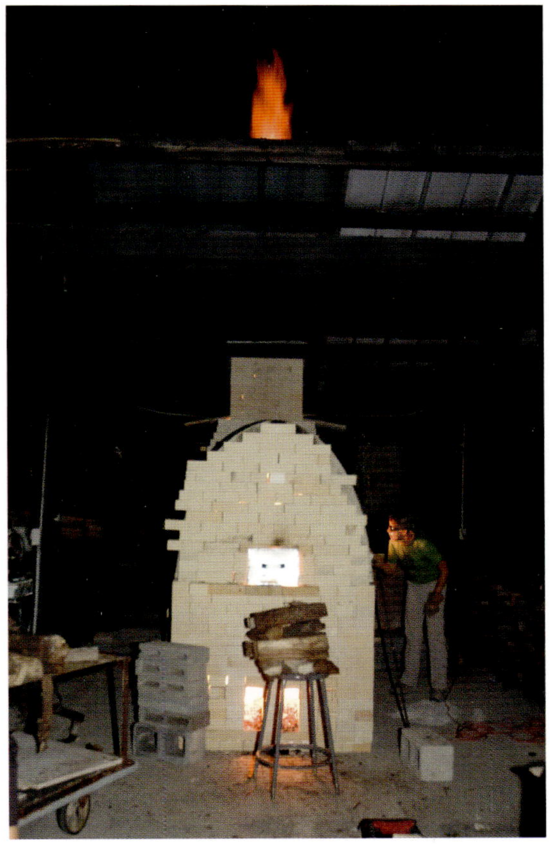

Manabigama Firing at Arrowmont School of Arts and Crafts in Gatlinburg, Tennessee. *Photo Credit: Lindsay Oesterritter.*

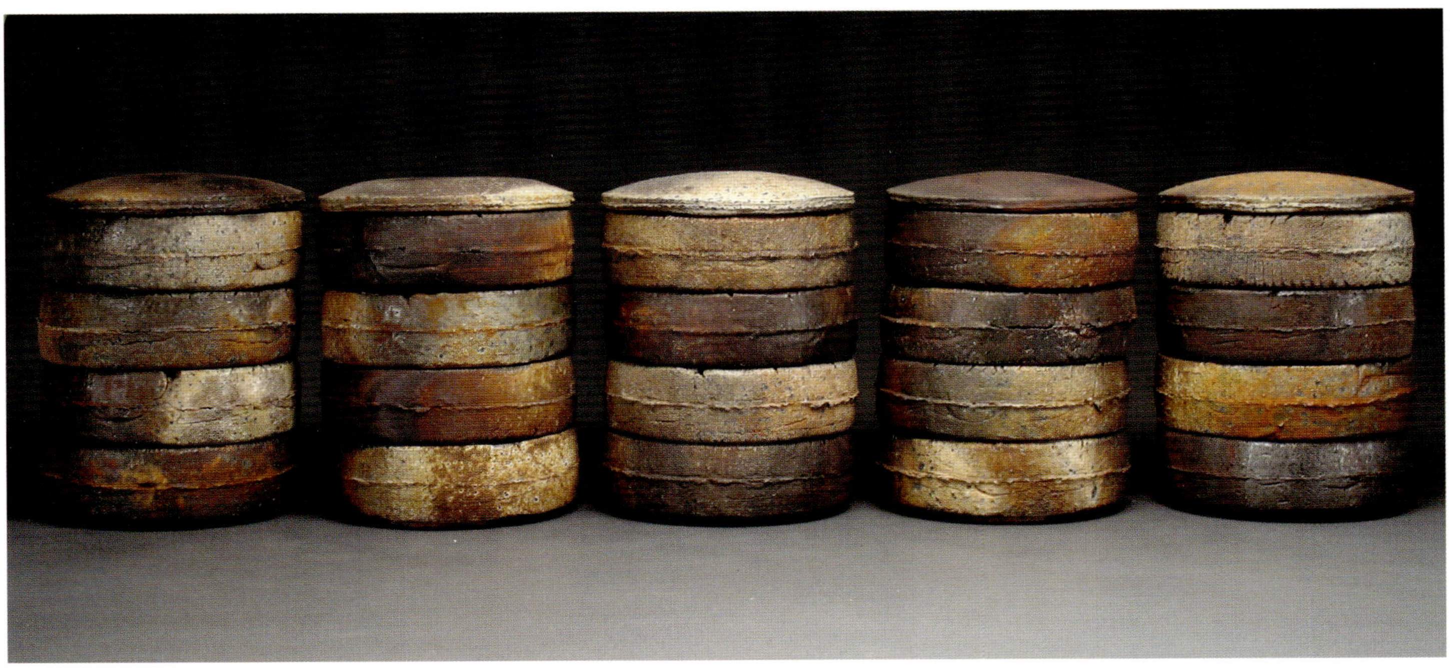

Lindsay Oesterritter, *Stacking Boxes*, 2009. Iron rich stoneware, press mold, 8" x 30" x 5". *Phcto Credit: Lindsay Oesterritter.*
Wood-fire reduction cool. Built with multiple part press mold, and purposely scattered throughout the kiln to get varied results on the individual box. Inspired by the ceramic roof tiles in China and old saggers. No glaze.

FREDERICK OLSEN

Mountain Center, California
olsenkilns@prodigy.net, www.olsenkilns.com

Artist's Statement

I use my twenty-eight year old *hollyweird* anagama, a two chamber Olsen wood-fired kiln, a Olsen fastfire wood. I fire with wood because I like the results from the various wood-fired kilns I have. I have been firing with wood for over fifty years and have found no other firing techniques that I can change methods, techniques, and time schedule to achieve the results I love. Also I like the challenge.

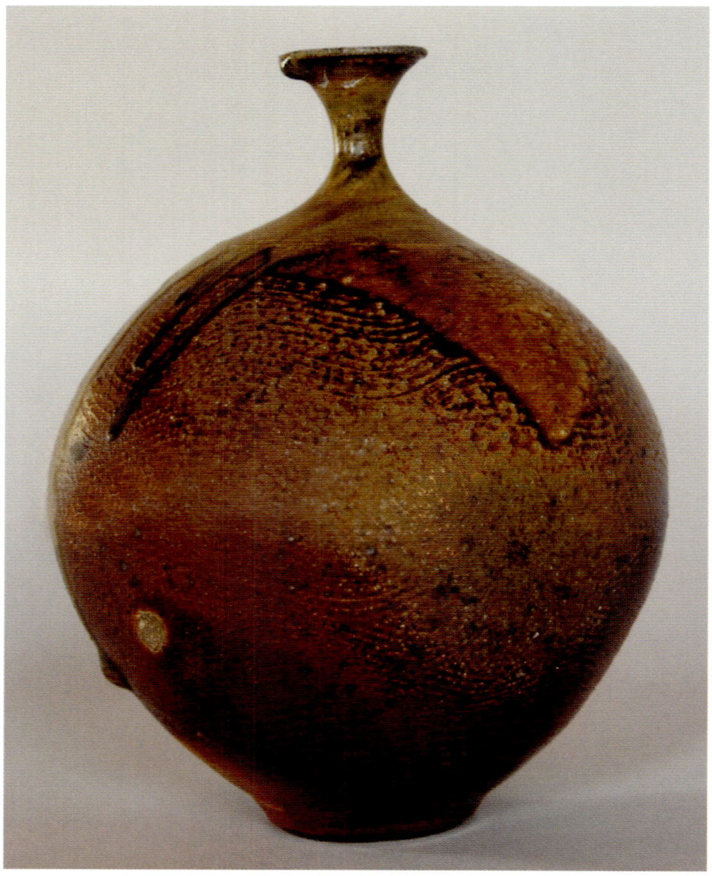

Frederick Olsen, *Bottle*. Anagama fired stoneware, 12" x 9". *Photo credit: Frederick Olsen.*

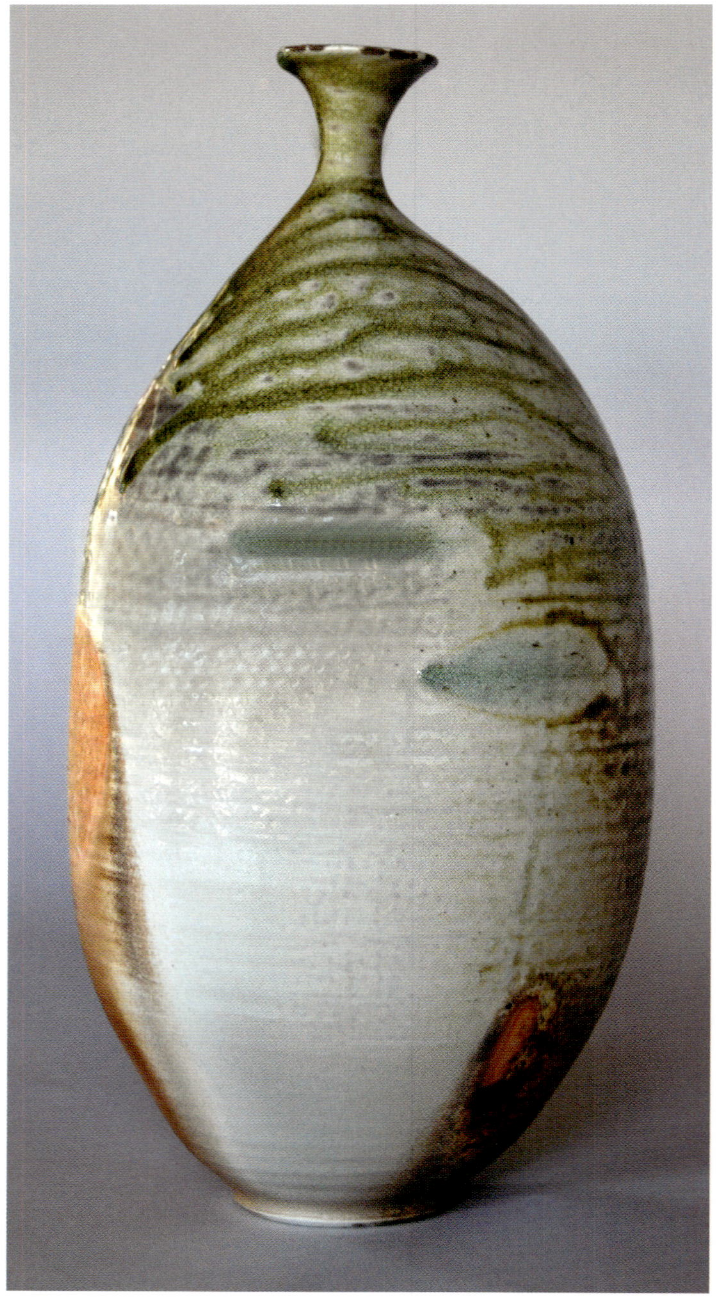

Frederick Olsen, *Bottle*. Anagama fired porcelain, 15" x 8". *Photo credit: Frederick Olsen.*

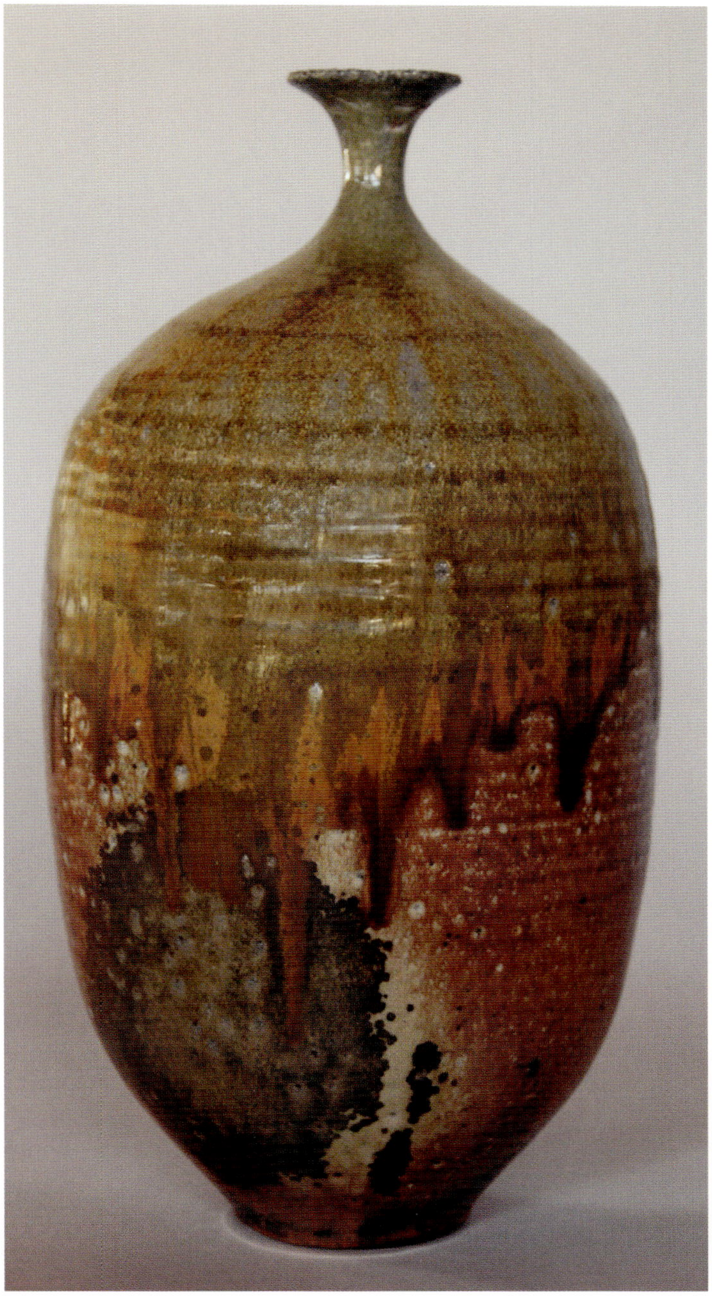

Frederick Olsen, *Bottle*. Anagama fired stoneware 12" x 6". *Photo credit: Frederick Olsen.*

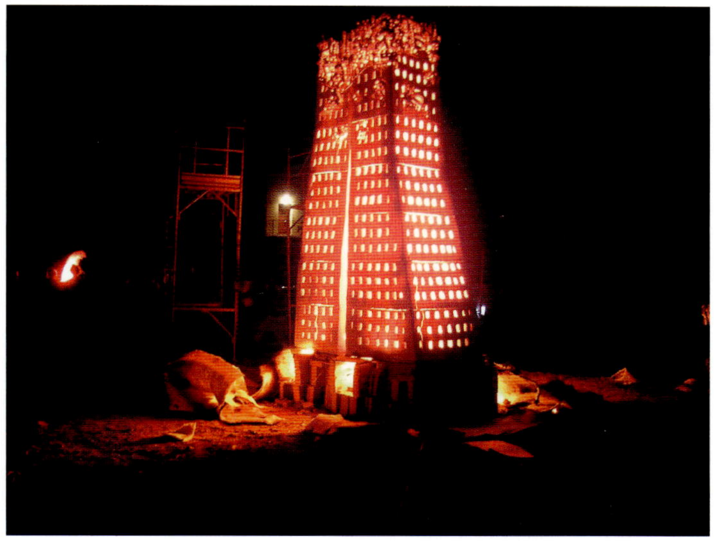

Fired in place 17 ft. tall sculpture in Nove, Italy, 2003. *Photo credit: Frederick Olsen.*

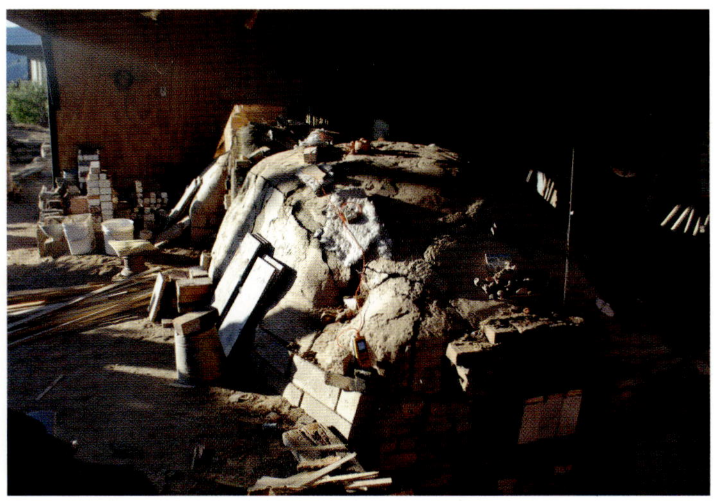

Hollyweird tube kiln built in 1984 and is fired in about 60 hours with wood. *Photo credit: Frederick Olsen.*

MICHAEL PONESS

Takoma Park, Maryland
mponess@wmcrparchitects.com, www.michaelponess.wmcrponline.com

Artist's Statement

As an architect and ceramist, I am continuously fascinated by how one discipline informs the other. As an architect, I deal with form, closure, structure, penetration, finishes, hardware, etc. The basic forms at the potters' disposal, the cylinder, sphere, cube, cone, and slab, are the basis of many buildings as well. The notion of form and function similarly is applicable to both disciplines. The creative process and the rigor of architectural design are aspects of my work that I bring to creating ceramic vessels. In architecture, a statement of requirements, a program, usually precedes the building design. The process begins with rough sketches. The design is developed with more detail, models, computer-aided-design, and finally technical documents for construction. It is typical for a building to take two or three years from the conceptual design to occupancy. The making of pots follows a similar linear process from start to finish, albeit in far less time. Virtually, all of my more complex vessels begin with a sketch of series or sketches. I try to work out the shapes and proportions before sitting down at the wheel with the clay. My pots typically have a strong base, middle, and top. This is the tripartite vertical organization found in buildings. I'm particularly drawn to lidded pieces, which correlates to designing roofs for buildings.

Another approach I use is to combine slabs with thrown forms. This combination of the plane or wall and the pure form has architectural precedents as well. A typical example of a vessel is a slab bisecting a cylinder. The slab then extends down to form one leg of a tripod base and up to form the lid and knob.

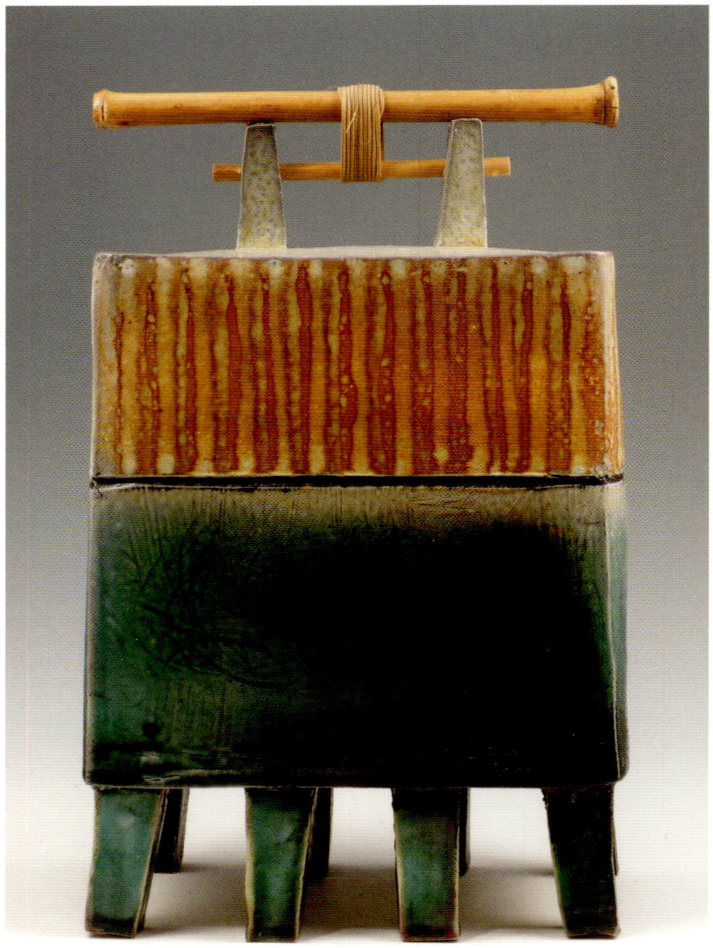

Michael Poness, *Eight Legs a Week*, 2009. Wood-fired stoneware, wheel-thrown, altered, 10" x 7" x 4". *Photo credit: Michael Poness.*

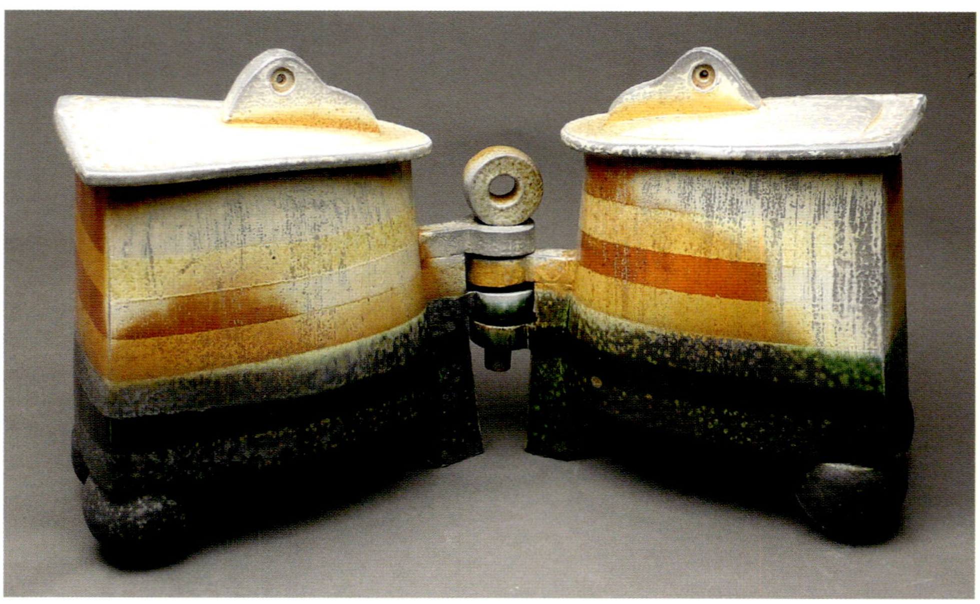

Michael Poness, *Linked for Eternity*, 2010. Wood-fired stoneware, wheel-thrown, altered, 10" x 18" x 5". *Photo credit: Michael Poness.*

I also enjoy lifting the piece with legs or feet, a gesture to columns in buildings. I draw on fine arts as an inspiration for surface decoration and texture."

"I choose to wood-fire my work for many reasons. I look for variegated surfaces and textures to compliment the forms. Firing in a wood kiln gives me that. I employ flashing slips, glazes, wood ash, and stains, inviting the flame to create the finish. I fire in a two-chamber norborigama style kiln at Baltimore Clayworks under the supervision of Jim Dugan. The first chamber is fired with wood only. The second chamber is used for atmospheric firing affects through the introduction of soda and salt at cone 10. This gives me an additional set of tools to explore surface textures and finishes.

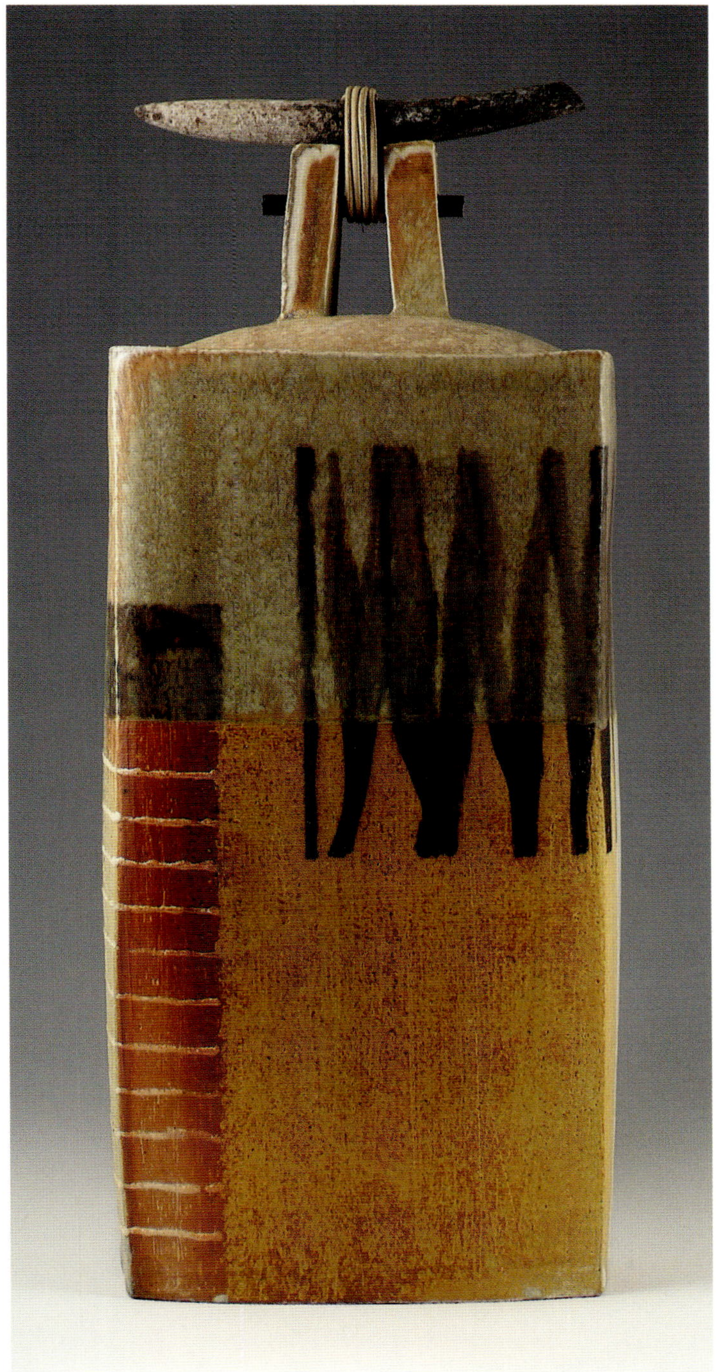

Michael Poness, *Tall Box*, 2010. Wood-fired stoneware, wheel-thrown, altered, 11" x 5" x 4". Photo credit: Michael Poness.

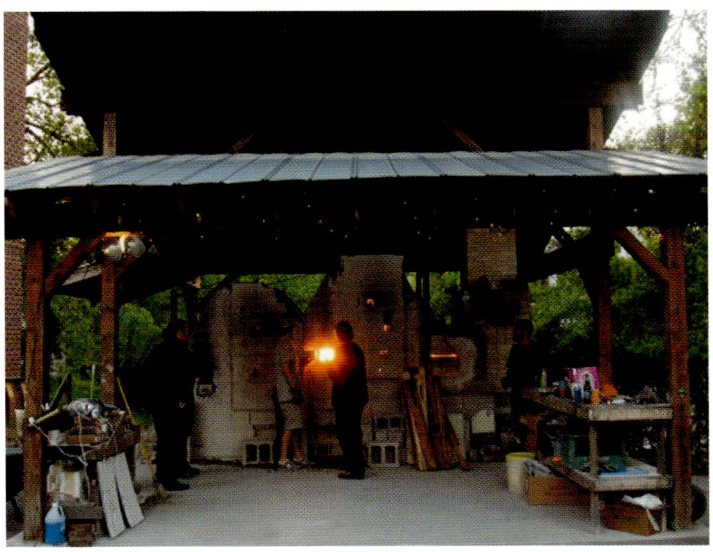

Firing the Norborigama at Baltimore Clayworks. *Photo credit: Michael Poness.*

DAVIE RENEAU

Glasgow, Kentucky
dreneau@campbellsville.edu

Artist's Statement

A friend from Arizona once commented to me that she missed home because the landscape out West was very male while the landscape in the Southeast was very female. I had traveled in the West so I understood that she was speaking of, flat and straight lines as opposed to rolling hills and curves. At least I thought I understood at the time ...

Several years ago a horrible accident forced me to slow down. I was on crutches for half a year and had more than enough time to reflect on different aspects of my life and work. It was only then that I truly came to understand her statement and process information that had been with me all my life and bring it to my pots.

I began to notice the belly-like hills and sensual textures that had surrounded me, growing up in a farming community. Raked hay, plowed fields, and graceful curvy roads took on a new visual meaning from her analogy that made other elements of my work and life make sense.

For the last twenty years, I have fired my pots primarily in wood kilns, specifically anagama kilns. There is an insane amount of work involved in the process of firing and there are many easier processes of finishing work. I am attracted to the spontaneity of ash deposits, flashing, and the beauty of naked clay exposed to raw flame. The mystery of unpredictable surfaces on the work intrigues me. I enjoy being physically involved in the firing process, not just doing passive turn-ups on a gas or electric kiln. I even like spending days cutting, hauling, and stacking wood. They are as comfortable and familiar to me as the long, hot days I spent as a child working in my dad's tobacco and hay fields. I realized this is one of the reasons I am so attracted to wood-firing. There is an honesty and purity in intense physical labor directed toward an ultimate goal. This is the spirit I want my work to exhibit.

Although many aspects of my pots have changed, the one thing that has remained constant is my adamancy about making utilitarian pots. I have always had an immense respect for the history and tradition, required mastery of skill and intimate relationship with the user commanded by functional pots. The exploration of the epiphany inspired by my friend's statement changed my considerations in making pots to still include, but go beyond, how a rim feels against the lips or takes the blow of a spoon. I began to look at the work in more sculptural terms.

Historically, my pots have always involved the rituals of food presentation and storage. I began to make vases that were more concerned with surface textures and relationships of forms that were directly inspired by the lush femaleness of Kentucky's farmland. I also began to make utilitarian forms that stack when not in use to form blossom-like sculptural objects. The textures on cut jars came to look like tobacco leaves. The straight lines in the liquor sets are reminiscent of the vertical planks in barns juxtaposed against the sensual curves of hills. It seemed that images from my past had begun to trickle into my work like Marc Chagall's painting, *I and the Village*.

I was an anomaly in graduate school because of my background (I once had a professor on my committee ask to look at my hands to acknowledge that I had calluses from working). I was also an anomaly to my family and friends in rural Kentucky, because I had become a serious, academic artist. The darkest year of my life after the accident brought an illumination that merged these two worlds in my pots. The clarity is inspiring. My work now not only reflects my sensibilities as an artist, but my own personal history as well.

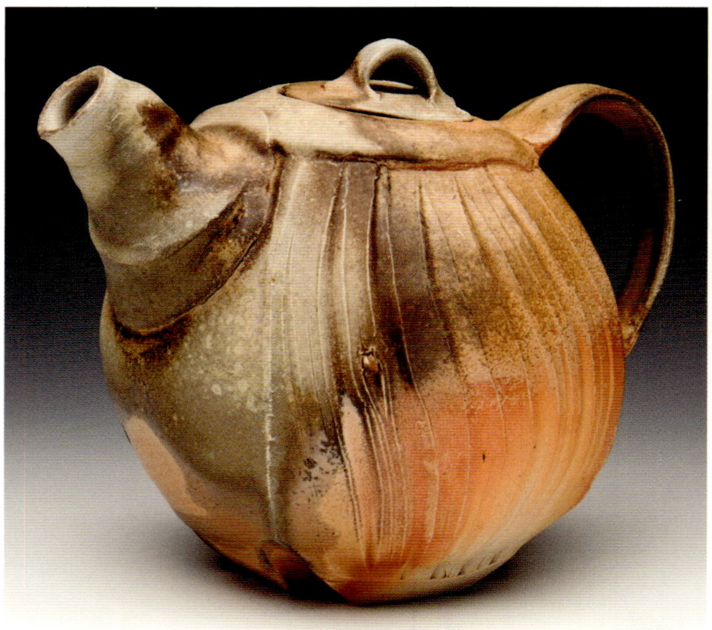

Davie Reneau, *Teapot*, 2010. Anagama fired porcelain, 8" x 7" x 7.5". *Photo credit: Bob Payne.*

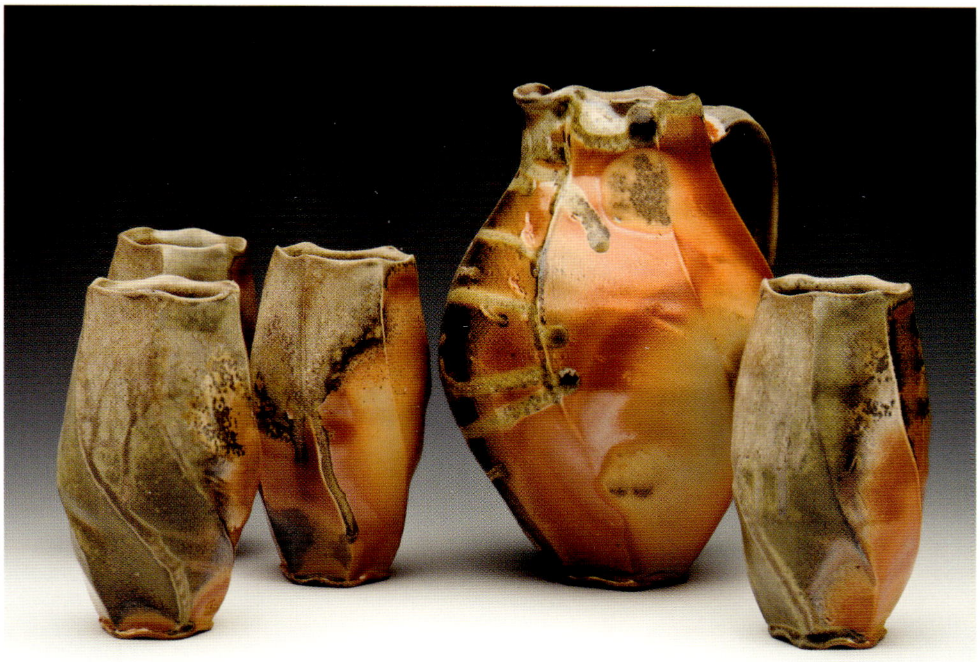

Davie Reneau, *Pitcher and Tumblers*, 2010. Anagama fired porcelain, 13" x 8" x 12.5", 8" x 3" x 2.5". *Photo credit: Bob Payne.*

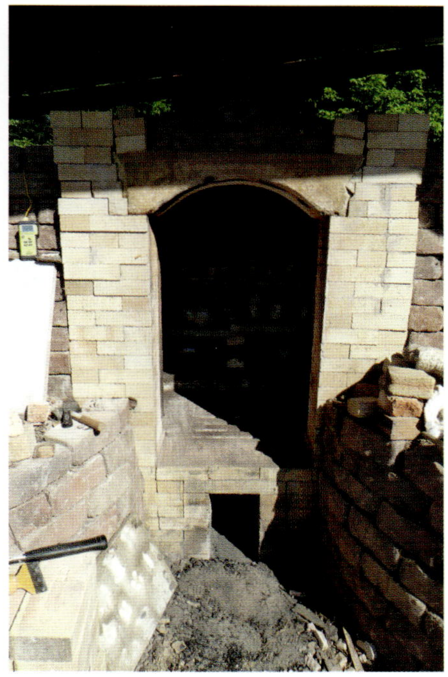

Anagama Kiln Firing, June 2011. Kiln after firing. *Photo credit: Renee Renfro.*

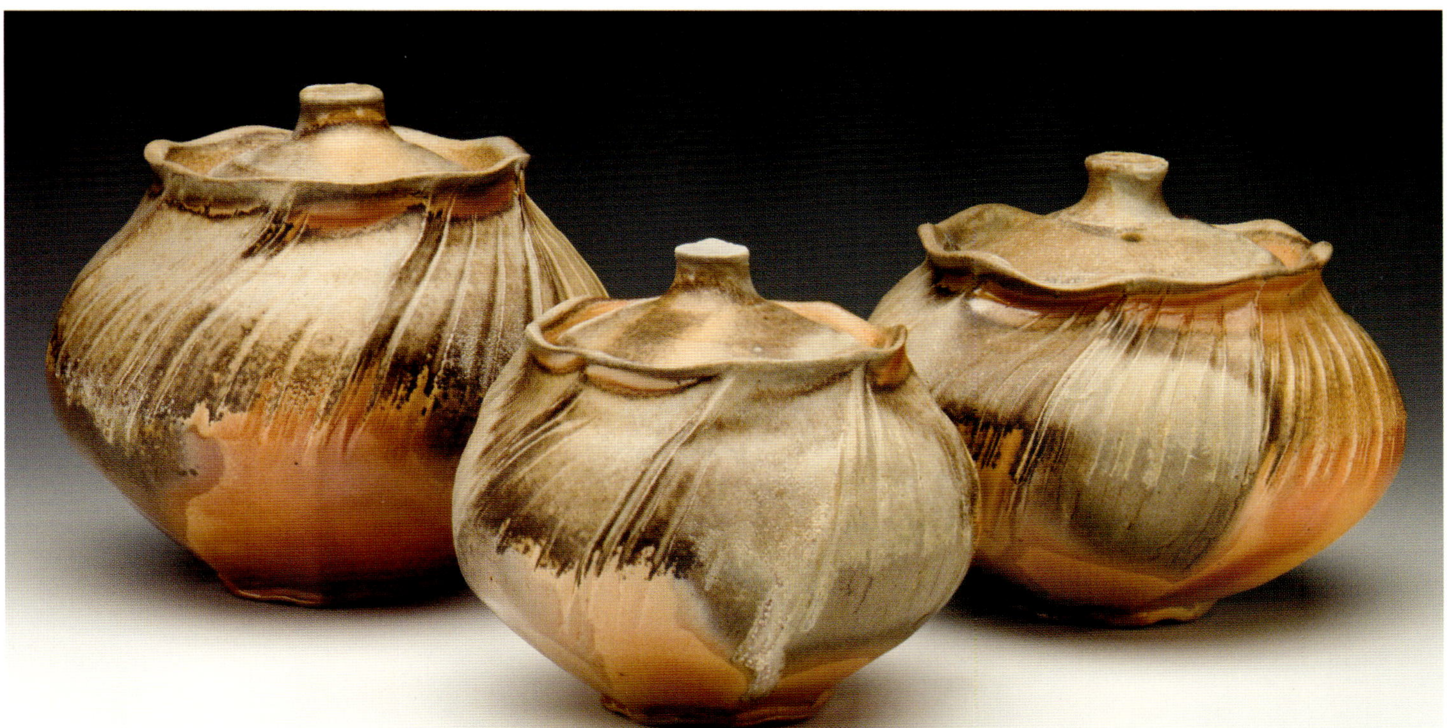

Davie Reneau, *Jars*, 2010. Anagama fired porcelain, 10" x 8" x 9.5", 8" x 7" x 7.5", 6" x 6" x 5.5". *Photo credit: Bob Payne.*

BEN RICHARDSON

Tasmania, Australia
ridgeline@bigpond.com, www.benrichardson.com.au

Artist's Statement

My work is a search for a maker's response to place, in this case the southeastern edge of Tasmania, the island state of Australia. On the edge of the Southern Ocean, this is a place of eroded cliffs, refracted ocean swells, rock stacks, and the linear natural and human patterning of the tidal zone. In my work there is a journey in place—to collect clay, to gather rocks to grind for glaze making, and to cut and stack black wattle for wood-firing. To work at maximum depth I have to start with materials that connect me directly to where I live, work, and play. They place me. Working with these materials is an approach not constrained by industry but driven by discovery as I search for the fired potential of simple combinations of clay, stone, shale, ash, and ochre. That my work can only really be appreciated in the fullness of use involves another journey of discovery, as in the creative act of seeing, sensing, and using potency can be revealed, explored, and renewed.

The techniques I use grow from the rich history of craft making. That past sustains and challenges me as I shape my local materials to express a maker's response to contemporary culture. I make my clay bodies from combinations of local clays, mostly with added feldspar and silica, and my glaze materials are gathered from around Tasmania, then processed and tested in the workshop. I see this testing as an ongoing research project to create the foundation for future making. The wood I use is native black wattle, gathered from our land on a sustainable natural rotation that allows for several kiln firings a year. Wherever possible, I am using what is available nearby, then building technique, process, and form on that. Soda rich glazes are now made up using seawater and marram grass from nearby beach dunes with local fat sand for wads used in kiln packing.

The kilns I design, build, and use are hybrid kilns, influenced by Oriental anagama but now blended with Western influences. Their shape allows the flame to be seen moving through the kiln; this direct visual connection with the fire/work interaction I now see as crucial to my engagement with the process. My firings are relatively short—less than two days—and do not depend heavily on natural fly ash deposit for surface and color development, except in the firebox and

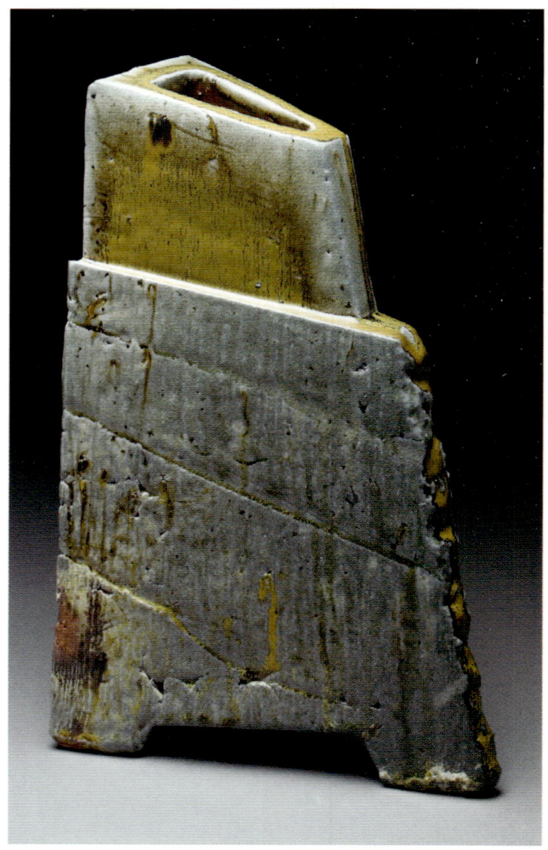

Ben Richardson, *Wedge Foliage Vase*, 2010. Off-form handbuilding, wood-fired stoneware, glaze of sandstone, dolerite, ochre, 15.5" x 10" x 5.5". *Photo credit: Robin Roberts.*

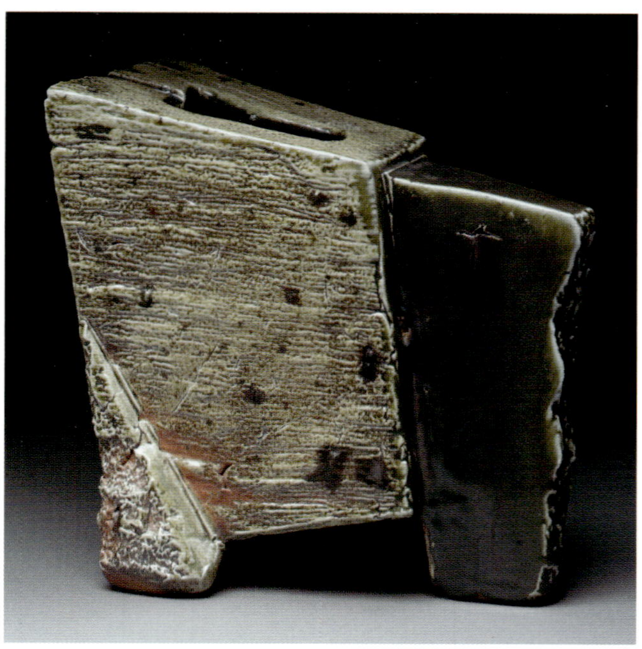

Ben Richardson, *Wedge Foliage Vase*, 2004, off-form handbuilding, wood-fired stoneware, glaze of sandstone, dolerite, ochre, 8" x 8" x 5.5". *Photo credit: Robin Roberts.*

side stoking areas. I develop work that combines the fired potential of the raw materials and explores the possibilities found through past firings for each zone in the kiln. The kilns are packed to create a flame-dividing pillar that splits the flame, encouraging it to travel around the kiln pack as well as through and over it. During the firing, firebox embers are shoveled to spread charcoal effects through as much of the base and front of the kiln chamber as possible.

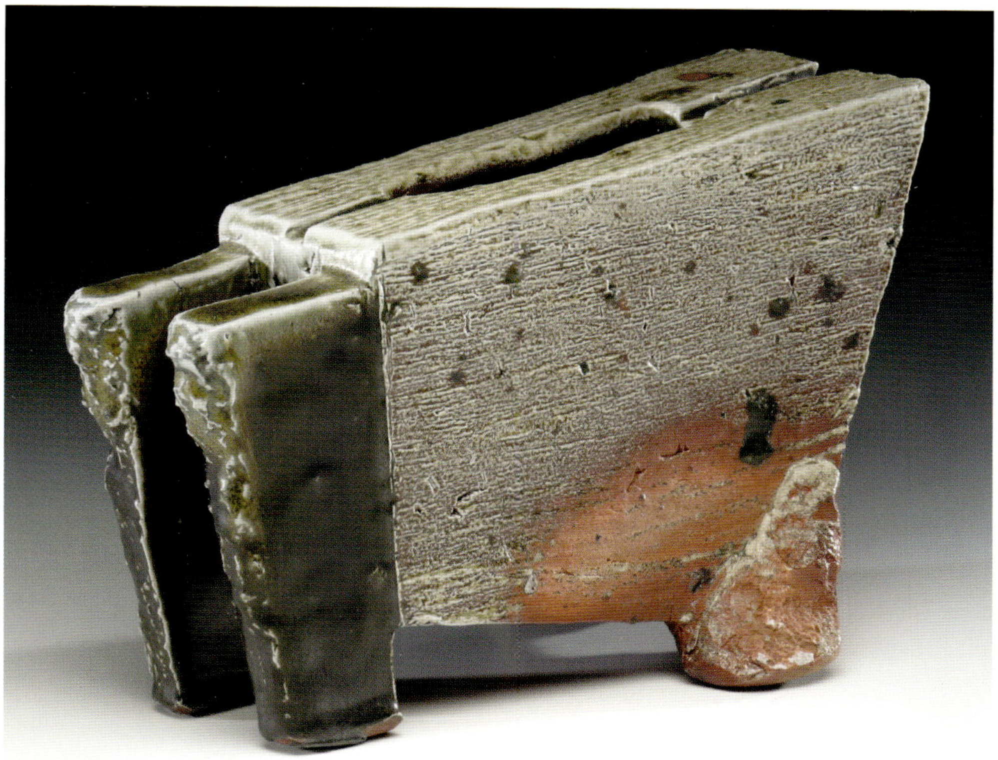

Ben Richardson, *Channel Foliage Vase,* 2006. Off-form handbuilding, wood-fired stoneware, glaze of sandstone, dolerite, ochre, 8" x 9" x 5.5". *Photo credit: Robin Roberts.*

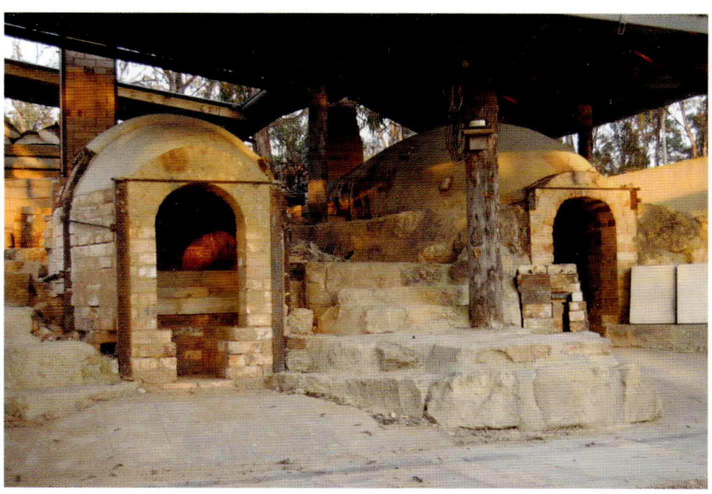

Kilns, 2011. *Photo credit: Ben Richardson.*

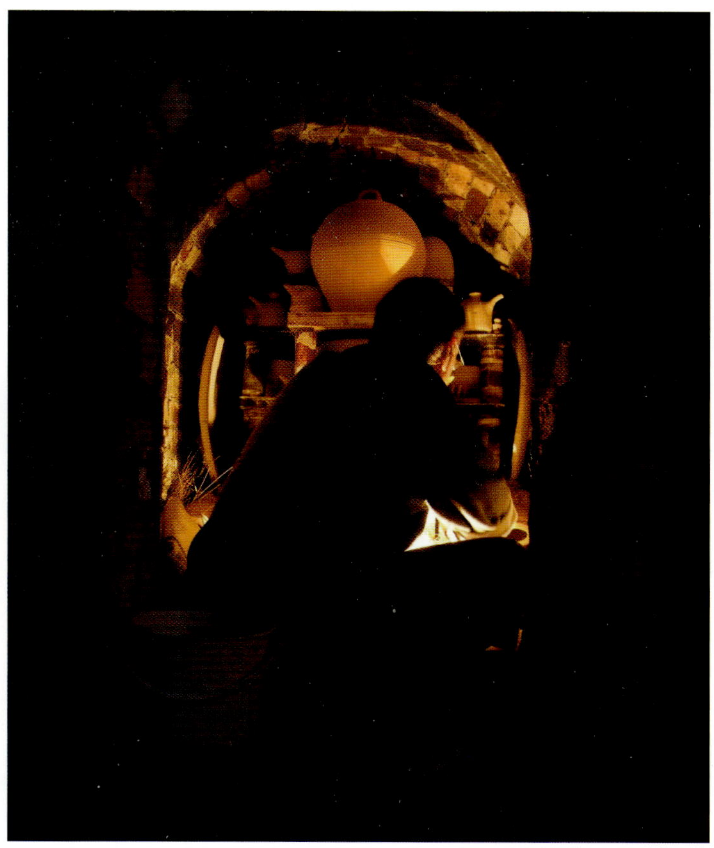

Night loading at Ridgeline Pottery. *Photo credit: Bec Tudor.*

STEPHEN ROBISON AND KATHLEEN GUSS

Ellensburg, Washington
robisons@cwu.edu

Artist's Statement

Objects of use and domesticity have a common language, which a diverse audience can appreciate and relate to. This may be the initial draw to our work but appreciation of the concepts and esthetics may seep into the viewer after further investigation. A firm belief in the connection of the mind to the hand and the hand to the media drive us to make objects with an inherent humanity. Like the lips to a reed, technology cannot replace or even come close to the sensitivity that the artists has with his or her material. Our intent is to create tactile qualities in these objects that offer an intimate relationship with the user and provide the objects with an innate value. Generating a pleasurable and possibly a reflective experience when being used and viewed creates new challenges with each object made. Visual balance by using proportional perspectives, physical balance within the weight, and pivot points of the piece along with tactile qualities are issues that are addressed to achieve these goals. Imagery and concepts that filter into our work in its forms and surfaces relate to diatoms, viruses, sea life, holes, openings, caves, and most recently using the bridge as a metaphor. This work is still meant to function as boxes, containers, vases, pouring vessels, or

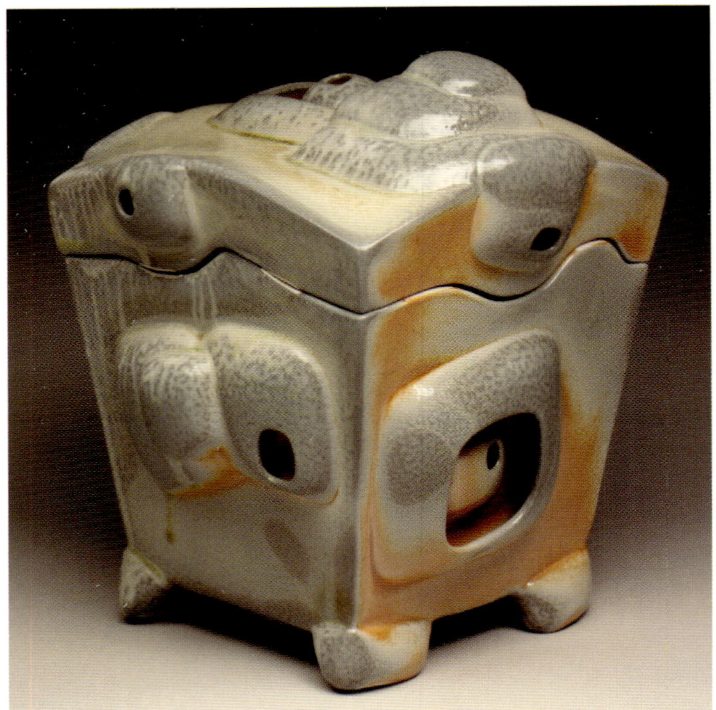

Stephen Robison and Kathleen Guss, *Box form*, 2011. Wood/soda fired, handbuilt porcelaneous stoneware, 16" x 8" x 7". *Photo credit: Stephen Robison.*

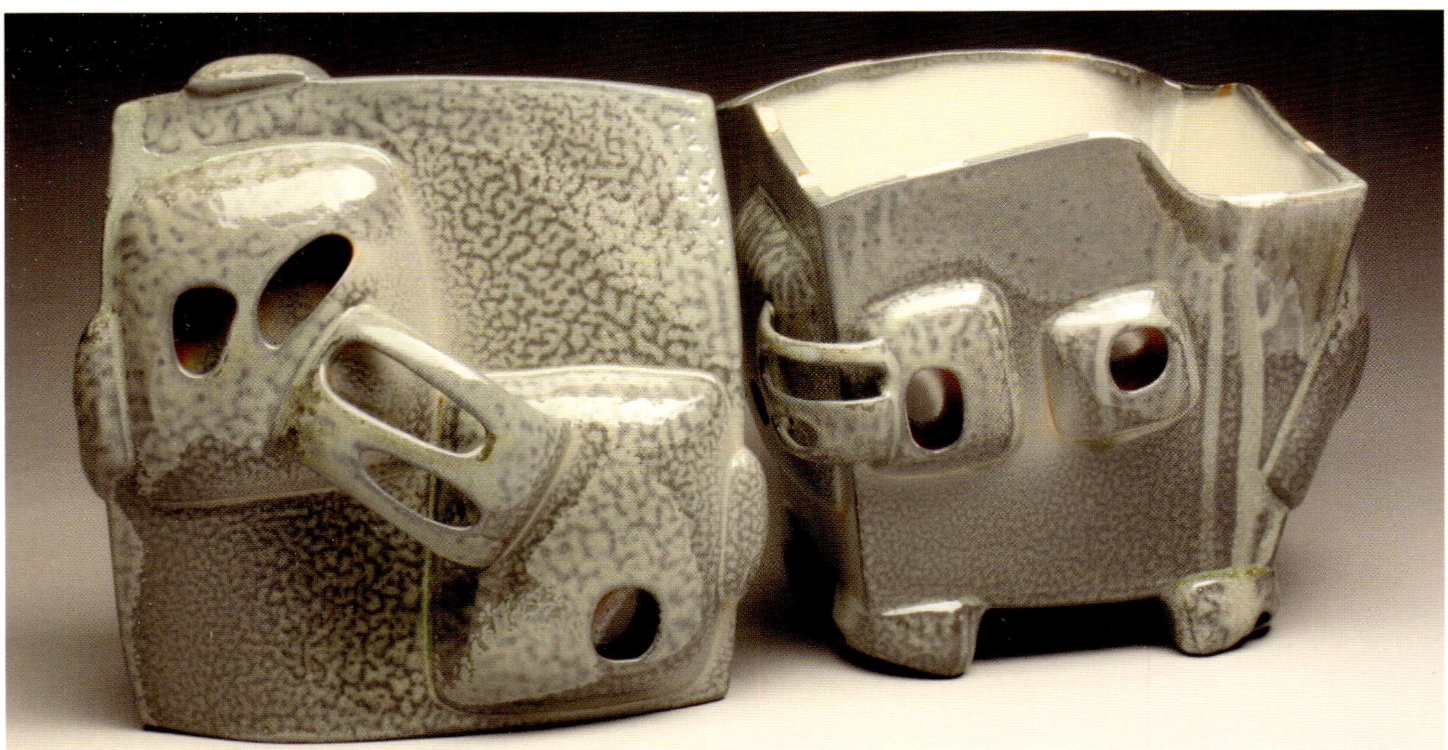

Stephen Robison and Kathleen Guss, *Box form with lid off*, 2011. Wood/soda fired, handbuilt porcelaneous stoneware, 15" x 8" x 7". *Photo credit: Stephen Robison.*

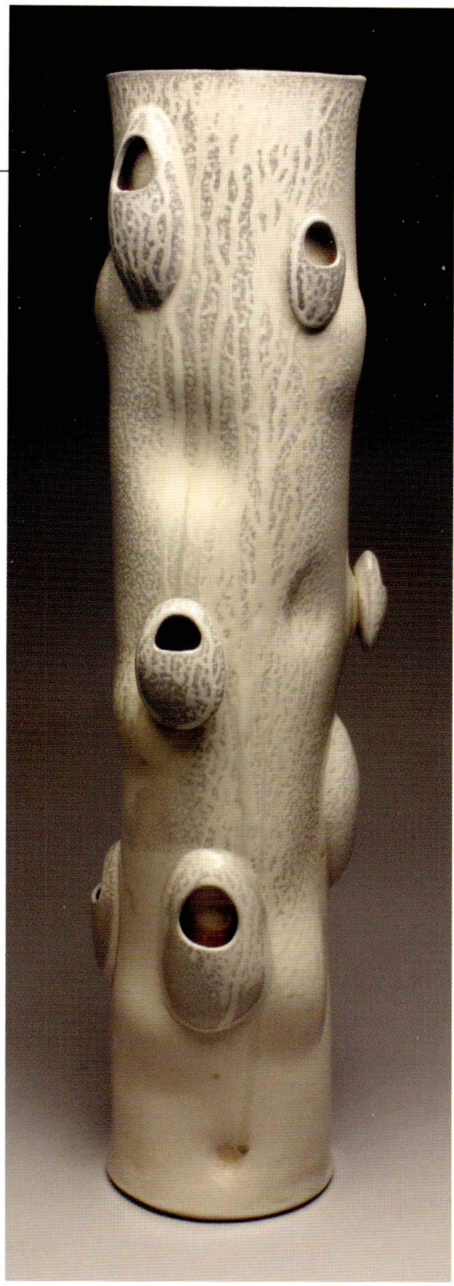

Stephen Robison and Kathleen Guss, *Vase form*, 2011. Wood/soda fired, handbuilt porcelaneous stoneware, 36" x 10". *Photo credit: Stephen Robison.*

drinking vessels, but some of that function may be sacrificed for form and concept.

Wood/Soda firing and straight wood-firing quite often are the means to the end and using certain types of firing, such as soda, salt, and wood-firing, adds subtleties to the surface of the forms. These natural subtleties and variations offered by the art of the loading process and the variety of different types of clays used open up an esthetic to our audience that is never garish, nor ever reflective of the plastic qualities of objects that most people are so overtly exposed to. The process of wood-firing is so enticing in an addictive way that it is also something to expose an audience to. Slowing down and working hard are somewhat embodied in the process of wood-fired work.

Building techniques are on and off the wheel. We use throwing, altering forms, slump molding, molds, and other additive and subtractive techniques.

We work with porcelain and stoneware clay bodies. Using a variety of techniques with slips, underglazes, terra sigillata, glazes, and atmospheric effects to achieve our surfaces. Primarily working with high fire temperatures and using some low temp techniques on top of the high fire surfaces and sometimes using sandblasting to achieve a surface.

Our two main kilns are a train kiln and a large catenary wood/soda kiln.

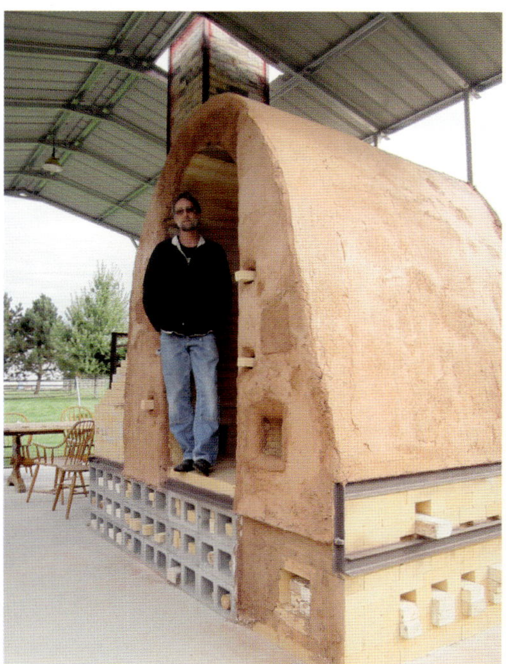

Wood/soda cat, seven feet at top of arch, built summer 2011. *Photo credit: Stephen Robison.*

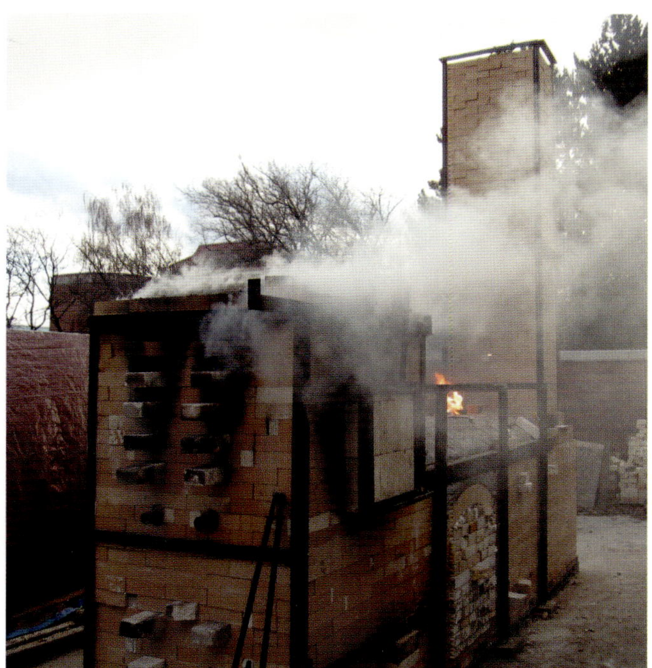

Train kiln at Central Washington University. *Photo credit: Stephen Robison.*

PETER ROSE

Knoxville, Tennessee
potterspond@bellsouth.net, www.peterroseceramics.com

Artist's Statement

My animals are not realistic, but are characters and portraits that express the quality of animal life—something large, great, and mysterious, remote from human civilization. I try to capture the form and attitude of these animals to convey them in a human meditative state, each with their own individuality.

However, form and shape is my primary interest in clay. Glazing to me covers up form, deceptive to beauty. Wood-firing enhances form, so from an aesthetic point of view, the sculptures I produced from the kiln would be strengthened from the lack of glaze and the fire markings from the wood. The problem with this is I also love color! I wanted to incorporate it to compliment the subtle surface that the fire and ash has on the form of my animals. I moved towards using stoneware glazes on the inside of sculpture that were intended for use, playing of the muted interiors with the palette of browns, grays, and blues of my clay on the exterior. The next step was to get color to enhance the form, but not overpower the piece, to use the build-up of ash on the surface as a compliment. I hit upon a series of black and rutile glazes I am still working and am happy to explore this further.

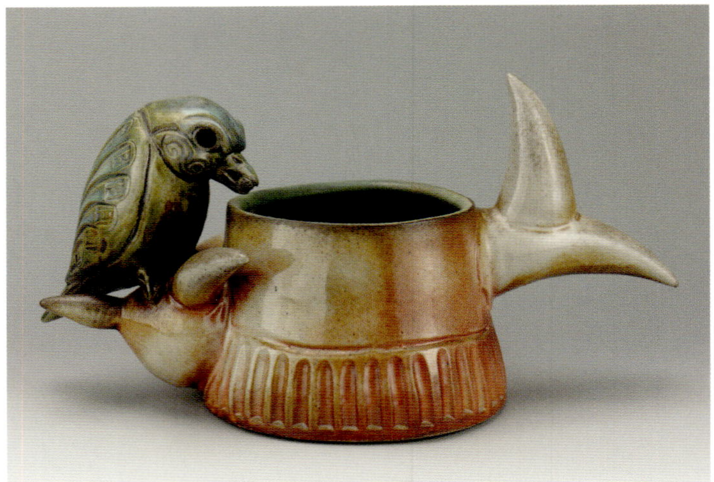

Peter Rose, *Thorn Tree Bird*, 2010. Wood-fired stoneware, wheel-thrown assembled sculptured, 15" x 9" x 9". Photo *credit: Dan MacDonald.*

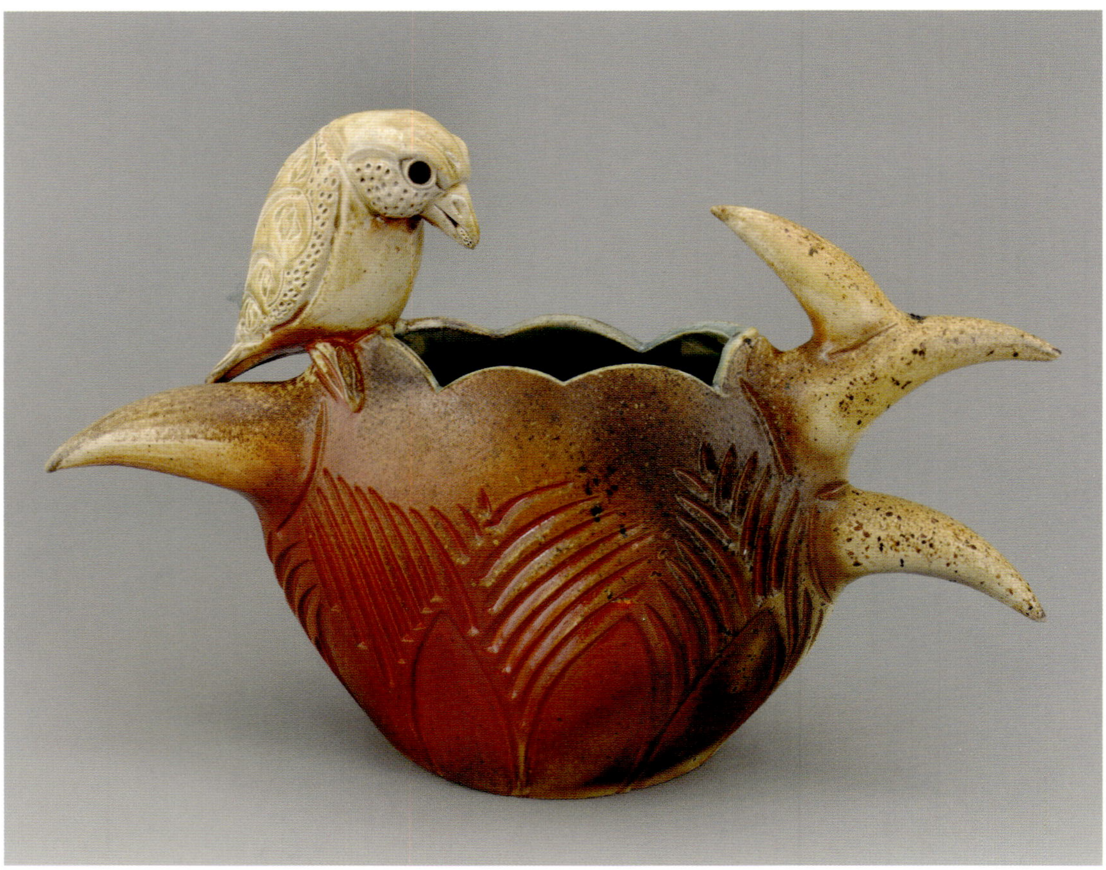

Peter Rose, *Thorn Bird*, 2010. Wood-fired stoneware, wheel-thrown, assembled sculptured, 14" x 14" x 7". *Photo credit: Dan MacDonald.*

The process of wood-firing is meditative. Collecting wood, splitting wood, stacking wood, loading the kiln, and then firing for three days, the whole experience I enjoy. I know the Zen of the process has been denigrated from the articles I have read over the last few years, but I have no illusion that I am creating art; to me the satisfaction stems from creating the work for myself. Yes, it's for purely selfish reasons that I wood-fire. My relationship with my kiln is most important. Every firing reveals knowledge and results I would have never imagined. This is also true for myself. The mystery, intimacy, the deepening understanding of the process, and the kiln never disappoints. So the question should be, why would I give up wood-firing? It's hard work but it always gives me the unexpected and makes the day!

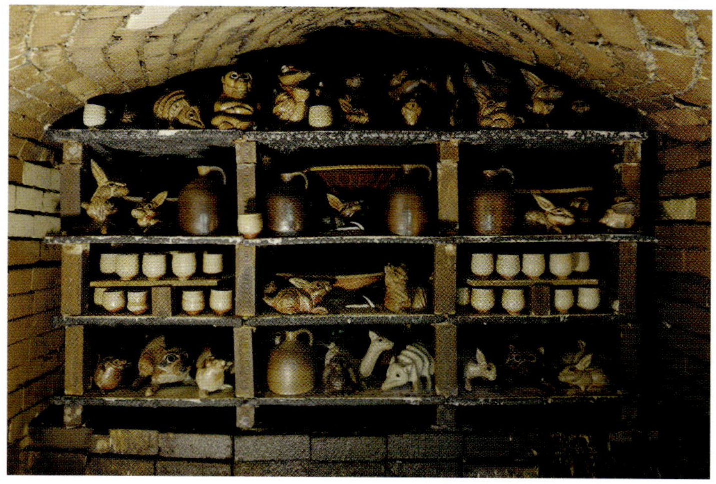

Inside kiln, 2011. *Photc credit: Dan MacDonald.*

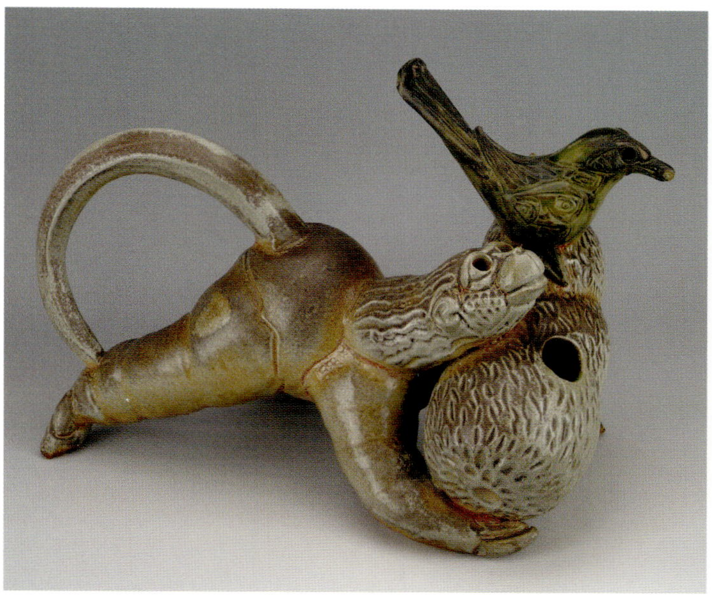

Peter Rose, *Monkey with Gourd*, 2010. Wood-fired stoneware, wheel-thrown assembled sculpture, 20" x 14" x 10. *Photo credit: Dan MacDonald.*

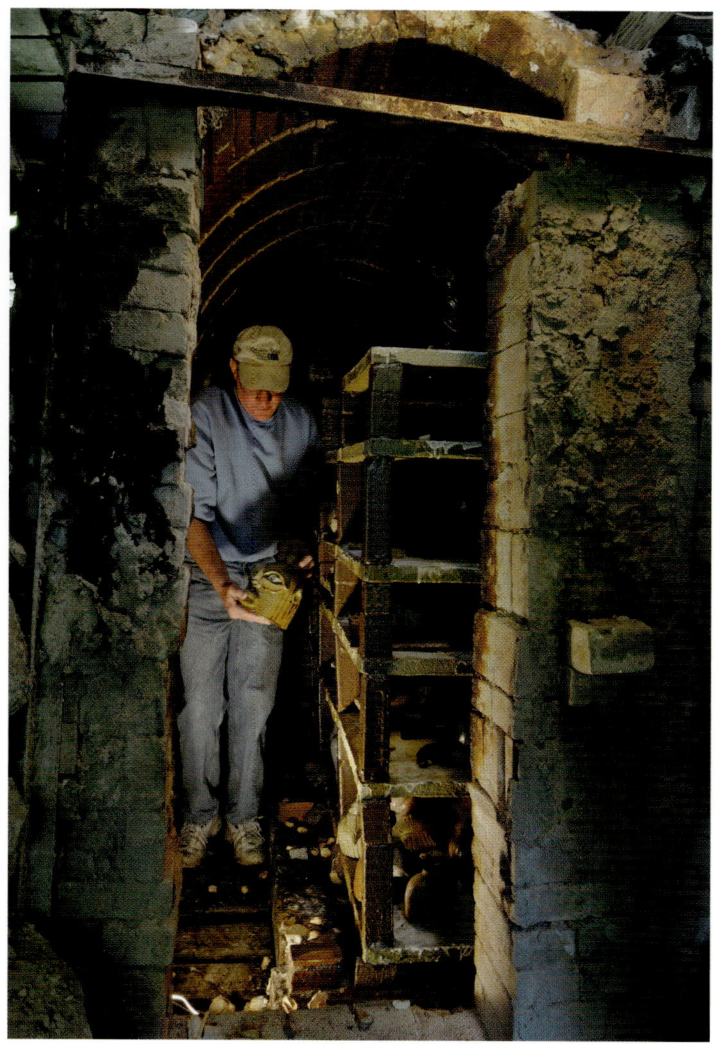

Peter Rose, inside of the back chamber of the kiln, 2011. *Photo credit: Dan MacDonald.*

MAT RUDE

Iowa City, Iowa
mathew.rude@gmail.com, www.artaxis.org/ceramics/rude_mat/mat_rude.htm

Artist's Statement

Throughout our lives we discover secret places that give us comfort, escape, or just a place to hide. As time passes, we may lose or out-grow a particular space, just as the relationships we develop that take us to or make us leave those spaces transform as well. Whether it's a playful energy, scared energy, or perhaps a sanctuary one finds to be alone, my current body of work is representative of those places. The forms are also reflective of my conversation with the material and a record of the process.

Wood-fire for me, as with most wood-fire people, started out with physicality and process. Through the act of firing, camaraderie and community emerged. In turn, information was passed on from one artist to another. This sequence became an important part of my art making. As time passed, it was evident that process and kiln space became the major factor in the creation of the work. The work was more collaboration with these elements, the firing playing perhaps the most crucial role. In recent years, I have taken a more active role, though still collaborating with the process; the work has become more focused on the presence and meaning of an object than the firing techniques employed to complete it. The wood-fire surface still compliments the new textures and forms, but most don't need it in order to be successful. I have found that loosening the shackles we put on ourselves with process, medium, and philosophical dogma is crucial for growth and healthy discourse.

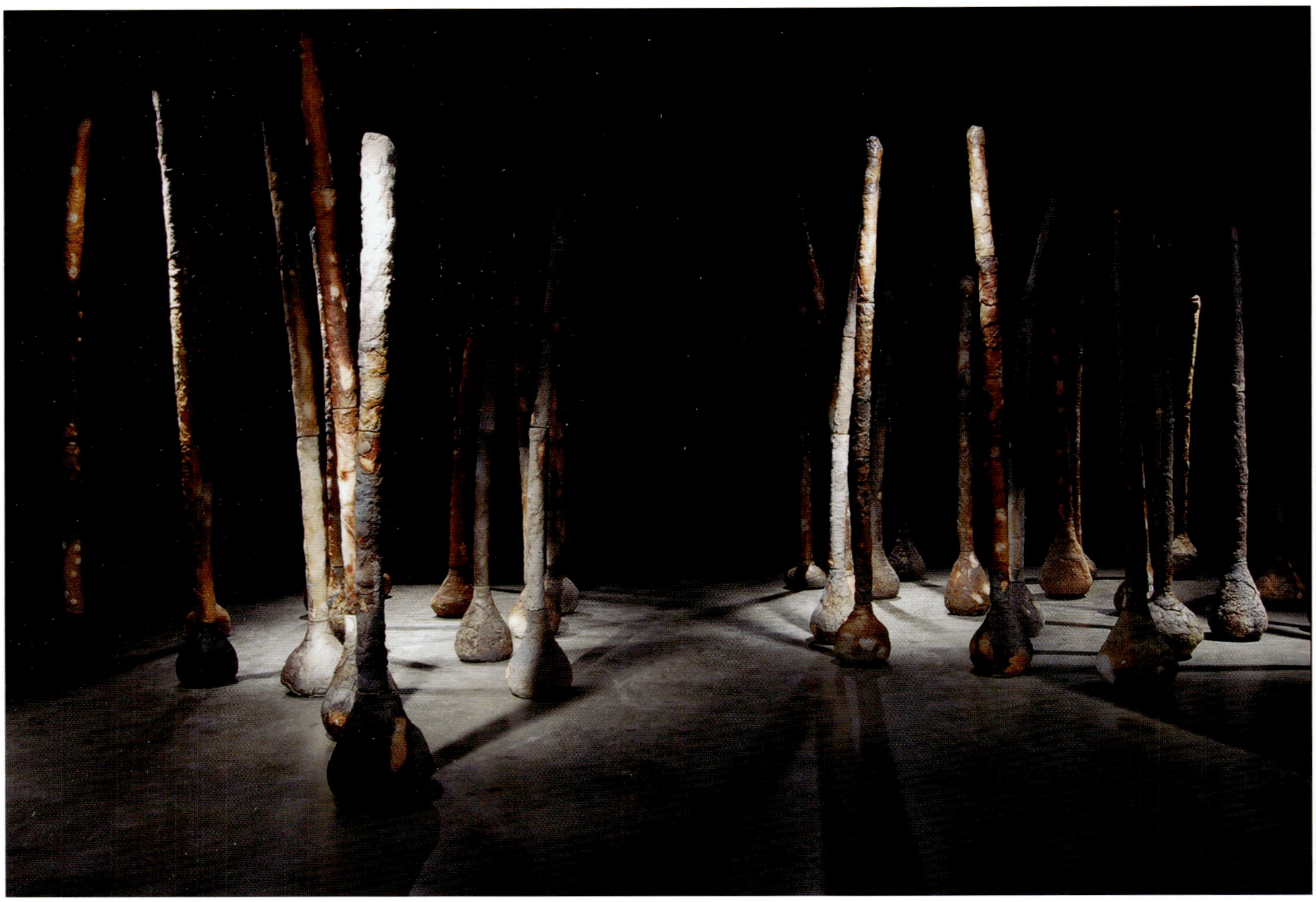

Mat Rude, *Translations*, 2010. Wood-fired stoneware and white stoneware, wheel-thrown and extruded, 11.5' x 20' x 30'. *Photo credit: Mat Rude.*

The kiln I use most frequently is a train kiln built with my students in a kiln-building course. I also use a wood/soda cross-draft and a chubby anagama, also built with students. Our train is similar to most trains, with a few modifications. Most notably, the throat arch is a sprung arch with three openings six inches above it to help eliminate the "triangle of death" in the front of the ware chamber and to give the kiln more breathing room in the area between the firebox and ware chamber. We also use three stainless steel bars, instead of two, as a grate system in the upper firebox to help keep the wood in place. The kiln is a side loader with a sprung arch over the ware chamber. The throat arch, ware chamber arch, door arch, and firebox arch are all the same arch in terms of length and rise. The responsiveness of the train, its relatively short firing time (forty hours is my preference), and the minimal amount of wood used, make it a wonderful teaching tool, perfect for academic institutions as well as studio artists.

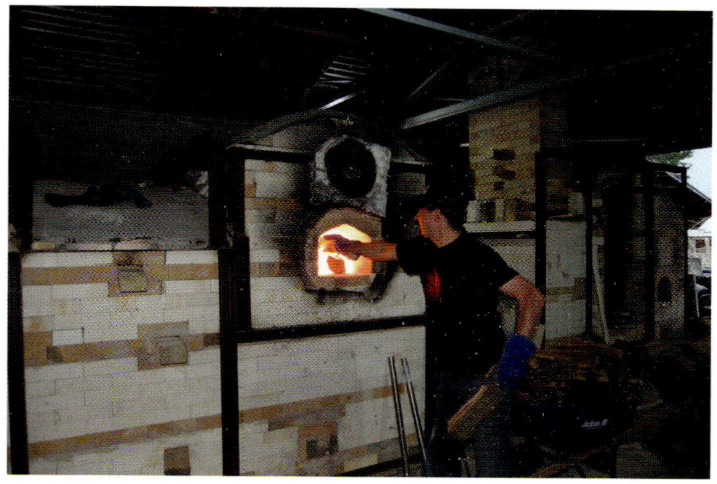

Mat Rude, Firing Train Kiln. *Photo credit: Mat Rude.*

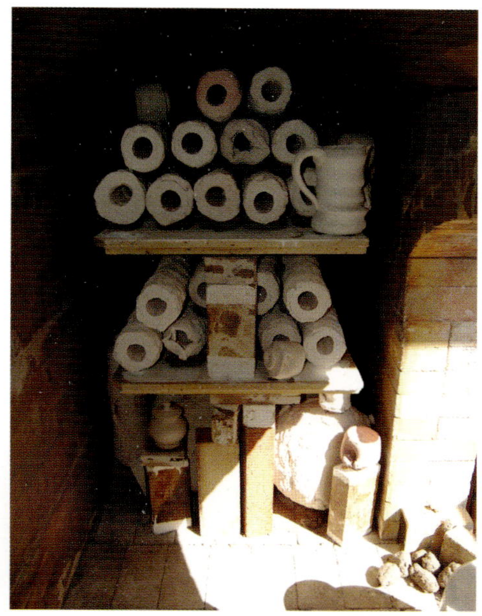

Loading Train Kiln. *Photo credit: Mat Rude.*

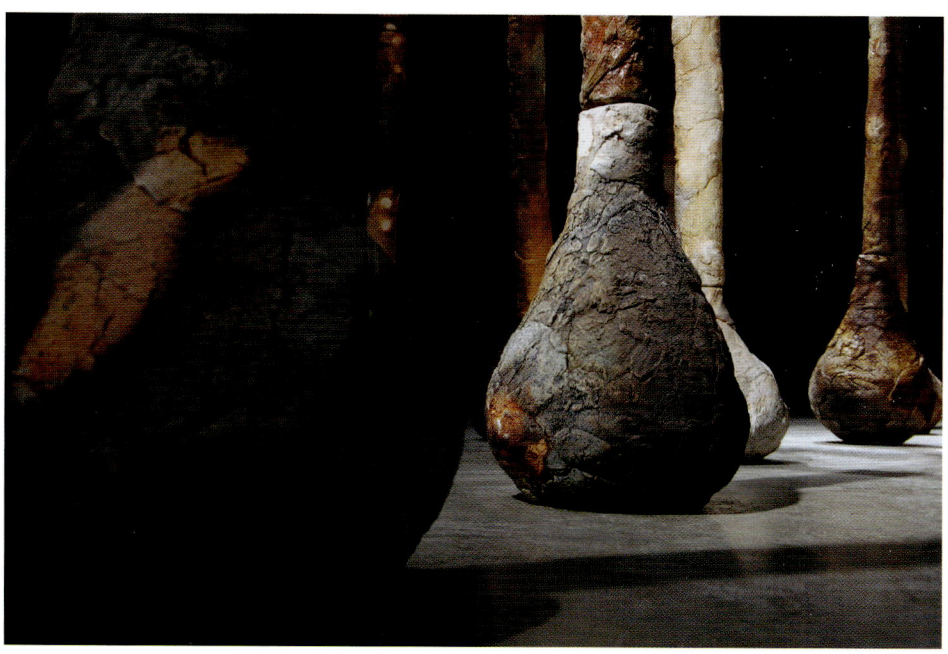

Mat Rude, *Translations* (detail), 2010. Wood-fired stoneware and white stoneware, wheel-thrown and extruded, 11.5' x 20' x 30'. *Photo credit: Mat Rude.*

RURI

Sheridan, Oregon
RuriClayWork@aol.com, www.ruriclaywork.us

Artist's Statement

I still clearly remember the strong impact I felt when I saw a yakishime yohen vase coming out of the kimitsu room of Mr. Shimaoka's noborigama in Mashiko, Japan, in 1977. Through this vase, I saw the whole Universe with multiple layers of matt colors and natural textures in such depth. I thought this is the direction I would like to follow in my clay work.

I have been always intrigued by the textures, weathering effects, and the natural contours that trees, clouds, mountains, etc. present against the vast sky, and also each other. I also observe natural harmonious compositions created by rocks in the woods, on the beaches, in the mountains, in the rivers and streams. All these natural elements resonate with each other, orchestrating a beautiful symphony.

After spending sixteen years of wheel throwing and alteration of forms fired in a reduction gas kiln at cone 11, the shift came after my mother's passing in 1989. I wanted to create something I was feeling deep within myself, and it was taking place in creating abstract clay sculpture using the coiling technique without thinking whether it was functional or not.

My inner urge for the yakishime yohen effect had been kindling all the time deep within me, and the clay sculpture I was creating was telling me not to be glazed, but to be christened by fire by wood burning itself, thus becoming a part of nature.

This gut feeling of wanting to recreate natural weathering on my clay sculpture, and making it become a part of the whole of Nature was getting stronger and stronger.

Thus, after so many years since the experience at the Shimaoka Pottery, *FuuKooGama* came into reality. And I have been even more strongly intrigued by the whole process of wood-firing—very spiritual and metaphorical, especially in yakishime yohen, all the natural elements and the artist's spirit come together to give birth to a new life.

I choose to utilize the technique of yakishime yohen, since it is based on pure interaction between clay, fire, falling ash, and the kiln atmosphere without any applications of glaze chemicals, slips, etc. Here the anagama is acting like an alchemist creating jewels of microcosm.

Each firing has been presenting such a wide range of yohen effects, which makes each firing more challenging based on the previous firings, since the size and type of work, clay used, and how they are loaded affect so much how the flames travel through the kiln, and results in creating variety of yohen. So many variables are involved, and just like in Nature, they all seem to be working harmoniously, sometimes with unexpected storms.

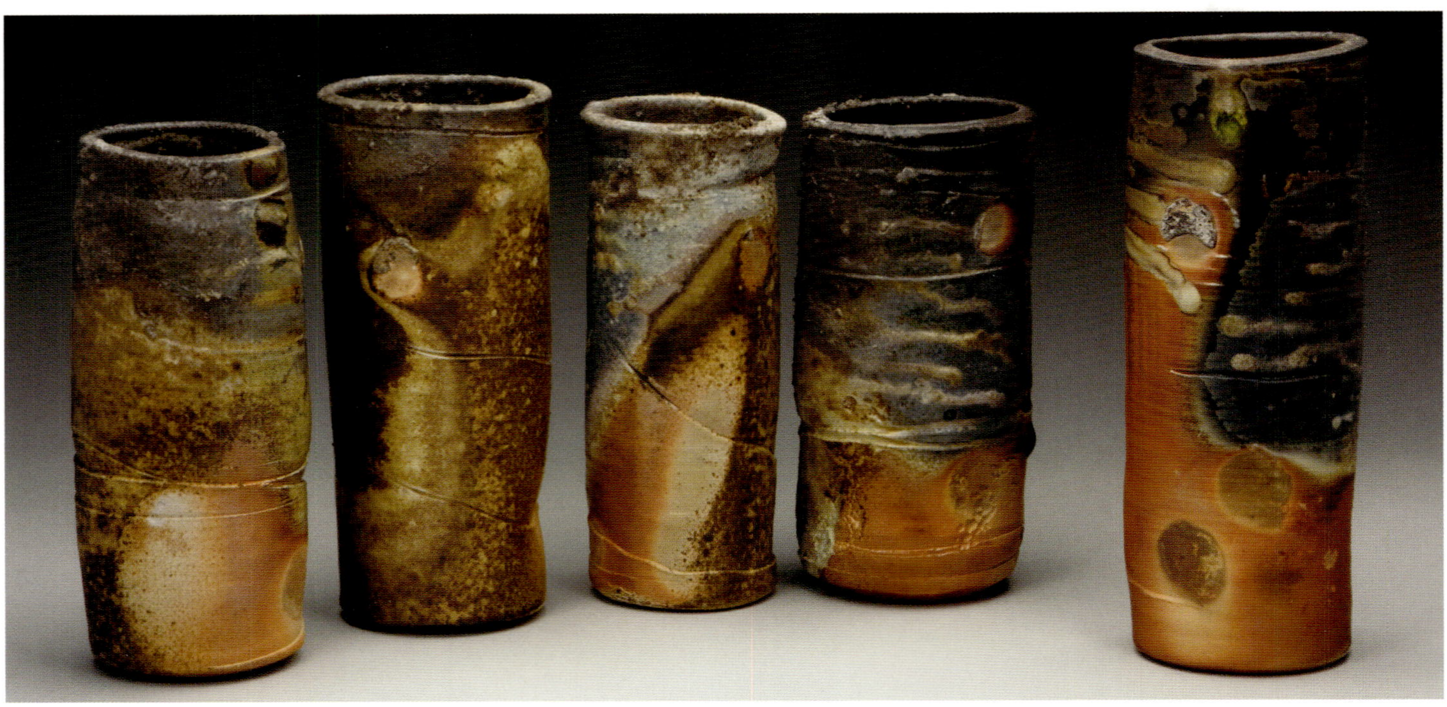

Ruri, *Cosmic Play Series*, 2009. Yakishime yohen, anagama fired porcelaneous stoneware, series of thrown and altered vessels with dancing and singing marks left by flames, hot ember, and ash, creating a microcosm, reminding us we are all a part of the whole, 8" x 3.5" x 4.5" (average size of each work). *Photo credit: Bill Bachhuber.*

All my works are fired in *FuuKooGama* (Wind & Light Kiln), which is a modified anagama with a small chamber added behind the anagama section. They call it an ananobori kiln in Japan.

I normally fire for over seven days, with three to six people in the firing crew. Using about eight cords of two foot long split firewood in three to five different sizes in diameter, using mostly Douglas fir, some maple, and oak.

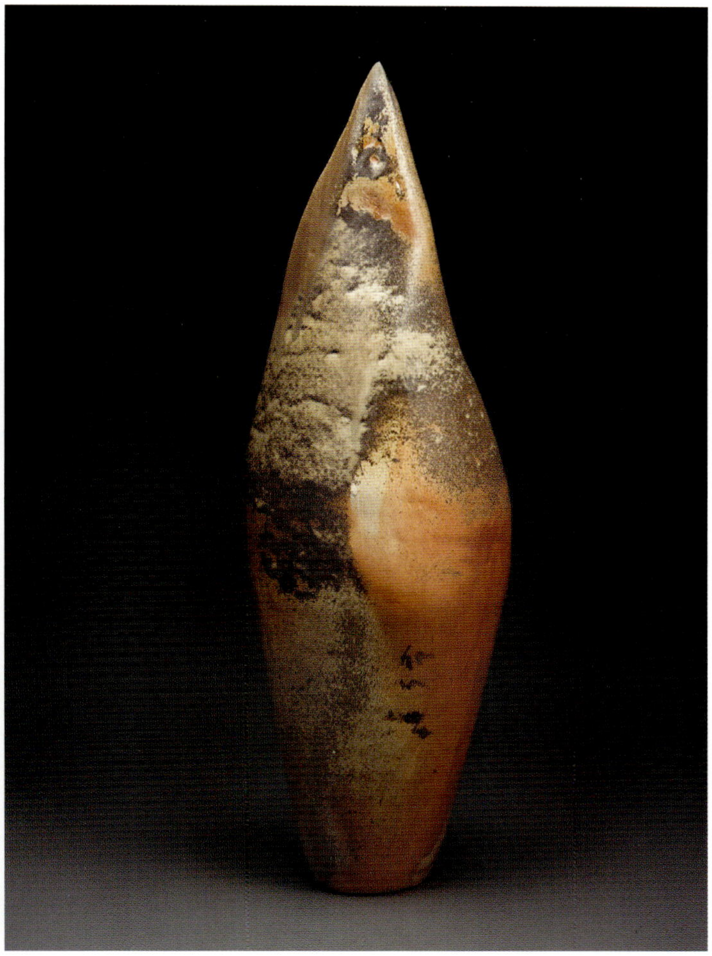

Ruri, *Yearning*, 2009. Yakishime yohen, anagama fired porcelaneous stoneware, handbuilt with coiling, longing for peace and unity, 39" x 11" x 15". *Photo credit: Bill Bachhuber.*

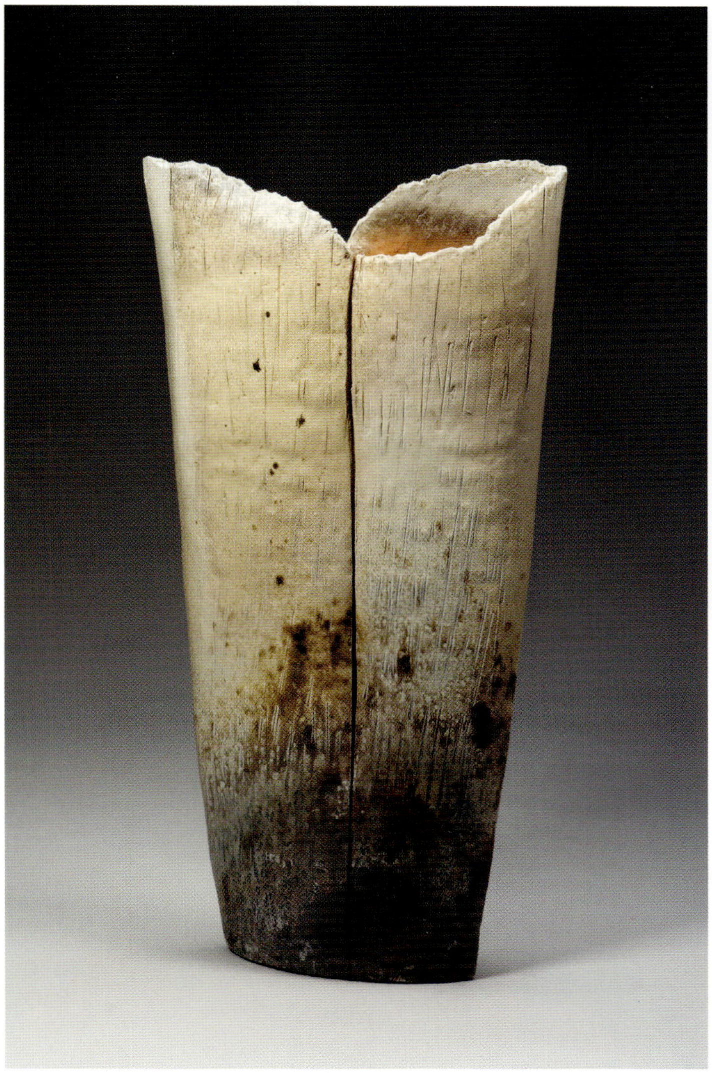

Ruri, *Prayer*, 2005. Yakishime yohen, anagama fired porcelaneous stoneware, handbuilt with coiling, created for peace within and without, 37" x 6" x 16". *Photo credit: Bill Bachhuber.*

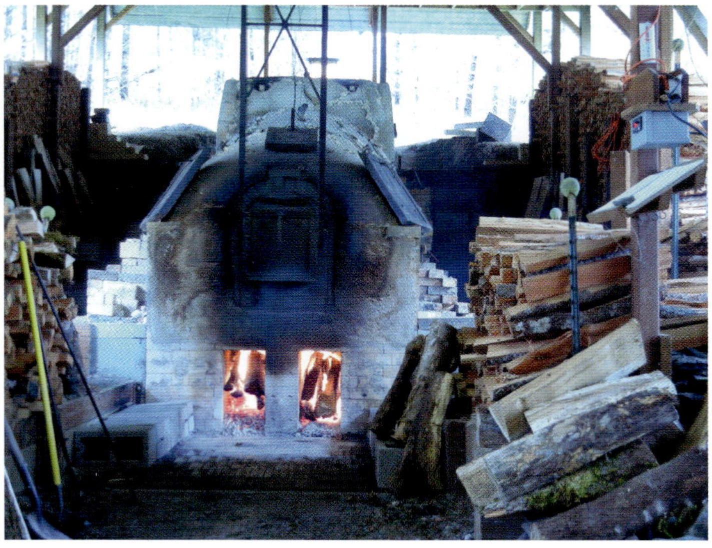

FuuKooGama kiln, view of the second day kindling. *Photo credit: Ruri.*

JEFF SHAPIRO

Accord, New York
jeffshapiro62@gmail.com, www.jeffshapiroceramics.com

Artist's Statement

I have been firing with an anagama single chamber tunnel kiln since returning from Japan twenty-nine years ago. I also fire a small down draft wood kiln.

I have chosen to fire with wood as an aesthetic choice first. The surfaces that are produced from a wood kiln are very different from those produced in other kilns: gnarly, crusty, with the potential for drips that range in color from green jade to amber to mustard yellows and an occasional opalescent blue. Sometimes the effect of melted ash on a dark iron rich clay background can give the appearance of comets and constellations in a night sky. On other occasions the work may appear to have been dug up from a volcano or after having been buried beneath the ocean for millennia.

Another aspect of wood-firing that drew me to it was the lifestyle, in that the work and the daily life are directly connected. In a way, it is a little like farming; everything has its time and cycle to be done. And the eight day firing, though arduous and demanding, may feel like a marathon that once successfully completed gives a real feeling of satisfaction.

The techniques I employ to create my artwork are varied and numerous. The loading is a key factor in acquiring a wide range of firing results. The placement of work, whether on shelves or tumble stacked, directly affects the final work. The loading creates the patterns, and the firing creates the color and texture. I have become more of a pragmatist in recent years, rather than a romanticist. I use whatever technique is necessary to produce the required results. Slips and glazes have been added to the techniques of building a wood-fired natural ash deposit.

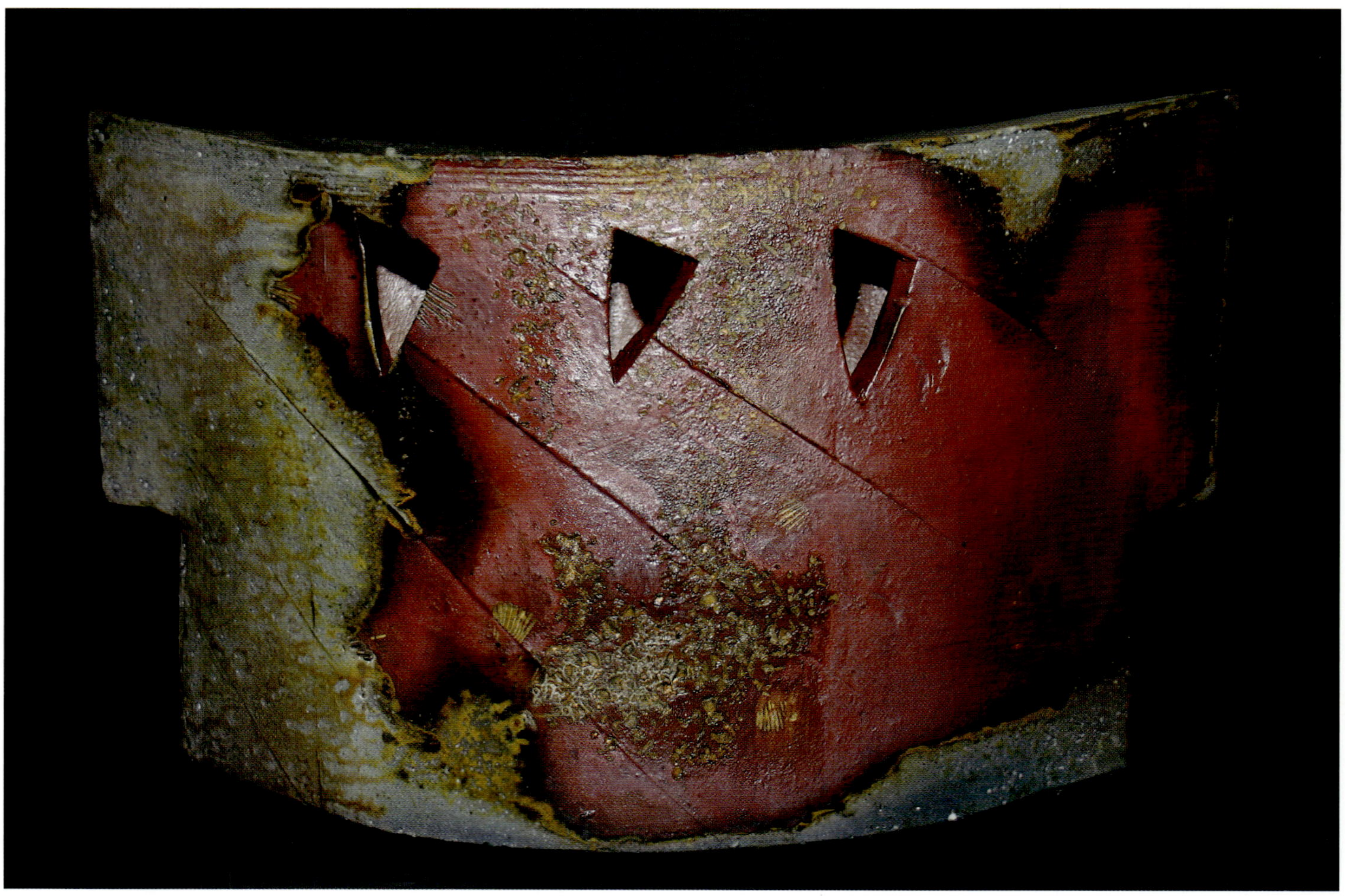

Jeff Shapiro, *Abstract form*, 2010. Wood-fired stoneware, natural ash deposit, slab construction, 22" x 9" x 13".

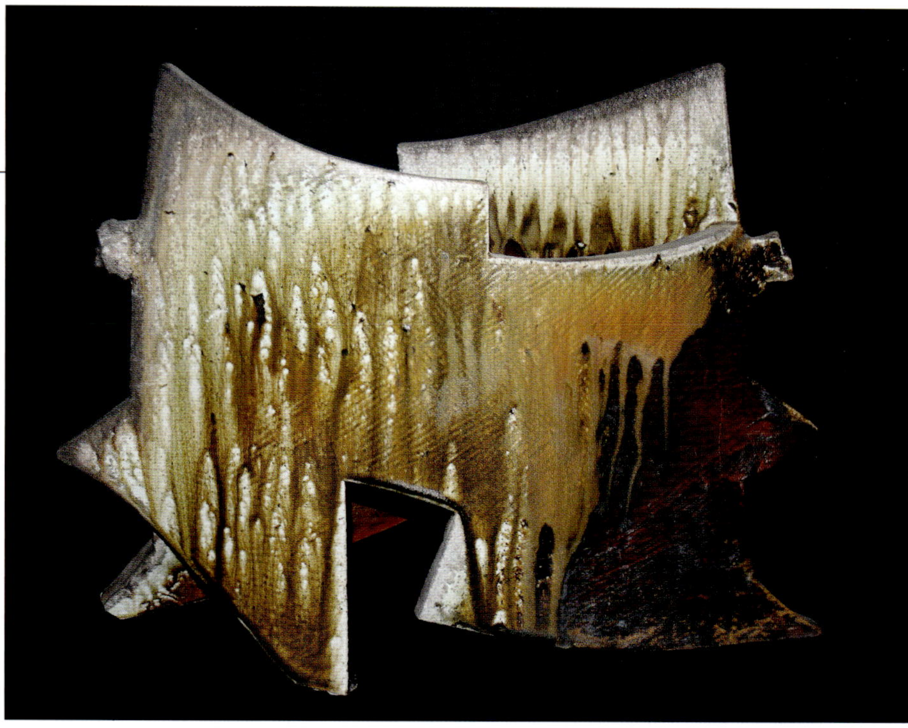

Jeff Shapiro, *Abstract Vessel*, 2010. Wood-fired stoneware, natural ash deposit, slab construction, 20" x 9" x 15".

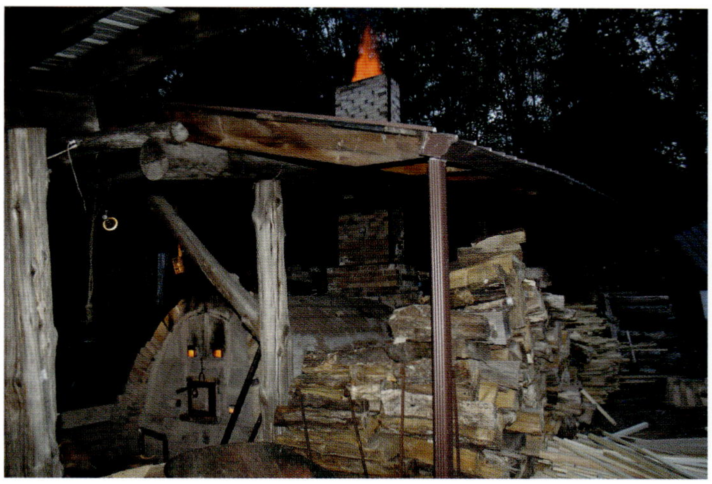

Anagama tunnel kiln at night.

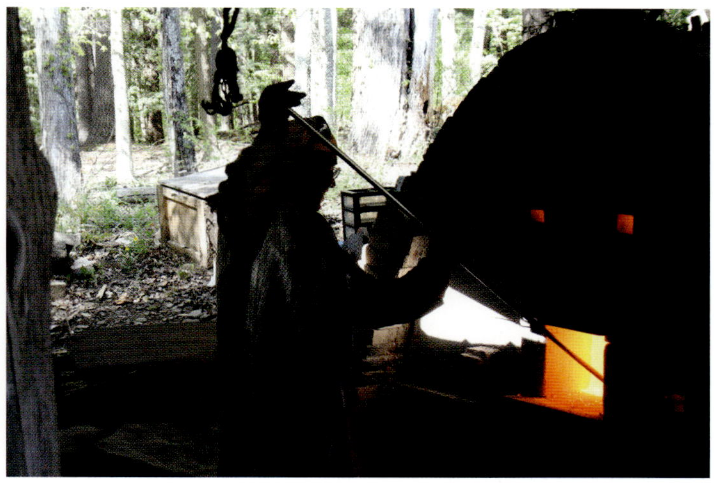

Jeff Shapiro, Front stoking anagama kiln.

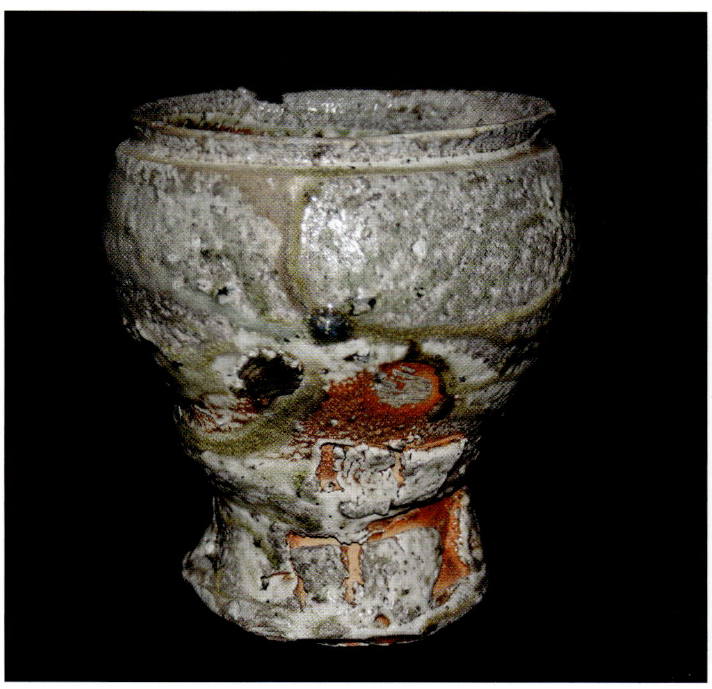

Jeff Shapiro, *Altered Vessel*, 2010. Wood-fired stoneware, natural ash deposit over thin shino, wheel-thrown and altered, 9" x 8.5" x 9.5".

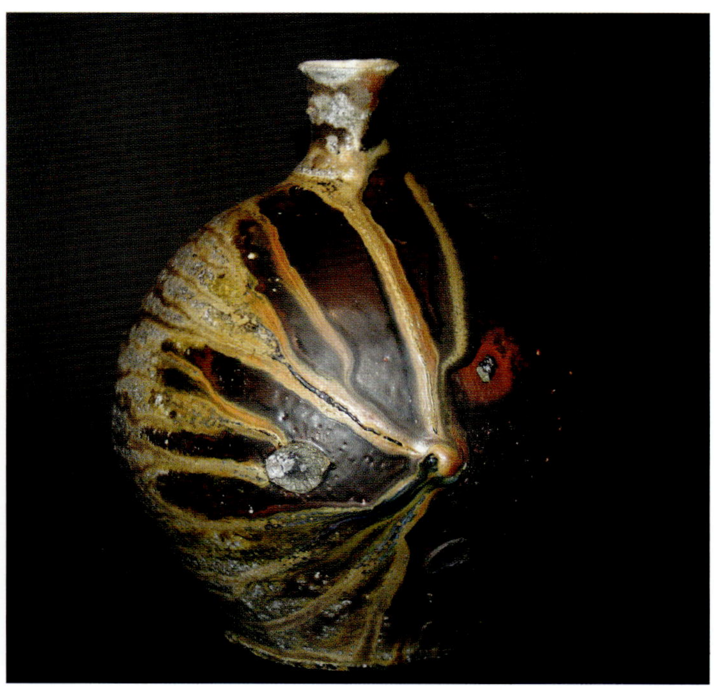

Jeff Shapiro, *Round Jar*, 2011. Wood-fired stoneware, natural ash deposit, wheel-thrown, 7" x 7" x 8".

WILLI SINGLETON

Kempton, Pennsylvania
info@willisingleton.com, www.willisingleton.com

Artist's Statement

I was introduced to high temperature wood-firing during an apprenticeship in Tamba, Japan, in 1982. The simplicity and power of the Tamba wood-fired climbing kilns awed me. From that point on I have focused on firing with wood. After I left Tamba, I traveled extensively within Japan, investigating possibilities of further study. I found an amazing variety within Japanese wood-fired ceramics. I was eventually drawn to Mashiko, where I learned how to throw on the kickwheel and participated in kiln building projects.

After two years in Mashiko, I returned to the United States in 1987 to establish a pottery on family land in eastern Pennsylvania and built a four chamber climbing kiln. I have been working here by myself since 1990, but occasionally get some assistance from my daughter, Sami, and the many friends that arrive to help me fire the kiln several times a year. The recently rebuilt kiln and I have now survived more than fifty firings.

My pots are mostly wheel-thrown on a kickwheel and fully glazed. I make my glazes mostly from materials coming from the surrounding area—creek clay, wood ash, corn stalk ash, and bamboo ash. I also use feldspar, kaolin, silica, and two oxides from the ceramic supply house. Most of my pots are made with a mixture of a fine grained red clay from the Stancill's clay mine located where the Susquehanna enters the Chesapeake Bay, with another clay from the top of Hawk Mountain just up the road from my pottery. For large jars and large plates, I use the same clay mixed with industrially mined clay. For some other pots, especially for tea ceremony, I use just the Hawk Mountain clay. I don't consider myself a "purist" (i.e., restricting myself solely to local materials) but I find it interesting and workable to use primarily locally obtained materials.

Making pots is like cooking. You have to start with good ingredients to get a flavorful, satisfying result. By using local clays and various ashes I am trying to make pots that have a pleasing flavor. Using a wood flame to fire these pots enhances the character of these materials. Although there are not many obvious ash deposits on my pots, the variety of colors and textures in them are largely due to the wood flame interacting with the clay and glaze materials, just as food cooked over an open fire carries a taste of the flame and smoke.

Working with natural materials and with the unpredictability of firing in a wood kiln, I am always surprised when I open the kiln. Although the potter has a lot of control over the processes leading up to the firing, the kiln ultimately decides how each piece will be transformed. My glazes consist mainly of mixtures of plant ashes (wood, corn stalk, bamboo, and rice hull) with feldspar and creek clay, and have a narrow range of successful firing, so manipulating the fire to achieve the correct temperature is critical.

Although wood-firing potters are viewed by some as anachronistic masochists, I believe wood-firing becomes increasingly relevant in this age of incredible technological advances. The growing popularity of wood-firing shows the seductive allure of this labor intensive approach. In contrast to anagama firing, many of my pots don't appear to have been wood-fired at all, but upon close inspection, they are each altered somehow by their encounter with the fire. The unidirectional flame, the slight flashing due to ash deposits, and the always fluctuating kiln atmosphere make interesting things happen. Wood-firing won't turn a bad pot into a good pot, but there's a chance that it will turn a good pot into a great pot. By using a wood flame and natural clays and ashes, I believe there is a possibility of attaining a deeper and more subtle beauty.

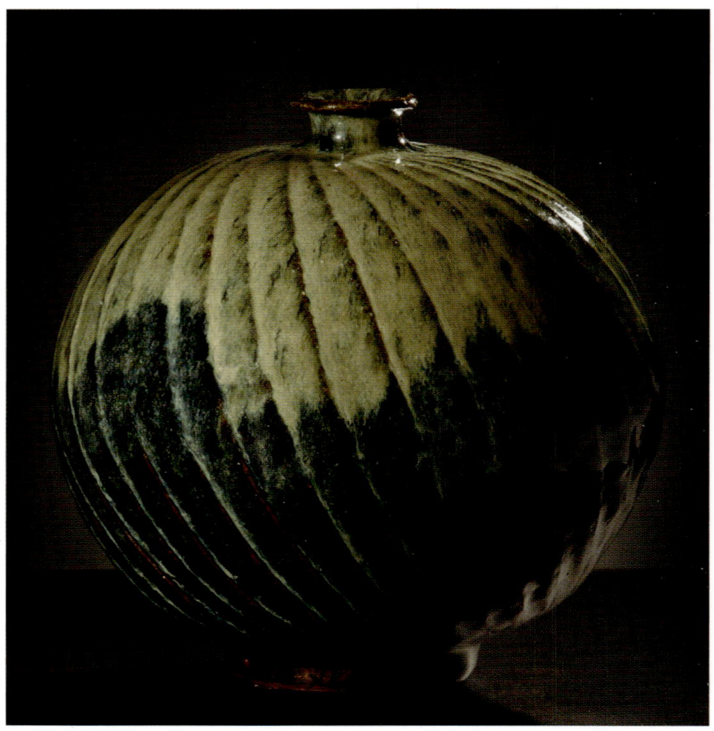

Willi Singleton, *Carved Vase*, 2011. Wood-fired local clay, kickwheel-thrown and carved, bamboo ash glazed upper, creek clay glazed lower, creek clay glaze on lip, 8.5" x 8.5" x 8.5". *Photo credit: Kenek Photography.*

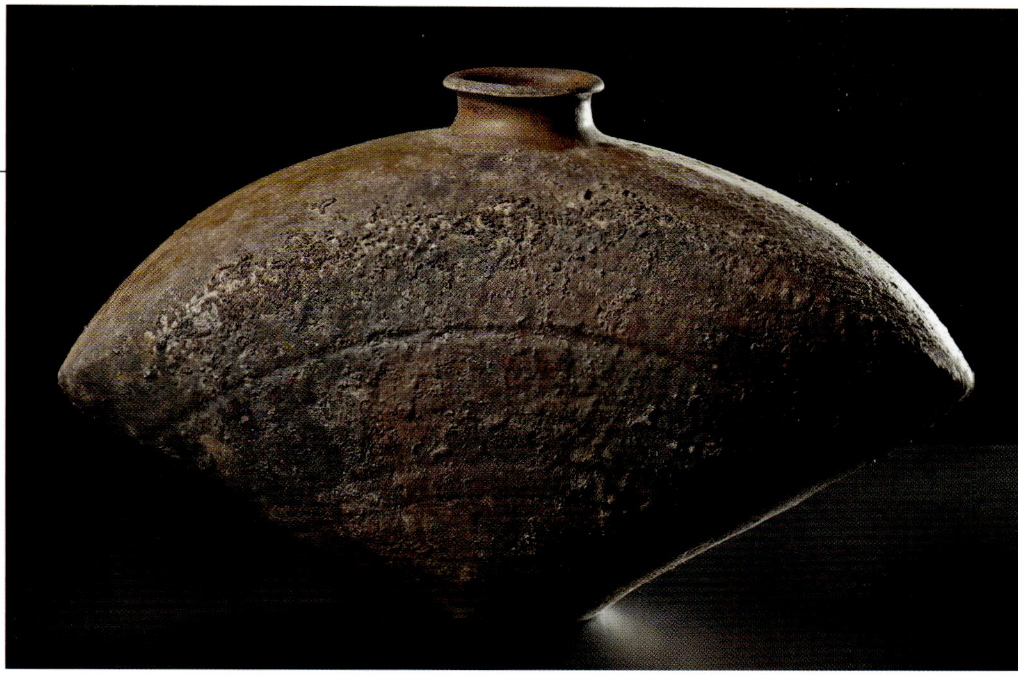

Willi Singleton, *Ash-encrusted Vase Form*, 2011. Wood-fired local clay, kickwheel-thrown with slump formed addition, thin coating of creek slip, covered with ash deposits from flame and coals, 19" x 11" x 9". *Photo credit: Kenek Photography.*

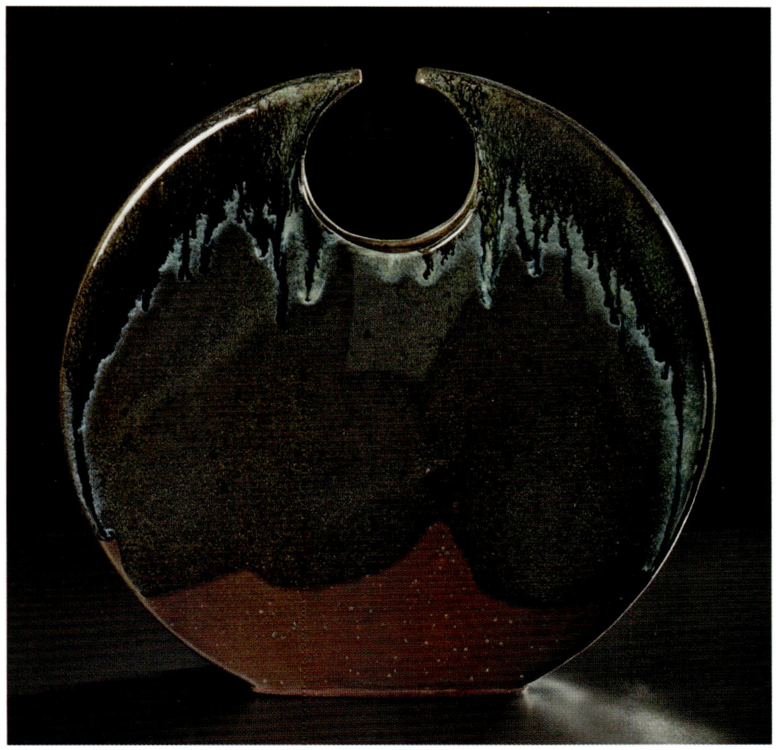

Willi Singleton, *Moonlight Landscape Vase*, 2011. Wood-fired local clay, kickwheel-thrown, cut and assembled, bamboo ash glazed upper, creek clay glazed lower, creek clay glazed lip, 14" x 14" x 4". *Photo credit: Kenek Photography.*

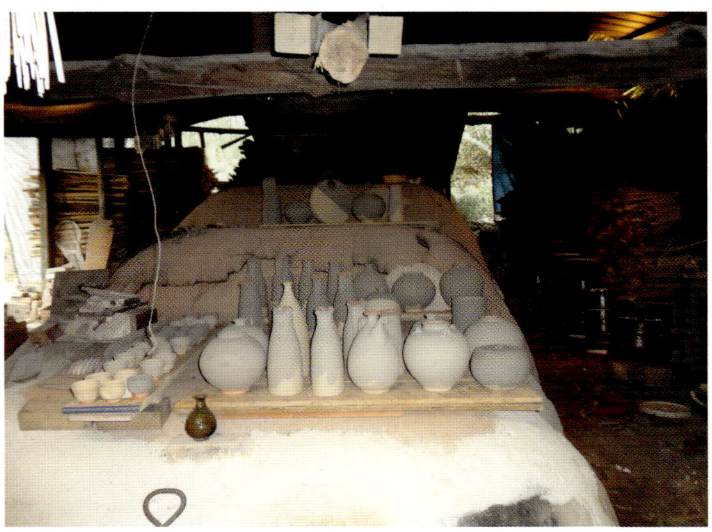

Singleton kiln, with pots atop the kiln during loading. *Photo credit: Willi Singleton.*

CONIFER SMITH

Cedar Rapids, Iowa
conifersmith@yahoo.com, www.newarts.com/wildfire

Artist's Statement

To create the sculptural pieces, a substructure of very coarse stoneware with sand and grog is used to withstand the heavy layer of porcelain slip applied when the piece is dry. Porcelain is chosen for its silky texture, purity, and reaction to the wood-fire process. The application of porcelain slip on the base of dry sculpture clay results in the porcelain cracking as it shrinks and dries. This texture enlivens the surface. What some view as a negative drying effect has a positive result in the sculptures.

Minerals in the clay and the type of wood fueling the kiln create the color palette I choose to work with. My challenge is to capture subtle differences in colors and vibrancies. Common colors are greens, yellows, and browns. At times subtle blushing pinks, flesh tones, rich cabernets, and warm violets are possible. An anagama style kiln helps to achieve these results. Firing with wood, the flames search out a path through the kiln. This path is recorded on the pieces. At times it leaves a mark reminiscent of a quiet caress; other times the marks evoke a sense of pulsating urgency and power.

Reduction cooling, introducing a few sticks of wood into the kiln at approximately 1450 degrees Fahrenheit during the cooling process, deepens and enriches the colors. I am continually seeking stochastic results in the surfaces of my sculptural work.

The teapots are hand-built from one slab of porcelain. Repeated stretching of a spiral wedged lump of clay produces the slab's texture. This texture resembles human skin in some areas, waves or tree bark in others. Joining is difficult without removing this texture. A fine layer of slip is applied before they are fired. I enjoy the subtle results achieved in firing a wood fueled hybrid *anagama*.

Intuition has become a large part of my creative process. While working in clay, I often attain a mindset where I am unconscious of what is going on around me and become unaware of the passage of time. Everything seems effortless, and for brief periods I dance with clay.

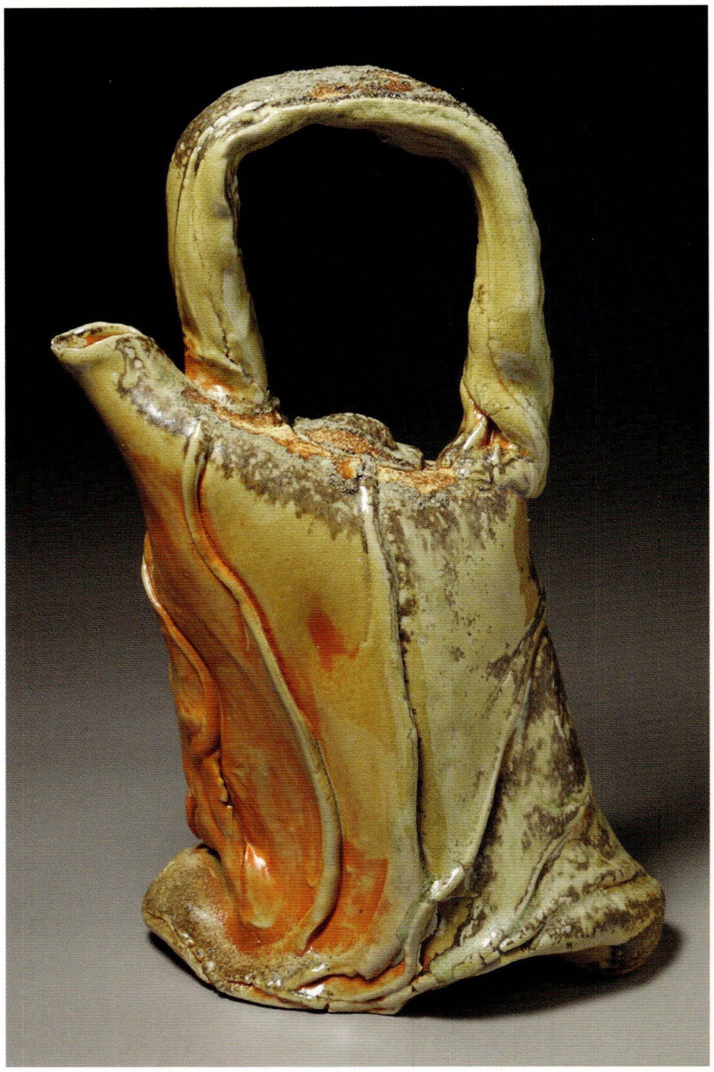

Conifer Smith, *Art Nouveau Teapot*, 2007. Wood-fired porcelain, handbuilt, reduction cooled, 13" x 6" x 3". *Photo credit: Vernon Windsor.*

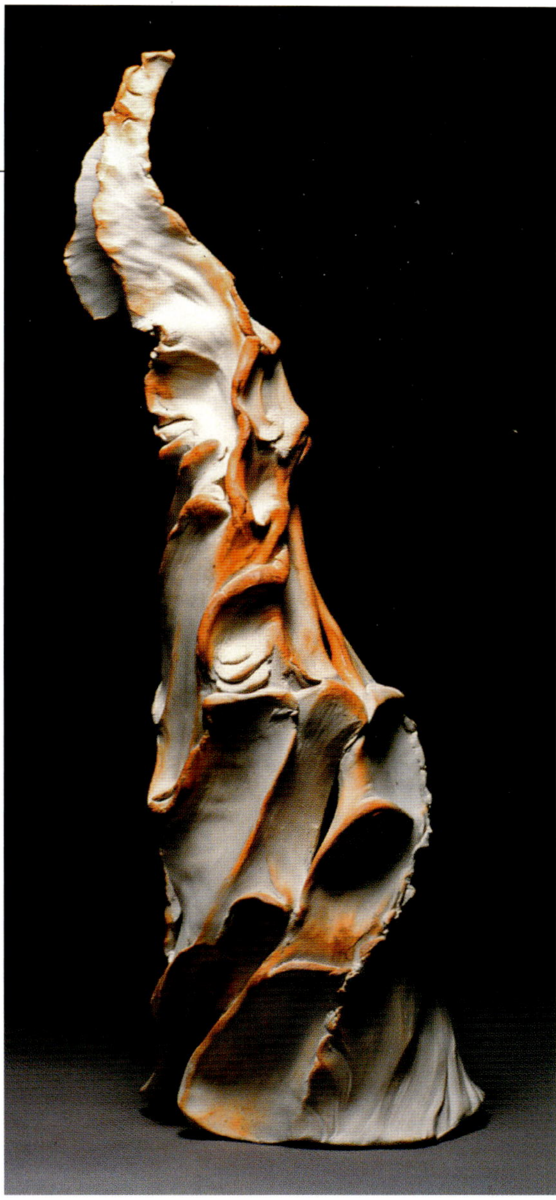

Conifer Smith, *Winter Fire Form*, 2003. Wood-fired porcelain, soda ash wash, reduction cooled, 27" x 7" x 7". Photo *credit: Vernon Windsor.*

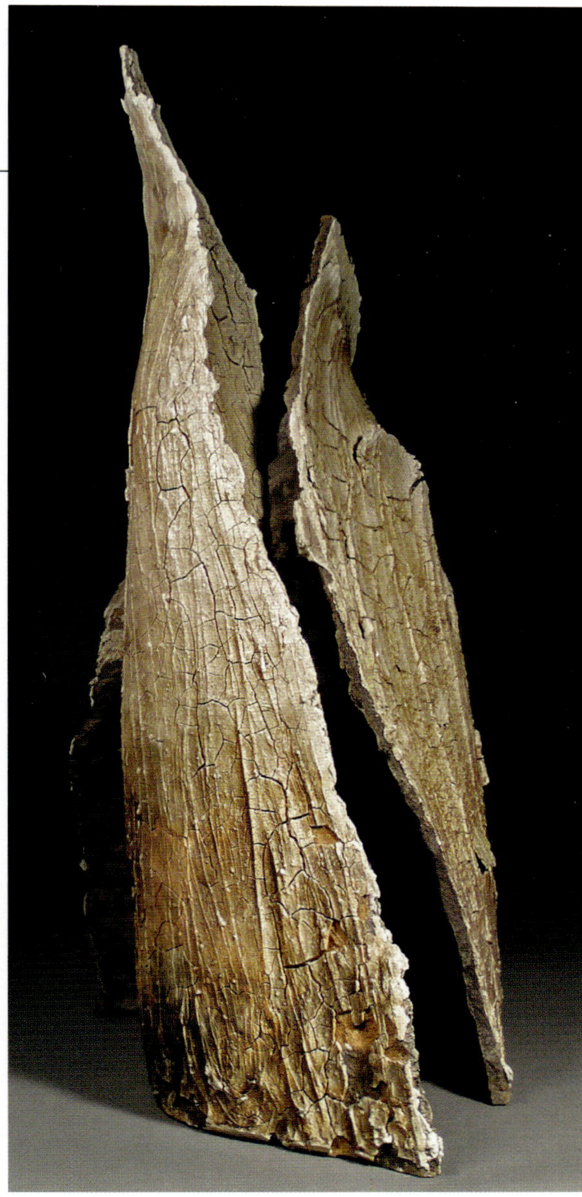

Conifer Smith, *Canyon Conversations*, 2005. Wood-fired stoneware, with porcelain slip, reduction cooled, 37" x 20" x 10". *Photo credit: Vernon Windsor.*

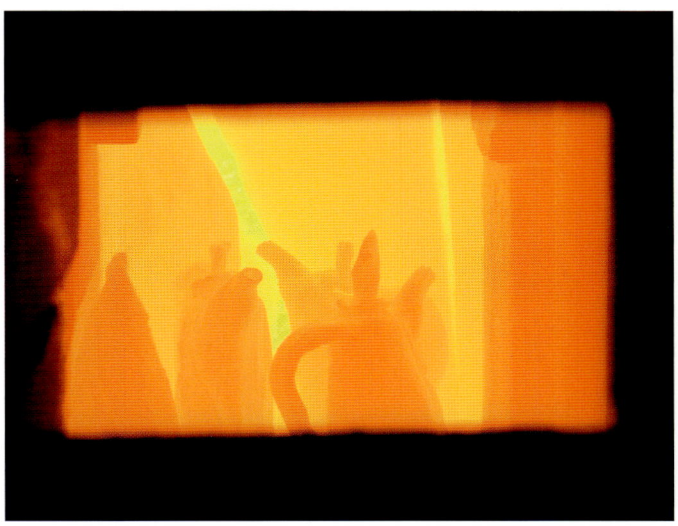

Conifer Smith, *Firing Image*, 2008. *Photo credit: Conifer J. Smith.*

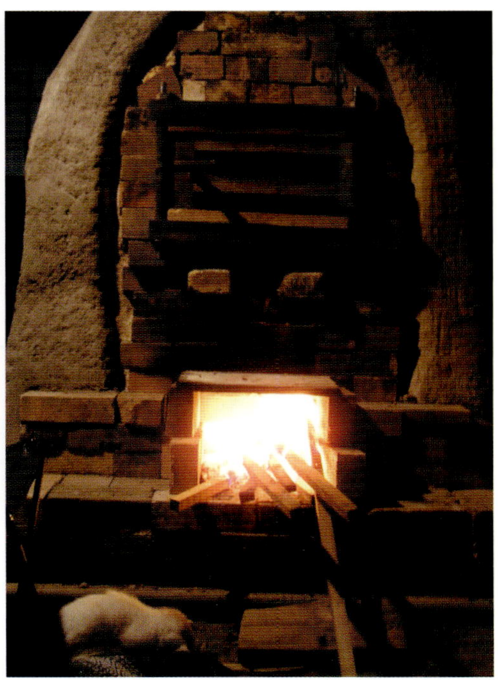

Conifer Smith, *Kiln Image*, 2008. *Photo credit: Conifer J. Smith.*

SAMANTHA STUMPF

Rochester, New York
slstumpf@gmail.com, www.slstumpf.com

Artist's Statement

My work focuses on the forms of action and reaction that occur in natural environments, and I see my work as a visual investigation of these natural forces. A sense of simultaneous deconstruction and construction—such as what occurs when a mountain slowly weathers or the way that water erodes a streambed—informs the process of my ceramic vessels.

Given the primary influence of nature, I see my pieces as metaphoric landscapes. I created these clay pieces by exploiting the responsiveness of the clay. I pushed, pulled, and tore the clay in an attempt to create a physical dialogue between the material and myself.

I have fired in Baltimore at Baltimore Clayworks and in Avon, New York, at Studio Sales. Both kilns are two-chamber Noborigama kilns. I chose to fire this body of work in a wood kiln because I wanted varied surfaces within the body of work. I really enjoy the results of the Noborigama kiln because of the endless variations of ash, flame, and airflow that it creates on my surfaces. Each piece in his body of work was built using wet slabs. While handbuilding each piece, I knew that this firing method was going to accentuate the textures and physical marks on the surface.

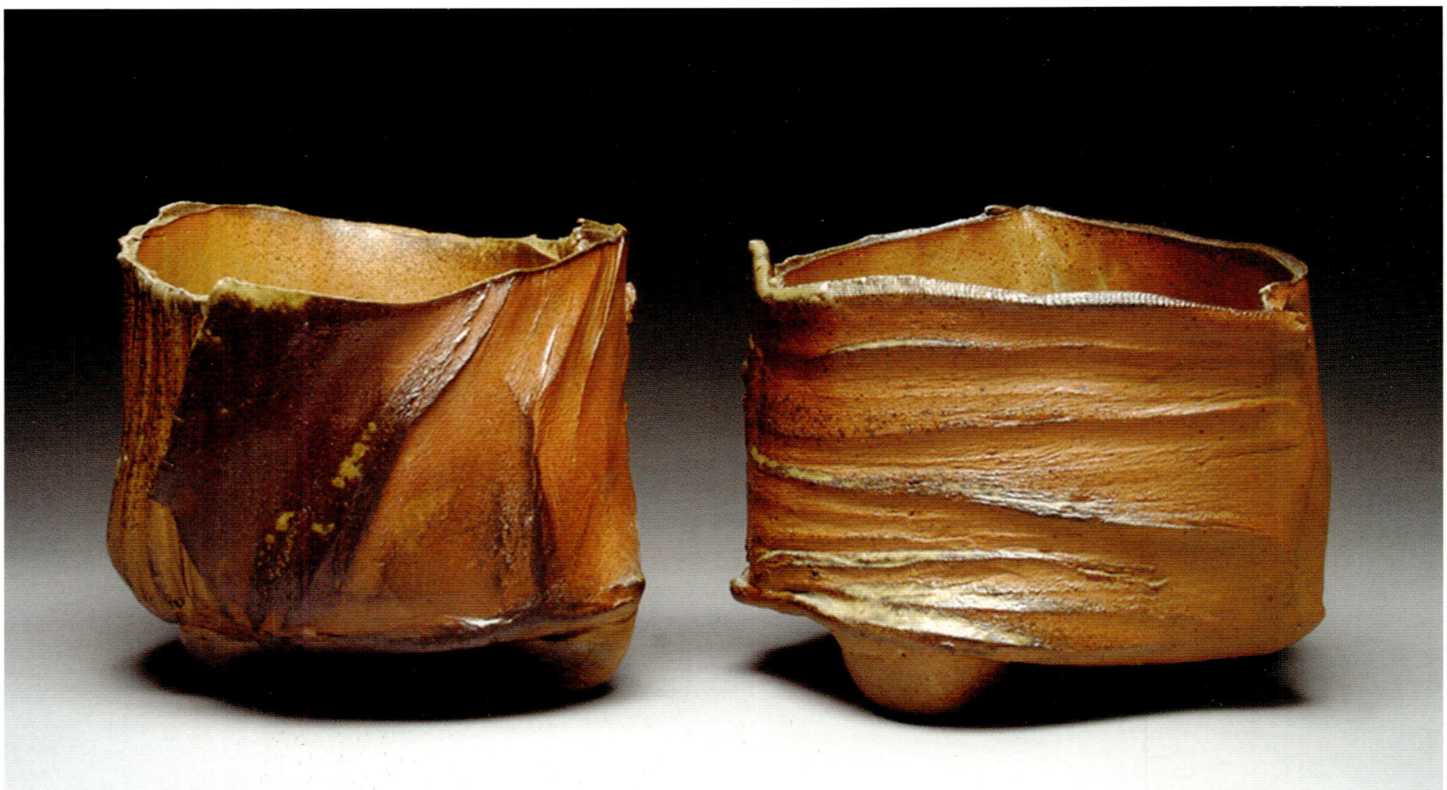

Samantha Stumpf, *Landscape Grouping #7 and # 8,* 2009. Wood-fired stoneware, handbuilt both are 5" x 8" x 6".
Photo credit: Samantha Stumpf.

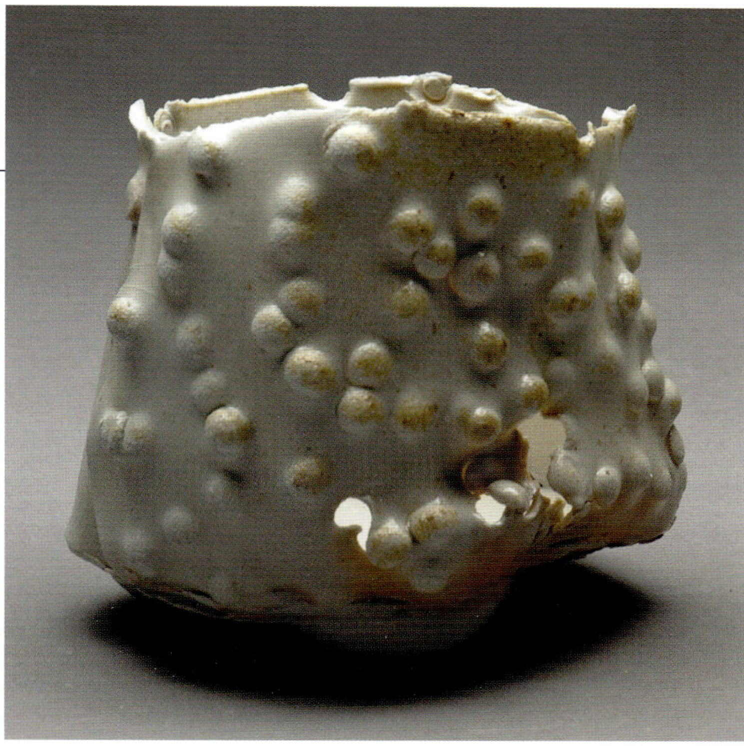

Samantha Stumpf, *Bumps*, 2010. Wood-fired porcelain, handbuilt, 5" x 7" x 4". *Photo credit: Samantha Stumpf.*

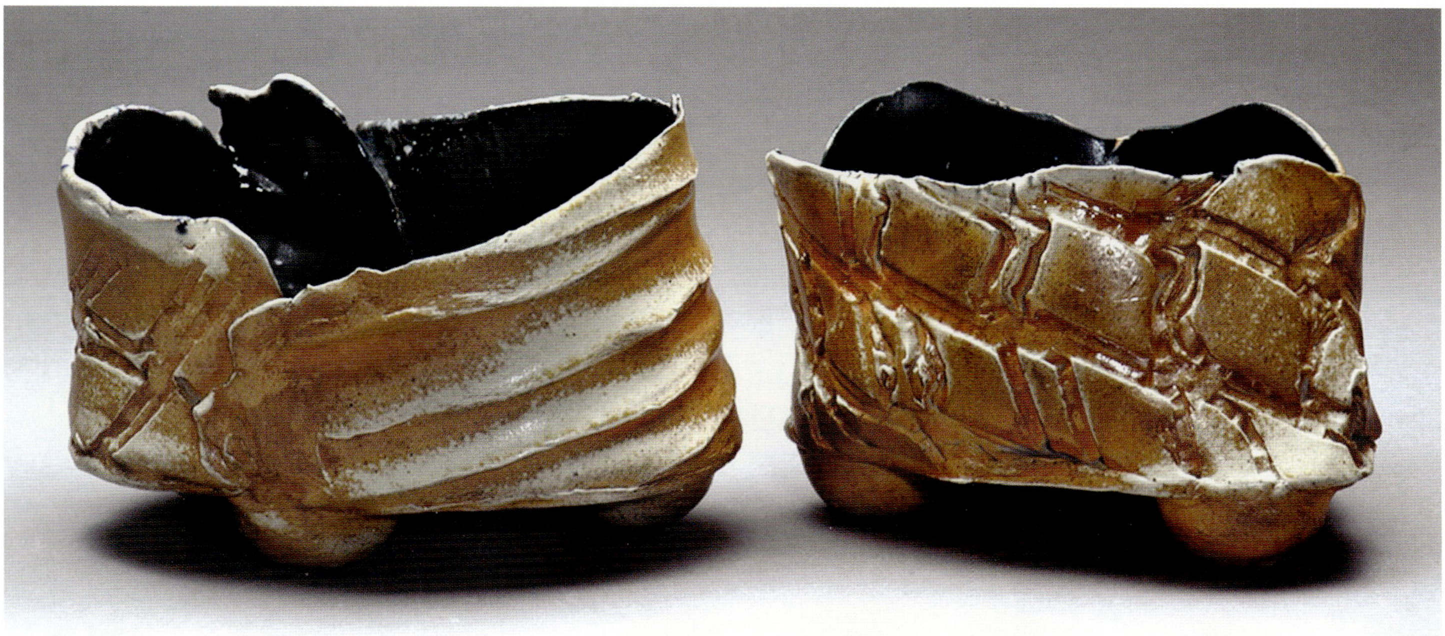

Samantha Stumpf, *Layers*, 2010. Wood-fired stoneware with colored slip, handbuilt, both are 5.5" x 7.5" x 6.5". *Photo credit: Samantha Stumpf.*

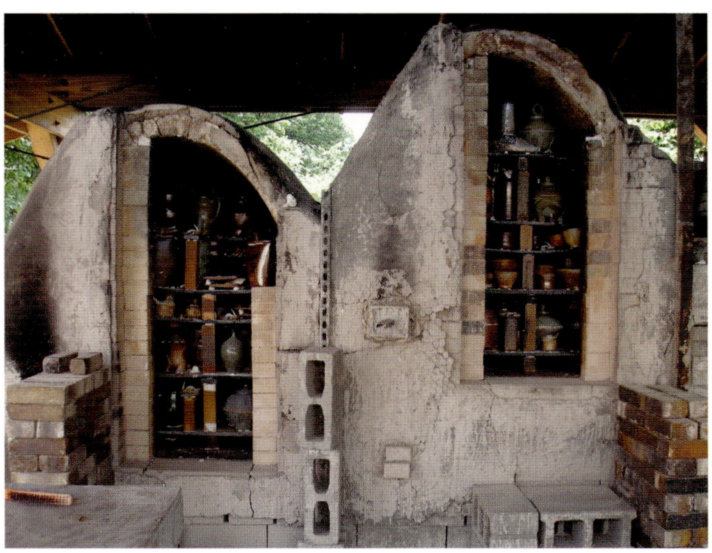

Baltimore Clayworks Kiln. *Photo credit: Samantha Stumpf.*

VINCENT SUEZ

Upland, California
vsuez@fullerton.edu

Artist's Statement

My work has its basis in traditional pottery. First and foremost I consider myself an artist/potter. I draw and paint, using traditional ceramic processes to achieve particular affects on my work in order to exploit and develop form. My concern with nature is revealed through my use of animal and bird imagery. Marks of stamps, inscribed lines, and the touch of brush emulate this fantasy in nature.

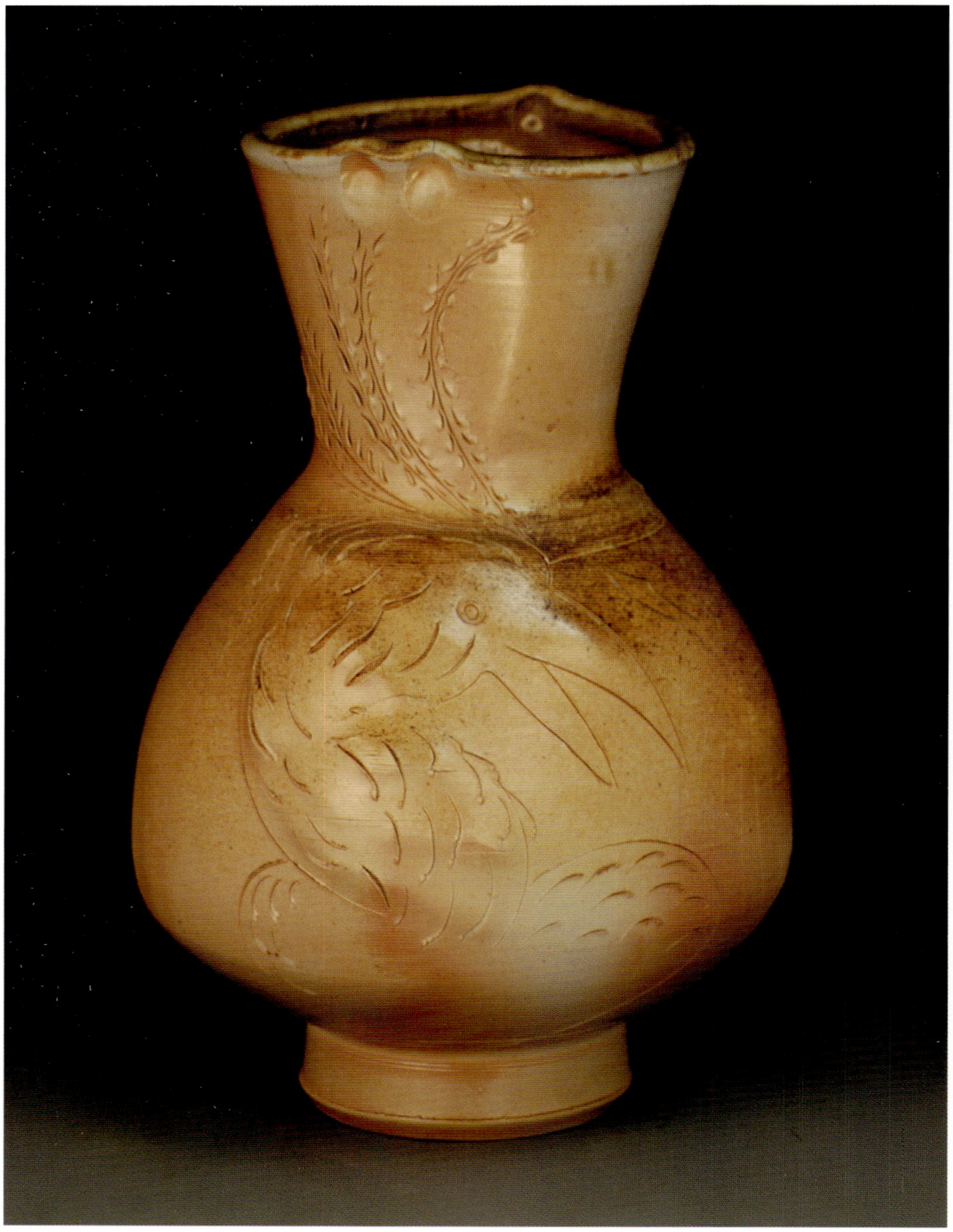

Vincent Suez, *On the Fly Pot*, 2009. Olsen's kiln Shino and fly-ash, wood-fired B-mix wheel-thrown, incised drawing relief, 17" x 10" x 10". *Photo credit: Vincent Suez.*

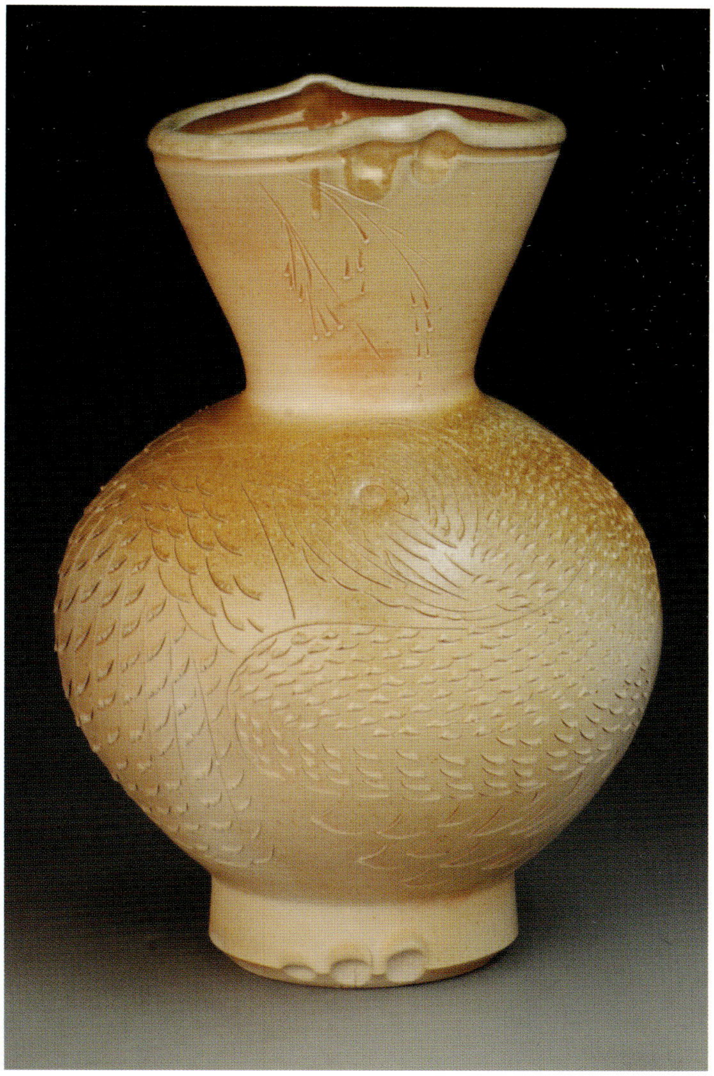

Vincent Suez, *Birds of a Feather Pot,* 2007. Olsen's kiln, fly-ash exterior/shino thin interior, wood-fired B-mix, wheel-thrown, incised drawing, 16" x 10" x 10". *Photo credit: Vincent Suez.*

In 1985 Fred Olsen invited me to be part of his wood-kiln building workshop. As the years passed I have become friends with Conrad Calimpong in Ferndale, California. He also has a wood anagama and has invited me to fire with him. Prior to this, my work had been painting and drawing figurative landscapes with a Majolica attitude at cone 6. At first I simply made marks in the pots and found that I needed something more. Thus the human condition and my relationship with nature jelled together and I am where I am with my work no matter the type of firing. There is also the camaraderie and the labor and quiet time. They are all very appealing to me. I feel fortunate that I have friends that also relish this process. As my friend Joe Bennion says, sometimes its "... Christmas and some times it's Halloween ...". For me the process is always enjoyable, the friendships are rewarding, and, who knows, sometimes you get a real winner.

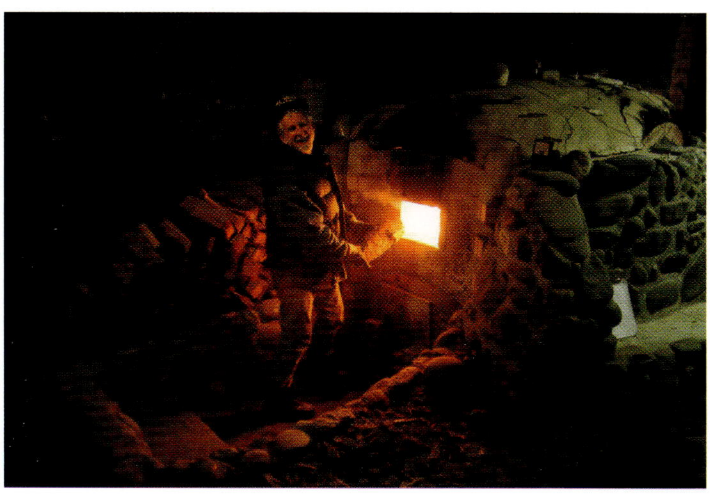

Conrad Calimpong's Kiln. *Photo credit: Conrad Calimpong.*

JOY TANNER

Burnsville, North Carolina
joy@joytannerpottery.com, www.joytannerpottery.com

Artist's Statement

The current focus in my wood and soda fired pottery is the interplay between form, texture, detail, and variations of surfaces. I am intrigued with the repetition of patterns I find in nature and I incorporate traces of these textures into the clay surfaces. Because of my interests in the environment and sustainability, I fire my work in a wood kiln using waste wood and strive to bring this awareness of conservation into my work and my lifestyle everywhere I can. The kiln is a one chamber sprung arch wood kiln with an internal firebox and is designed to burn wood pallets discarded from local industry and headed for the trash pile. By reclaiming these materials and using them as energy, I am preventing this wood fuel from becoming a wasted energy source. The wood flame and ash that form during the firing create striking variations and earth tones on the surface of my pottery and is integral to the success of my work.

Cradling a cup or bowl in hand, I hope people connect with my functional pots through the subtle textures I have impressed or hand carved into the clay. At the wheel, I consider the fullness of a form, the division of contrasting surfaces, and rhythms that occur in the textures. These individual elements are treated similarly to how I observe nature. At first glance, my eyes are unable to focus on the variety found within nature that surrounds me. Yet the little details are what I am drawn to, such as a leaf connecting to its stem, the folds of a mountain range, or the softness of a seed enclosed in its shell. These small components in nature that reveal such variations are what I feel bring vitality and energy to my life. By bringing these details into closer view, I hope each pot possesses the energy and insights I gather from my natural surroundings. My admiration of nature comes full circle, directly into the hands using my pots. From drinking coffee out of one of my carved mugs or serving a fresh green salad in one of my bowls, it is my hope that each detail I've revealed in my pottery brings a sense of awareness and joy to the daily rituals in our lives. Perhaps by seeing my impressions of detail, these people too, will realize the importance of our environment and make sustainable choices within their lives.

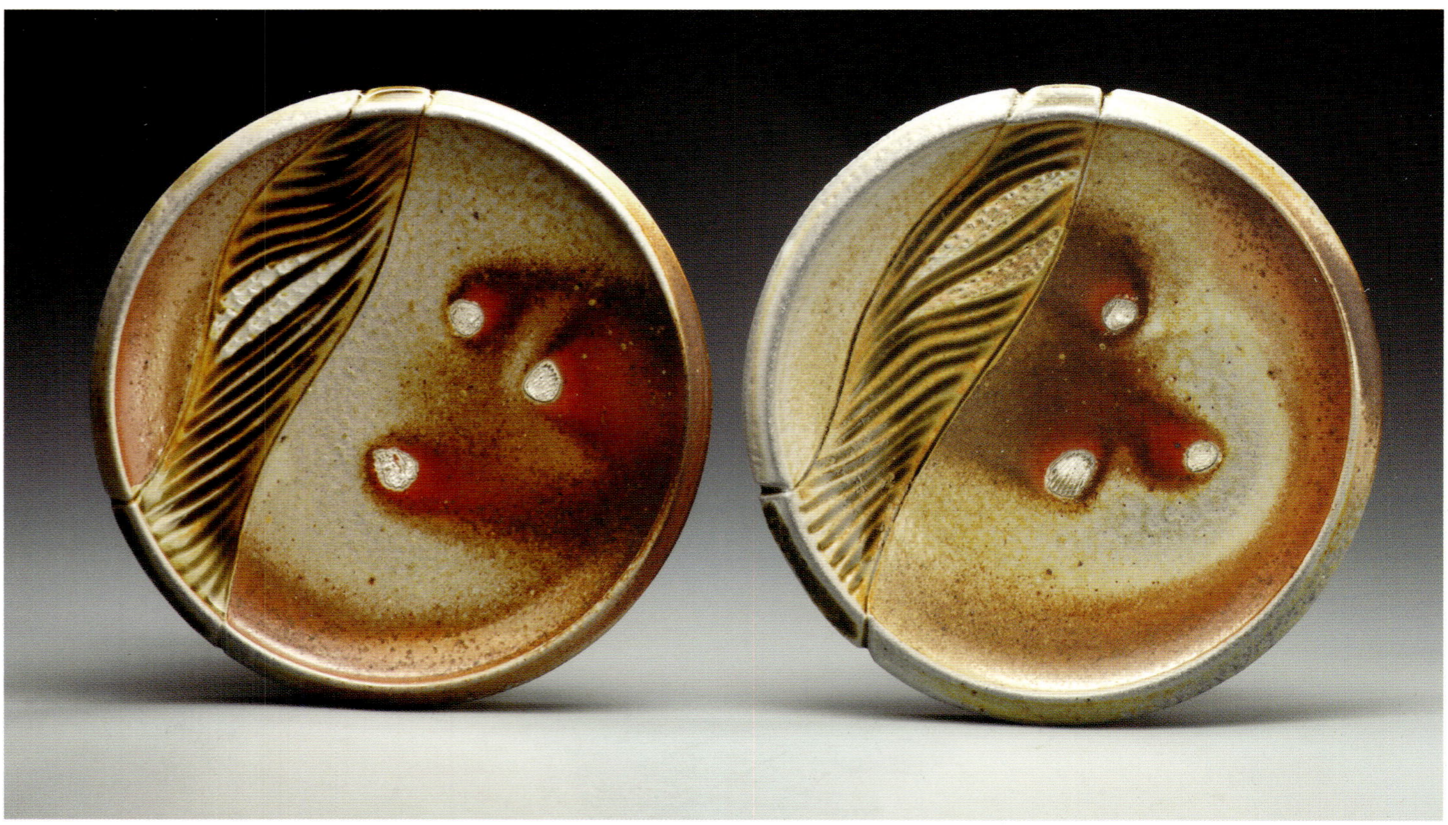

Joy Tanner, *Carved Plates*, 2011. Wood and soda fired stoneware, wheel-thrown and altered, 1.5" x 7.5" each. *Photo credit: Joy Tanner.*

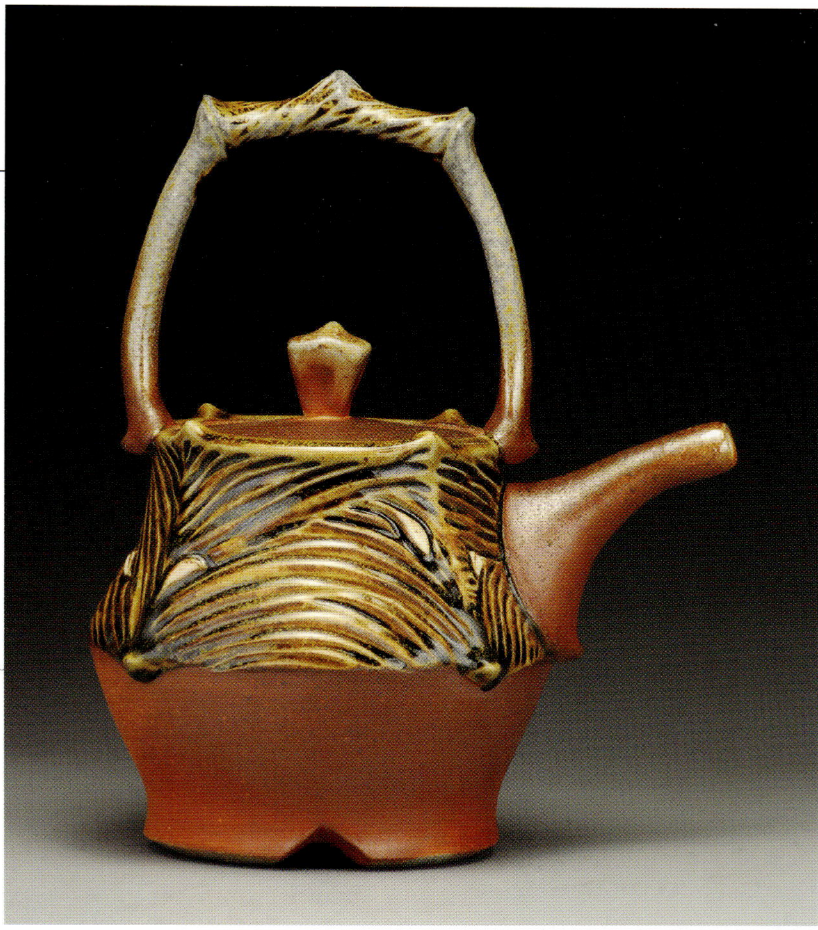

Joy Tanner, *Carved Teapot*, 2011. Wood and soda fired stoneware, wheel-thrown and altered, 9" x 7" x 5.5". *Photo credit: Joy Tanner.*

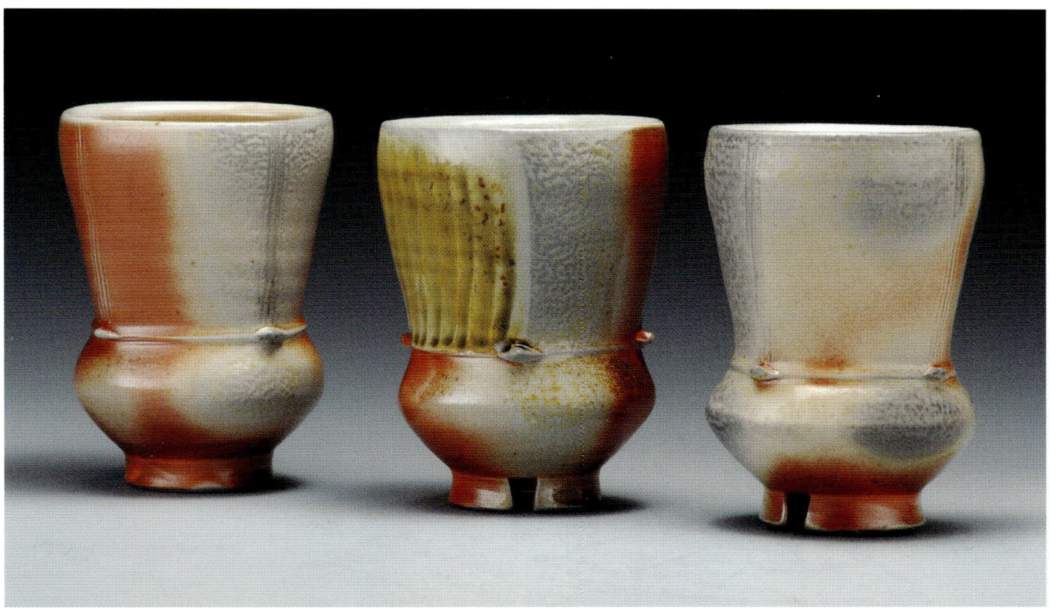

Joy Tanner, *Cups*, 2011. Wood and soda fired porcelain, wheel-thrown and altered, 5" x 3" x 3" each. *Photo credit: Joy Tanner.*

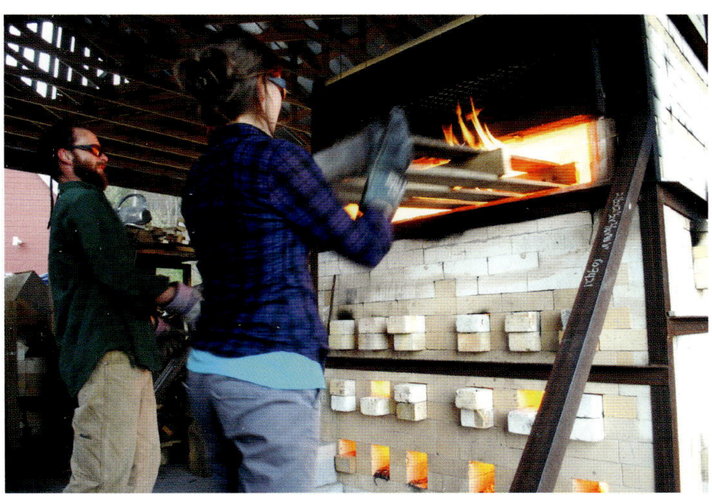

Joy Tanner and William Baker, Stoking wood kiln, 2011. *Photo credit: Joy Tanner.*

JORDAN TAYLOR

Chapel Hill, North Carolina
jordantaylorclay@gmail.com, www.jordantaylor.us

Artist's Statement

The weighty history of ceramics gives the most attention to what is or what remains, even though there is much more meaning invested in what is not or is no longer. In archeology shards are clues to the larger whole not only of a vessel, but of an entire culture. We are left with little choice but to deal with the incomplete material record that ceramics offers. Even work based on documentation of ephemera using unfired clay is still based on that documentation of "what is," no matter how briefly "it was."

I have become interested in what escapes material recognition, in ceramic absence. The two clearest examples of this in my work both come from stone objects I encountered in the backcountry of the Guatemalan jungle at the remote archeological site of the Mayan ruin *El Mirador.* One is a *molino,* a hand mill for grinding corn. The Mayas believed that their food was literally their body (you are what you eat predates Columbus) and so implements used in the preparation and eating of food were links of the spirit to the physical world. As such, a person would use a bowl, or a *molino,* their entire life. Part of burial rituals included a postmortem perforation of an individual's food utensils to free their spirit from the soon buried body. The other objects were *stelae,* standing stones carved with glyphs documenting a king's rule. These documents have largely been washed blank once unearthed because of limestone's vulnerability to acid rain. The ones that remain intact remain buried or, in a few cases, a shelter is built around them so they can remain on display.

Pictured are "untitled (square)," 2008 and "untitled (*stelae, 1, 3, 4, 5*)." While both are dense and massive, both pivot on absence.

With (square), the perforation at the center is the most obvious, but the ring made by the wad marking, which echoes both the shape of the perforation and of the outer edge of the piece, indicate interfaces with what is beyond the piece and what precedes the piece. Wad marks reference the near umbilical relationship to a firing environment, whereas the border of the piece acts as a horizon between the object and its new (post firing) context. Further, the "square" label is intentionally misleading, calling attention to the triangles suggested in the corners, the circle/cone of the perforation, and the sphere implied by the concavity of the piece.

In the case of *(stelae),* the series is numbered though 6, but stele 2 is installed on private land, and images of it will not be published. Stele 3 was buried at the site of the kiln that fired it after the kiln was dismantled. All were fired to a slight porosity of 6–8%, where 5% porosity is what the brick industry considers frost proof. The *stelae* will disintegrate over a period of two to three hundred years. Freeze thaw will eventually free this series to follow the watershed back to the mouth of the Susquehanna River where the clay was mined. While massive and an imposing presence, they also mark an absence, a divot in their geologic deposition that can never be truly replaced, even when the material returns to the same location (the hole will have filled in by then). Lastly, the *stelae* relate to the dimensions of the windows of the Everhart Museum of

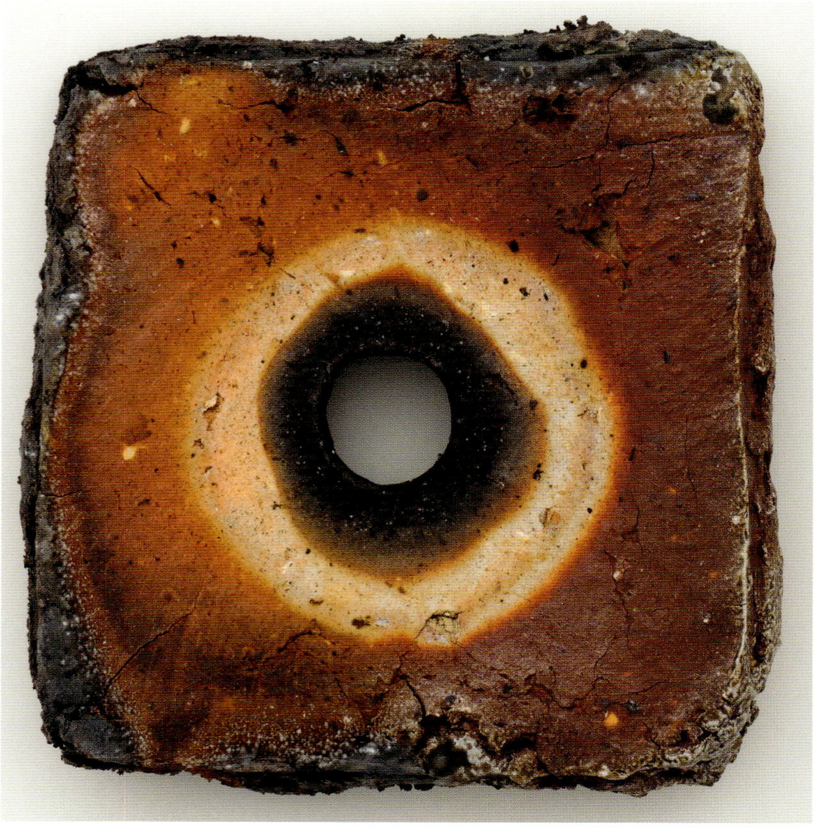

Jordan Taylor, *Untitled (square),* 2008. Indigenous brown clay and industrial salvage kaolin, wood-fired, 14" x 14" x 4". *Photo credit: Jordan Taylor. Collection: Museum of Fine Arts Houston, TX, Garth Clark and Mark Del Vecchio Collection.*

Natural History, Science and Art. While the piece consists of four ton geometric solids, it functions primarily as a window.

Understanding absence depends on object permanence, developed between eight and twelve months of age, and is the essential marker for development of memory and of sensory-motor cognition. Object permanence is the beginning of proprioception in the flow of space. Memory, later coupled with—if not indistinguishable from—imagination, form our compass in the flow of time.

Some of my work in clay demands long firings to ensure even firing of large work, and so wood is a natural choice for mercenary reasons. It is cheaper to fire large kilns with wood than with gas, oil, or electricity. More important to my creative aims is that wood demands an attentive human participation around the clock, and long firings involve high risk. My best firing lasted 220 hours and had a nearly 80% loss rate. In a 10 day firing I experience a sleep deprivation beyond what my wife and I experienced during the infancy of our daughter, and I begin to lose my memory of previous firings around day 5. I used to keep careful notes, but as my firings grew longer I embraced the idea that the firing process was a deliberate stripping away of my intellect, leaving all decisions affecting the work to my emotion, intuition, and to the small community of people I had built around letting me rest briefly during parts of the firing. The experience of my work and the experience of making my work are both about two explorations: the absence offered by object impermanence and timelessness offered by the erosion of memory.

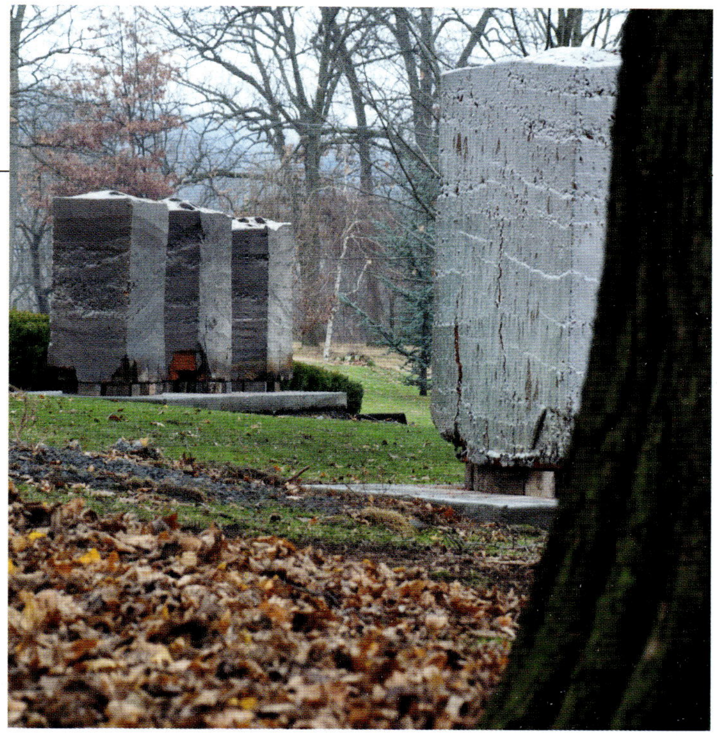

Jordan Taylor, *Untitled (stelae 1, 4, 5, 6), 2009*. Susquehanna River clay, wood-fired, 76" x 48" x 32" each component. *Photo credit: Jordan Taylor. Collection: Everhart Museum of Science, Natural History and Art, City of Scranton, PA.*

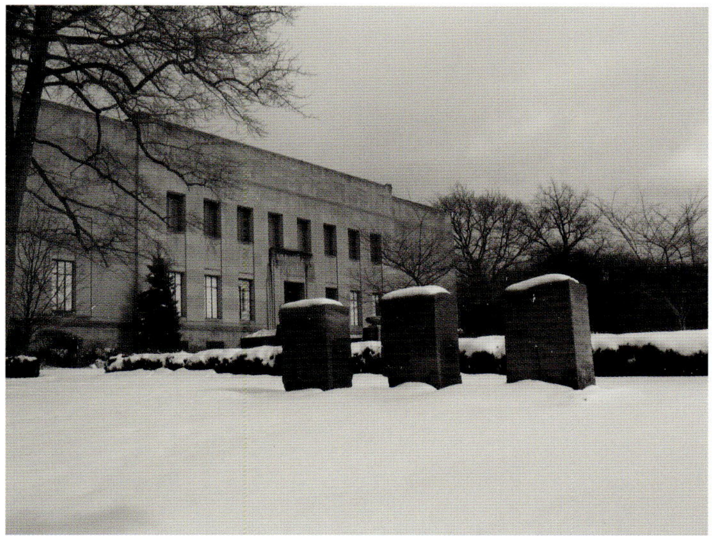

Jordan Taylor, *Untitled (stelae 4, 5, 6)*, 2009. Susquehanna River clay, wood-fired, 76" x 48" x 32" each component. *Photo credit: Jordan Taylor. Collection: Everhart Museum of Science, Natural History and Art, City of Scranton, PA.*

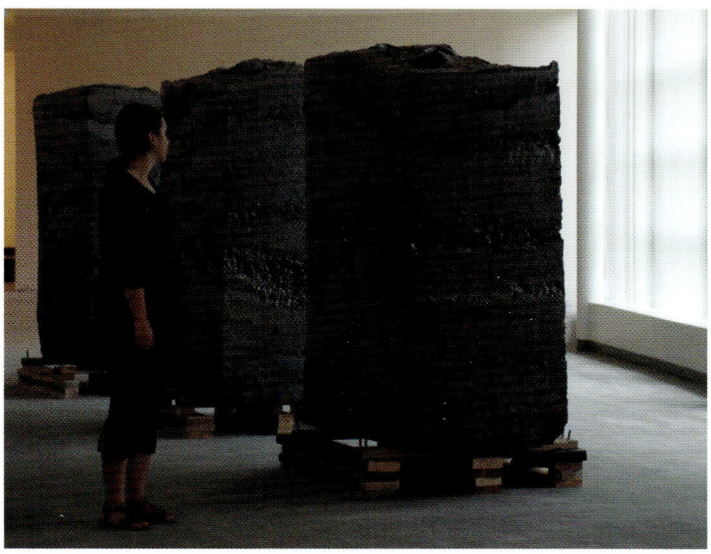

Jordan Taylor, *Untitled (stelae 4, 5, 6)*, 2009. Susquehanna River clay, wood-fired, 76" x 48" x 32" each component. *Photo credit: Jordan Taylor. Collection: Everhart Museum of Science, Natural History and Art, City of Scranton, PA.*

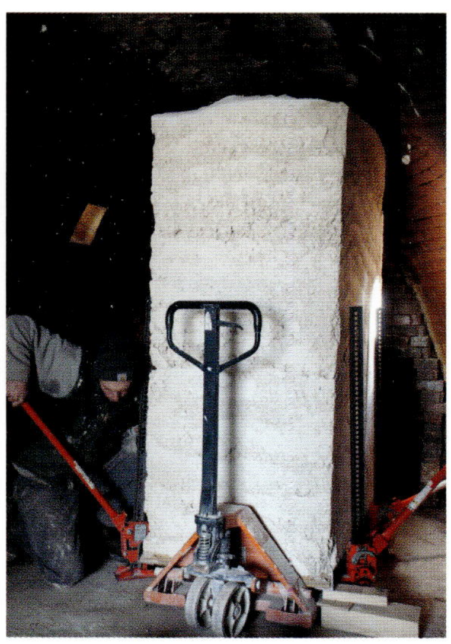

Jordan Taylor, loading *untitled (stele 3)*, Second Kiln, tapered catenary tube, catenary combustion chamber, Union Dale, PA, 800 cubic feet, 6 firings 2005–2009. *Photo credit: Jordan Taylor.*

SKEFFINGTON THOMAS

Monroeville, New Jersey
thomass@rowan.edu, www.skeffthomas.com

Artist's Statement

Each morning as my fingers play across the body of my coffee cup, and my lips are welcomed by the roundness of a rim, I am reminded about the subtle quality of form and surface that the ceramic arts provide. I choose to use the wood-firing process for finishing much of my artistic output because there is that great level of subtlety in the ware that emerges from each firing.

I engage in my artistic practice using strong form, bold color, and a predestined unique outcome. This approach provides me the incentive to design and make functional pottery. I hope the thought or care with which each pot is made impacts the users' ability to embrace the translucency or other subtle qualities of color, texture, line, or pattern hidden within the form. Ultimately, the distinction between functional craft and art is blurred, and the viewer/user is granted an experience of a visceral nature simply by using an artistically designed, made, and wood-fired piece of functional pottery.

As a believer in the importance of daily practice, I continue to examine and contemplate function in the context of use in serving versus the context of function as decoration or ritual support. In this era where functional objects of ceramic art are overshadowed by the plethora of commercially available cooking and serving utensils, it is only the connoisseur of the handcrafted and designed object of the studio artist that looks at, and tries to see, the design vision of the artist. I enjoy wood-firing, because I enjoy talking with people who look for the subtle intentions of the maker, combined with the unique contribution of the kiln, that gives each object its own personality.

These ceramic forms and pottery have been fired using a "fast-fire" kiln based on the plans provided by Fred Olsen. The kiln I built has been modified from the Olsen plans to add more insulation and larger fireboxes. The kiln, constructed with hard fire brick for the foundation and fireboxes, and K26 insulating fire brick for the chamber walls, has been coated with ITC coating to reduce corrosion from the 1–5 lbs of salt I introduce into the kiln between cone 5 and cone 10. The kiln can reach temperature in seven hours, but I prefer a slower firing that lasts ten hours or more. The firings have been continued until the last of the six cone packs in the kiln has achieved cone 10. Cone 12 has been achieved for most of the kiln. The stoneware or porcelain clay body is used to provide the opportunity for the glaze choice and surface finish to be as rich and full of depth as can happen during the wood-firings. When working with a kiln fueled by wood I consciously give up some control of the outcome to the vagaries of the firing. This technical process of firing ceramic art allows for personal creativity within the transformative process and yet still allows for the serendipitous or unique outcome that makes each piece unique, be it a humble cup or serving bowl.

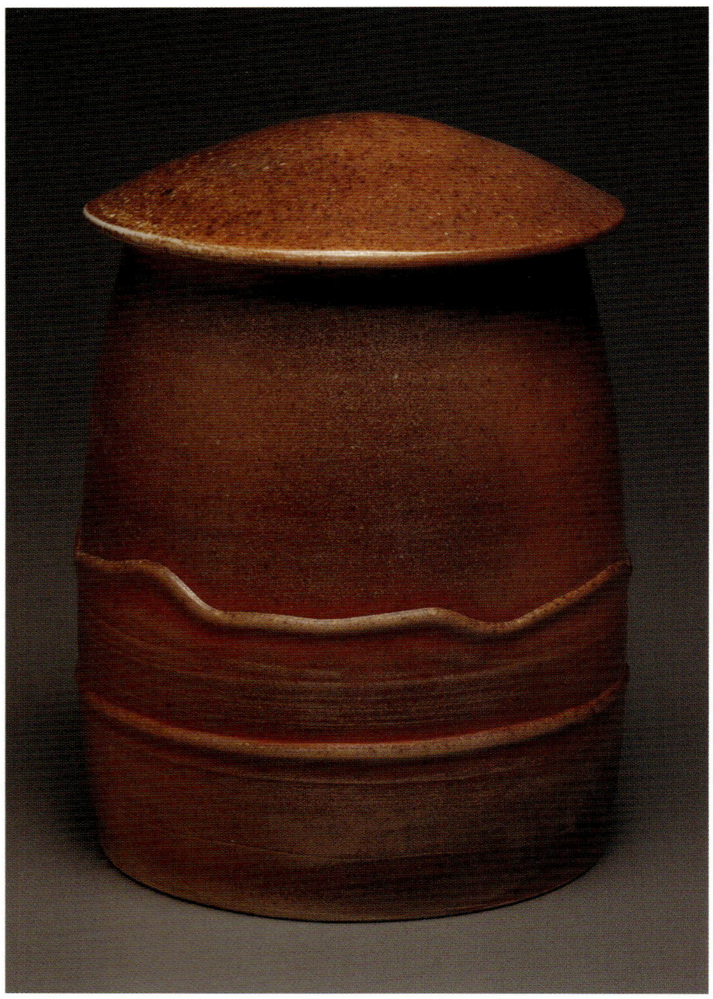

Skeff Thomas, *Storage Jar*, 2011. Wood-fired porcelain with flashing slip, salt glaze, 10" x 5" x 5". *Photo credit: John Carlano.*

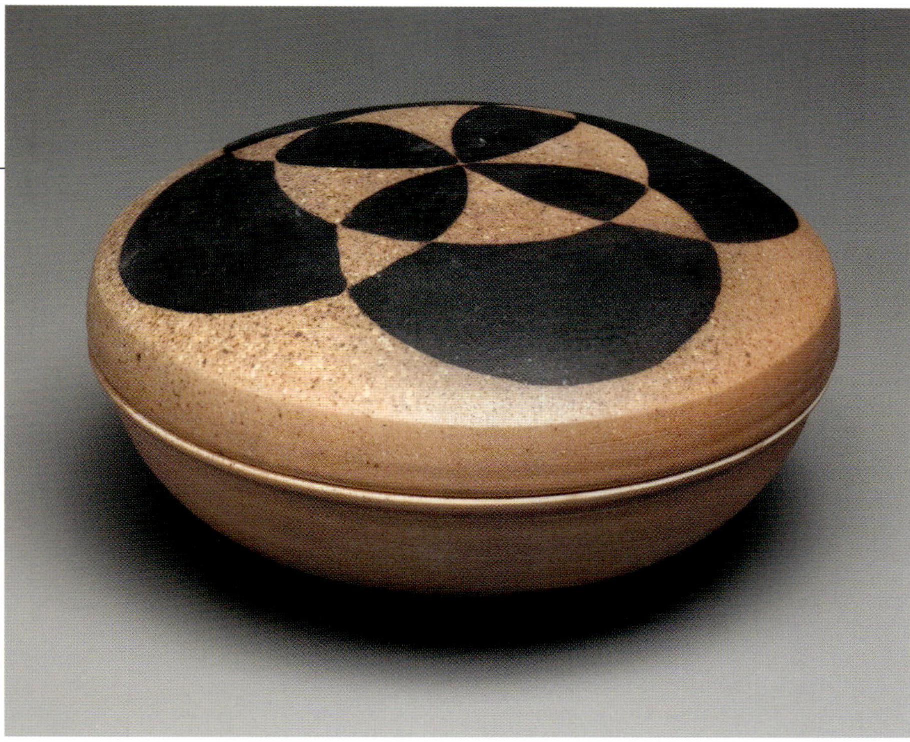

Skeff Thomas, *Covered Serving Dish*, 2011. Wood-fired stoneware with black underglaze, salt glaze, 7" x 11" x 11". *Photo credit: John Carlano.*

Skeff Thomas, *Vase and Bottle with Blue Spots*, 2011, Wood-fired porcelain with flashing slip and salt glaze, 13" x 4" x 4". *Photo credit: John Carlano.*

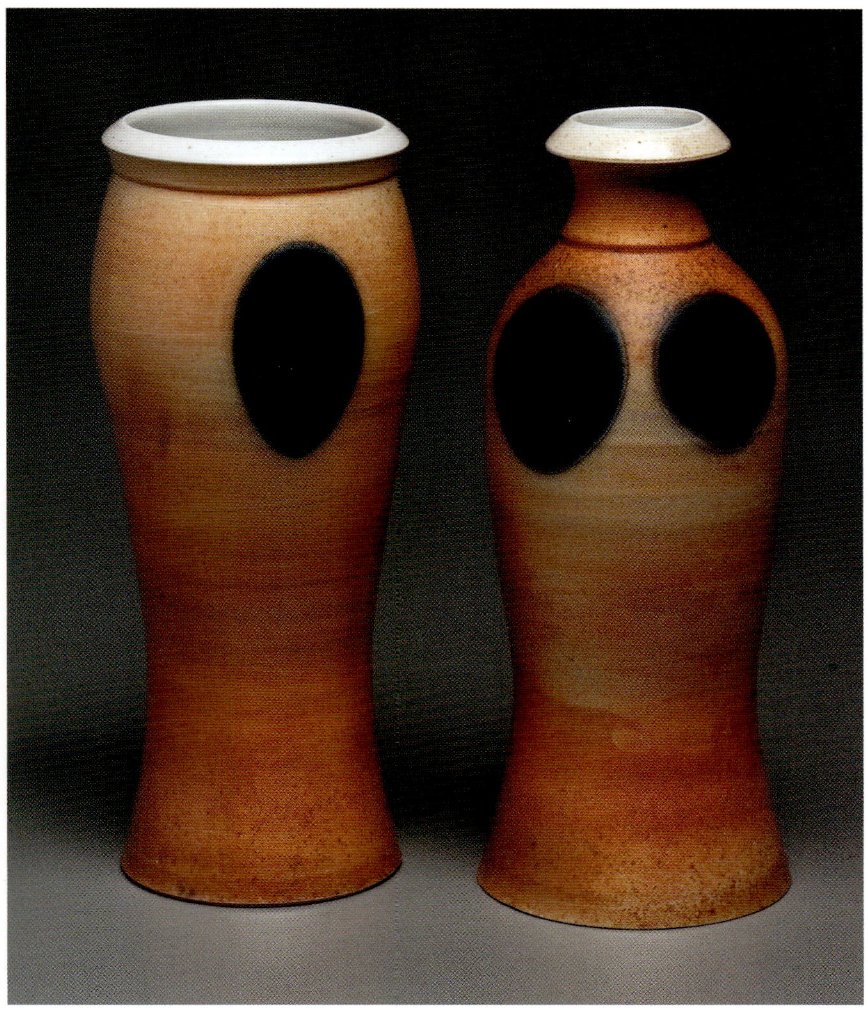

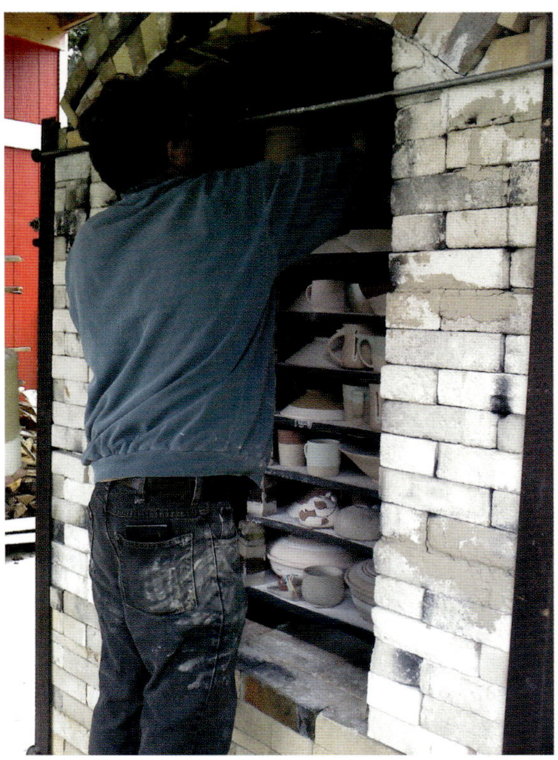

Skeff Thomas, Loading the kiln, 2011. Olsen Fast-fire kiln with 26K IFB.

SHANE TIDMORE

Houston, Texas
harupots@sbcglobal.net

Artist's Statement

As a clay artist, I am interested in providing unique, interesting, and usable objects to elevate the daily rituals of life. The majority of my work is functional ceramics to be used in the preparation and service of food and drink. This would include cups, bowls, tea sets, baking dishes, etc. My five-year tenure, as both apprentice and professional working artist, in South Korea and Japan solidified my foundation in and appreciation for this type of ceramic art.

The remainder of my work deals with sculpture. These pieces often include or start with found objects, such as old wood, metal, or rocks. The sculpture is experimental in terms of texture, size, color, materials, procedures, and firing techniques. I enjoy the challenges and results of wood-firing. This process seems to fit the finished look of my work using little or no glaze, which to me seems to cover the most important element of the work that is the clay. My work in ceramics deals with gesture, asymmetrical balance, line, color, function, content, mood, volume, and ritual. These thought- provoking objects are personable and usable, both on a daily basis and/or on special occasions.

Currently I am the owner and operator of the Lone Pine pottery located in the woods of San Jacinto County, Texas. It is my desire to build large kilns modeled after those I helped construct in Korea and Japan, which will enable greater possibility for the types of ceramic work I make. The kiln I currently use is an anagama that is approximately twenty feet long, five feet wide, and four feet tall. I have been firing with an equal mix of oak and pine from four to five days, reaching cone 12 in the front and cone 9 in the back. I use a variety of both commercial and natural clay bodies ranging in color from light to dark, which yield a wide range of results.

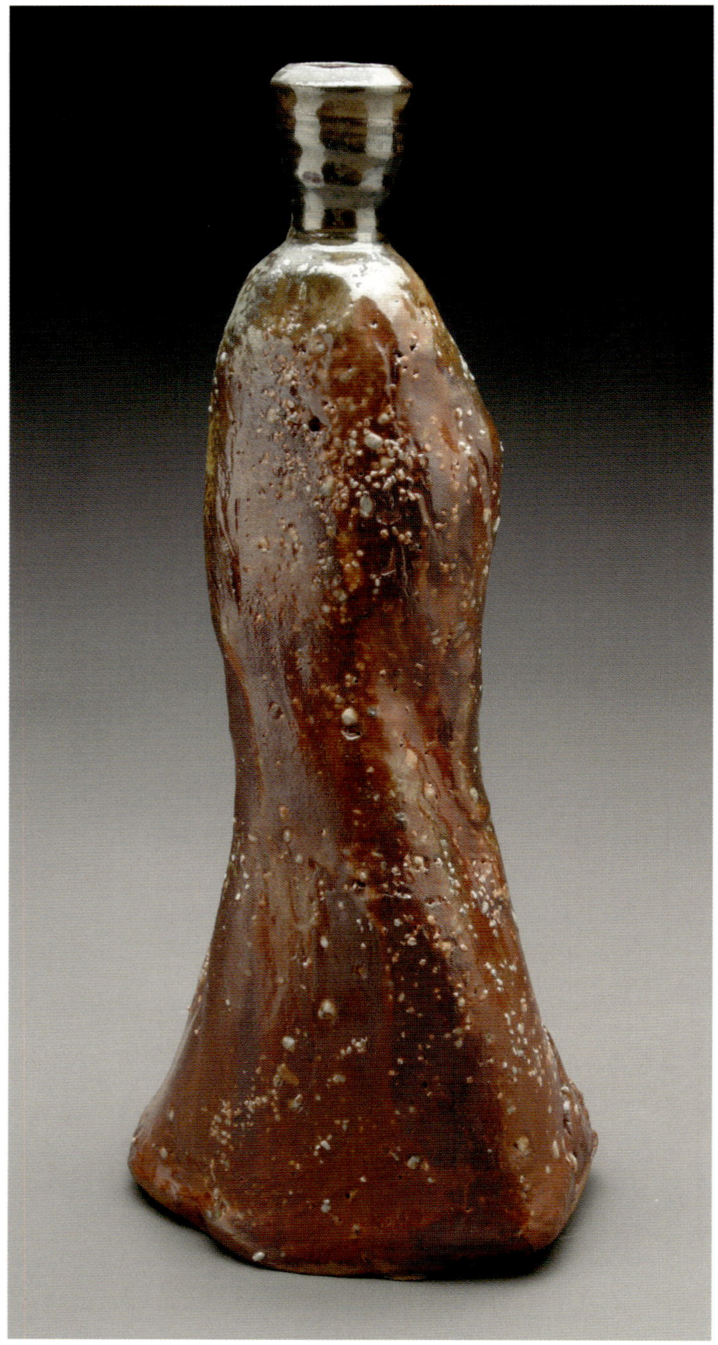

Shane Tidmore, *Bottle*, 2010. Wood-fired stoneware, handbuilt and wheel-thrown, 20" x 7" x 7". *Photo credit: George Ramirez.*

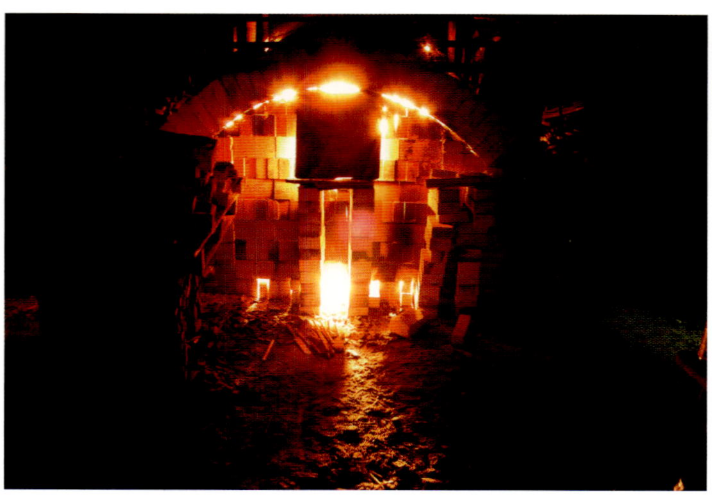

Shane Tidmore, Front of anagama, 2010.
Photo credit: Horacio Rodriguez.

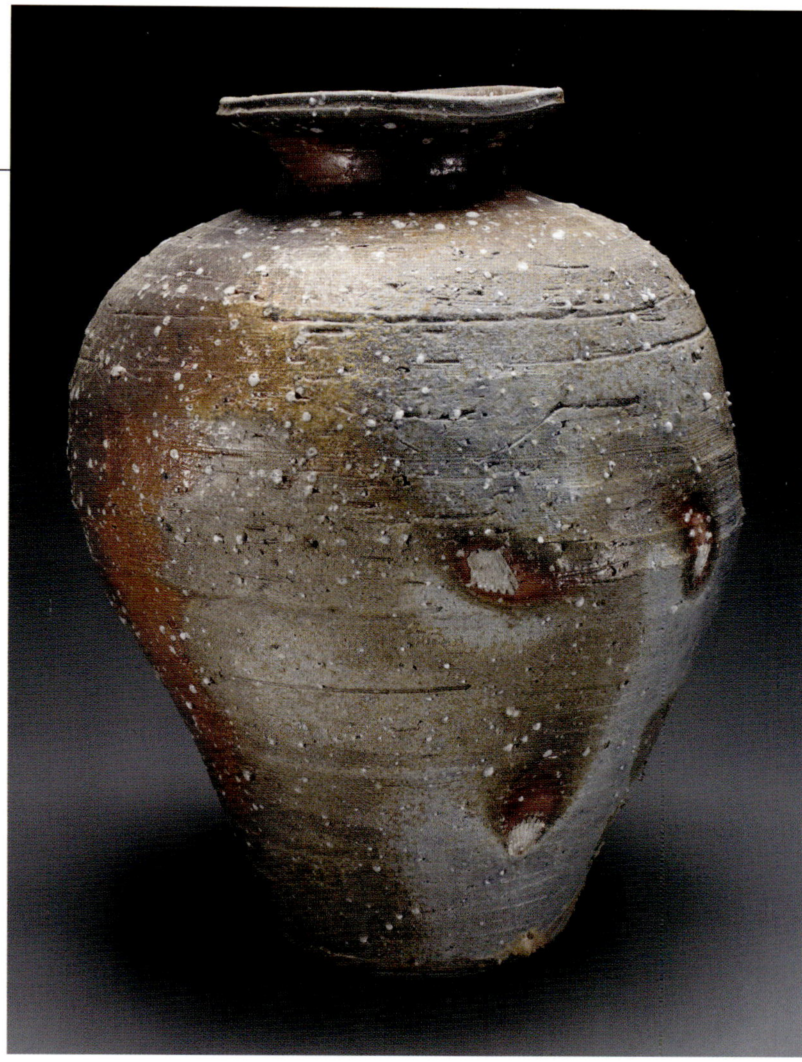

Shane Tidmore, *Jar,* 2011. Wood-fired stoneware, wheel-thrown, 28" x 18" x 18". *Photo credit: George Ramirez.*

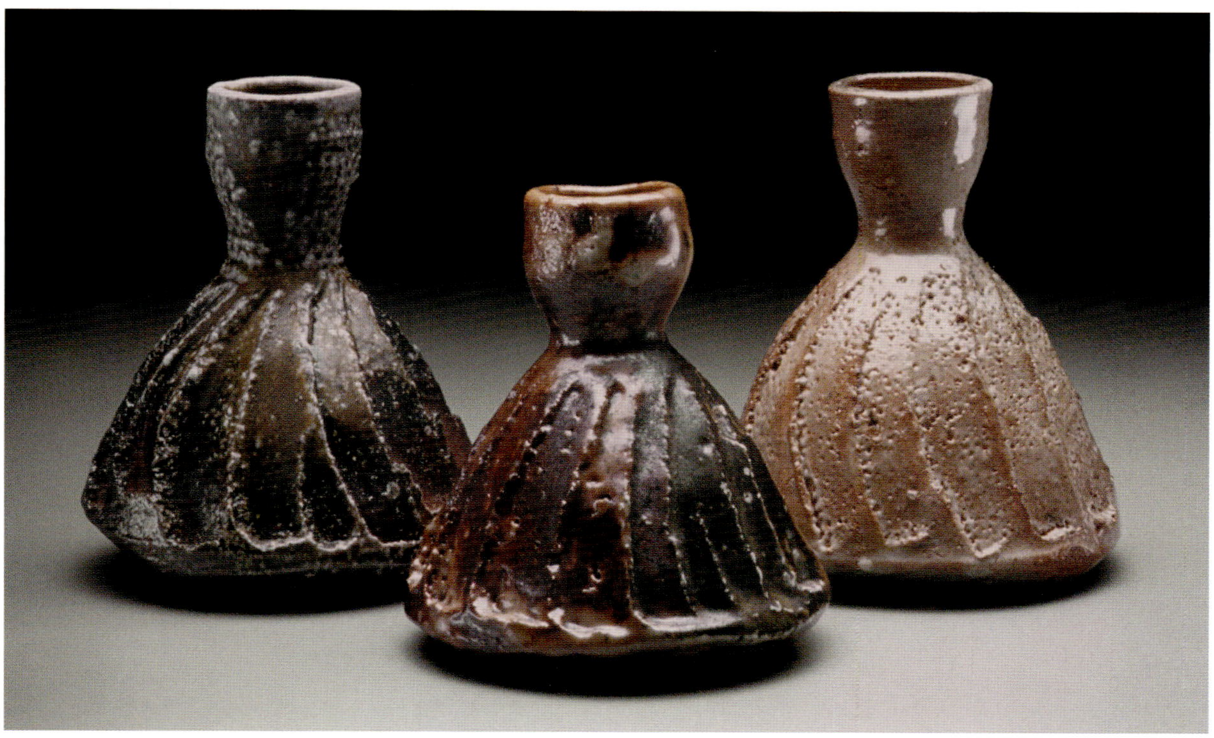

Shane Tidmore, *Three bottles*, 2011. Wood-fired stoneware, wheel-thrown and carved, 6" x 5" x 5" x 3. *Photo credit: Brad Flowers.*

JAMES TINGEY

Athens, Ohio
jamestingey@gmail.com, www.jamestingey.com

Artist's Statement

My body of work explores relationships between functional considerations and process; an engagement that relates the hand to environments. My intention is to engage the space of the viewer, exploring how bodies relate to objects. Through rich and varied surfaces, I invite the user to investigate and explore surfaces and form. I am interested in the ways that pottery is used to punctuate our lives and becomes integrated into our experience of the everyday.

Integral to my studio practice is utilizing the wood-kiln in the development of layers of information contained within the work. I am interested the interactions between pattern, chance, accumulation, and erosion to enrich surface. Through addition and removal, surfaces serve as a record of an object's history. In this context, the viewer begins to understand the relationship to landscape, process, and physical presence.

I am invested in the activity of mark making as a record, directly through the index of the hand and indirectly through the residues of firing. Traces and layers of process are evident through the residues of forming and firing, engaging in a visual and tactile dialogue that is discovered through use. Preserving the records of material change and contingent processes, these surfaces record layers of time. These residual marks relate to the viewer through the traces of touch and the history of the object. This engagement of the user through these layers feeds my ongoing passion of functional works.

Recent work focuses on press molded forms that explore the intersection of material and process, regularity and difference within a group of forms. Form is dictated through the regulating force of the mold, material and process becomes the vehicle to explore difference. Through the contingent environment of firings, the individual is increasingly differentiated from the group. Through these processes I am complicating the notion of the regularity of the multiple. The work allows the viewer to transcend functional references of the individual objects and to consider the work in relationship to landscape via material, process, and its physical presence.

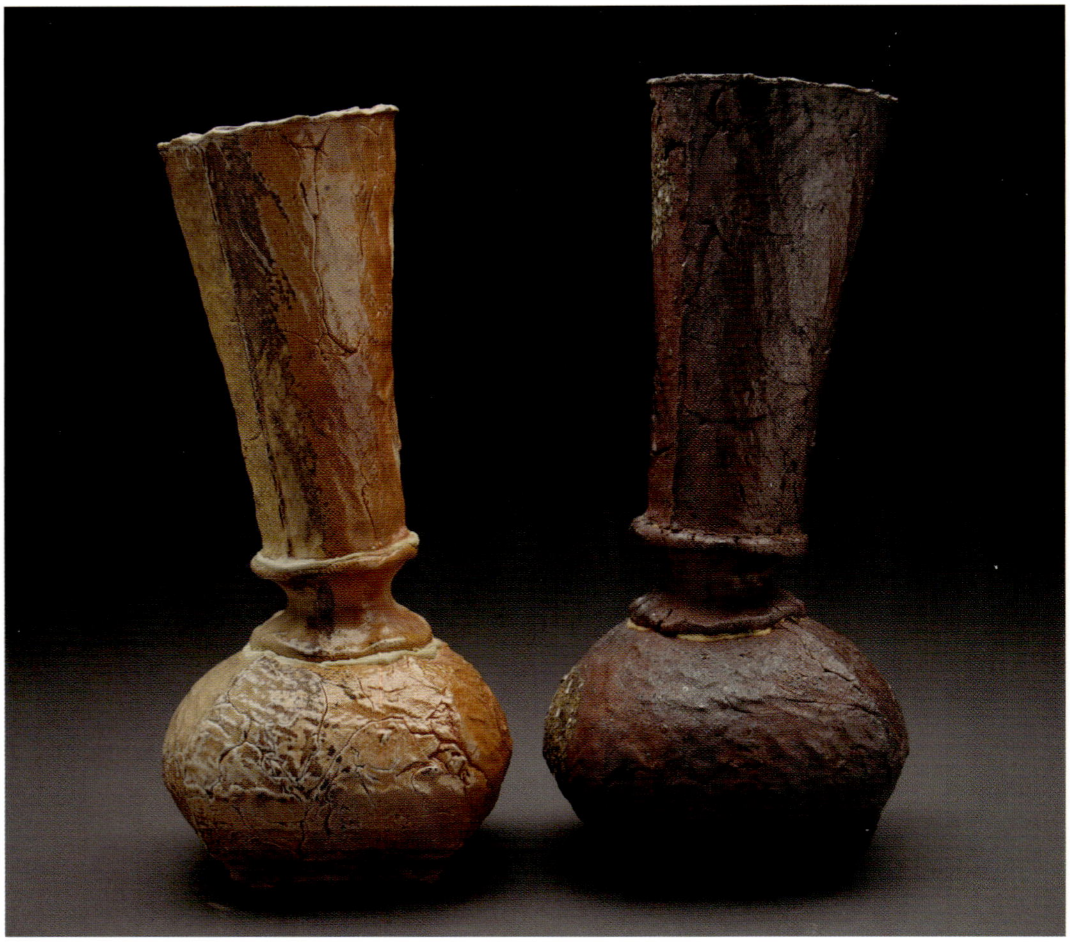

James Tingey, *Two Bottles*, 2011. Wood-fired, press molded stoneware and porcelain, 18" height. *Photo credit: James Tingey.*

I utilize a variety of kiln types to produce my work. I view myself as nearing the end of my "journeyman" years as a wood-fire artist; firing my work in a variety of wood-fired kilns for more than ten years. For the past three years as a graduate student at Ohio University I have selected to fire my work in a "tube" style kiln. I fire this kiln typically for 48 to 60 hours to cone 12. Though the kiln holds approximately 60 cubic feet of work, this smaller size allows for a quick turn around of work and a smaller crew to fire easily. A variety of clays are used from porcelain to heavily grogged, iron-rich stoneware. Firing utilizes a combination of hardwood and pine, augmented by a supply of hardwood slats from a local small-scale sawmill and campus recycling. I fire slowly through the initial stages of firing for a slow, even increase in temperature, to accommodate larger work and to develop low velocity ash deposits. To encourage ash deposits, approximately two bales of straw are introduced throughout the firing. The design of the kiln also promotes several areas of direct ember contact with the work, pieces in these areas are selectively buried or pulled into the embers to develop surface.

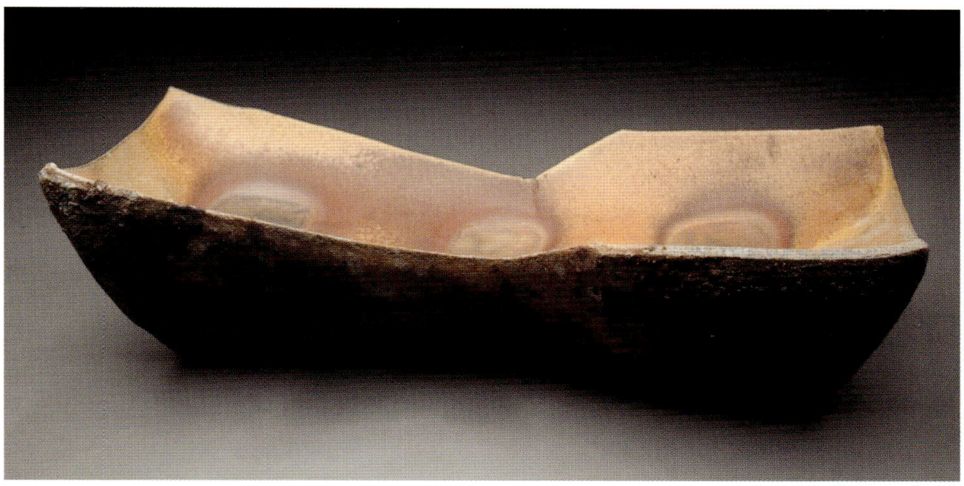

James Tingey, *Trough*, 2011. Wood-fired, handbuilt stoneware, 9" x 12" x 30". *Photo credit: James Tingey.*

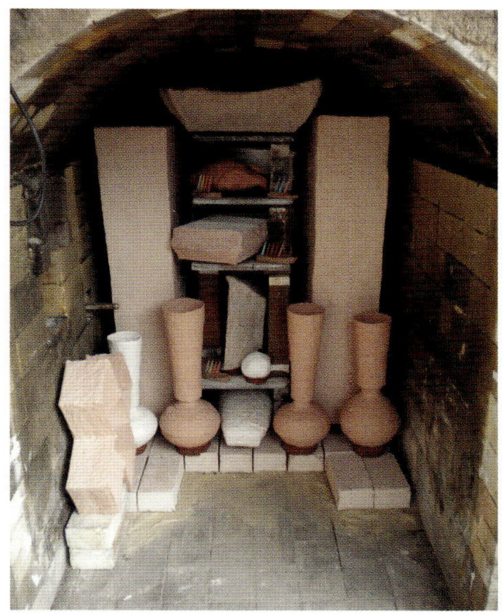

Loading the Tube Kiln at Ohio University in Athens, Ohio, 2012. *Photo credit: James Tingey.*

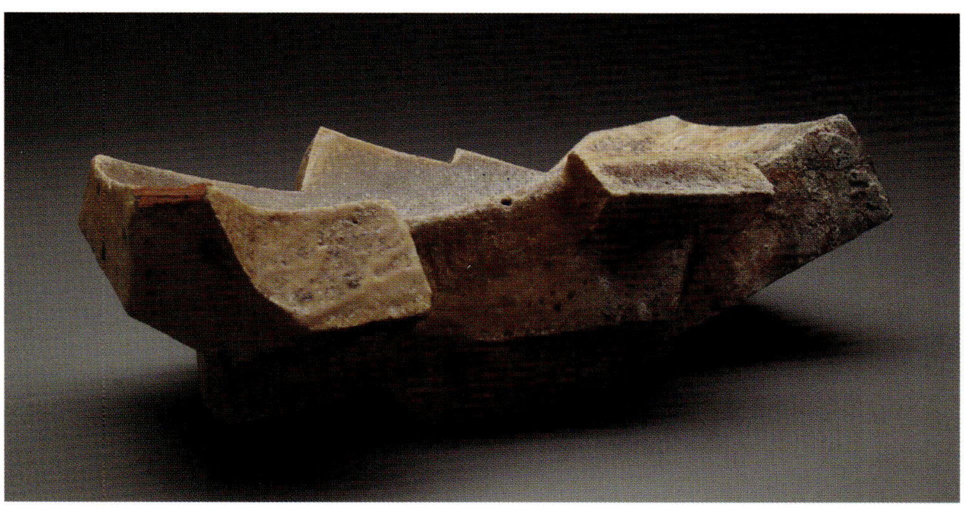

James Tingey, *Untitled*, 2011. Wood-fired, handbuilt stoneware, 5" x 6" x 18". *Photo credit: James Tingey.*

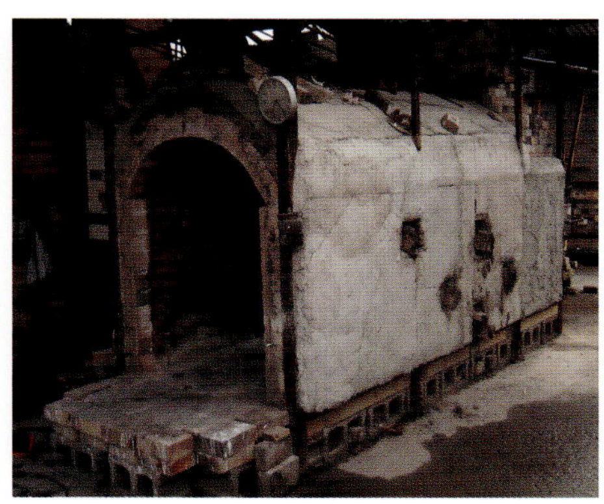

Side view of the Tube Kiln at Ohio University in Athens, Ohio, 2012. *Photo credit: James Tingey.*

TRI TRAN

Fairfield, California
teetran@aol.com, www.tritranstudio.com

Artist's Statement

There are so many words that come to mind when one describes wood-firing effects. Chief among them: unique, natural, organic, complex, rare, understated, mysterious, subdued, tranquil, earthy, elemental. Wood-fire offers the beauty of extraordinary textures and colors that embody some or all of the above qualities.

In addition to the physical beauty that it offers, the demanding process of wood-fire brings people together as a community in this age of individualism and electronic isolation. Thus, it enhances the humanity of the work we do.

I fire all of my work in anagama kilns. One I built with six colleagues in Napa and for the other I participate with colleagues in Pope Valley, California. My work is fired for approximately ten days.

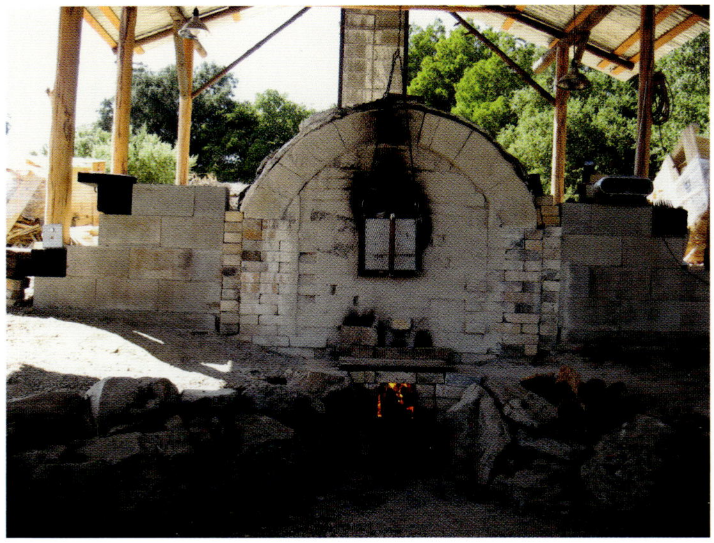

Front of the kiln, Napa, California. *Photo credit: Tri Tran.*

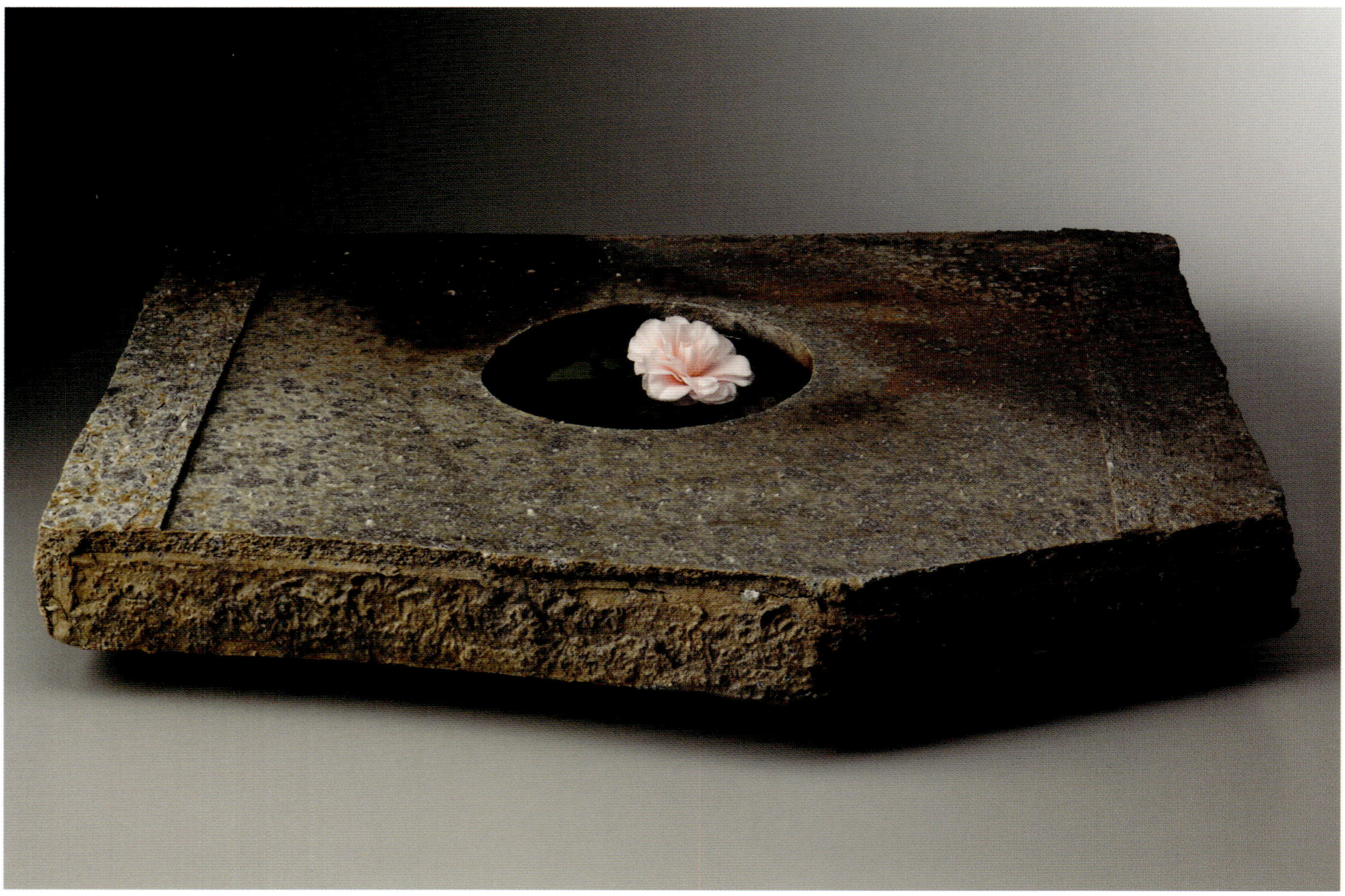

Tri Tran, *Water and Rock Garden*. Large vessel, wood-fired stoneware, slab / coil built with ultra heavy grog, 28" x 22" x 5".
Photo credit: Bill Schwob.

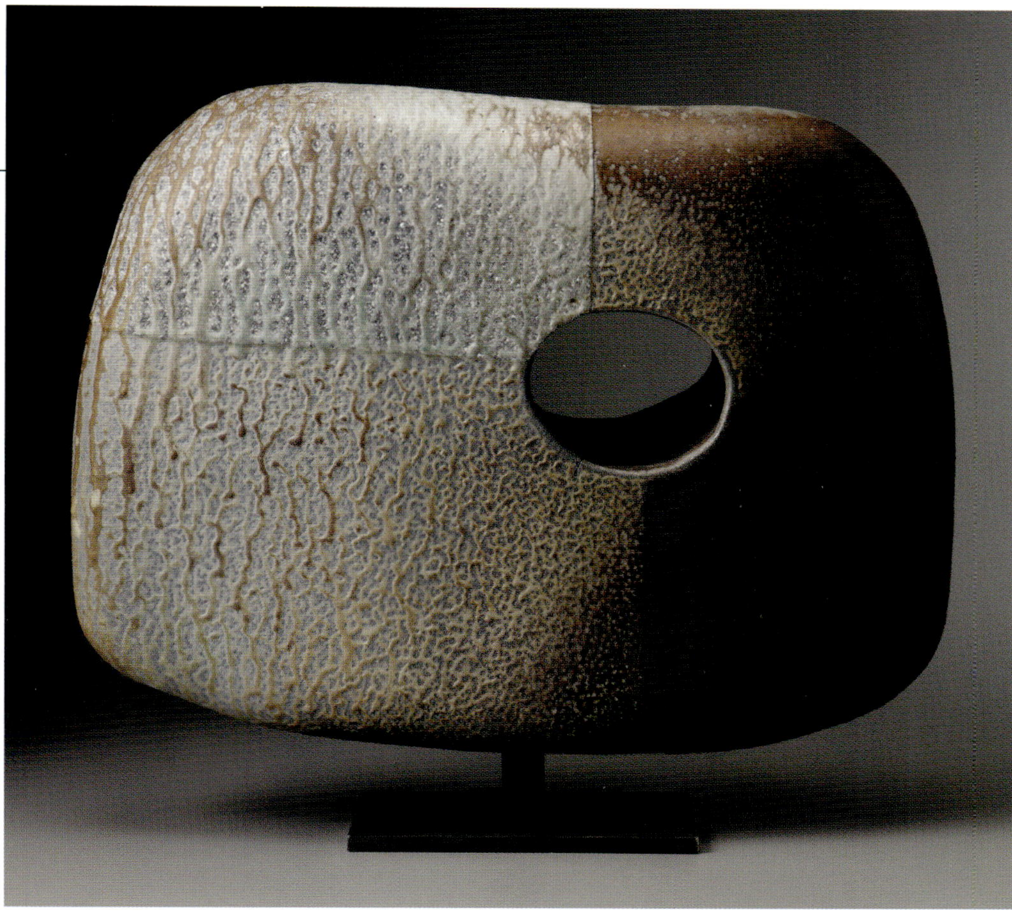

Tri Tran, *Four Seasons*. Sculpture, wood-fired stoneware.
Technique: slab and coil built on metal stand. Top left with porcelain slip, bottom right with manganese slip,
creating four distinctive sections, 20" x 18" x 6". *Photo credit: Bill Schwob.*

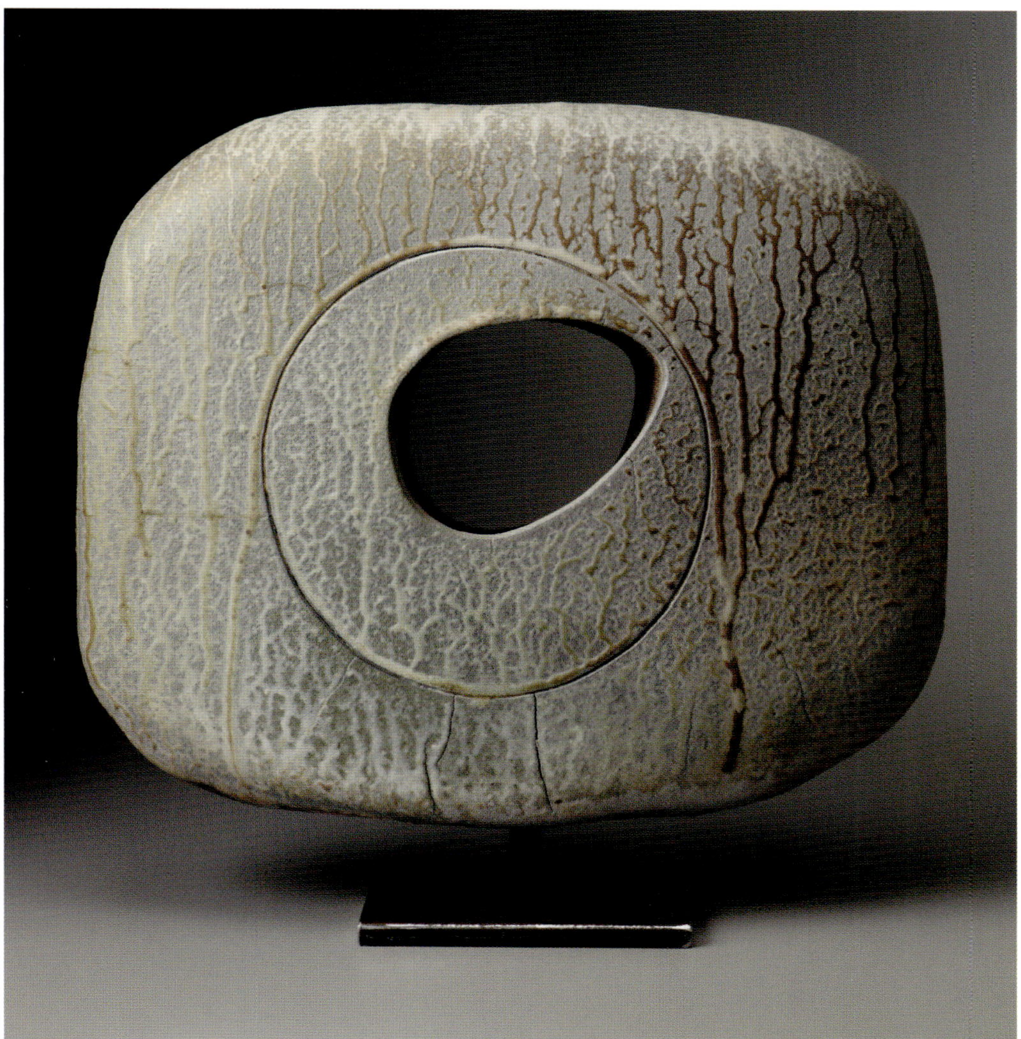

Tri Tran, *Gateway*. Sculpture, wood-fired stoneware, slab / coil built on a metal stand, 20" x 20" x 6". *Photo credit: Bill Schwob.*

JACK TROY

Huntingdon, Pennsylvania
jgaylordt@yahoo.com, www.jacktroy.net

Artist's Statement

After first firing a small anagama a few of us built at Juniata College in 1978, there was no going back; no resisting the unquantifiable magnetism. There has been only a persistent, ravenous, curiosity about how to understand and apply what we began to learn by striking that first match: how to yield to humbling beauty that doesn't give itself up easily or to everyone.

Wood-firing is an amalgamation of chain saws, airplane travel, alluring books, cremating botanical corpses, waking up in new places, transmogrifying earth-flesh, log-trucks, bricks, skews, wedges, arches, soaps, and splits by the thousands, shirts holding smoke smell, divine and calamitous kiln-openings, splitting wedges, museum visits, pots revealing fiery origins, axes and splitting mauls, aesthetic theories both elegant and fatuous, flames—surging, eddying, and vanishing. You can't look the adventure in the eye; you just try to keep pace with it.

Daniel Rhodes once described anagama firing as like trying to fill a burlap sack with liquid—only a huge quantity of fuel/fluid can accomplish the goal. Although it is perhaps the least "efficient" way to fire, it presents a superb learning opportunity for interacting with others in a non-competitive relay for a common purpose. Fuel gleaning necessitates interacting with people for whom contact with artist/ceramists may be rare. Ultimately, aesthetic choice making in anagama firings involves a variety of people committing as much as several weeks of working together.

The beauty of anagama-fired work sometimes contradicts and displaces culturally acquired aesthetic values, all to the good. In the long run, my preference for firing an anagama lies in the simple, reductive process offering, on a good day, a happy ratio between challenges and rewards. Local trees provide the fuel, delivered by a friend. The participants know one another, work together for a common purpose, and learn the vanity of claiming to be fully responsible for the work we make. That we take nothing for granted and count our blessings when we are less than cursed comes with the territory.

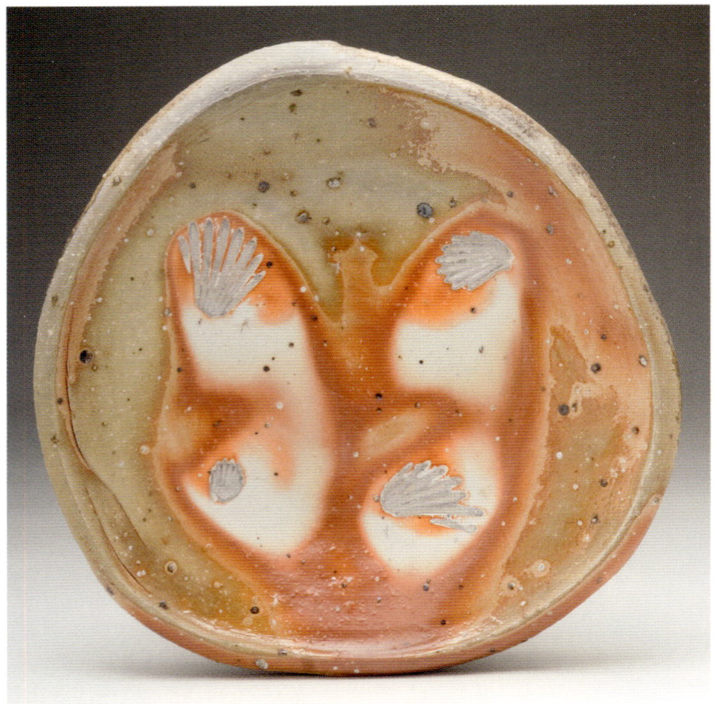

Jack Troy, *Butterfly Platter*, 2011. Porcelain, natural ash glaze, anagama fired,16" x 2.5". *Photo credit: Paul Hazi.*

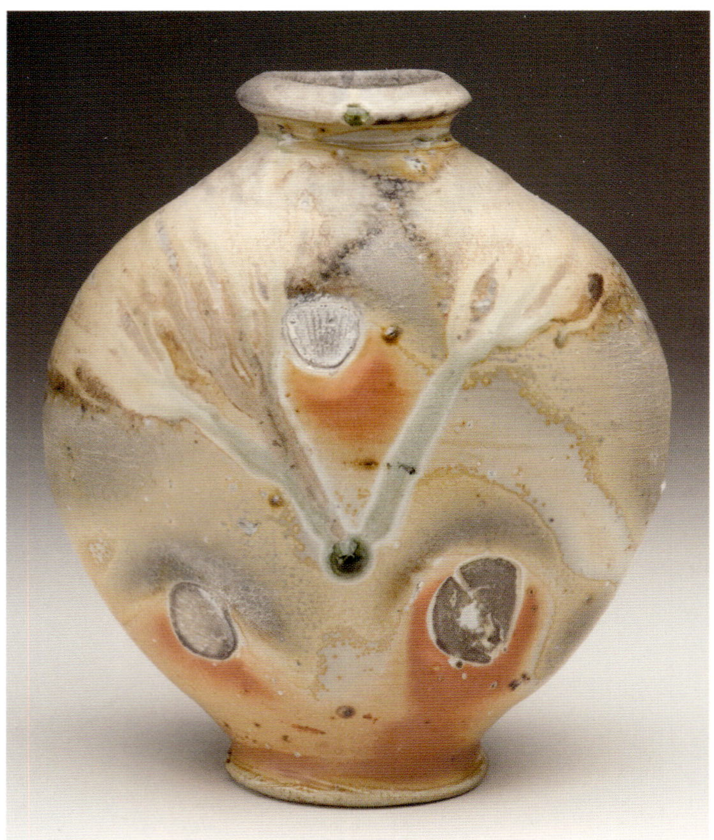

Jack Troy, *Paddled Porcelain Pot*, 2009. Natural ash glaze, anagama fired, 15" x 15" x 9". *Photo credit: Paul Hazi.*

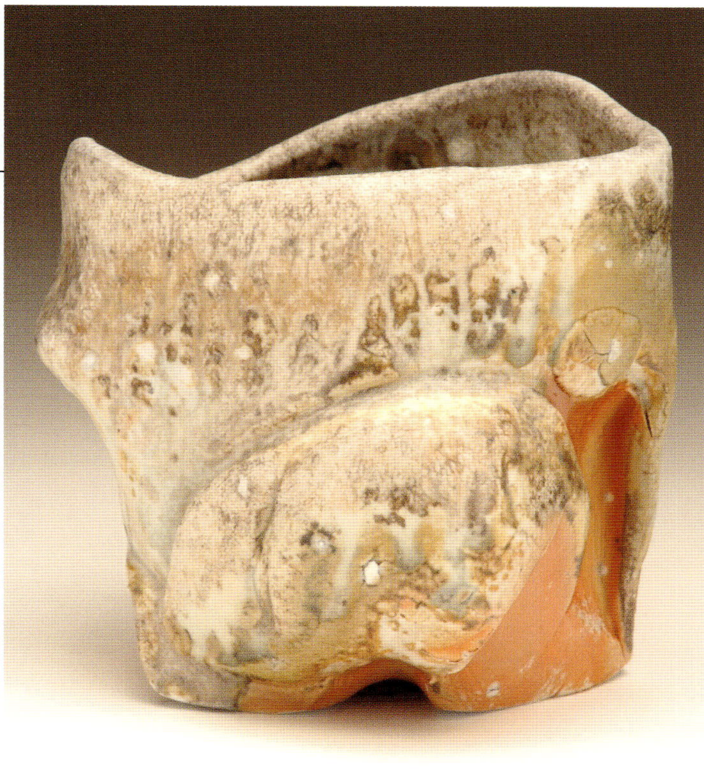

Jack Troy, *Porcelain Tea Bowl*, 2010. Thrown and altered porcelain, anagama fired, 4" x 5" x 3.5". *Photo credit: Paul Hazi.*

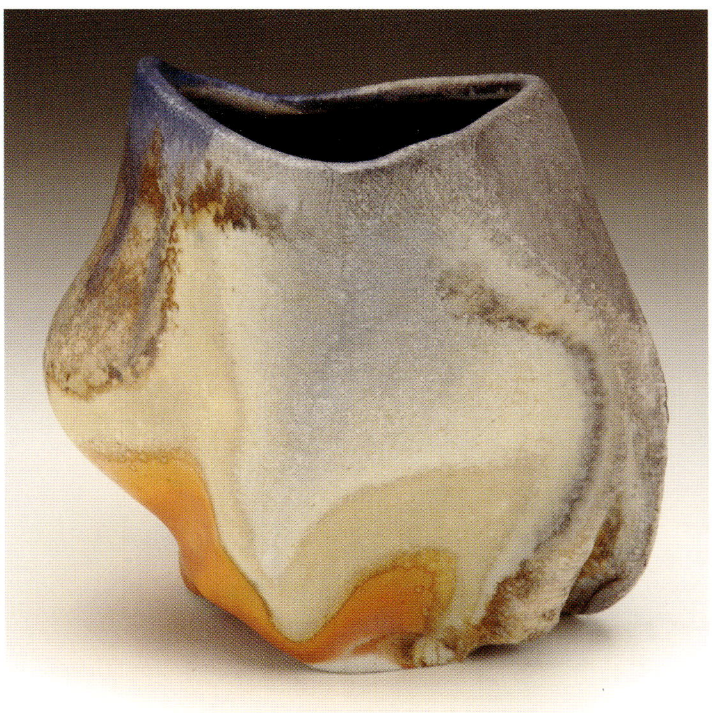

Jack Troy, *Porcelain Tea Bowl*, 2007. Applied and natural ash glaze, anagama fired, 4.5" x 5" x 4". *Photo credit: Paul Hazi.*

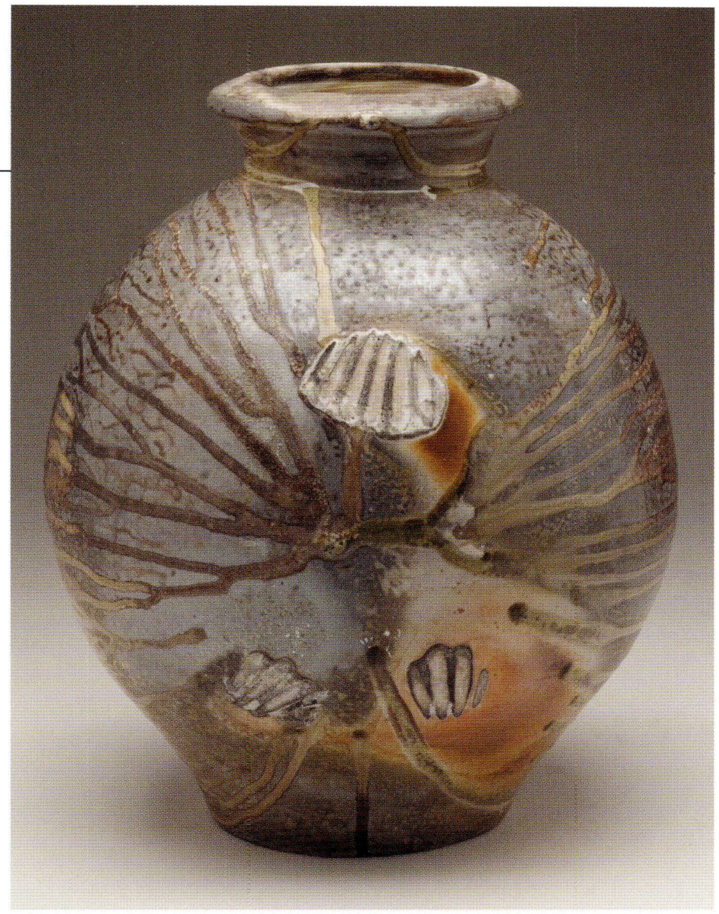

Jack Troy, *Porcelain Jar*, 2011. Natural ash glaze, anagama fired, 16" x 14" x 14". *Photo credit: Paul Hazi.*

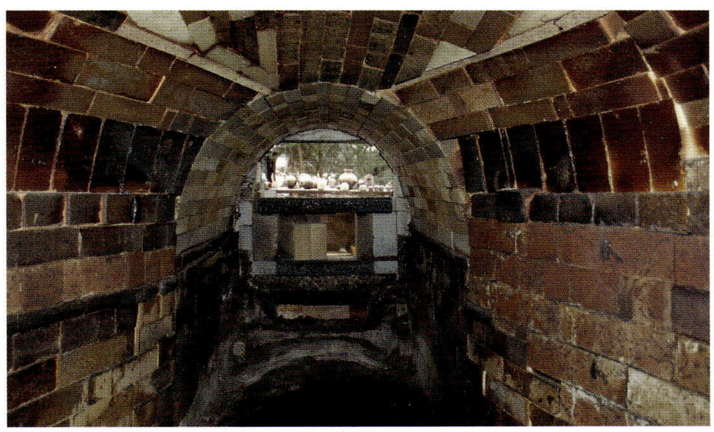

Pixiegama firebox. *Photo credit: Jack Troy.*

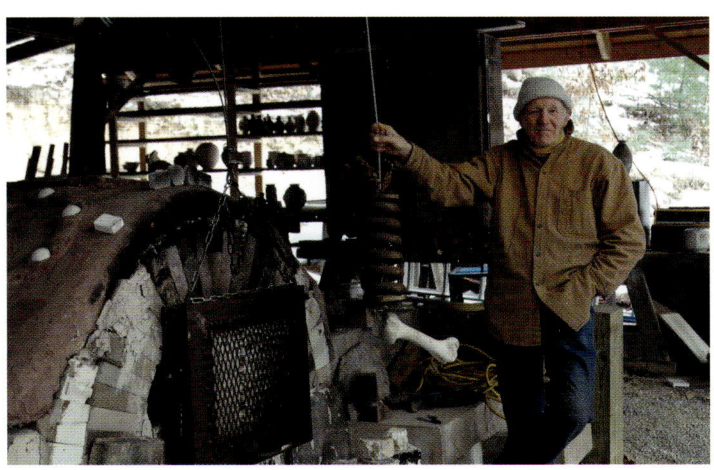

Jack Troy at Pixiegama, 2007. *Photo credit: Carolanne Currier.*

RENATA WADSWORTH

Spencer, New York
renata@nataspots.com, www.renatawadsworth.com

Artist's Statement

Working with clay develops its own rhythm. Wedge, weigh, center, pull ... and repeat. Often, I pause during this dance to alter, trim, add a handle, or just contemplate a new form. Pots line up on the studio shelves full of hope and promise.

My training included influences from the folk, Chinese, and Japanese traditions. These potentially opposing methods of working have actually helped me strike a balance between technical precision and creative intuition. I use simple lines and gentle asymmetry to create elegant pots that invite a little fresh humor.

My pottery is made predominantly from porcelain ... a clay that encourages the intense carbon trapping I look for in the shino surfaces. A light soda spray at about cone 8 further enhances these effects. The kiln is then brought to cone 10–11 and cooled slowly over two to three days.

During each stage of making, there is at least one moment of anticipation ... as there are so many ways things could go wrong. My moments of truest happiness often occur when things go wonderfully right and I find myself holding a beautiful piece of pottery.

Three and a half years ago I drove a twenty-six foot rental truck to Shickshinny, Pennsylvania, where I moved about 40,000 lbs of fire brick from a bed of poison ivy to the bed of the truck. It took three trips in the peak of summer and I managed to blow the transmission on one of the trucks from over-loading it.

This was the beginning of my right-of-passage into the world of wood-firing. When I decided to build my first wood kiln (a 60 cubic foot bourry box), I had no previous experience with wood kilns or wood-firing. I figured I'd learn as I go, save some money, and reduce my carbon footprint along the way. The effects on the pots themselves did not become part of the conversation until later.

I know I could have done things a bit differently, but as a serial risk-taker I have thoroughly enjoyed the journey, and when I pull a piece out of the kiln that takes my breath away, it all seems worthwhile.

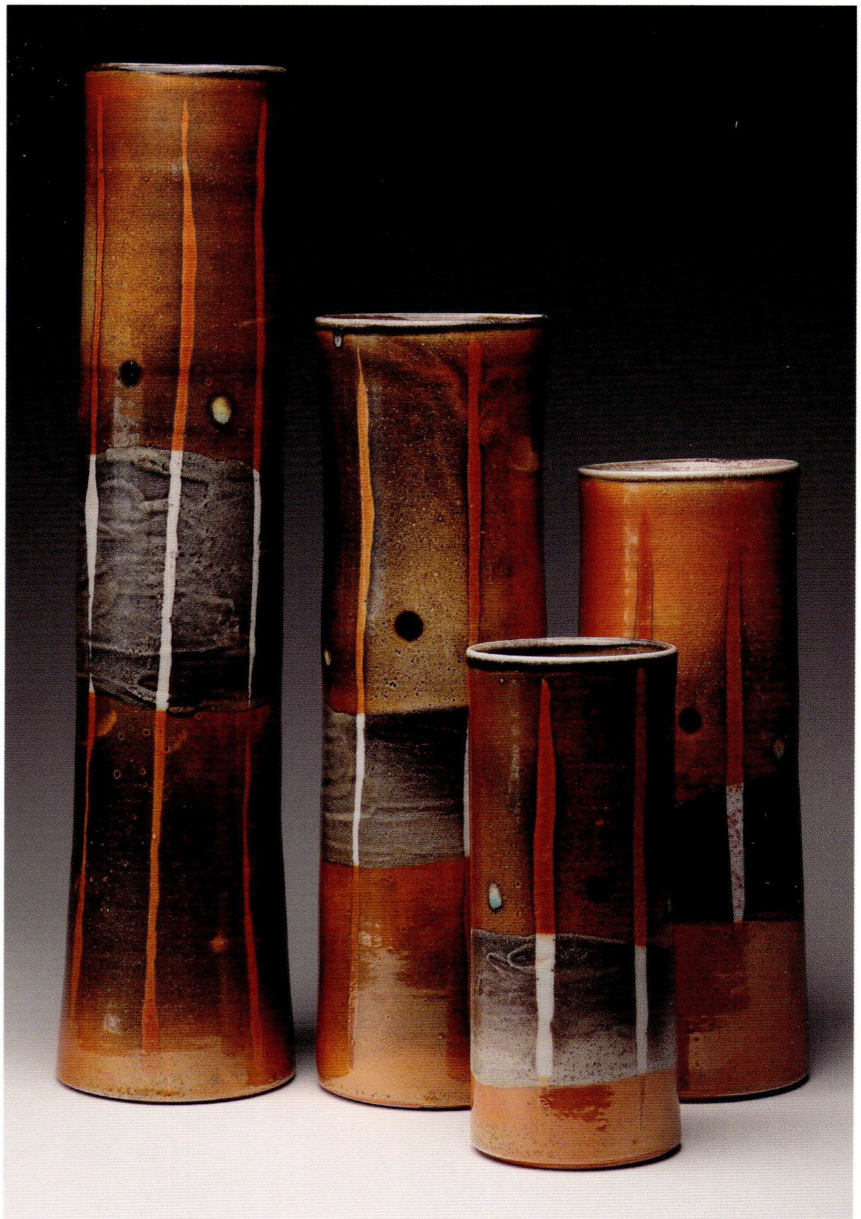

Renata Wadsworth, *Cylinders*, 2011. Wood-fired porcelain, wheel-thrown, 20" x 4" x 4" (tallest). *Photo credit: Alex Solla Photography.*

The first design of my bourry box kiln included a primitive oil drip system (I utilized used veggie oil) in an effort to reduce the total firing time and make it feasible for me to fire the kiln on my own. After months of tinkering with stinky oil and destroying countless steel plates to the fire, I invested in a manufactured waste oil burner that would do the work of atomizing for me. This $2,800 burner didn't end up carving time off my firing ... so back to the drawing board I went.

In August of 2011, my apprentice and I built a new, smaller, more efficient wood kiln that either one of us could fire on our own. It is a small down-draft with the fireboxes (2) under the floor of the ware chamber. After about 10 firings we can get this baby to cone 10–11 in about 15 hours with 3 or more hours of heavy reduction and a light soda spray towards the end.

For me, the making and the firing of my pottery have married and become my teaching partners. I cannot quite picture one without the other. I make pots that will come alive in the kiln and help the kiln bring the pots to life.

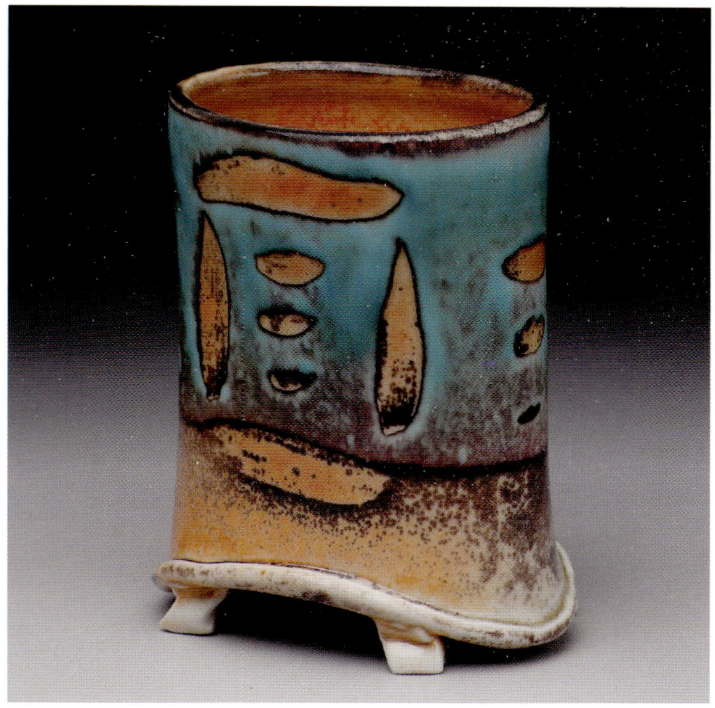

Renata Wadsworth, *Footed Vase,* 2011. Wood-fired porcelain, wheel-thrown and altered, 5" x 3" x 2.5". *Photo credit: Alex Solla Photography.*

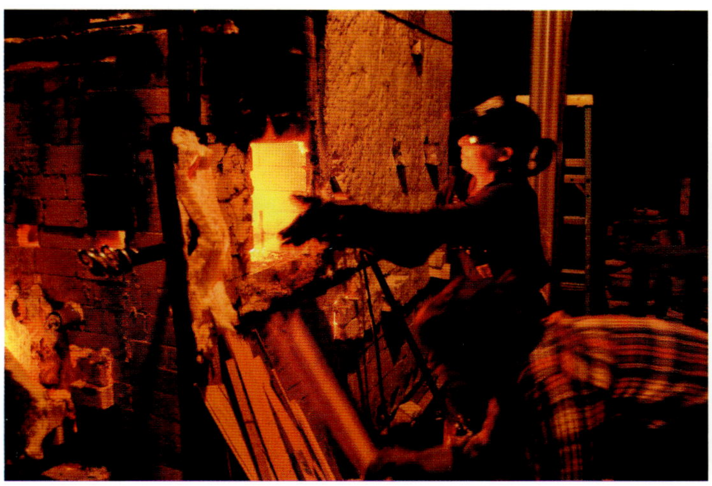

Renata Wadsworth, Stoking the Bourry Box. *Photo credit: Alex Solla Photography.*

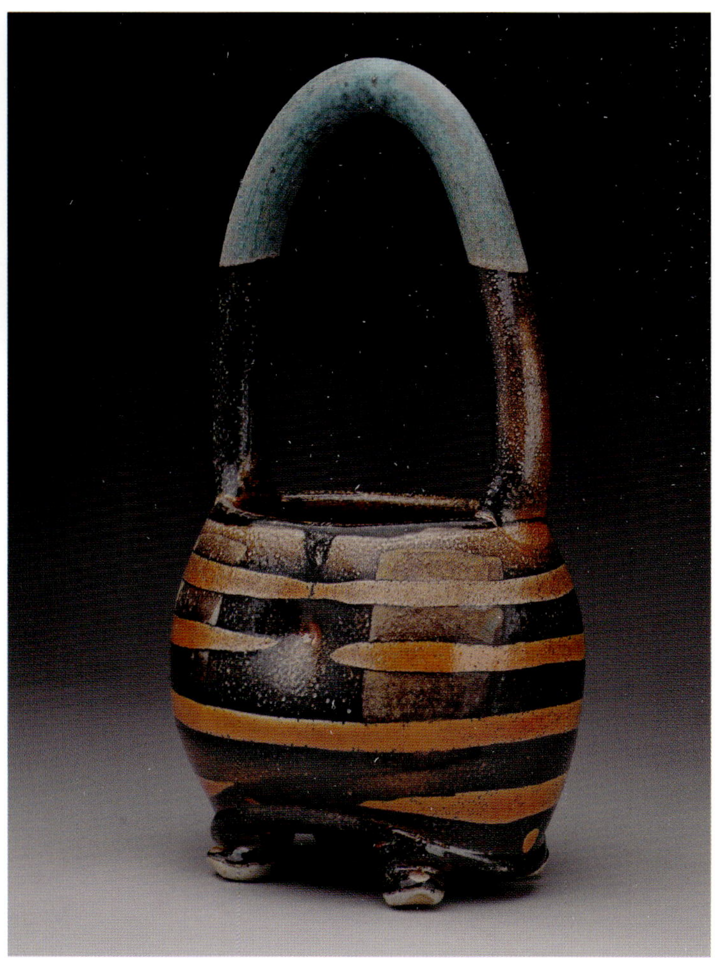

Renata Wadsworth, *Basket,* 2011. Wood-fired porcelain, wheel-thrown and altered, 15" x 8" x 6.5". *Photo credit: Alex Solla Photography.*

HONG-LING WEE

New York, New York
HL@ceramicus.com, www.ceramicus.com

Artist's Statement

I like to give a contemporary freshness to a traditional craft by specializing in one-of-a-kind handmade ceramics that transform everyday utilitarian pottery into delightful, decorative objects.

I enjoy exploring simple objects that are often taken for granted. I see successful designs as the union of function and aesthetics. I would like my work to occupy a living space and be engaged in daily use to fulfill ideas of the useful and the beautiful.

Once a year, I get invited by the Lewis Central High School in Iowa as a visiting artist to teach a workshop with Clay Cunningham, the ceramic instructor at LCHS. There, we spend the first week making work and the second week loading, firing, and cooling the anagama.

This single-chamber wood kiln is fired steadily over two days to reach cone 10, then allowed to soak for a few hours at that temperature before being sealed to cool very slowly for three days. The firing uses just under two cords of wood. I usually line my functional pots with a shino and leave the porcelaneous stoneware exterior unglazed to let the fire do its magic. Living in New York City limits my firing options; hence this annual wood-firing ritual is a real treat for me and my work.

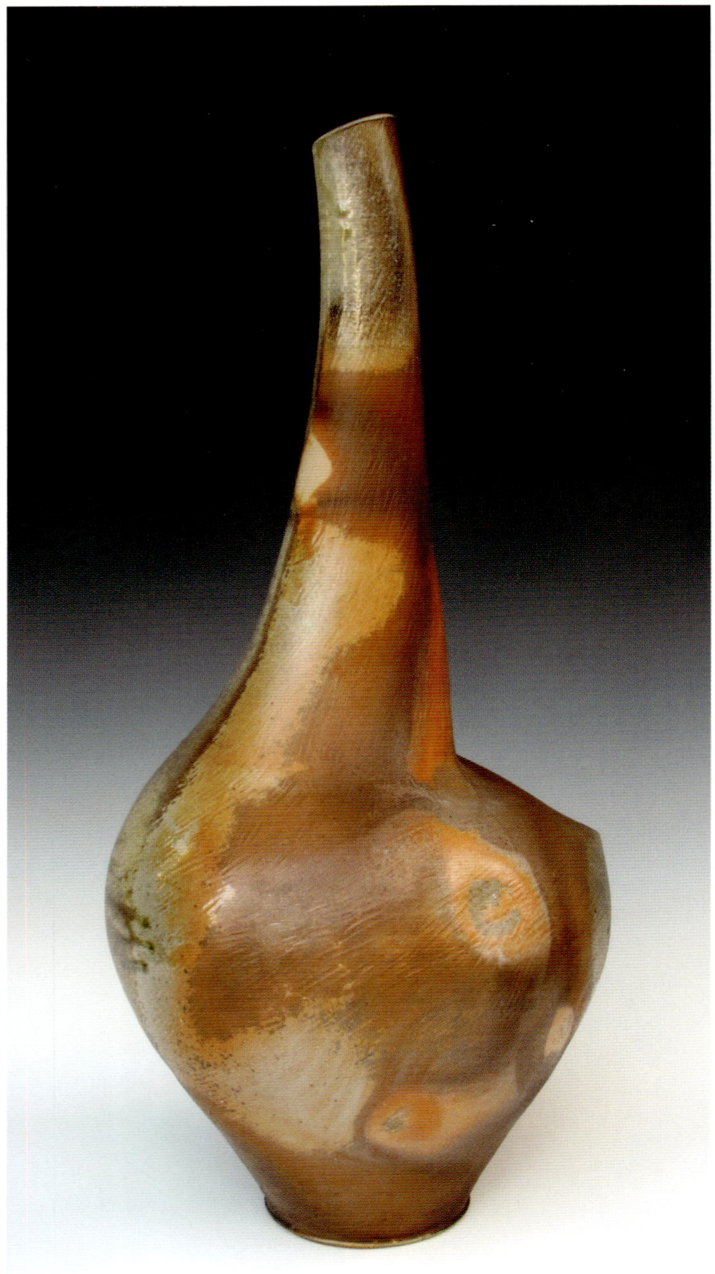

Hong-Ling Wee, *Bird Vase – Nina*, 2009. Stoneware, wheel-thrown and altered, wood-fired without glaze, 24" x 11" x 1". *Photo credit: Hong-Ling Wee.*

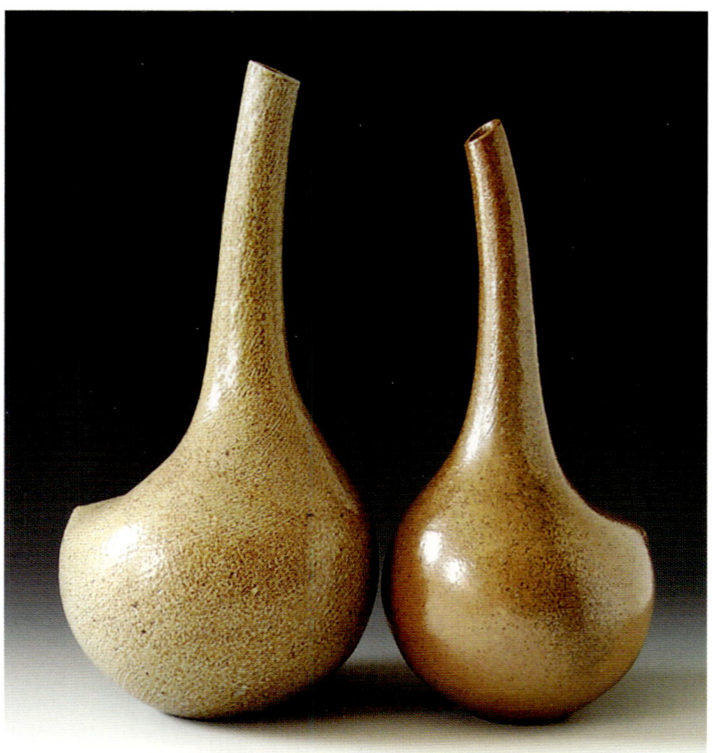

Hong-Ling Wee, *Together*, 2009. Stoneware, wheel-thrown and altered, wood-fired without glaze, 17" x 6" x 7" each. *Photo credit: Hong-Ling Wee.*

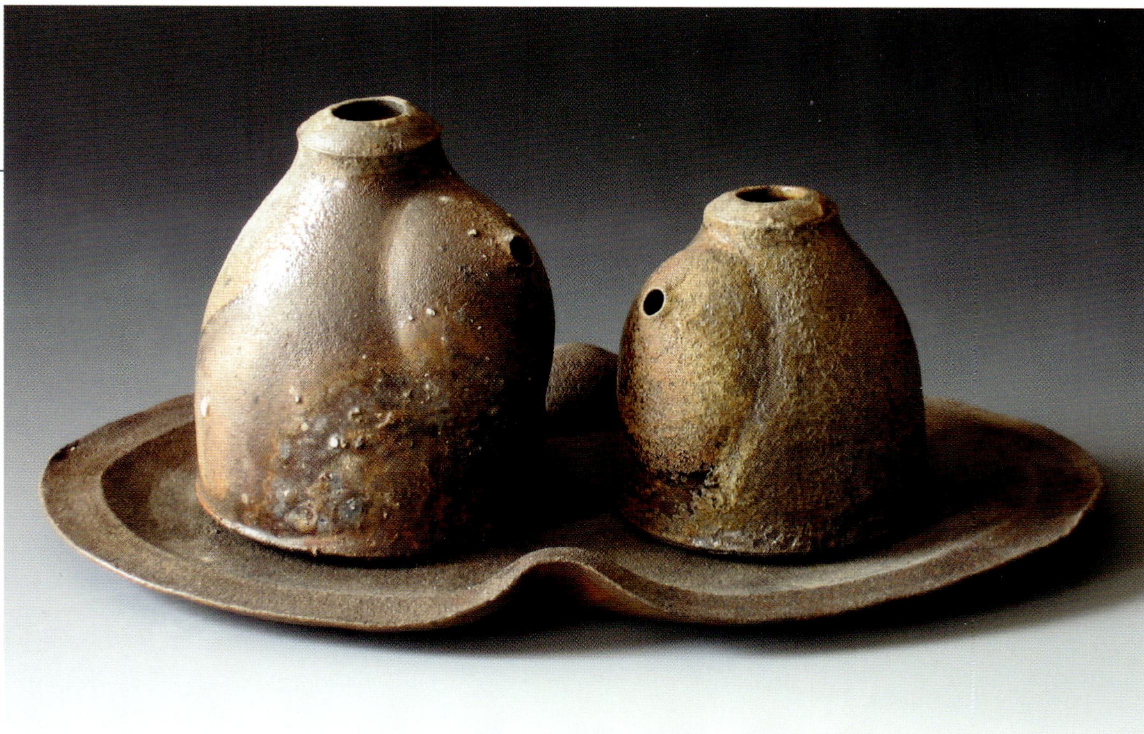

Hong-Ling Wee, *Oil & Vinegar Ewer Set*, 2008. Stoneware, wheel-thrown and altered, wood-fired without glaze, 4" x 5" x 9". *Photo credit: Hong-Ling Wee.*

Hong-Ling Wee, *Wood-fired Ovoid Bowl 2*, 2010. Stoneware, wheel-thrown and altered, wood-fired without glaze, 4" x 14 x 16" each. *Photo credit: Hong-Ling Wee.*

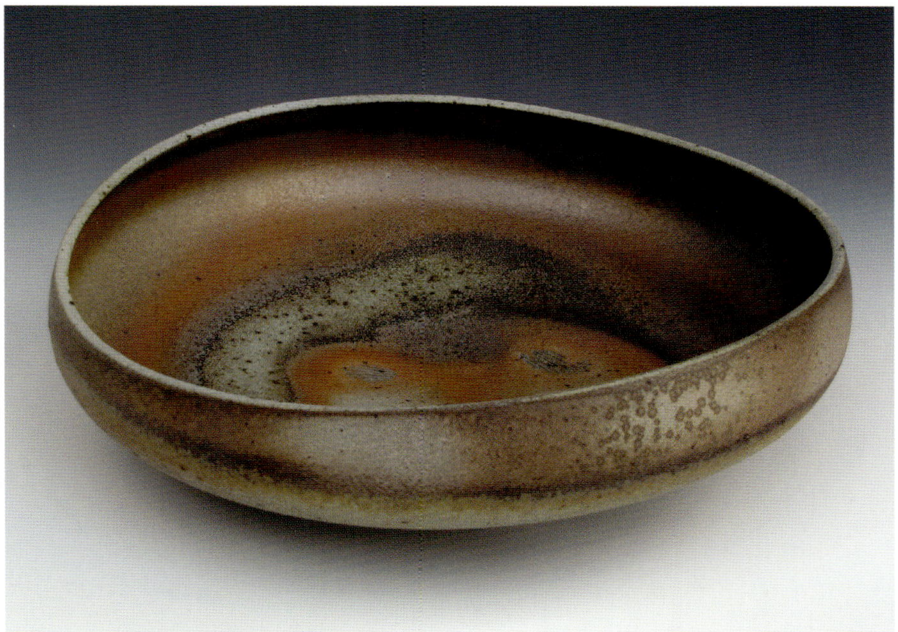

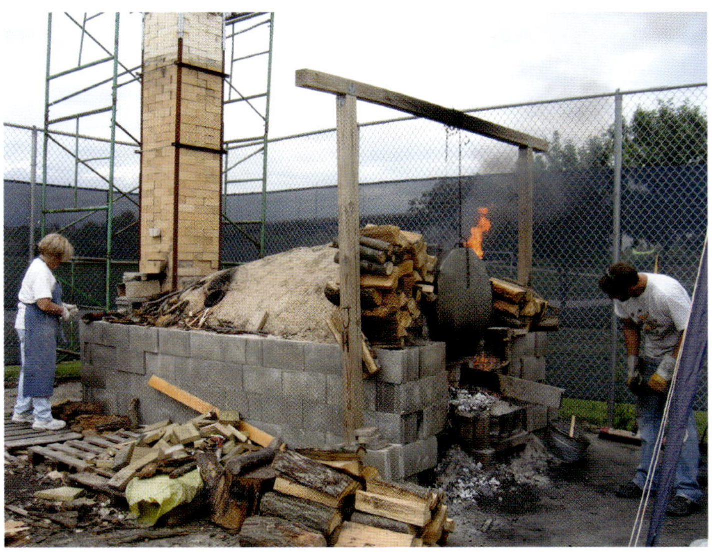

The anagama, 2009. *Photo credit: Hong-Ling Wee.*

JINNY WHITEHEAD

Vancouver, British Columbia
vwhitehead@shaw.ca, www.jinnywhitehead.com

Artist's Statement

I am a self-taught potter living and working in Vancouver, British Columbia. I first started working with clay in 1983. During the next fifteen years, I attended workshops and courses, learning techniques and working intuitively to develop my own style. In 1998, with four other potters, I established a cooperative studio that, twelve years later, is still thriving.

From 1999 to 2004, I participated in the firing of an anagama kiln located in Washington State. It required a team of twelve people working eight-hour shifts. Progressively expanding my mastery of firing techniques, I moved to a train kiln, which only required four people to fire. In January 2009, I celebrated the inaugural opening of a train kiln built and fired by three colleagues and myself.

Since 1999, I have been actively involved in the Potters Guild of British Columbia and in 2003 I was elected President. The Guild is over fifty years old, an organization with over 300 members and a successful gallery displaying the work of over 100 BC artists. In the same year, I was elected as Vice-President of the North-West Ceramics Foundation, a charitable society fostering education and the appreciation of ceramics.

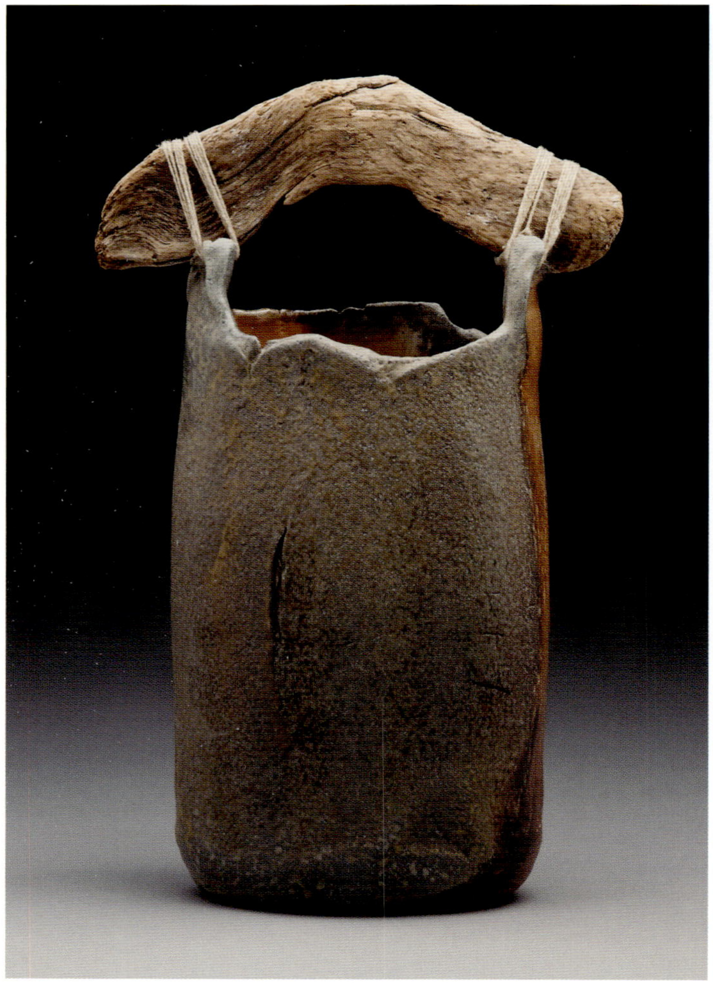

Jinny Whitehead, *Water Carrier,* 2009, Handbuilt vase with driftwood handle, wood-fired stoneware, 11" x 7" x 3". *Photo credit: Goran Basaric.*

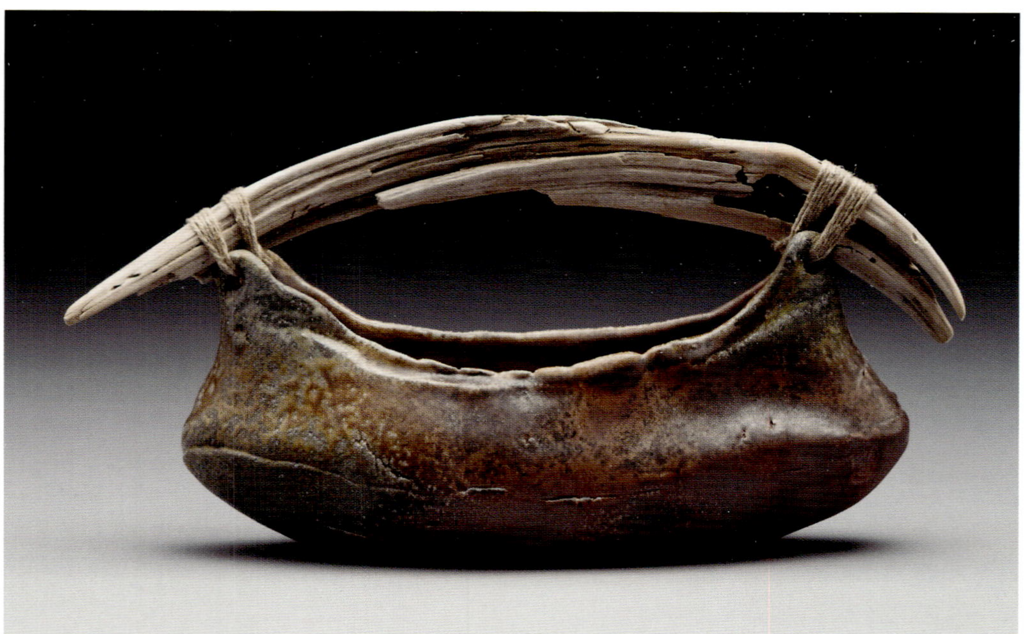

Jinny Whitehead, *Raincoast Life Contained,* 2009. Handbuilt vessel with driftwood handle, wood-fired stoneware, 15" x 6" x 8". *Photo credit: Goran Basaric.*

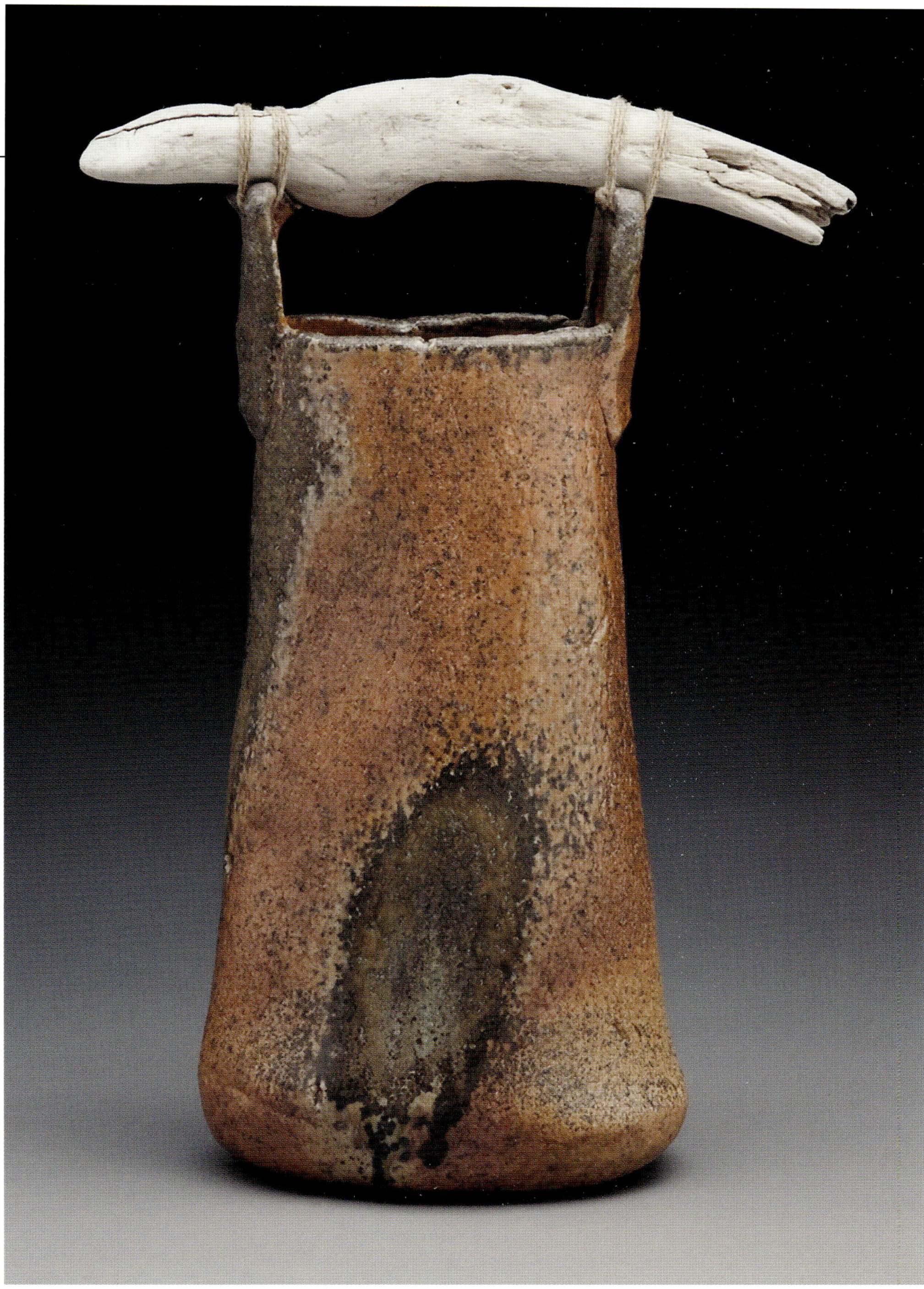

Jinny Whitehead, *Water Carrier,* 2009. Handbuilt vase with driftwood handle, wood-fired stoneware, 14" x 7" x 4". *Photo credit: Goran Basaric.*

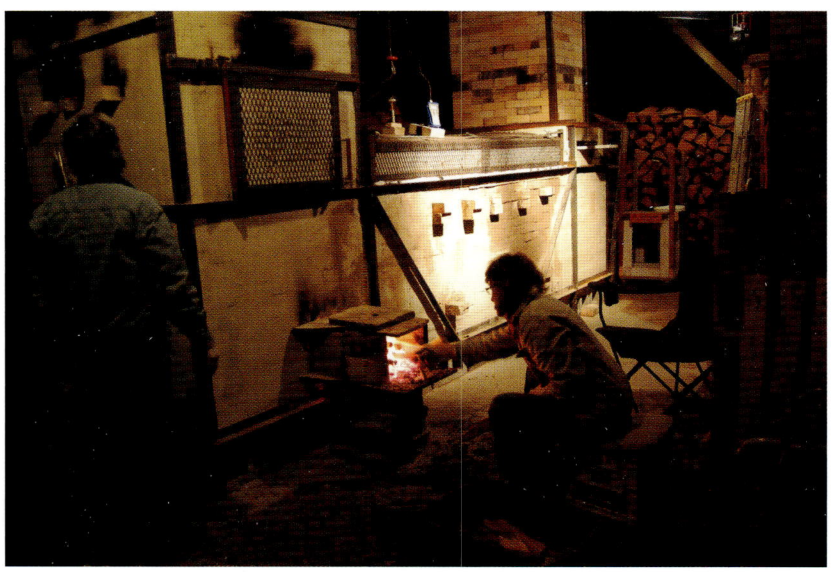

Stoking the mouse hole. *Photo credit: Ron Robb.*

BETSY WILLIAMS

Dixon, New Mexico
riftgallery@yahoo.com, www.enbistudio.com

Artist's Statement

About 95% of my work is wood-fired. While most Westerners associate wood-firing with large climbing kilns fired by groups of people for days, if not weeks, and the resulting ware as ash-laden brown ware with flashes of color, the wood-firing that I do stems from a different tradition in Japan. I studied in an apprenticeship with Mr. Yutaka Ohashi of Karatsu, Japan, from 1994 to 1999. The ware of Karatsu, on the western-most island of Kyushu, is closely related to the Korean ceramic tradition, and incorporates glazed ware fired in noborigama style kilns.

I glaze all of my ware, using glazes that I make using things like ash (from previous firings), and I load the kiln using wadding, but I fire only until I reach temperature (somewhere between cone 8 and cone 11, depending on position in the kiln), and do not rely on heavy ash buildup on the ware.

Much of my work is coated in white slip at the leather hard stage, and then decorated and glazed with a clear glaze, a technique that originated in Korea and was subsequently adopted by the Japanese. This technique, called "*kohiki*" (or "powdered" as in make-up) yields a soft and vivid decorating surface with some blushes of color when fired in reduction, but also a toasty warm clay body underneath.

A sense of directness and calm in my work, together with an overall visual rhythm among the pieces, are important to me. My work includes multiple-cup compositions, installations of cups, which can be used and rearranged. I am interested in work that presents a concept and/or visual image without losing any of its usefulness.

At the center of what I do is the conviction that good pots have the capacity to communicate something essential and ineffable about the human spirit, and that they do this by bringing an element of the sublime into the everyday. I search for a balance between subtlety and self-expression, minimalism and warmth, delicacy and usability, tradition and originality.

My kiln construction is based on a design which, in Japanese, is called "*itte-koi gama*," meaning simply "go and come back kiln" and referring to the circulation of the flame. The proportions of my kiln, including the chimney height, are based on the principles outlined in Fred Olsen's *The Kiln Book*, including the adjustments for high altitude. (I am at 7320 feet above sea level.) The kiln is built with hard brick and refractory mortar, and the kiln chamber arch includes a layer of insulating fiber. The firebox is below the kiln chamber, with its arch perpendicular to the direction of the kiln chamber arch. The kiln is stacked with about six levels of small to medium ware and then the kiln chamber is bricked up with firebrick, sealed with clay, and the threaded rod is screwed firmly onto the corner angle iron. During the firing, the flame first moves to the back of the kiln and up through the inlet flue in the floor of the kiln chamber. Then the flame follows the shape of the kiln chamber arch toward the front of the kiln, finding the exit flues, moving past the dampers, and then exiting up the chimney, which is just above the stoke port. The kiln is fairly small (about thirty-six cubic feet) and requires about thirteen hours and a half cord of pine to fire. Upon reaching temperature, all ports are sealed and the kiln is left to cool for three days. I fire about six to eight times per year.

The possibilities and types of wood-firing are much deeper and far-reaching than many people imagine.

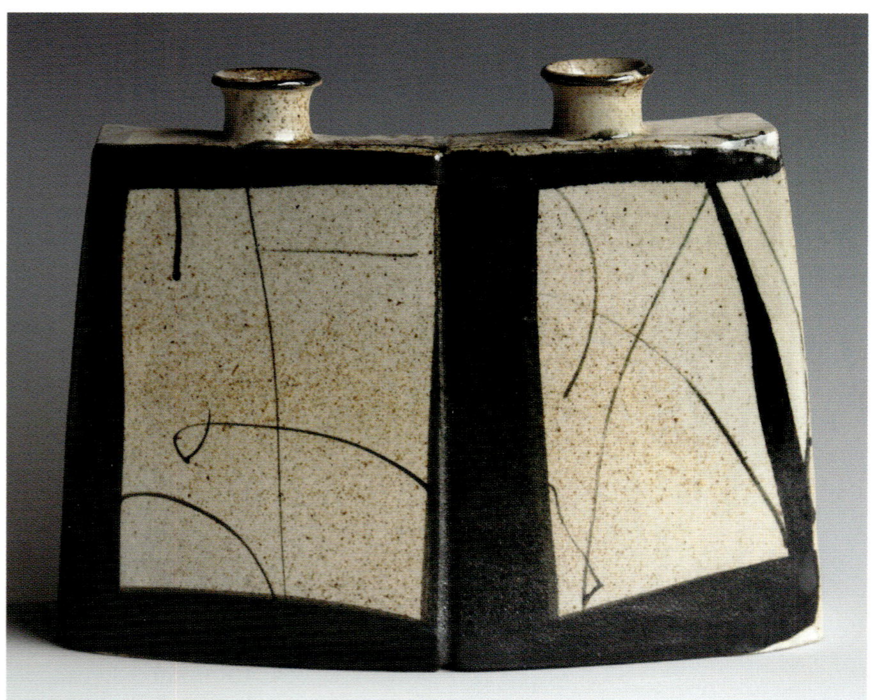

Betsy Williams, *Double Flask with Lines*, 2008. Slab built with wheel-thrown components, underglaze oxides, wood-fired stoneware, 9" x 8" x 3". *Photo credit: Betsy Williams.*

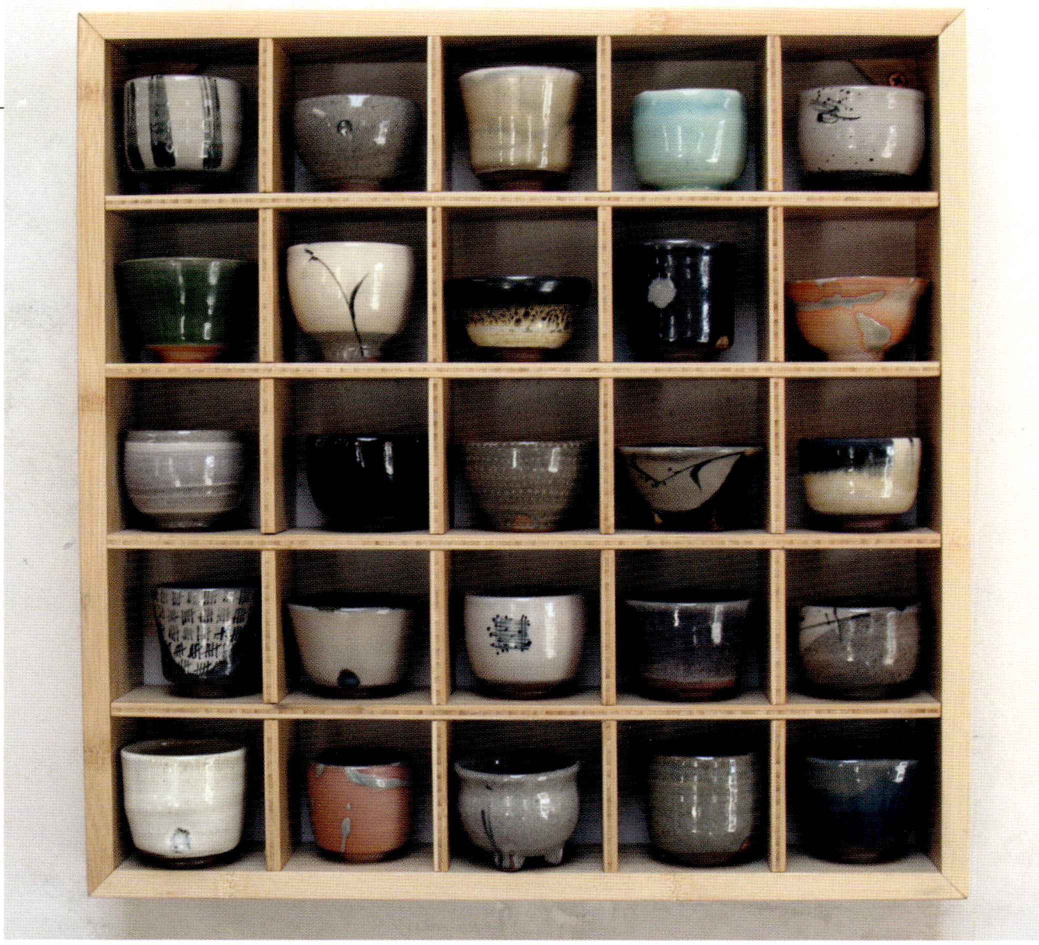

Betsy Williams, *Cupism 25*, 2011. 25 wheel-thrown cups in a bamboo grid, wood-fired stoneware (except for turquoise cups), 16" x 16" x 3". *Photo credit: Betsy Williams.*

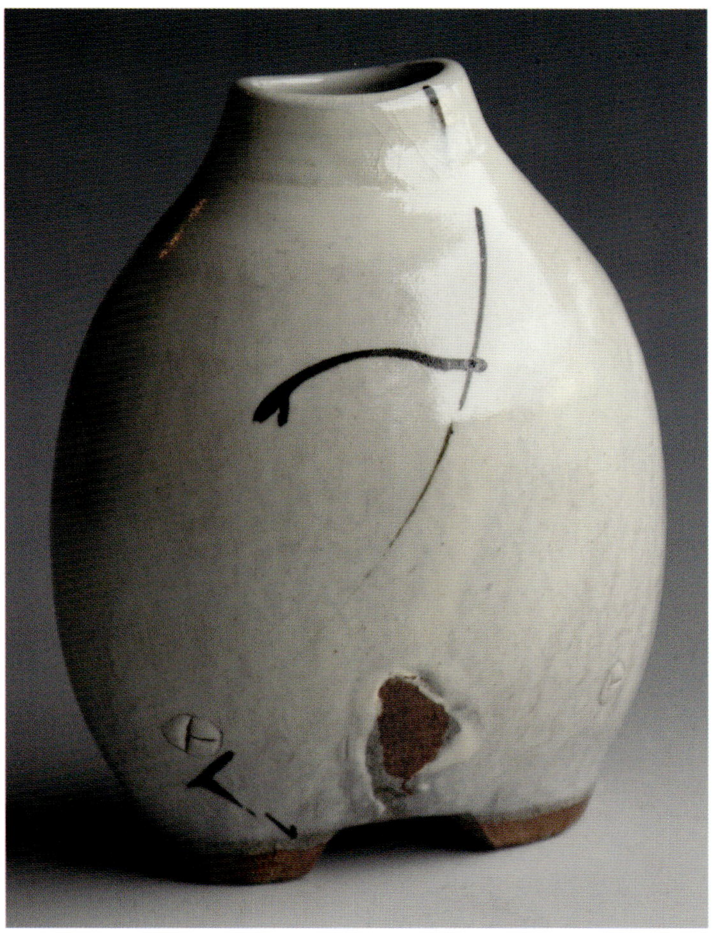

Betsy Williams, *Oval Flask with Lines*, 2010. Wheel-thrown, altered; white slip, underglaze oxides, wood-fired stoneware, 9" x 7" x 3". *Photo credit: Betsy Williams.*

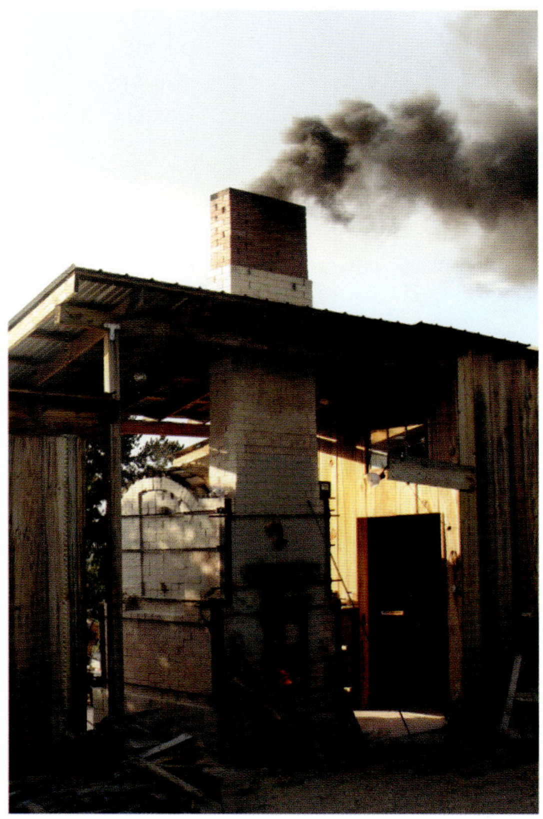

Betsy Williams, Kiln firing full view. *Photo credit: Betsy Williams.*

TARA WILSON

Montana City, Montana
tara@tarawilsonpottery.com, www.tarawilsonpottery.com

Artist's Statement

The pottery I make is thrown, altered, and fired in a train style wood kiln. I was drawn to the wood-firing process early on as an undergraduate student at the University of Tennessee. I enjoy the community aspect of the process and being intimately involved with the pot's transformation. I also love the physicality of the process. For some odd reason, I love physical labor; I don't have to work out if I chop wood. I feel satisfied by the exhaustion at the end of a firing. I also like the connectedness to the work as it's transformed into the finished state. I love and hate the unpredictability of the process. The way that I work in the studio is very controlled; the wood-firing process is a good balance for me, forcing me to give up some control to the kiln. Ultimately it's about the surfaces that I get on the work. The surfaces achieved through wood-firing compliment my forms. I choose to fire a train style kiln because of the fairly short firing cycle and length of firing required while still achieving surfaces that appear to be fired for a longer period of time.

My quiet pots initially speak softly, yet reveal themselves through continued investigation and use. Embodied in my wood-fired pottery is the serenity that I experience by surrounding myself on a daily basis with a rich natural environment. While the surfaces of the vessels represent the natural world, the forms often relate to the figure. My thrown pieces are altered in such a way that they appear natural, as if the form could possibly exist in nature. Like the figure, I want the parts of my vessels to feel like they belong together, the feet and arms emerging naturally from the body. I also want the viewer or user to have to investigate the piece in order to figure out how it was created, holding their attention, and revealing more details over time.

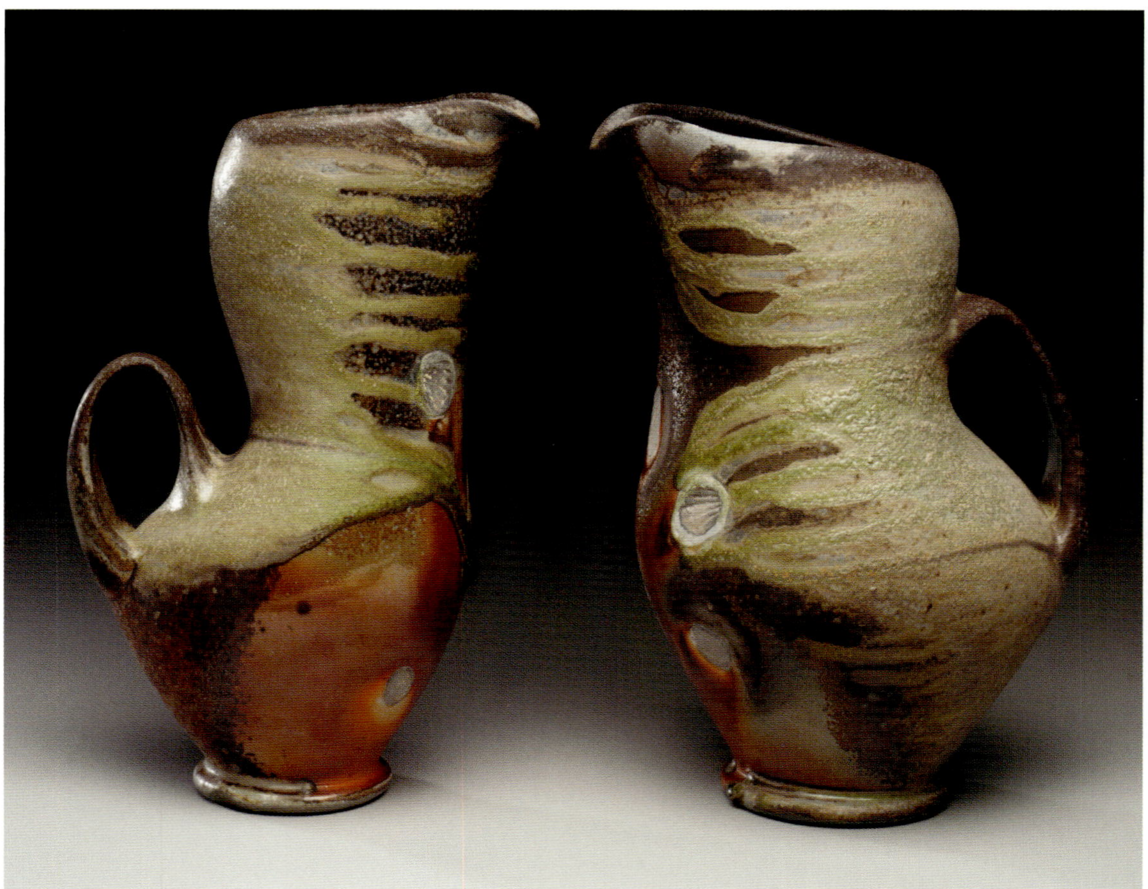

Tara Wilson, *Pitchers*, 2011. Wood-fired stoneware, thrown and altered, 12" x 15" x 8".
Photo credit: Chris Autio.

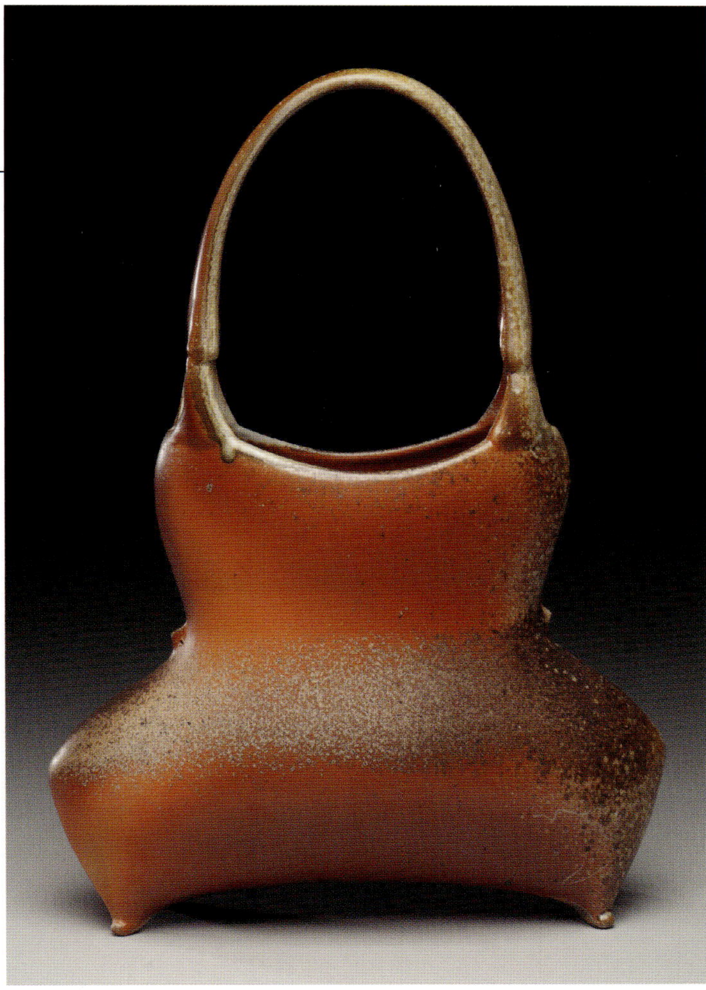

Tara Wilson, *Basket*, 2011. Wood-fired stoneware, thrown and altered, 12" x 9" x 5". *Photo credit: Chris Autio.*

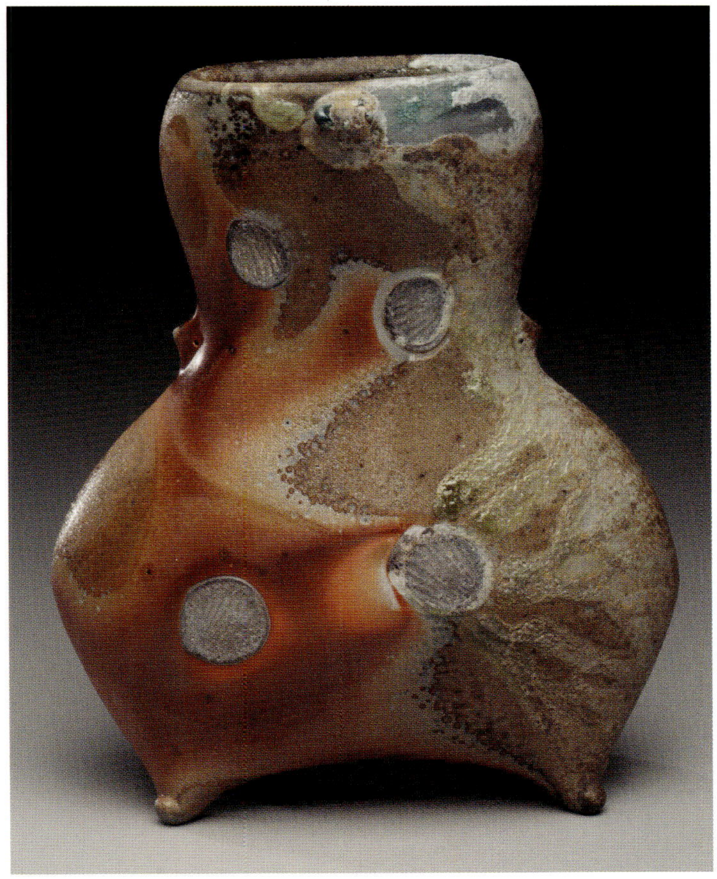

Tara Wilson, *Vase*, 2011. Wood-fired stoneware, thrown and altered, 7" x 6" x 4". *Photo credit: Chris Autio.*

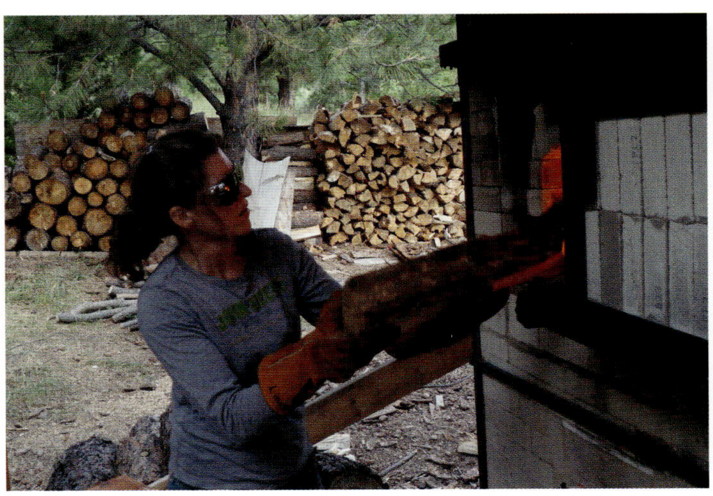

Tara Wilson stoking, 2011. *Photo credit: Chris Autio.*

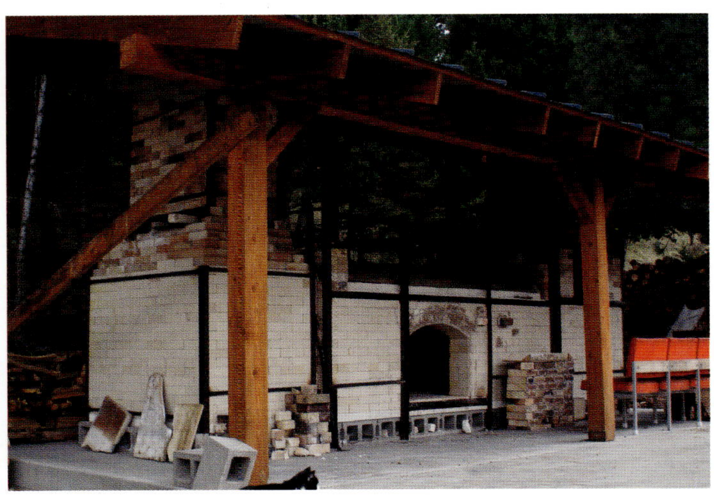

Tara Wilson's Train Kiln, 2012. *Photo credit: Tara Wilson.*

DAVID WRIGHT

Thrussington, Leicester, England
david.wright52@virgin.net, www.davidwrightpottery.co.uk

Artist's Statement

I prefer to build upwards with coils of clay. I enjoy the slower pace, manipulating the shape and form as the pot develops. I work on a theme, making three or four pots at a time, each one a development of the preceding vessel. Sometimes I revisit a theme at a later date, taking it in a different direction. Texture is always an important aspect of my work and smooth insides often contrast with rough outsides. My glazes are mainly made from wood ash and the naked clay is often left exposed. All the processes are designed to be as simple as possible.

My pots have been described as "quiet," they invite the viewer to look more closely to discover the myriad of subtle colors and nuances; wood-firing adds an extra dimension to the controlled forms. The vagaries of ash and flame have an unpredictable quality. The weather, type of wood, and duration of firing can give up different and subtle results. I use the wood kiln as a tool. Where the pot might go in the kiln and how it might be glazed are often considered as the form takes shape.

My present kiln is a small 12 cu. ft. down draught, a development of the kiln that I built before and similar to Jack Troy's Phoenix Kiln. Handbuilding is a slow process, and to develop my work, the smaller kiln means that I can fire more often and experiment more with glaze and fire.

I fire, mainly alone, using scrap wood from pallets, firings last about twelve hours or until cone 11 is over.

I am often invited to help fire other wood kilns and I am always surprised how each kiln can give its own signature to the same vessels and glazes.

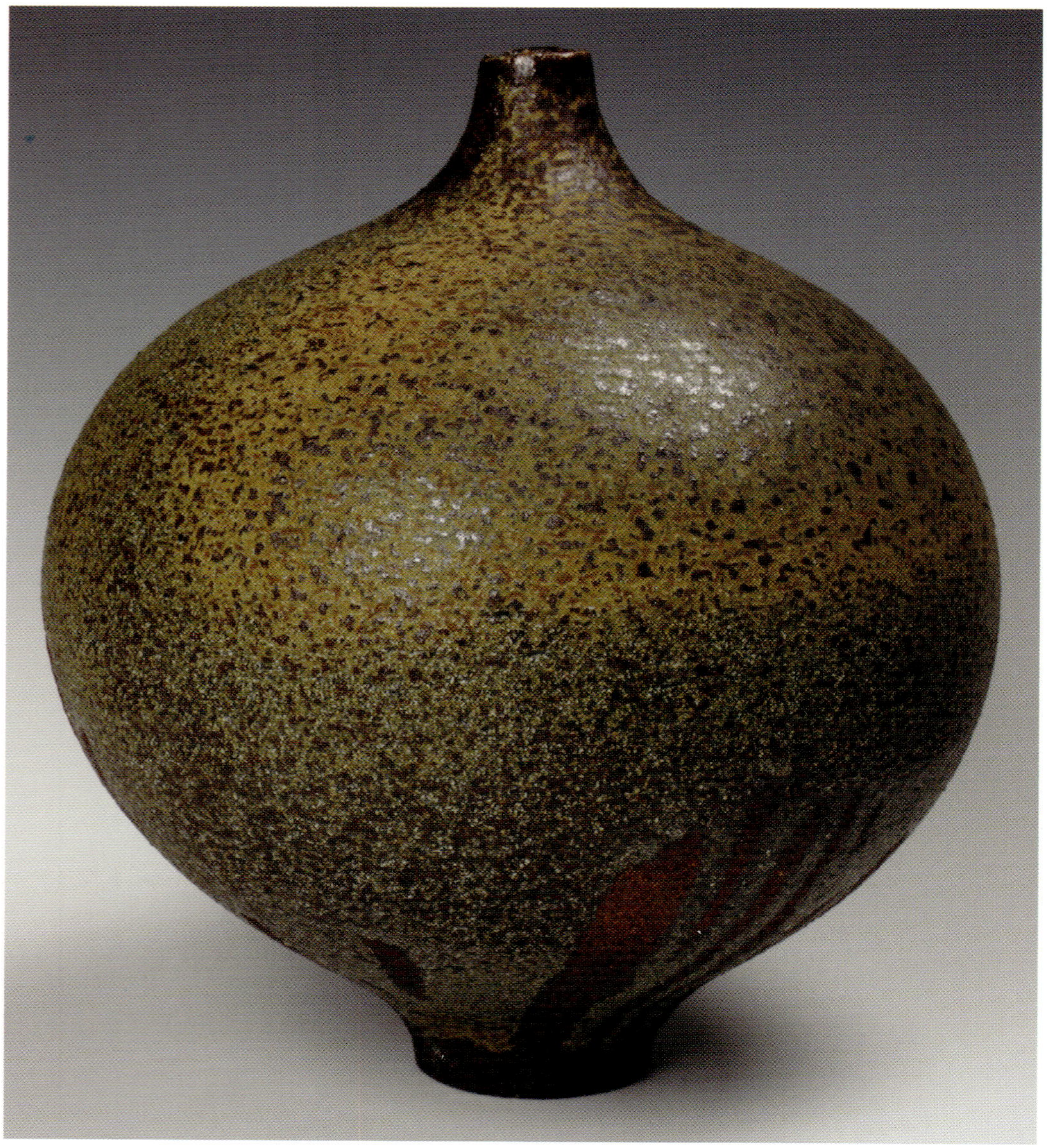

David Wright, *Fat Boy*, 2011. Stoneware with wood ash glaze, coil built, 9" x 8". *Photo credit: David Wright.*

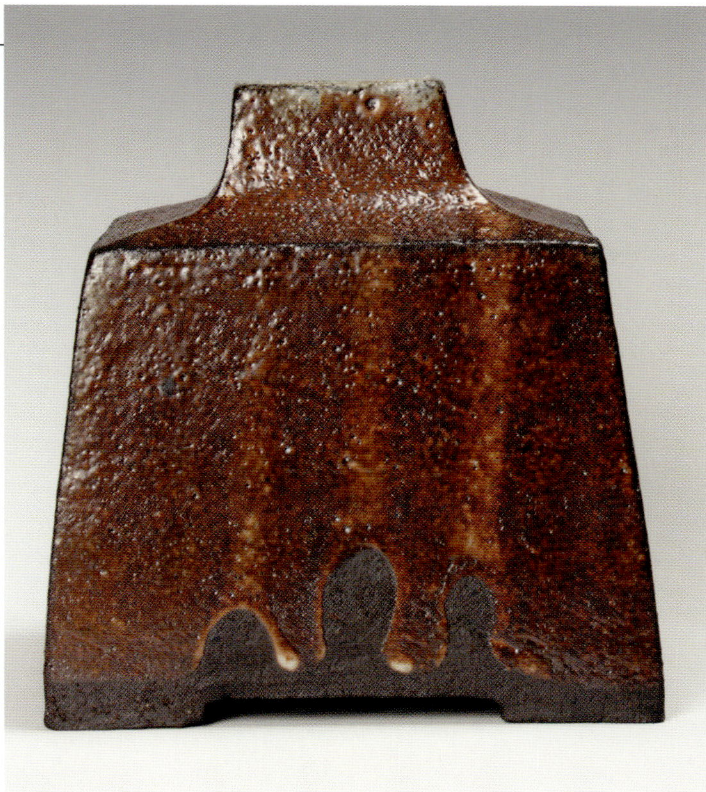

David Wright, *Square Bottle*, 2011. Iron rich stoneware with Shino glaze, coiled and paddled, 5.5" x 5.5" x 4". *Photo credit: David Wright.*

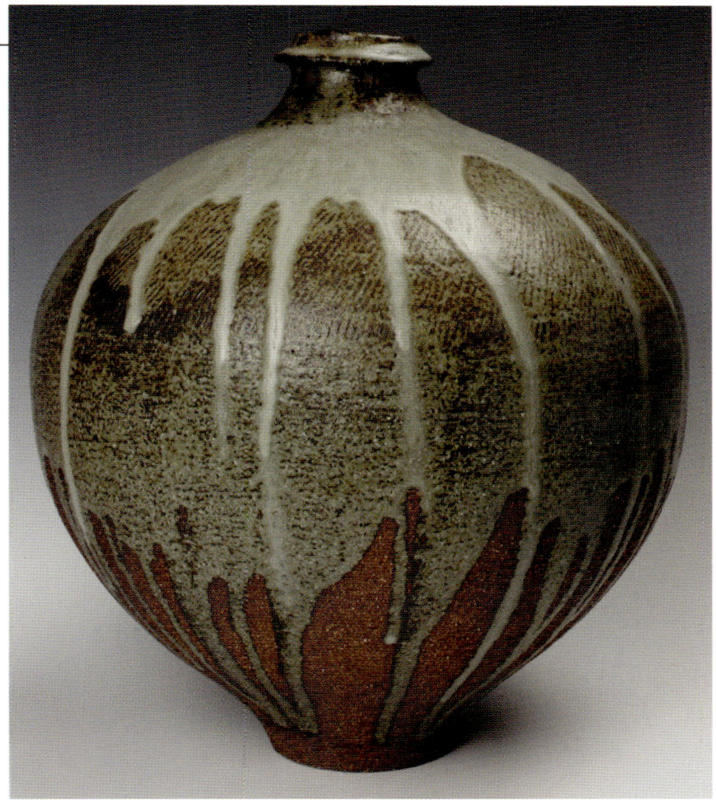

David Wright, *Bottle*, 2011. Stoneware with Sycamore ash glaze, coil built, 12" x 11". *Photo credit: David Wright.*

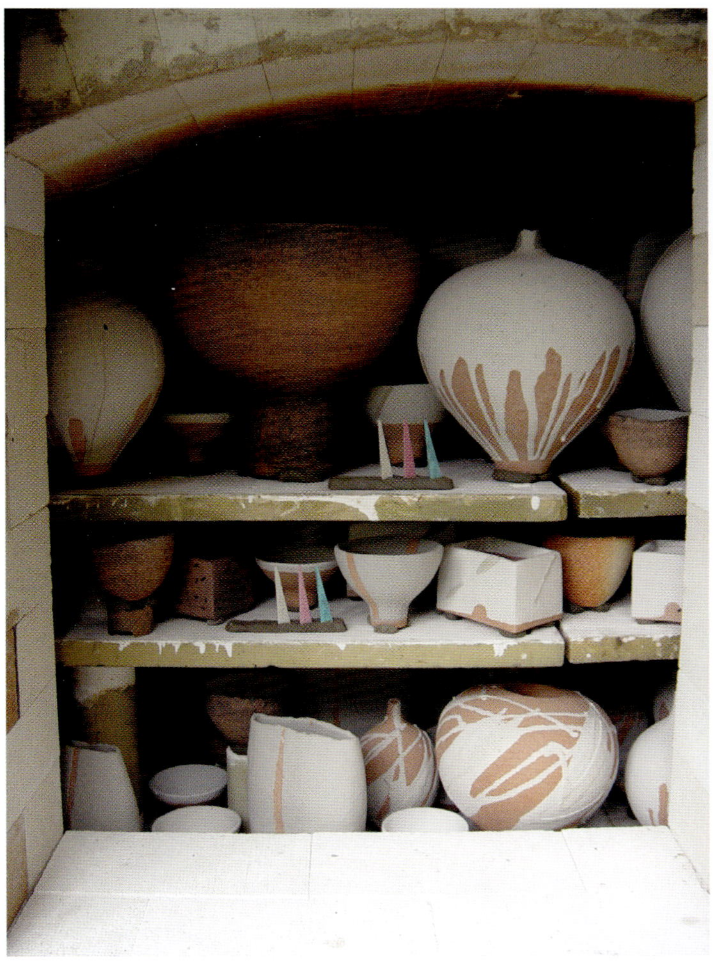

Packed kiln ready to close up. *Photo credit: David Wright.*

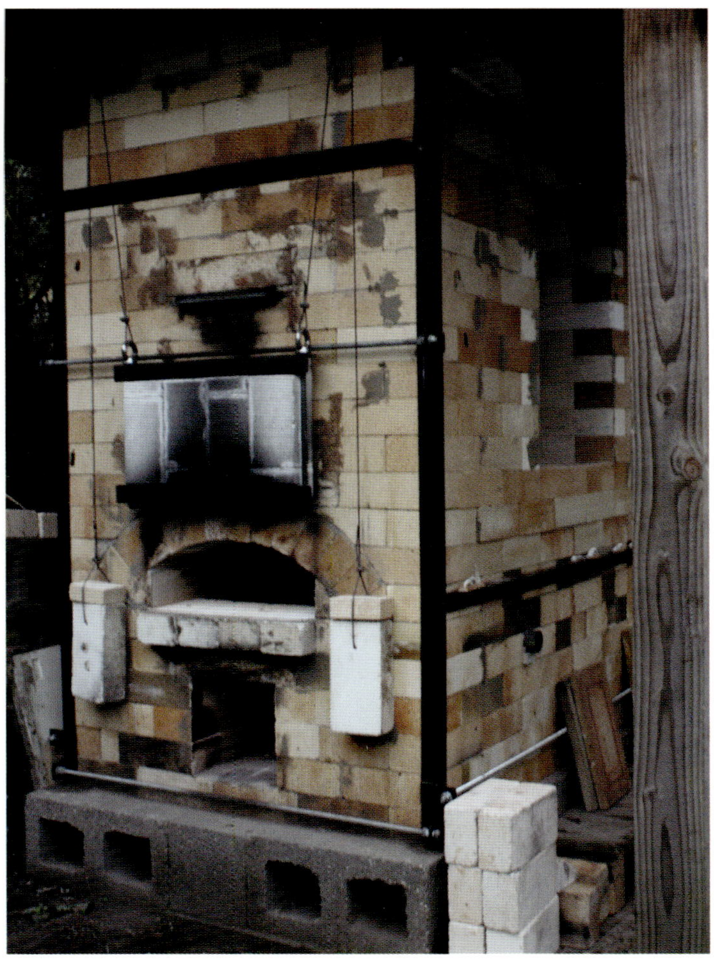

Kiln firebox end. *Photo credit: David Wright.*

The Author and His Work

AMEDEO SALAMONI

Quarryville, Pennsylvania
asalamoni@georgeschool.org, www.salamonipottery.com

Artist's Statement

I work in clay because it enables me to be expressive in ways I cannot in other mediums. I work on several different dimensions. On a practical level, I honor the craft itself, utilizing the skill and knowledge it takes to be a potter. As an artist, I use the process of wood firing to create a chance element in my finished work. The nuances of the flame, the drips of ash, happy mistakes—all this excites me. Finally, as a functional potter, I want people to use my work. For one to pick up a pot and feel it, to use it, to enjoy both the visual beauty and the functional aspect—only then is my work complete.

Most of my work has been fired with anagama-type kilns, but I have utilized a variety of kiln types over the years. Most recently, I have been firing my work in a single chambered catenary arch kiln, with my firing partner Kevin Lehman. A typical firing will last anywhere from twelve to sixteen hours, firing to cone 10 and adding salt or soda toward the end of the firing.

By concentrating on producing simple, yet classic forms on the wheel, I allow the process of wood-firing to give the pots their ultimate beauty. Manipulating the kiln environment—varying types of wood, forcing a rapid or slow temperature climb, or deliberately positioning the pots for effect—contributes to the individuality of each piece.

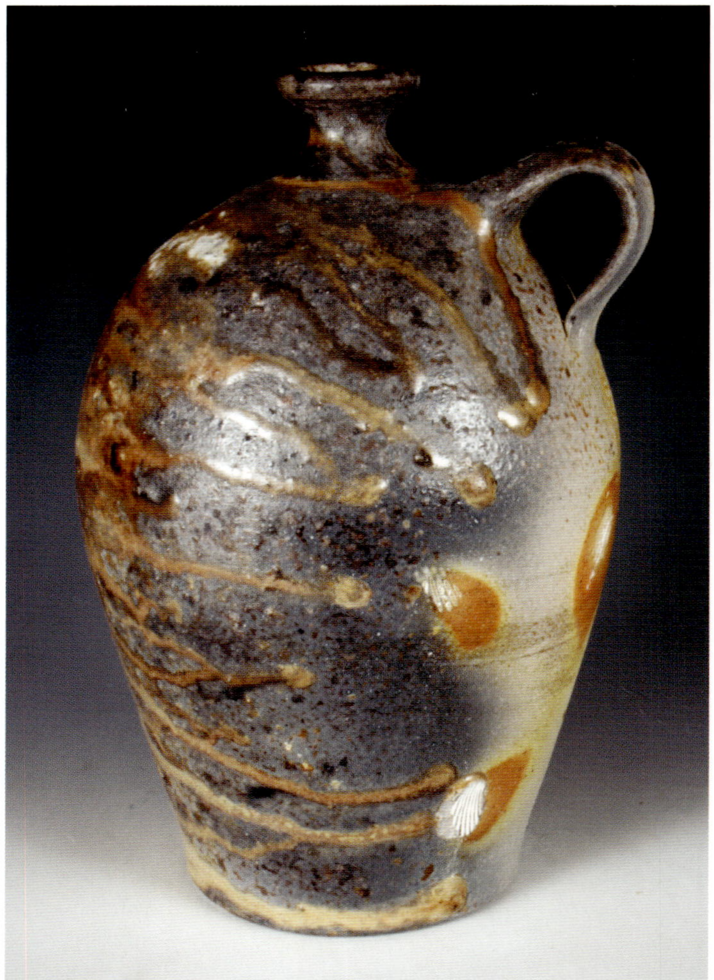

Amedeo Salamoni, *Side Fired Jug*, 2005. Wood-fired stoneware, fired in anagama kiln, buried in embers, natural ash deposits, wheel thrown, 9" x 5.5" x 5.5". *Photo credit: Amedeo Salamoni.*

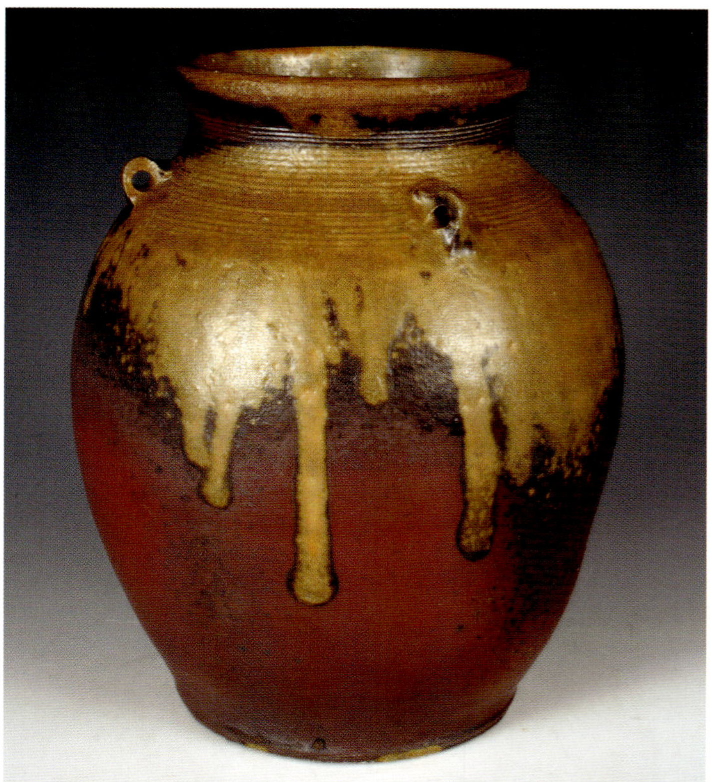

Amedeo Salamoni, *Jar With Handles*, 2001. Wood-fired high iron stoneware, fired in anagama kiln, natural ash deposits, anagama fired, wheel thrown, 8" x 6" x 6". *Photo credit: Amedeo Salamoni.*

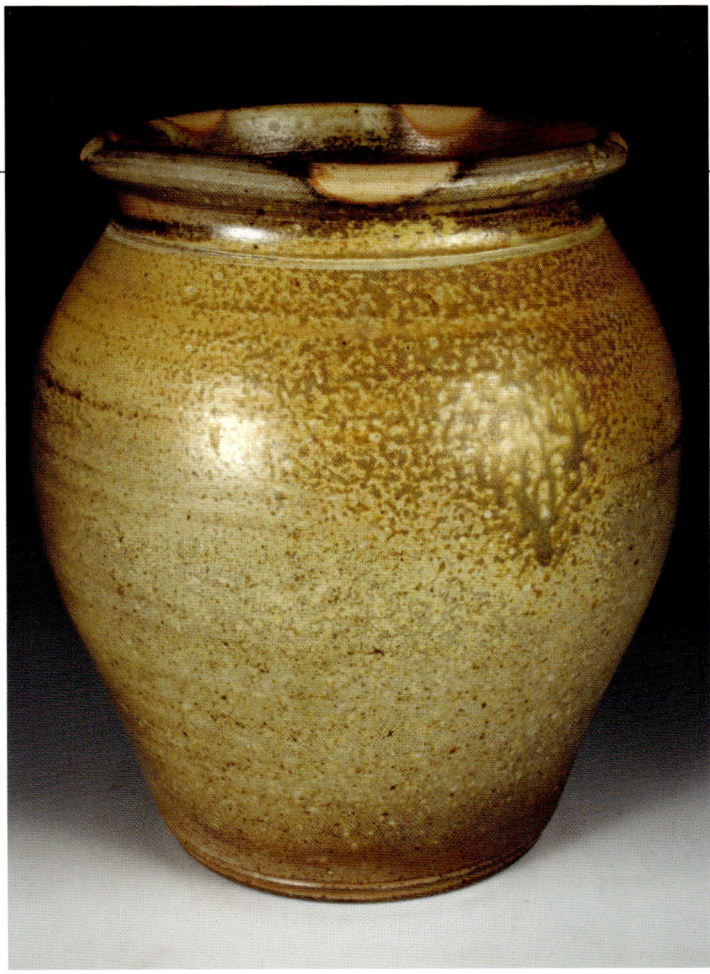

Amedeo Salamoni, *Jar,* 2005. Wood-fired stoneware, natural ash deposits, fired in anagama kiln, wheel thrown, 14" x 13" x 13". *Photo credit: Amedeo Salamoni*

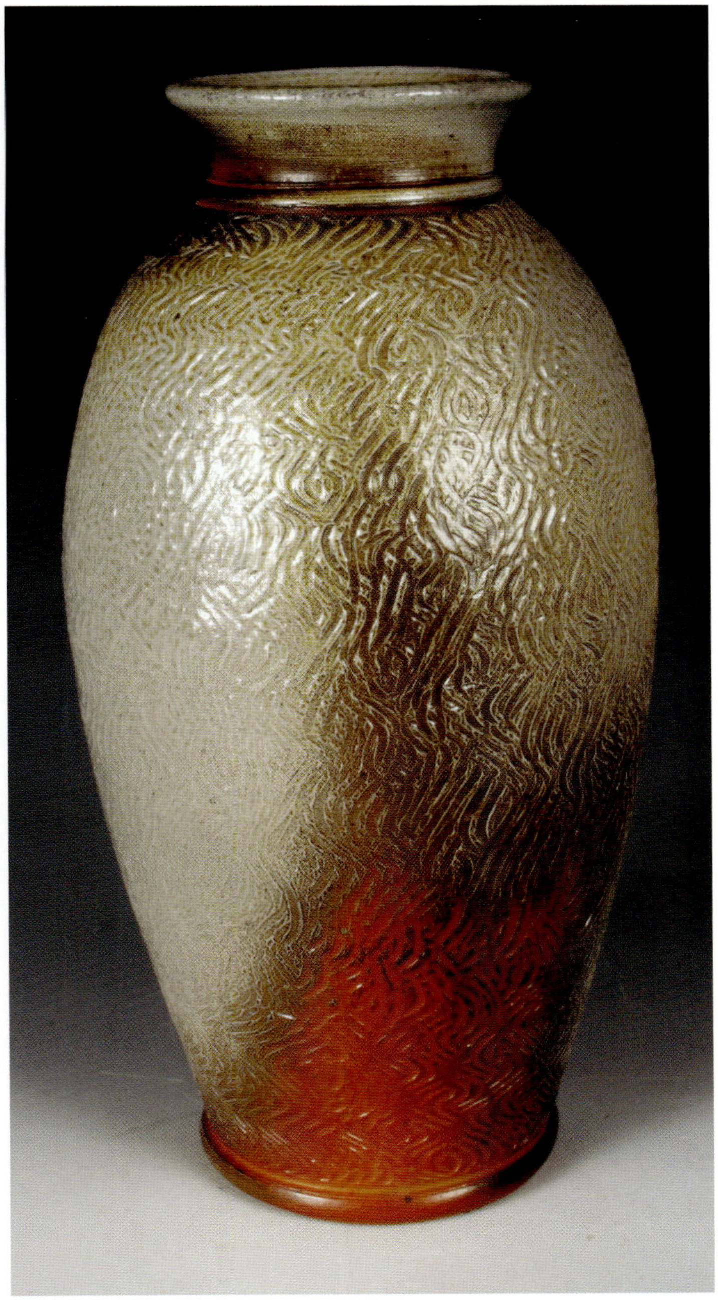

Amedeo Salamoni, *Carved Vase,* 2009. Wood-fired stoneware fired in cross draft type kiln with soda and salt, natural ash deposits, wheel thrown, from 12" x 6" x 6". *Photo credit: Amedeo Salamoni.*

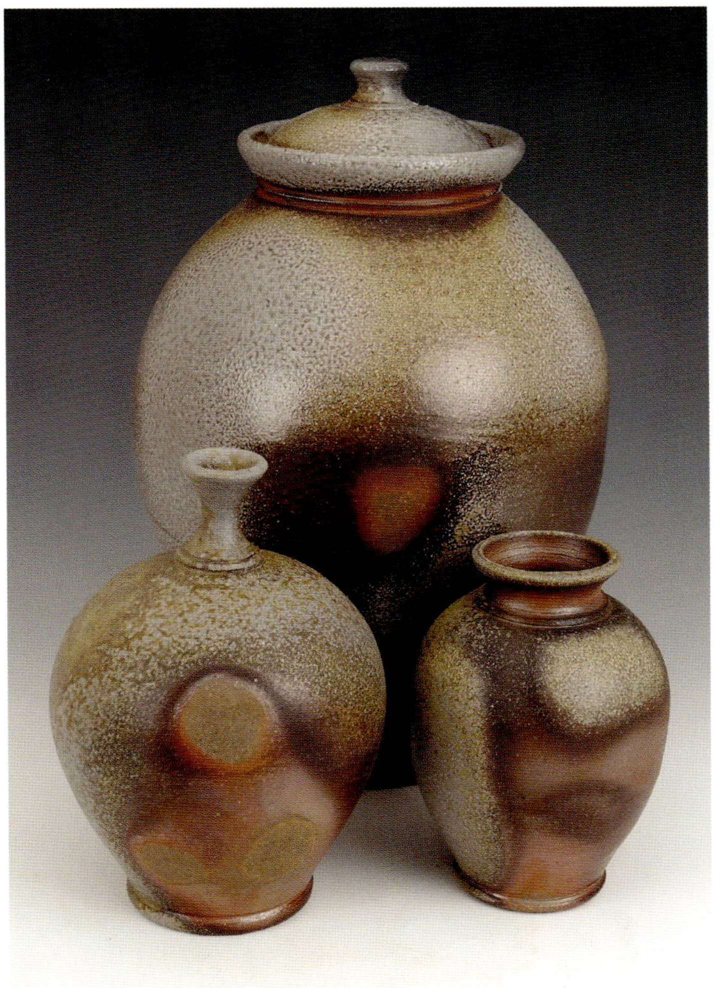

Amedeo Salamoni, *Covered Jar, Bottle, and Vase,* 2006. Wood-fired stoneware, fired in anagama kiln with soda, natural ash deposits, wheel thrown, from 7" h. to 13" h. *Photo credit: Amedeo Salamoni.*

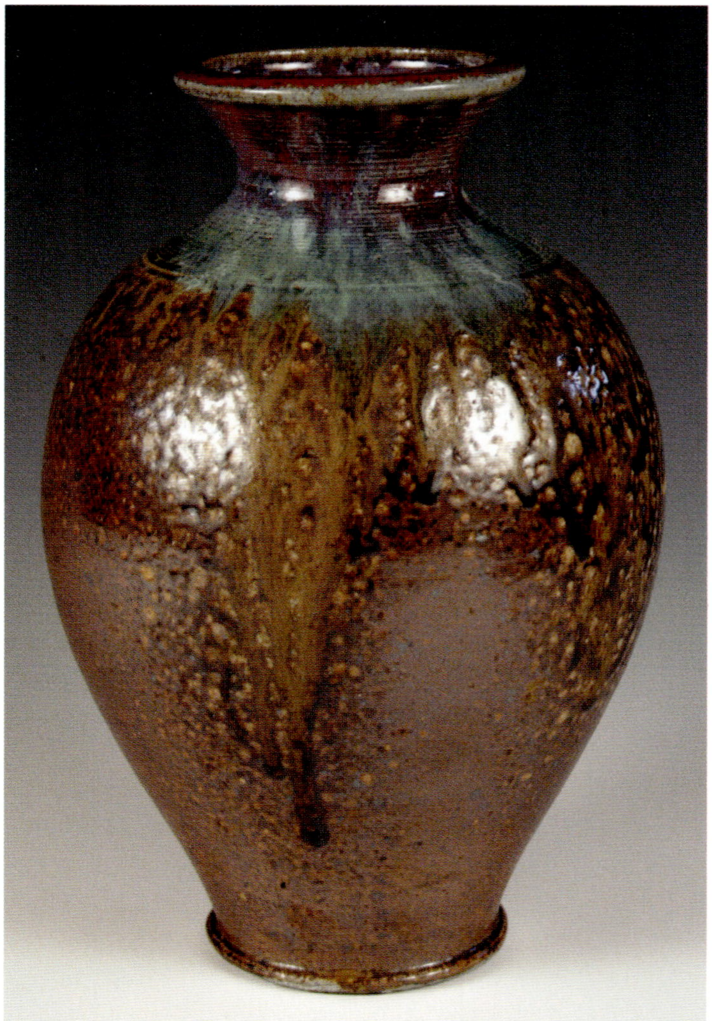

Amedeo Salamoni, *Vase,* 2011. Wood-fired stoneware, fired with soda, applied and natural ash glaze fired in cross draft type kiln with soda and salt, wheel thrown, from 10" x 6.5" x 6.5". *Photo credit: Amedeo Salamoni.*

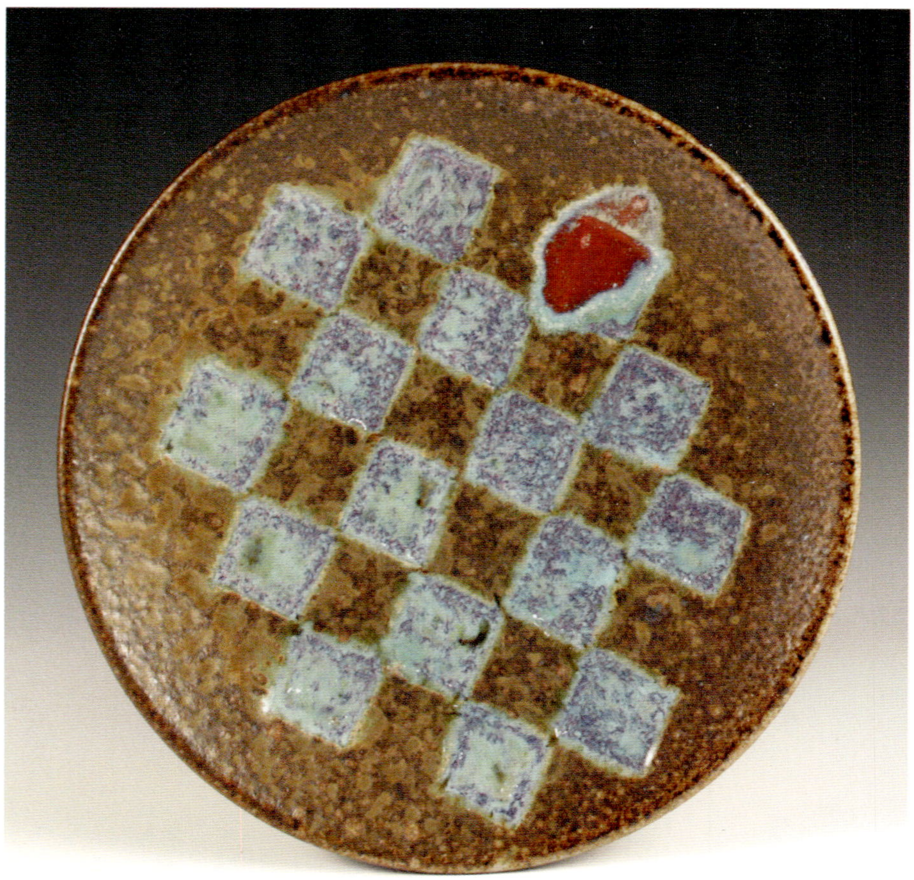

Amedeo Salamoni, *Checkered Plate,* 2012. Wood-fired stoneware, applied and natural ash glaze fired in cross draft type kiln with soda and salt, wheel thrown, from 9" d. *Photo credit: Amedeo Salamoni.*

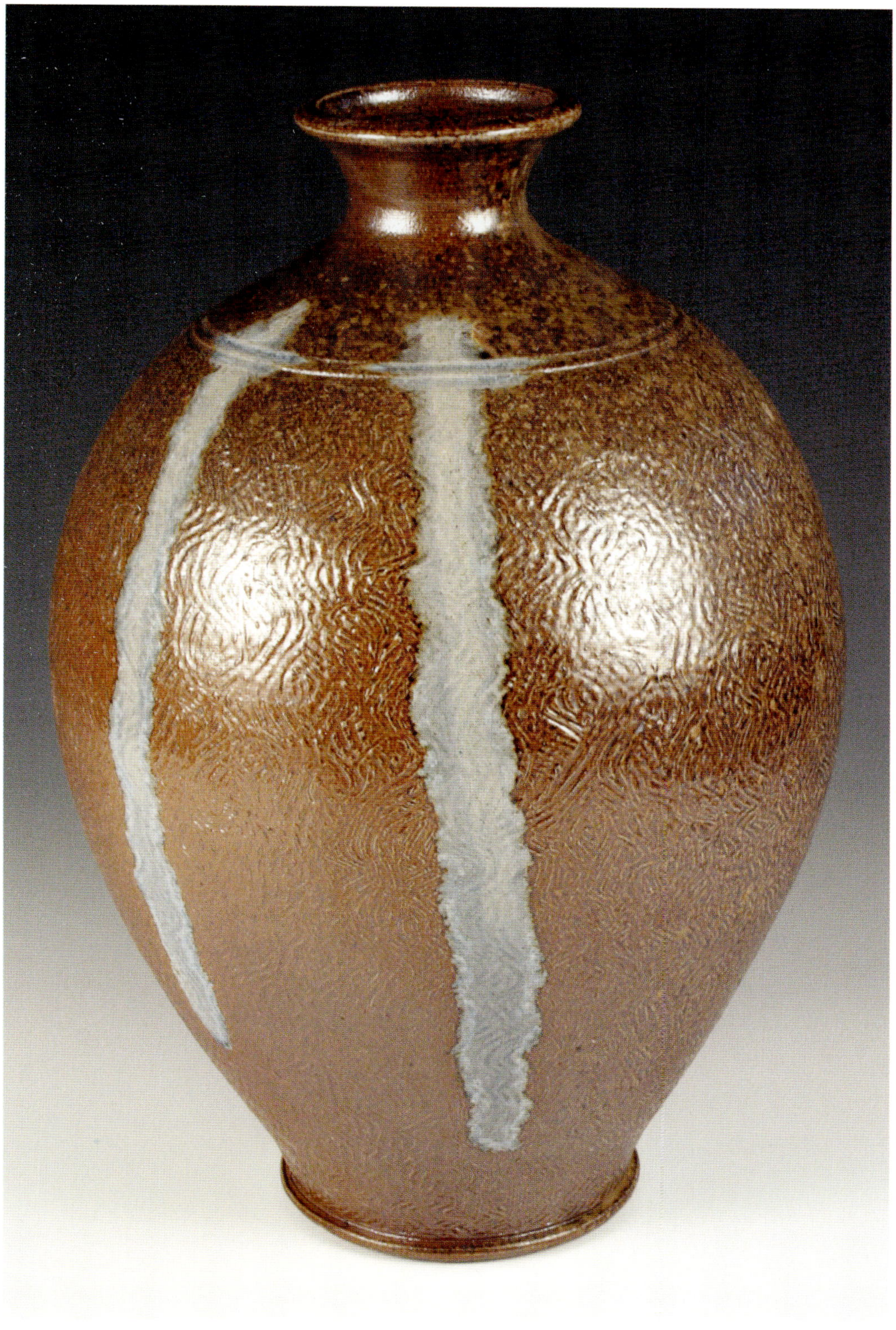

Amedeo Salamoni, *Carved Jar,* 2012. Wood-fired stoneware fired in cross draft type kiln with soda and salt, natural ash deposits, blue glass runs, wheel thrown, from 15" x 10" x 10". *Photo credit: Amedeo Salamoni.*

Glossary

anagama: translates "cellar kiln"—traditional Japanese kiln evolving from the bank-kiln, and featuring a long, swelling inclined tubular ware chamber, with lower extreme serving as firebox. Characteristically produces heavy flame-flashing and residual-ash effects.

alkaline glaze: this glaze was composed of clay and wood ash, as well as one of two additional silica sources: iron slag, or glass.

alumina hydrate: Al(OH)3—alumina source, rarely used in clay bodies or glazes, because all needed alumina comes from clay and feldspar. Used primarily for shelf wash and wadding—better adhesion and suspension than aluminum oxide (Al2O3). Small additions increase viscosity of glaze melt. Should not be used as matting agent in functional glazes—produces immature glaze.

ash-slagging: heavy deposition of fly-ash onto surface of wares, furniture, and kiln interior in a wood kiln.

Avery kaolin: well-known flashing slips for wood-fire, but no longer being mined.

back-pressure: in a fuel kiln, the internal pressure resulting from proper balance of combustion pressure and damper setting, resulting in more economical and efficient heatwork.

bag wall: refractory wall in some fuel-burning kilns to deflect heat and flames from direct contact with the wares.

bank kiln: early form of East Asian kiln, where a tubular kiln chamber was excavated in an earthen bank. Predecessor to anagama and tube kiln.

barrel arch: kiln arch that forms a perfect half circle, with the arch beginning and ending on a horizontal surface, using no skew bricks.

bisque-firing: initial kiln firing in which clay sinters without vitrifying, and though very porous, will no longer soften in water.

Bizen ware: a type of Japanese pottery most identifiable by its iron-like hardness, reddish brown color, absence of glaze, and markings resulting from wood-burning kiln firing.

Bourry box: type of wood-kiln firebox where primary air enters at top of firebox, passes down through wood, and combustion occurs at level of grates or hobs, and is supercharged by the heat of the coal bed. Heat and flames enter the firing chamber below grate-level. In a true Bourry box, wood can be piled above grates and is gravity-fed down to the combustion zone.

carbon-trapping: usually purposeful effect where carbon is trapped within the surface of the glaze, giving smoky shaded areas, especially in shino glazes. Encouraged by slightly early body reduction; can be promoted in high-fire by brushing saturated soda ash solution over glaze.

castable: short for castable refractory—a refractory mix that can be cast into molds to form kiln parts.

catenary arch: a parabolic kiln arch requiring no buttressing or steel frame; laid out by hanging a chain from two points and marking the resulting curve.

ceramic fiber: ceramic insulating material composed of spun kaolin fibers—available in blanket form, braided tape, rigid board, and tubular flue liners. Highest insulating rating of standard refractories, but can release carcinogenic fibers.

climbing kiln: any one of a variety of East Asian kiln designs featuring single or multiple chambers that climb up a slope, creating an adequate draft, often with little or no chimney.

climbing reduction; partial reduction: in a fuel kiln, atmosphere that is slightly reducing, but still allows increase in temperature.

cross-draft: fuel-burning kiln, usually downdraft, where heat enters at the floor level at one side of the ware chamber and exits at the floor level at the opposite side of the chamber.

cullet: crushed window or bottle glass, occasionally used as a glaze material.

damper: in fuel-burning kilns, adjustable refractory plate located in the exhaust flue, allowing control of the back-pressure and secondary air, regulating the kiln atmosphere.

downdraft: kiln where exhaust gases exit through the flue at the floor level. Requires a chimney to develop convection currents needed to draw off exhaust gases. Best system for reduction firing.

draft: the flow of exhaust gases out of a fuel kiln, affecting intake of flames and secondary air.

draw rings; draw trials: in vapor-glaze firing, small rings of clay, formed to stand vertically, which are placed inside a peephole and may be removed with an iron rod during the firing, quenched in water, and examined to determine the degree of glaze deposition.

earthenware clay: natural low-fire secondary clay—fluxed with iron, fires porous. Often called "common" clay, found almost everywhere, matures below 2000°F.

firebox: the part of a fuel-burning kiln where fuel gases combust before contacting wares. Gas kilns need little if any firebox, whereas wood and oil kilns produce long, hot flames that require a large firebox, unless flame-flashing and ash-slagging (with wood) effects are sought.

flashing: color change in fired clay or slip due to direct flame contact and residual ash deposition in wood-firing, or due to variable currents of vapor deposition in salt and soda firing. Flashing can occur on almost any light-colored clay body, but is most dramatic on porcelain bodies and slips.

flashing slip: slip that is painted or dipped onto wares in order to promote flashing effects in the firing.

flue: passages in kiln for flames or exhaust gases.

fly-ash: airborne ash in a wood-kiln.

forced-air: firing system in a fuel-burning kiln that uses power-driven blowers or other pressurized air source to entrain the primary air.

grog: crushed high-fired clay graded in sizes from 15-mesh (very coarse) to 150M (extra fine) added as a source of filler or tempering grit to claybodies to reduce shrinkage and give structure for throwing or handbuilding. Does not shrink in firing, so in medium and coarse grades will show texture through thin to medium glaze.

groundhog kiln: a long, low, horizontal sprung-arch or barrel-arch wood-fired kiln, often banked with earth, traditionally used by potters on the eastern slope of the Appalachian Mountains.

hibuse: hibuse is a term reserved for certain fire-effects that are produced in noborigama kilns. Noborigama kilns can be used to produce glazed pottery. However, because noborigama kilns are fired with wood, there is a risk that ashes falling on the glazed pottery could cause unwanted effects in the glaze. In order to protect the glazed pottery, unglazed pottery is placed at the floor level right next to the "firebox" The unglazed "guard" pots are placed in front of the glazed ware and stand between the burning wood and the glazed pottery, thereby protecting the glazed pieces. The unglazed "guard" pots are greatly affected by heat and ash from the wood burned in the chamber.

handbuilding: forming plastic clay by hand without the wheel, using pinching, coiling, and/or slab construction.

hardbrick: hard, dense firebrick generally used only in high-stress areas of kiln (floor, burner ports, flues, bag wall) and for corrosive firing processes (salt, soda, wood).

high-fire: high-temperature firing range, usually including cone 8 to cone 12, for firing stoneware or porcelain.

hikidashi guro: this method involves removing the pottery from the kiln during firing to make the pottery very black through sudden cooling.

hobs: steps constructed in walls of wood-kiln firebox in place of grates. Wood cut to proper length catches on hobs and remains suspended above the coal bed, insuring quick and complete combustion.

IFB; insulating firebrick; softbrick: porous firebrick with insulating values much higher than hardbrick.

inclusions: any particles or bubbles of material that remain suspended within the glaze-melt, affecting the glaze's appearance.

Jomon ware: the oldest prehistoric earthenwares in Japan received the name *Jomon* or "rope pattern" due to impressed rope and cord designs that ornament their surfaces. These wares were hand-coiled and fired in open kilns at low temperatures.

kaolin; china clay: Al2O3×2SiO2×2H2O—primary clay that fires pure white, very refractory, coarse particle size, low plasticity, high-temperature—major component of porcelain and some whiteware bodies. See EPK, Grolleg, Helmer, Tile-6.

kiln: a specialized oven or furnace used for firing clay for pottery.

kiln furniture: refractory shelves, posts, and stilts used in a kiln to support the wares.

kiln wash: refractory slip coating applied to the top surface of the kiln shelf to protect from glaze runs. For all but salt and wood-firings, 50-50 kaolin and silica. For salt and wood, 60% alumina, 30% kaolin, 10% ball clay.

kurinuki: a Japanese term relating to a particular hand-forming process. This technique employs carving out a shaped solid block of clay to obtain interior space.

low-fire: low-temperature firing range, usually below cone 02 (2048° F), used for most bisque-firing and for glaze-firing terracotta and whiteware.

mid-range: glaze-firing range, usually including cone 4 to cone 7, very popular with electric kilns.

mouse-hole: in some wood kilns, a small air-port and passage that admits air directly beneath the coal bed.

natural draft: firing system in a fuel-burning kiln that uses natural convection currents in the kiln (updraft) or in the chimney (downdraft), without any mechanical blowers, to draw in secondary air, circulate heat and atmosphere through the kiln, and exhaust combustion gases.

neutral atmosphere: in a fuel-burning kiln, atmosphere that is neither oxidizing nor reducing—most efficient atmosphere for efficient climb and even temperature.

noborigama: the Noborigama kiln is an evolution from the single-chambered anagama design as a multi-chamber kiln, usually built on a slope. The word "Noborigama" approximately translates from Japanese to mean "step-climbing kiln." With the initial firebox chamber at the bottom, each succeeding chamber is higher than the one before it. Wood is stoked in the front firebox at first, then through the side-stoking holes of the individual chambers. As the kiln is fired, each chamber preheats the next. The multiple chambers of a noborigama kiln allow the individual chambers to be fired in different ways. *See climbing kiln.*

oxidation firing: any kiln atmosphere with an abundance of oxygen to combust the fuel and oxidize the ceramic materials.

partial reduction; climbing reduction: in a fuel kiln, atmosphere that is slightly reducing, but still allows increase in temperature. *See reduction firing.*

peephole; spyhole: small holes in kiln door or wall, allowing viewing of cones, flames, wares, and/or atmosphere during firing.

porcelain: high-fired vitreous claybody containing kaolin, silica, fluxes, and often ball clay to increase plasticity, with total clay component not more than 50%. Usually pure white or "eggshell" in color; some porcelains may fire translucent where thin.

porcelaneous: white-firing stoneware claybodies closely related to porcelain.

primary air: in a fuel-burning kiln, air that feeds initial combustion; in a wood-burning kiln, air that feeds the base of the flames; in a gas kiln air that enters the back end of the burner tube in an atmospheric burner, or via a blower in a power burner. *See secondary air.*

primary clays; residual clays: clays that remain at the site of the parent rock, without being transported by wind or water, tend to be free of contaminants but have coarse particle size and therefore low plasticity. Includes the purest kaolins.

pyrometer: temperature gauge connected to a thermocouple, which indicates temperature within the kiln. Pyrometers provide good general reference, but only respond to temperature, whereas clay and glazes (and cones) are affected by temperature, duration, and atmosphere of firing. *See pyrometric cones.*

pyrometric cones: small slender pyramidal-shaped indicators made of ceramic material formulated to bend at a specific temperature—standard method for determining maturing temperature of firing. Like clay and glazes, cones respond to temperature, duration, and atmosphere of firing far more accurately than mechanical measurement.

pyroplastic: flexibility of clay or glaze when heated sufficiently, as a result of developing glassy-phase.

reduction cooling: in a fuel kiln, maintaining slight reduction atmosphere during cooling cycle, from maturing temperature down to approximately 1400° F, in order to minimize reoxidation of clay and/or glazes.

reduction firing: in fuel-burning kilns, firing atmosphere with insufficient oxygen to completely combust fuel, introducing abundance of unoxidized carbon and hydrogen, which extract oxygen molecules from the surface of wares, altering appearance of clay and glaze.

refractories: any materials highly resistant to the effects of heat. In kiln construction, high-temperature insulating materials, such as firebrick, ceramic fiber, etc. In clay and glaze chemistry, the neutrals or stabilizers that are resistant to melting and that combine with the fluxes (bases) and glass-formers (acids). Primary refractory material in ceramics is alumina, most often introduced in the form of clay (aluminum silicate).

residual ash: in a wood kiln, the surface effects created by fly-ash settling upon the wares. Effect is maximized by turbulence in firebox, considerable draft through kiln, and/or active stirring of the coal bed. In short, wood-firing's ash doesn't have time to melt, and leaves a rough, crusty surface.

residual salt/soda: in a well-used salt/soda kiln, the light salt/soda effect that may be obtained by firing wares in the kiln without introducing any additional salt or soda, due to the revolatization of salt/soda from the walls of the kiln.

saggar: a refractory clay enclosure in which wares are fired. Originally developed to protect wares from ash-slagging and flame-flashing in wood-firings. In contemporary use, with clean-burning gas firings, process used in exactly the opposite way.

saggar firing: contemporary firing process in which wares are placed within refractory saggars, along with chemicals and combustibles, in order to achieve certain surface effects.

salt firing; salt glazing: vapor-glazing process where salt (sodium chloride) is introduced into the kiln firebox at high temperature. Salt vaporizes, and sodium vapor combines with silica in clay surface, forming extremely hard sodium-silicate glaze. *See soda firing.*

secondary air: in fuel-burning kilns, air that enters after initial combustion and feeds flames as they leave the primary combustion source. In gas kiln, secondary air enters the burner port around the burner tip. In a wood kiln, secondary air enters above grates in conventional firebox, and below grates or hobs in Bourry-box. *See primary air.*

secondary clays; deposited clays; sedimentary clays: clays that have been transported away from their point of geologic origins by wind or water. Finer particle-size gives greater plasticity—ball clays, stoneware clays, fireclays, etc.

sequential firing: firing in a sloped tube kiln or noborigama, where as each chamber or zone reaches maturity, the fire is moved up to the next, and each one preheats subsequent chambers or zones.

Shigaraki ware: pottery and stoneware made in Shigaraki area, Japan. The wares for which the town is known were first produced in 1278; they have a crude shape and an oatmeal-like surface covered with various glazes.

silicon carbide: SiC—extremely refractory material used to form kiln shelves. Highly resistant to corrosive atmospheres, and therefore suitable for salt, soda, and wood-firing. Silicon carbide kiln shelves conduct electricity and should never be used in electric kilns

single-firing; green firing: process of firing glazed greenware without a bisque firing.

slip: clay suspended in water, usually the consistency of thick cream. May be colored and used to decorate surfaces, or may be cast into plaster molds to create ceramic forms.

soaking: during firing or cooling ramp, the act of holding the kiln at a steady temperature for a period of time to allow proper formation or maturation of certain clay and glaze effects.

soda firing; soda glaze: vapor-glazing process touted as modern-day nontoxic replacement for salt firing. Gives slightly less gloss and orange-peel. Instead of salt, soda ash (sodium carbonate) in water solution is sprayed into the kiln at maturing temperature, and sodium vapor combines with silica in clay to form sodium-silicate glaze.

sprung arch: kiln arch representing less than 180 degrees of curvature, and requiring buttressing to support the outward thrust of the arch.

stoneware: high-fired vitreous ware, literally as hard and durable as stone. Matures from 2200–2400° F. (cone 5–11). Also naturally occurring refractory clays with adequate fluxes to fire in stoneware temperature range.

thermocouple: temperature probe that produces minute variable electrical current dependent on degree of heat exposure—used in pyrometers.

tobacco-spit glaze: tobacco-spit glaze originated long ago among American folk potters. When the glaze is fired, it runs down the sides of the pot in golden brown rivulets that earn the glaze its name.

tube kiln: sequential-firing Korean kiln design evolving from Chinese bank kiln, with long sloping tubular firing chamber. *See anagama.*

underglaze decoration: process of applying any decoration to the bare (usually bisque-fired) clay surface directly before glazing.

vermiculite: a porous expanded mica product used as filler in clay bodies and in insulating refractory layer on kiln exterior.

vitreous; vitrified; vitrification: fired clay that has fused together completely, so that the pores between refractory particles are filled with glass and the body is impervious to water. Vitrification is sintering in the presence of a fully developed glassy-phase.

wabi: (Japanese) an aesthetic term, meaning quiet, solitary, unassuming.

wadding; wads: small balls or rolls of refractory clay mixture (40 alumina, 10 ball clay, 50 kaolin) placed under wares and posts, and between pots, lids, etc., in vapor-glazing and wood-firing processes, to keep surfaces from sticking together.

warping: distortion of clay forms caused by uneven stresses within clay due to forming method, uneven drying, uneven support in firing, or uneven or excessive heat in firing.

wood-firing: a firing technique used by necessity in many countries and by choice in others. In a properly designed kiln, wood is capable of delivering high temperatures so it is possible to make stoneware and porcelain. The kiln chamber in a wood kiln subjects the ware to a lot of ash and smoke and this profoundly affects its appearance. It is possible to fire pieces without glaze and the products of combustion of the wood will deliver enough fluxes to fuse the surface of the clay in a glaze-like manner.

yakishime yohen: high fired, unglazed, solely relied on the special effects coming from pure interaction of clay, flame, and wood ash.

yohen or youhen: literally "changed by the fire/flame," refers to changes in the kiln that cause the glaze to run during firing. Yohen also refers to the build-up of ash on the kiln floor and the natural glazing brought about by this ash, resulting in deep blues, browns, and reds—often seen on yakishime ware, like Bizen, and on Temmoku.

yunomi: (Japanese) is a form of teacup, typically made from a ceramic material, being taller than wide, with a trimmed or turned foot. Unlike the more formal chawan tea bowl that is used during the Japanese tea ceremony, the yunomi is made for daily (or informal) tea drinking.

Appendix 1

Kiln Drawings

The following illustrations are selected examples of a variety of kiln designs supplied by the contributing artists. Many of these designs are derived from kiln designs throughout the centuries. Each artist has modified the kiln designs to suite their work styles, and to get the desired results that they are after.

Hadi Abbas, Orlando, Florida

Anagama Kiln

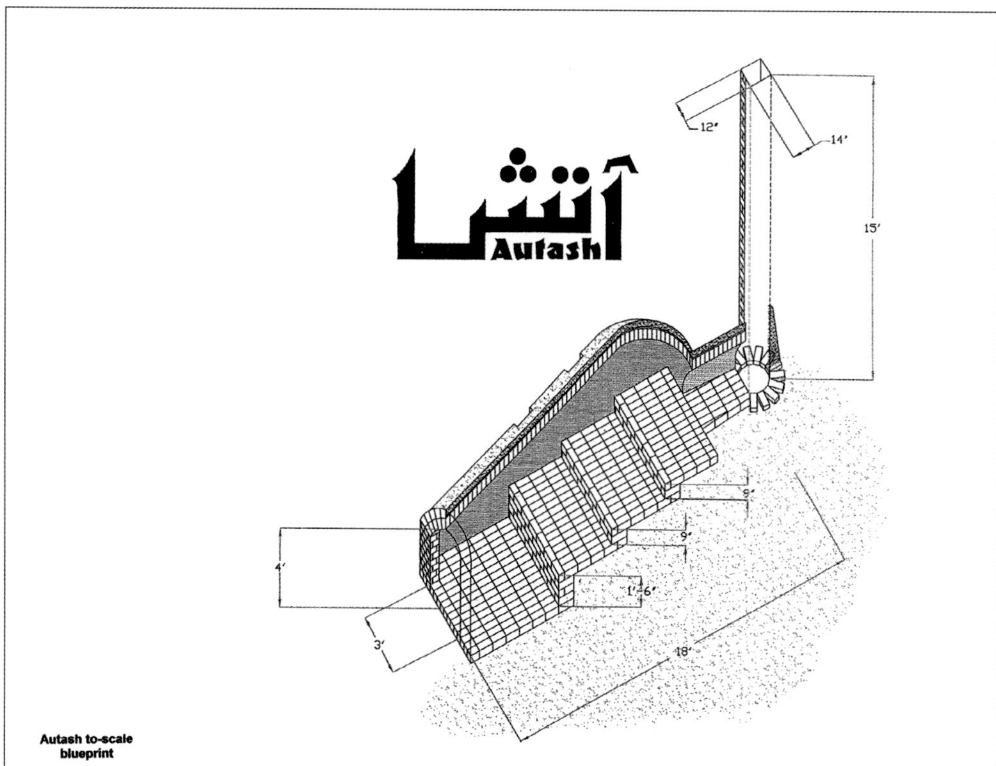

Hadi Abbas (USA). *Autash Kiln,* anagama kiln.

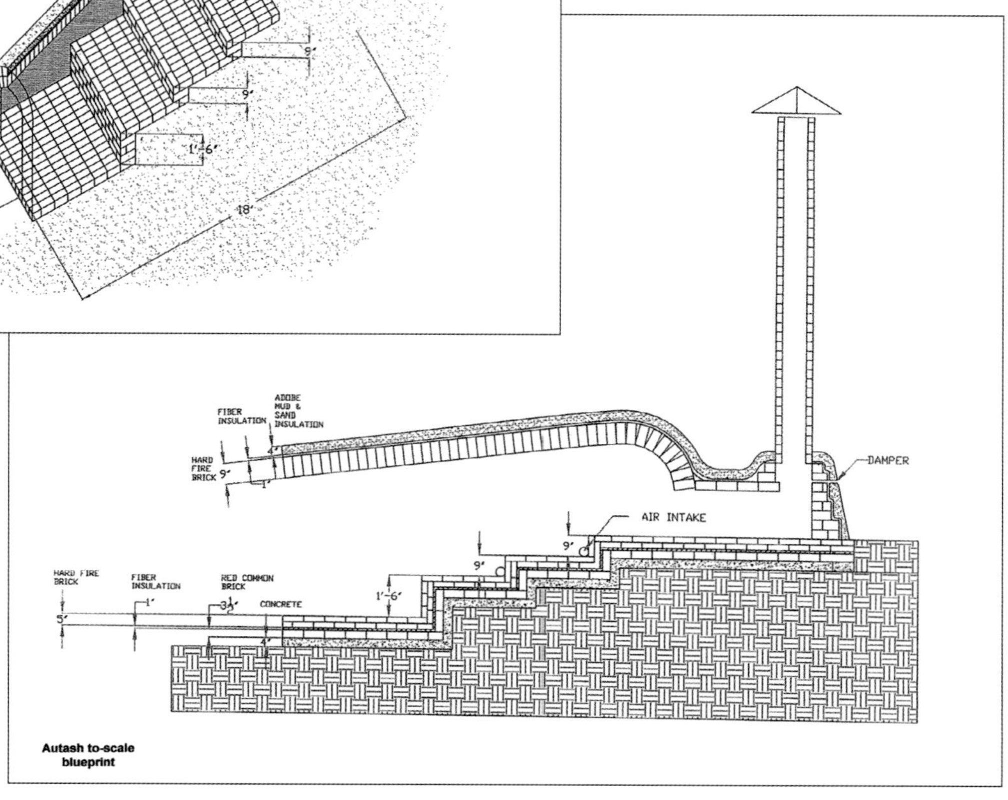

Hadi Abbas (USA). *Autash Kiln,* anagama kiln, side view.

Stefan Andersson, Tibro, Sweden

Modified Fast Fire Kiln

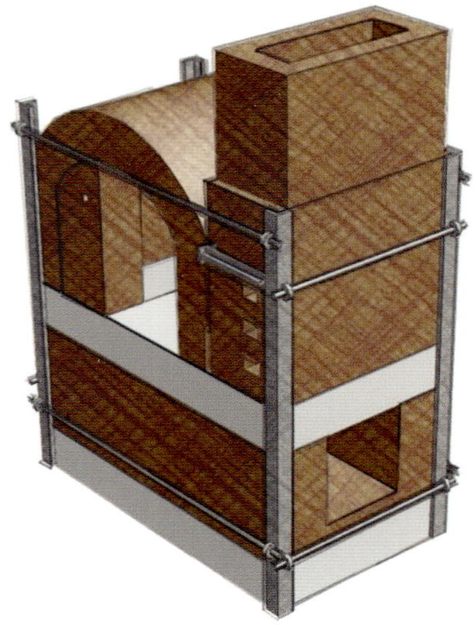

Stefan Andersson (Sweden). Computer rendering, Fast Fire kiln. The design of this kiln has been in use for a long time as a test kiln. I have worked on making it into a production kiln for small series and fast turn around speeds.

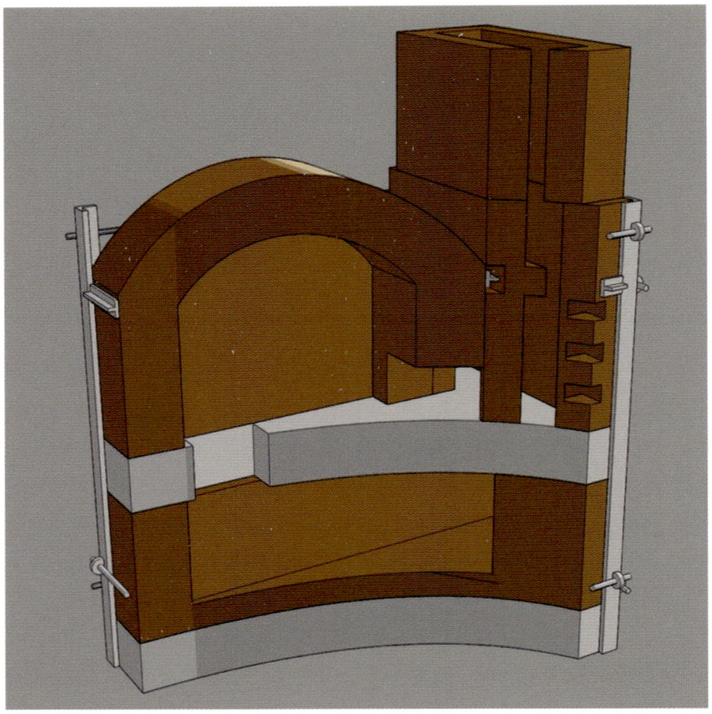

Stefan Andersson (Sweden). Computer rendering, Fast Fire kiln. The design can be described as a double cross draft with an oversized firebox. The firebox of this kiln is the key in making it very versatile with the power to fire in many different ways, with results you normally only get in primitive wood-kilns (like anagamas). The size makes it a good choice for use in schools.

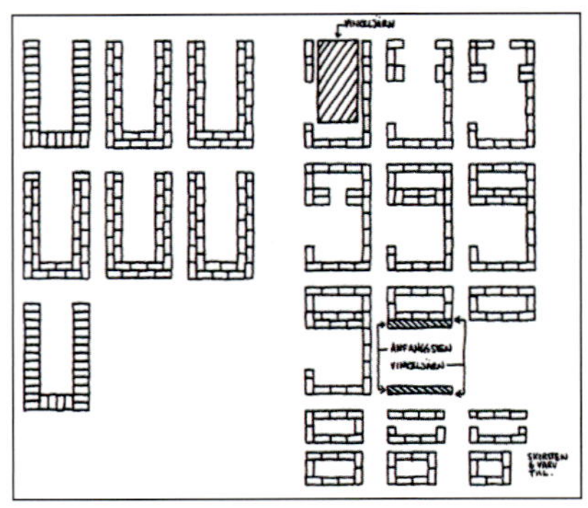

Stefan Andersson (Sweden). Building plan, Fast Fire kiln. The entire kiln can be built with hard firebrick making it possible to use salt or soda, for a faster and more efficient kiln the top half and outer wall of the firebox could be changed to IFB.

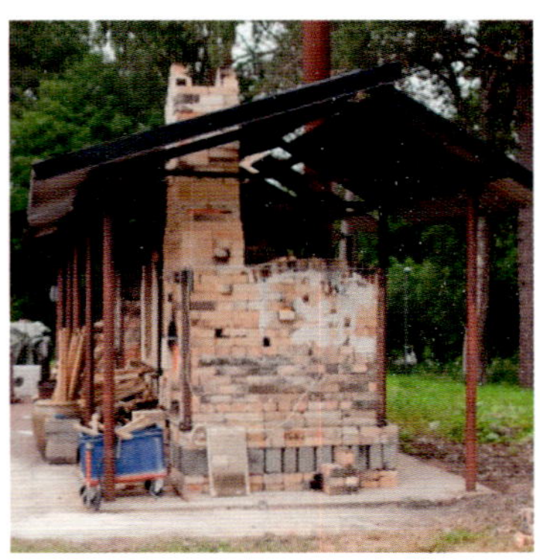

Stefan Andersson (Sweden). Fast Fire kiln.

John Benn and Colleen Gallagher, Shelton, Washington

Train Kiln, Modified Anagama Kiln

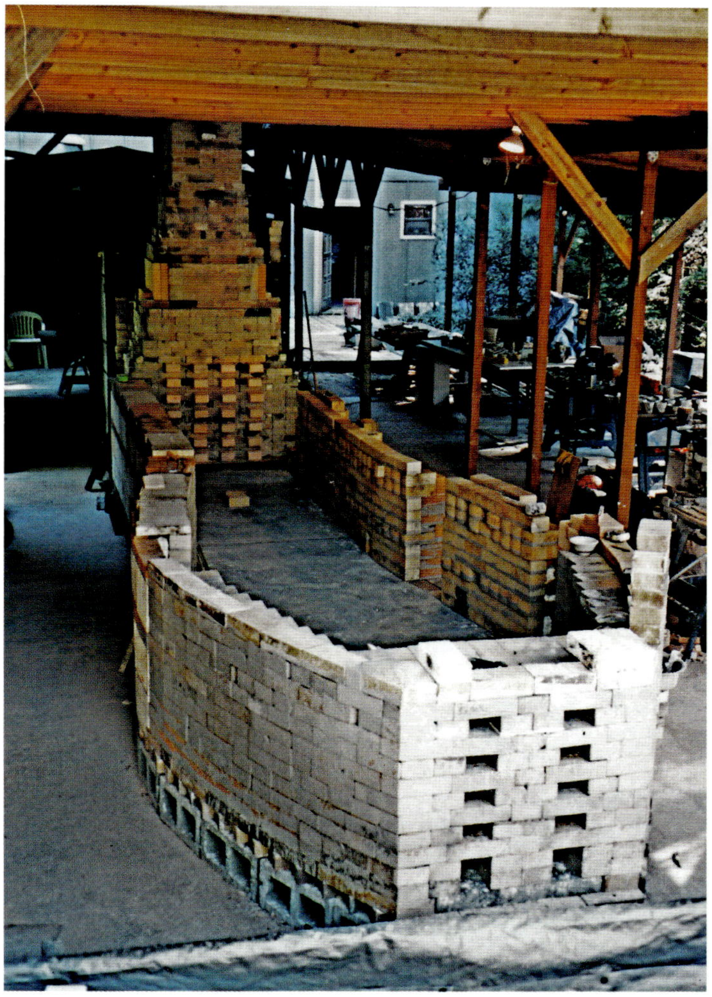

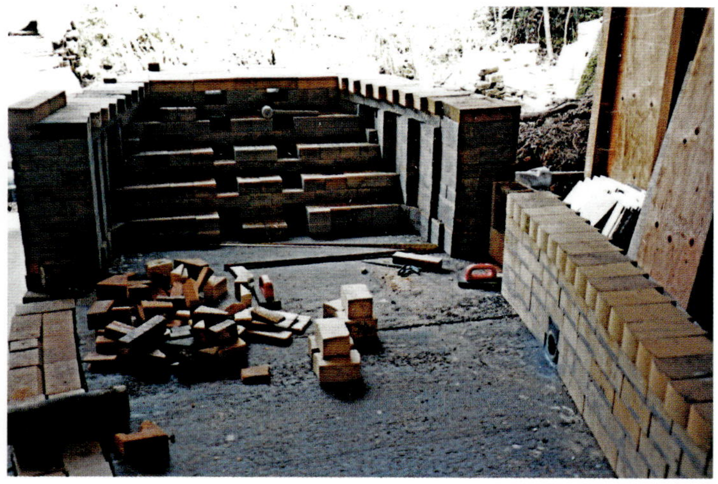

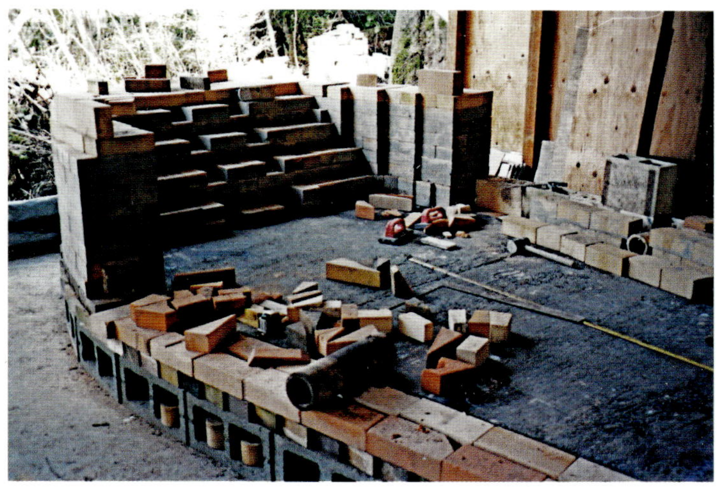

John Benn and Colleen Gallagher (USA). *Mongrelgama Kiln,* Train Kiln. These two photos show the step-down grate in the firebox of the *Mongrelgama.*

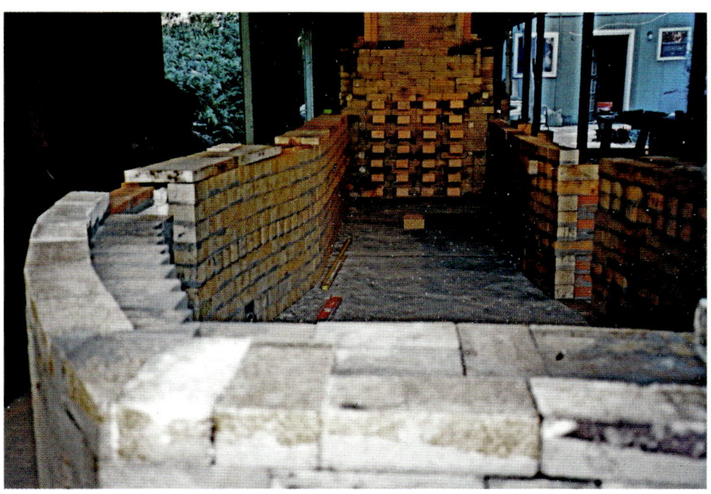

John Benn and Colleen Gallagher (USA). *Mongrelgama Kiln,* Train Kiln. The first two images show the *Mongrelgama* kiln's footprint. The chimney is shared with our salt kiln (on the far side) and the protruding bricks are a checkerboard flue. The floor is castable refractory.

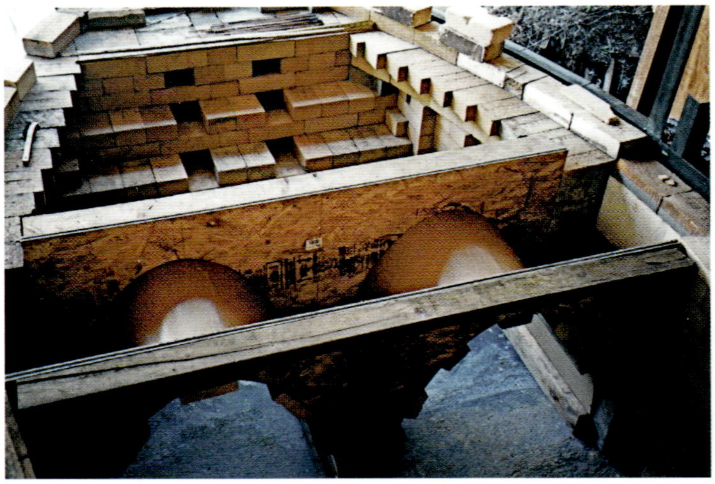

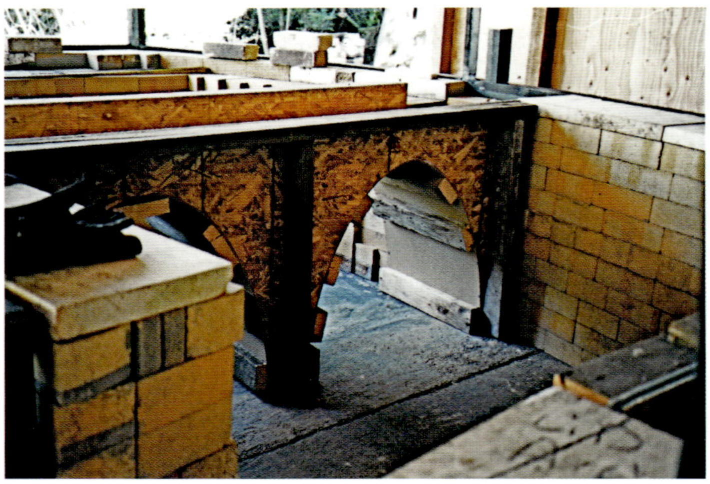

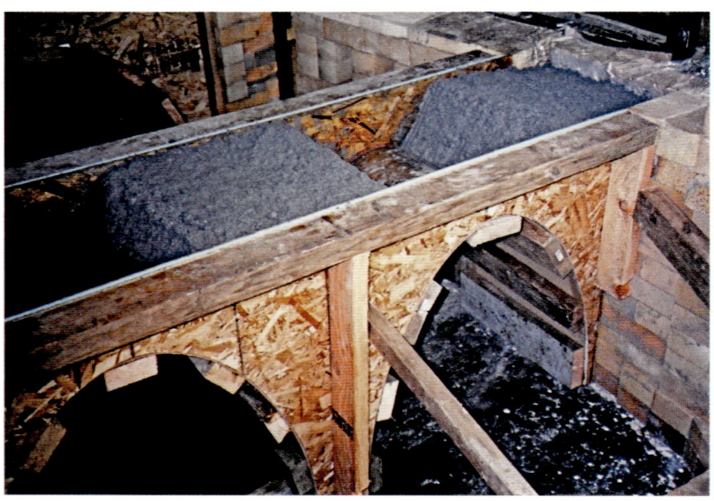

John Benn and Colleen Gallagher (USA). *Mongrelgama Kiln*, Train Kiln. These images show the plywood and hardboard forms for the throat arches between the step-down grate and the first chamber of the *Mongelgama*. The castable refractory is packed into the forms as we were building a brick arch.

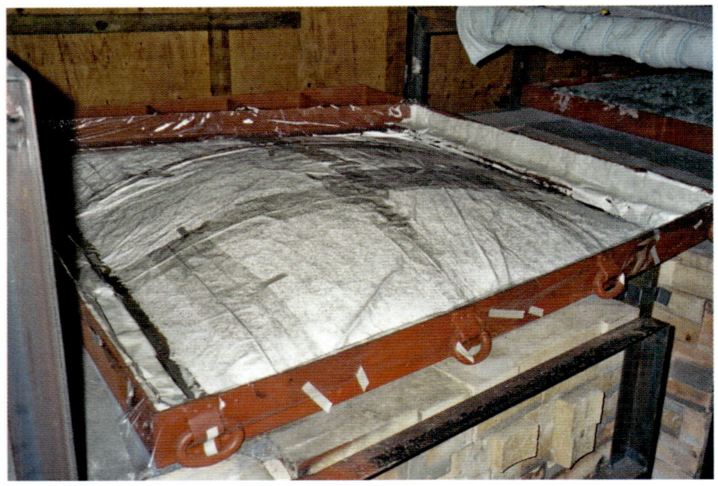

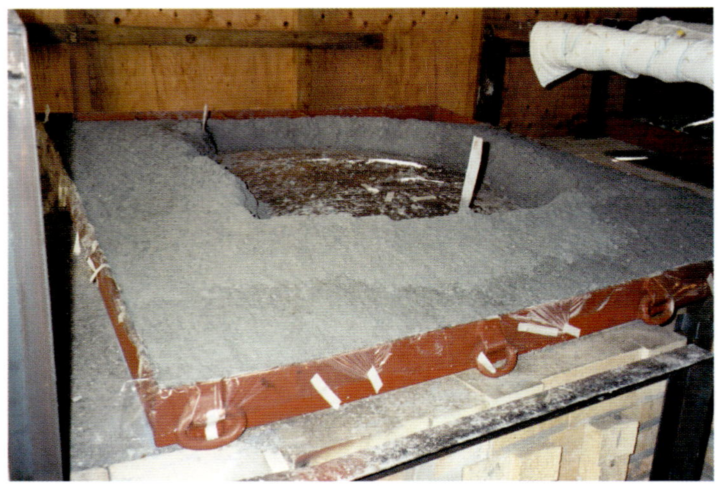

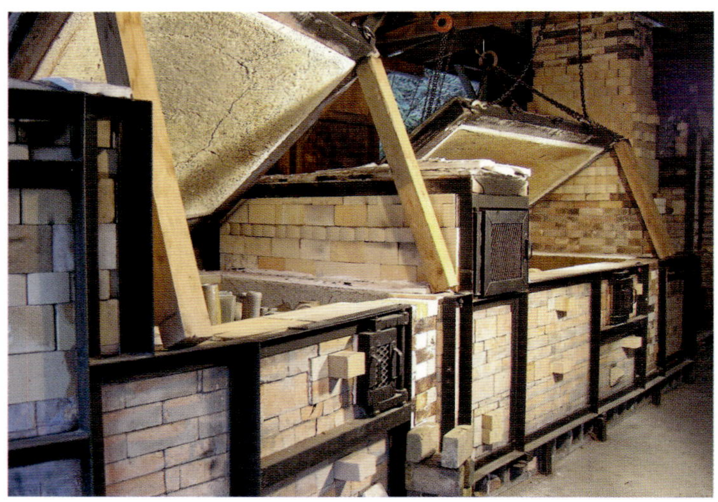

John Benn and Colleen Gallagher (USA). *Mongrelgama Kiln*, Train Kiln. The lids of the *Mongrelgama* being cast. A heap of sand is formed inside the welded lid frame, and covered with plastic sheet. Then the castable is packed over that as if we were building a dome of bricks.

Jim Busby, Iowa City, Iowa

Train Kiln, Modified Anagama Kiln

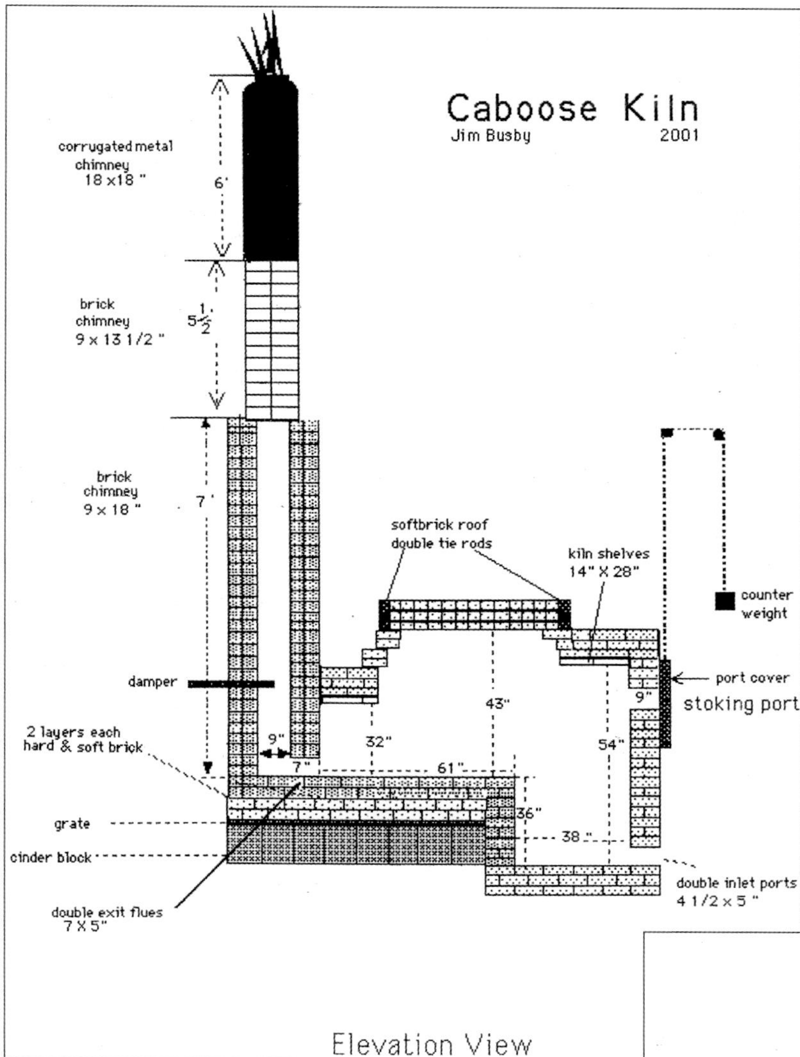

Jim Busby (USA) *Caboose Kiln,* modified anagama type kiln. The roof of the chamber is a "Flat Top" design, made of K-26 soft brick. The roof is coated in ITC 100 HT and can be raised or lowered with minimal effort.

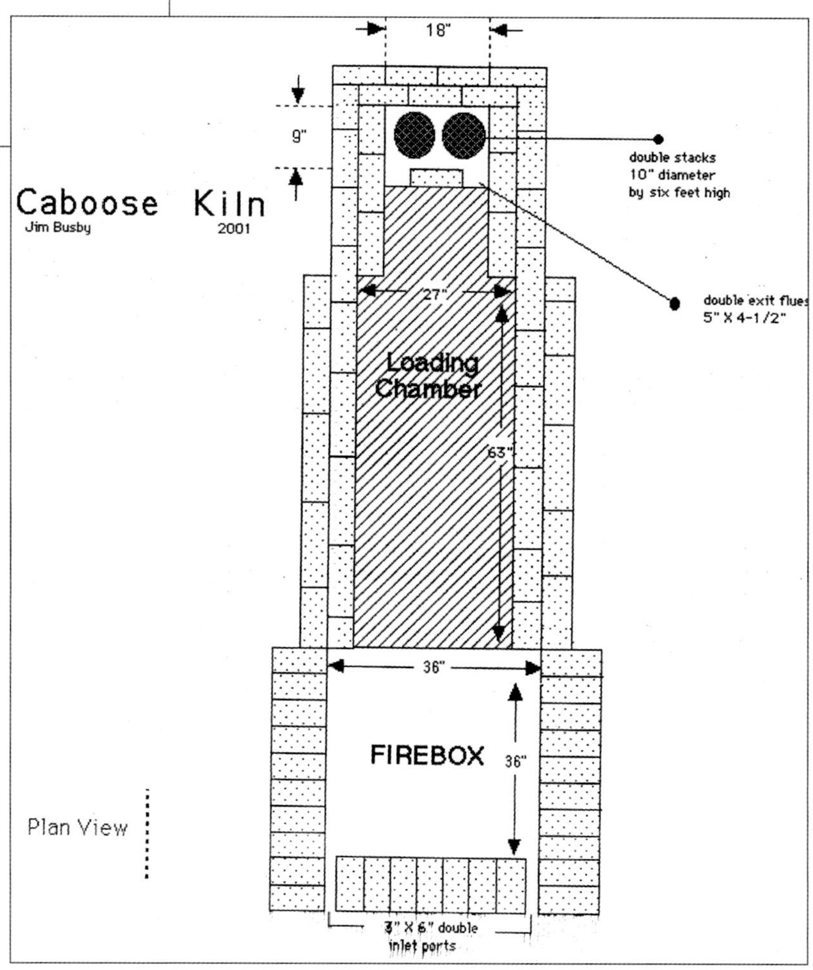

Jim Busby (USA). *Caboose Kiln,* modified anagama type kiln. Exit flues are partially below the floor level of the chamber. The "Hole in the ground" firebox: provides cool, oxygen-rich air to the coals, opens up pre-chamber affecting flow and efficiency of flame-stream, allows ideal placement of two "extra" platters (fired "on edge") down in the firebox.

Anagama Kiln

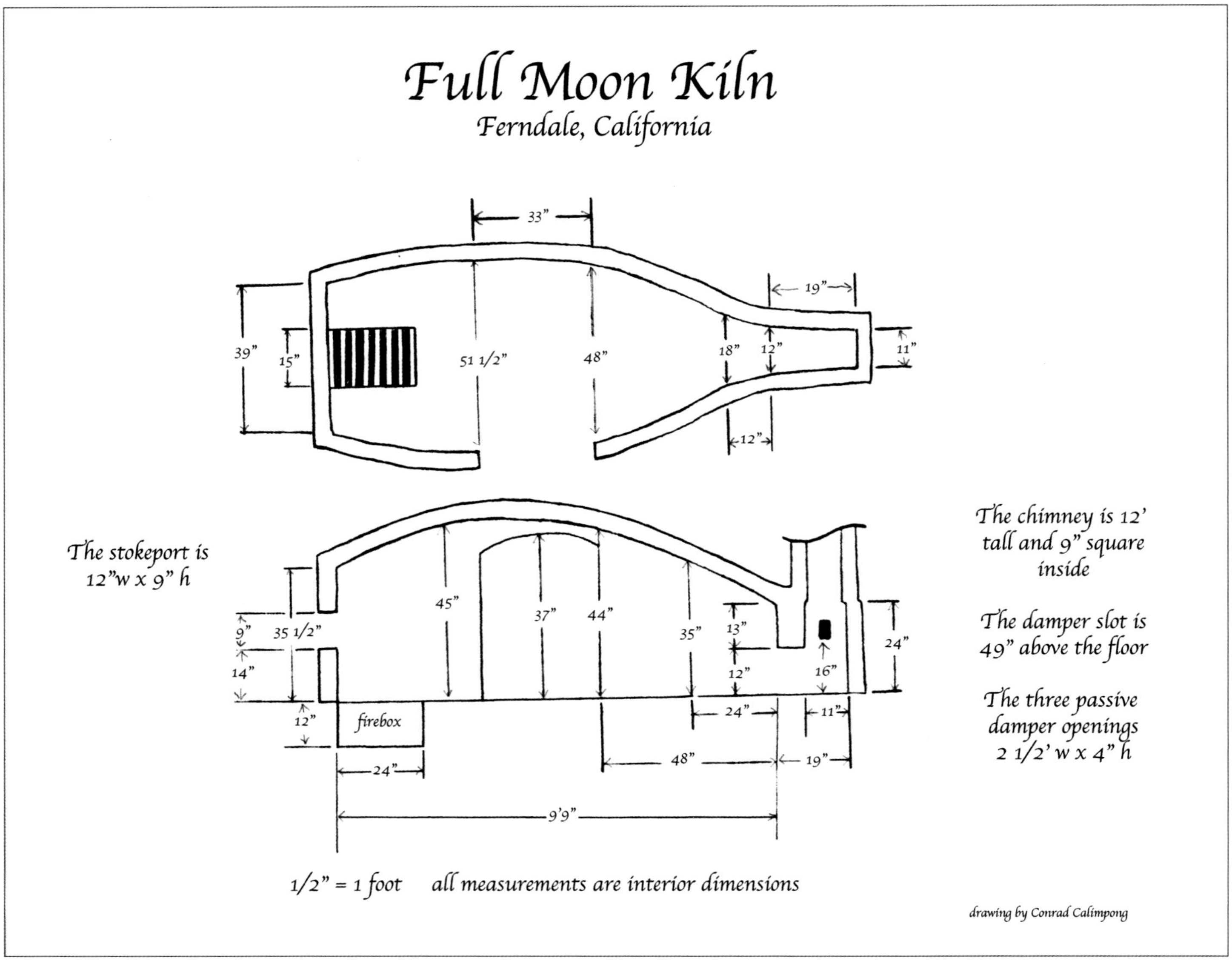

Conrad Campilong (USA). *Full Moon Kiln*, anagama kiln.

Peter Callas, Belvidere, New Jersey

Anagama Kiln

My kiln is an anagama. It sits on an eighteen degree slope with three side stoke holes and an upright chimney. We are able to reach cone 12 in the back of the kiln consistently over the eight day firings.

In 1976, I built an anagama in Piermont, New York. I believe it to be the first Japanese style natural ash kiln in the US. It was successful enough over ten years that it served as the prototype for the Belvidere, New Jersey, kiln, built in 1987. Together with my wife Chrissy and a friend, we completed the masonry in thirty working days. We then applied a 1400 lb. castable, including a chimney coat with about 4" to 6" of homemade mix in one day.

I designed a kiln floor with a mix of brick and sand. The brick frames out the rising steps and provides structural points for posting up; the sand areas are used for firing work without the need for wading material. Another improvement was elevating the bottom stoke hole about 2' above grade, so there is minimal bending when stoking it.

The big difference in the Belvidere kiln was the structural integrity, using large block salvaged from a smelter plant in Sayreville, New Jersey. It made a stronger kiln that has withstood 100 firings, and the arch shows no wear. The workflow around the kiln makes more sense now too. It sits on a level concrete pad with enough shed space to store wood, clay, and equipment too.

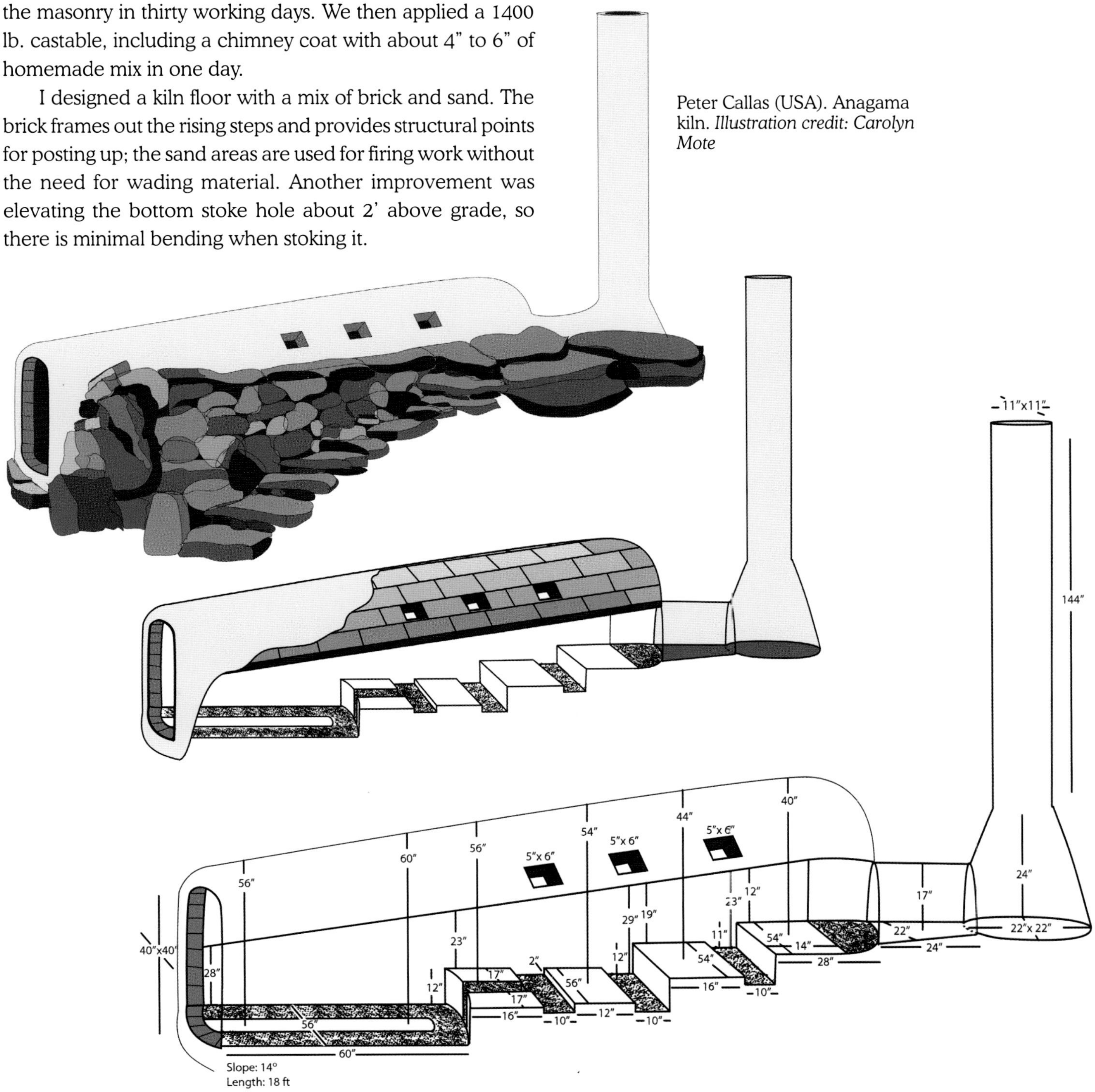

Peter Callas (USA). Anagama kiln. *Illustration credit: Carolyn Mote*

Multi-Chambered Kiln

Linda Christianson (USA). Multi-Chambered Climbing kiln

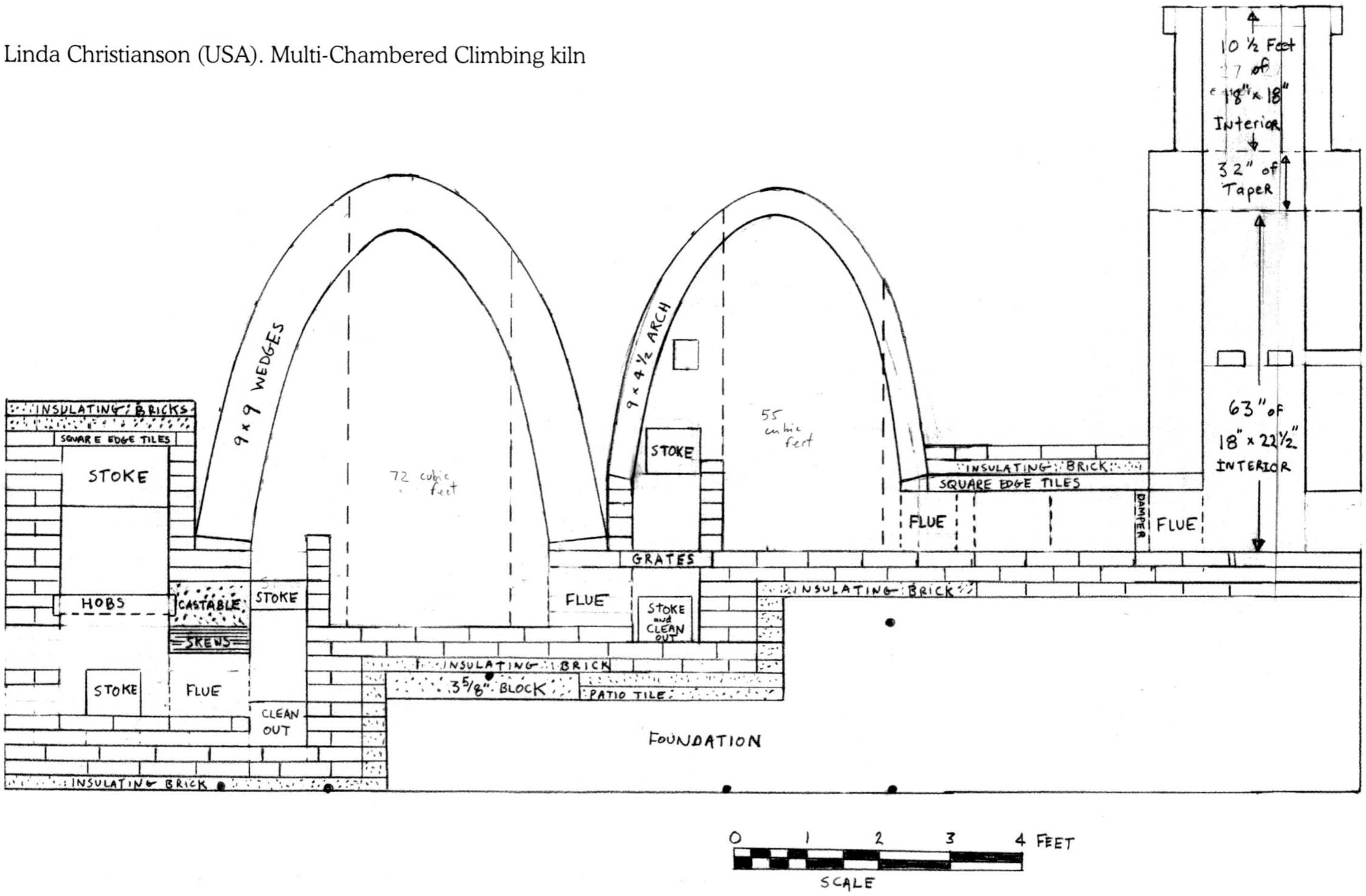

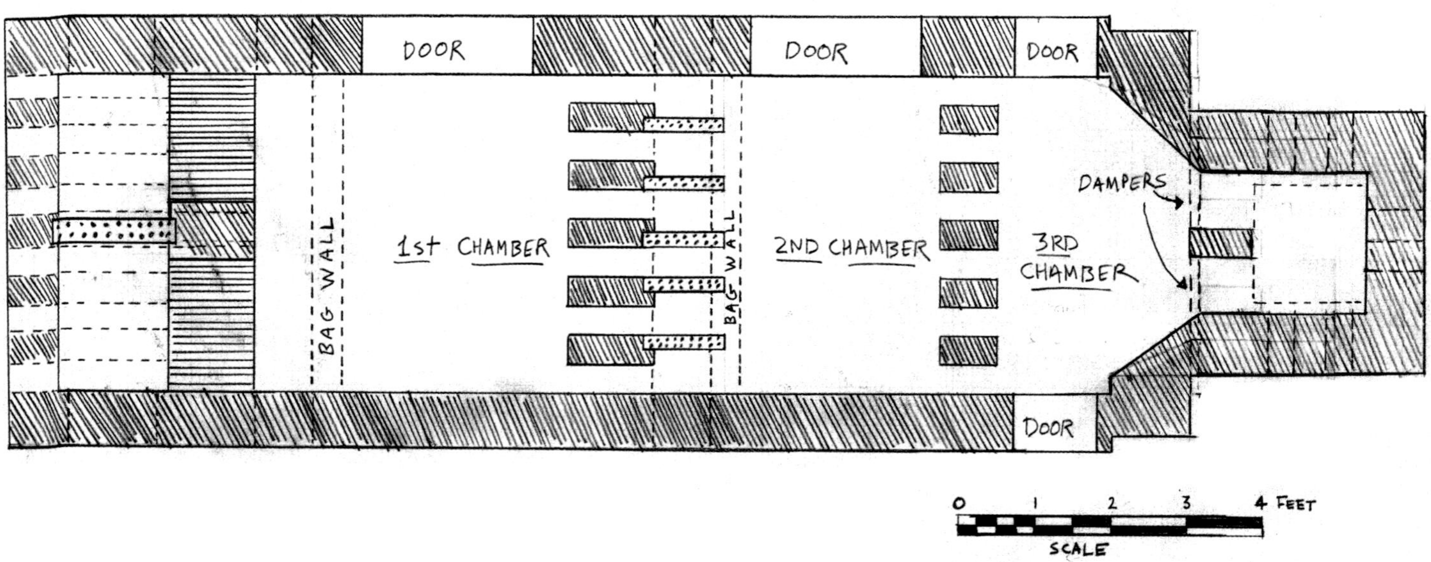

Makoto Hatori, Ibaraki-ken, Japan

Multi-Chambered Climbing kiln

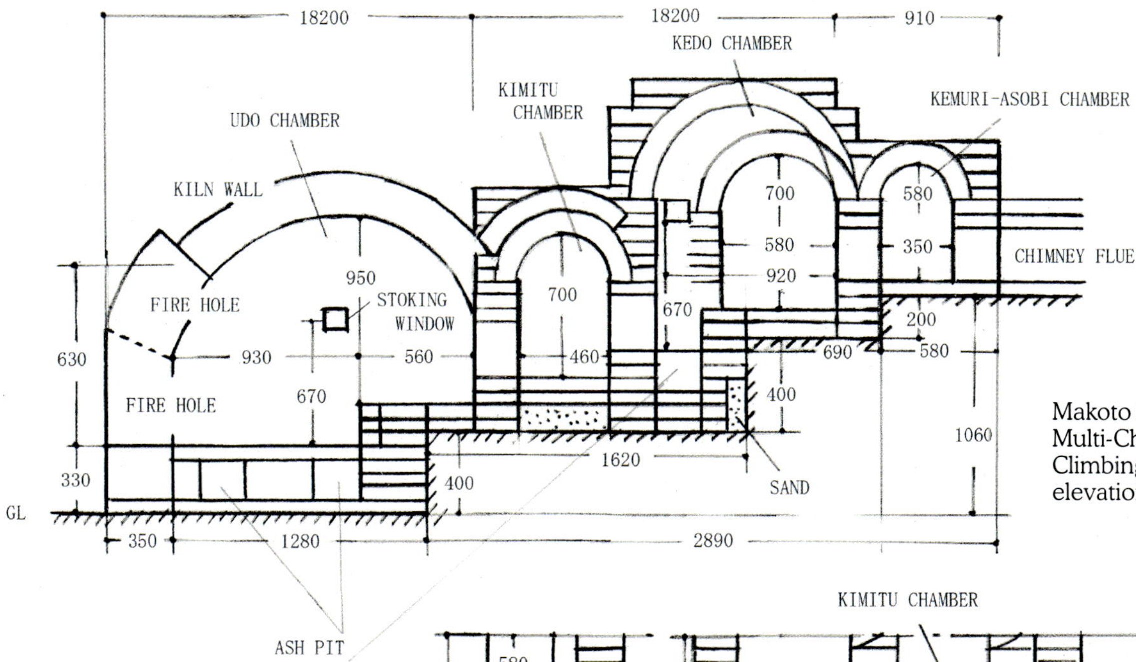

Makoto Hatori (Japan). Multi-Chambered Climbing kiln, side elevation

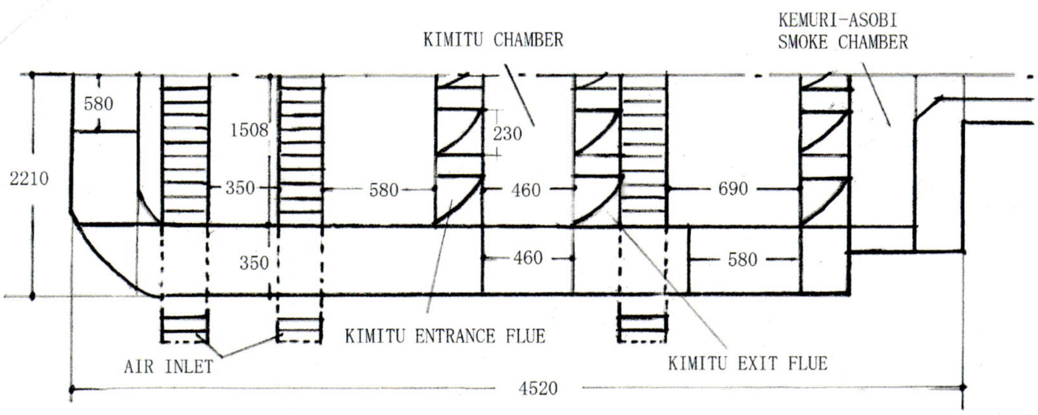

Makoto Hatori (Japan). Multi-Chambered Climbing kiln, floor plan

Dale Haufmann, Pittsburgh, Pennsylvania

Modified anagama kiln

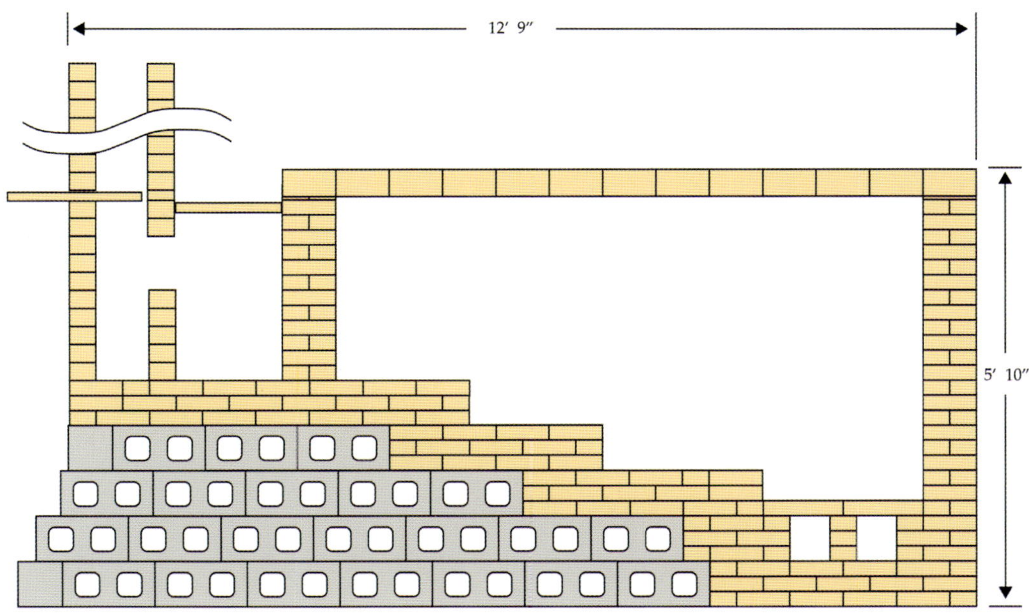

Dale Haufmann (USA). Computer rendering, modified anagama kiln, side elevation

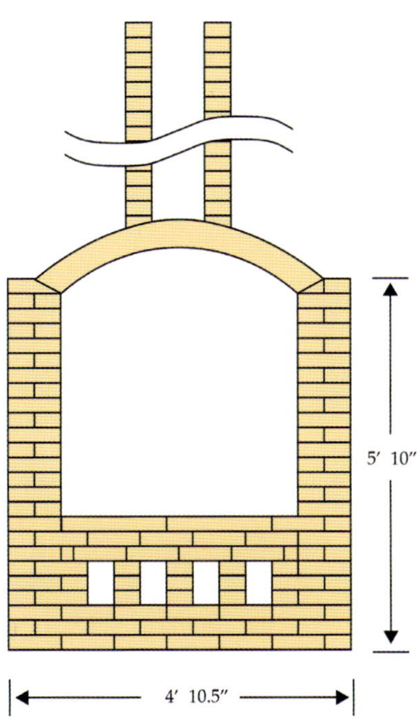

Dale Haufmann (USA). Computer rendering, modified anagama kiln, exterior measurements: width, 58.5 inches, length, 153 inches, height, 72 inches, Interior volume is approximately 80 cubic feet.

Jody Johnstone, Swanville, Maine

Anagama Kiln

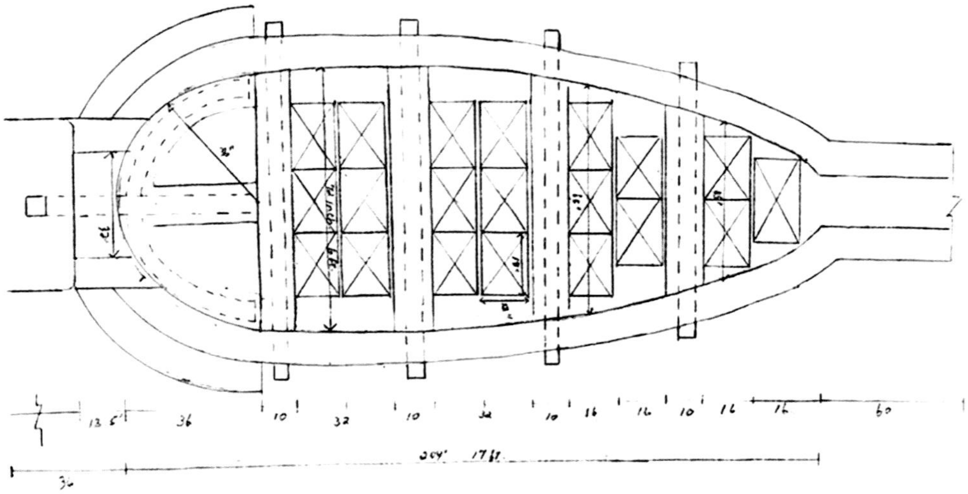

Jody Johnstone (USA). Anagama kiln

Kevin Lehman and Amedeo Salamoni, Lancaster, Pennsylvania

Single Chamber Cross-draft Kiln

Our kiln is a modified catenary arch style kiln. The peak of the kiln has a barrel arch instead of following the typical curve of the catenary style and we also placed the floor and bag wall closer to the firebox, both of which increased the stacking space. The kiln is about 54 cubic feet and holds about 150 pieces of various sizes.

Our firings vary depending on what look we are trying to achieve. However, our general firing method consists of firing to cone 10 flat in about 14 hours, holding the kiln at cone 10 for about 1 hour. We begin to salt around cone 8 and use about 6–9 pounds of salt and/or soda. When salting, we pour about 2 pounds of salt onto a flat board and place that in the firebox, letting it fume for one or two more stokes before we add the next batch of salt.

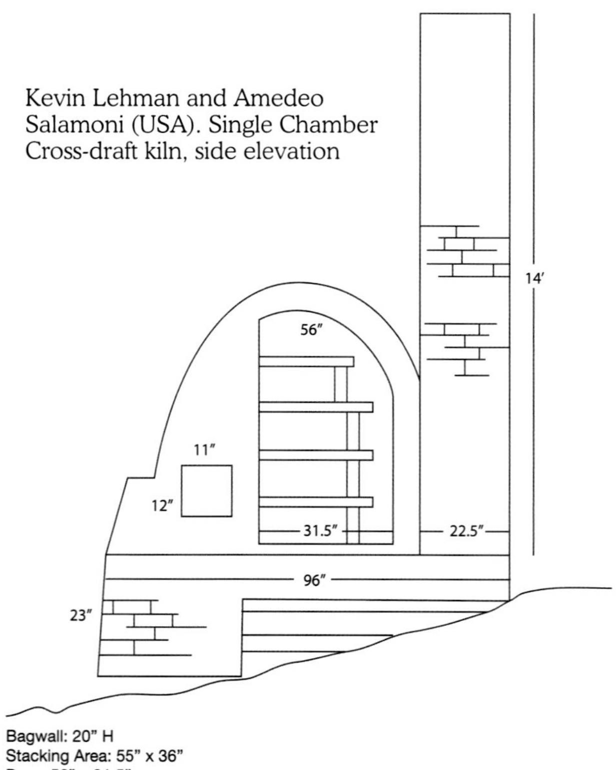

Kevin Lehman and Amedeo Salamoni (USA). Single Chamber Cross-draft kiln, side elevation

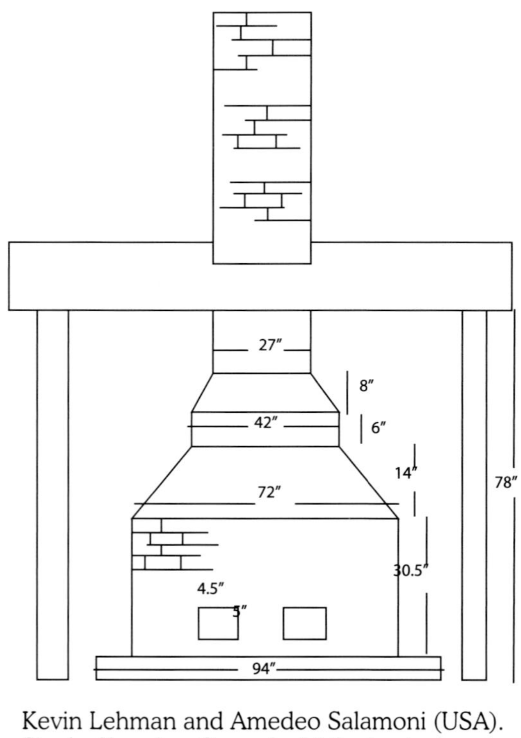

Kevin Lehman and Amedeo Salamoni (USA). Single Chamber Cross-draft kiln, back elevation

Simon Levin, Gresham, Wisconsin

Anagama Kiln

In August of 2009 I was commissioned to build an anagama for Tainan National University of Arts in Tainan province of Taiwan.

Ching-Yuan Chang the senior ceramics professor has been building a world-class department. He regularly brings his students to NCECA and arranges for them to study abroad for part of their degree program.

Ching-Yuan also has created a fabulous visiting artist program, allowing for a constant flow of new ideas and interpretations to enhancing the graduate studio.

I designed the kiln on Google Sketch up with a long climbing flue to allow for a lot of passive control of draft.

I also designed the pad for the kiln pad, which preceded my arrival in Taiwan.

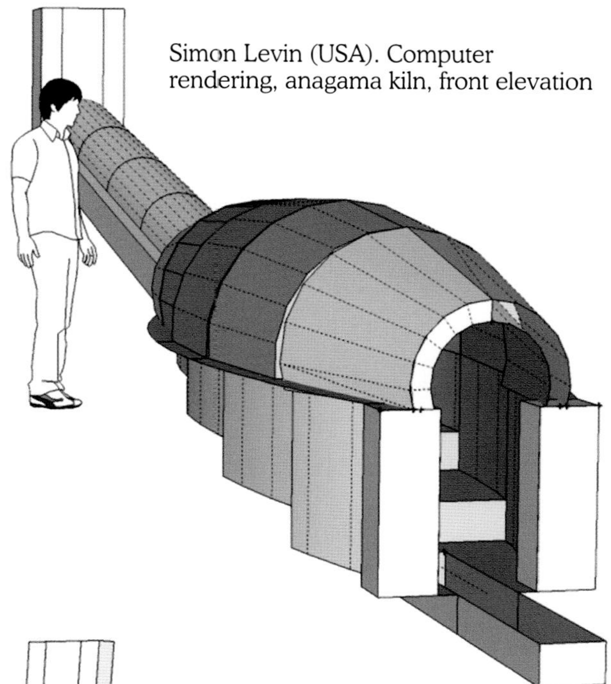

Simon Levin (USA). Computer rendering, anagama kiln, front elevation

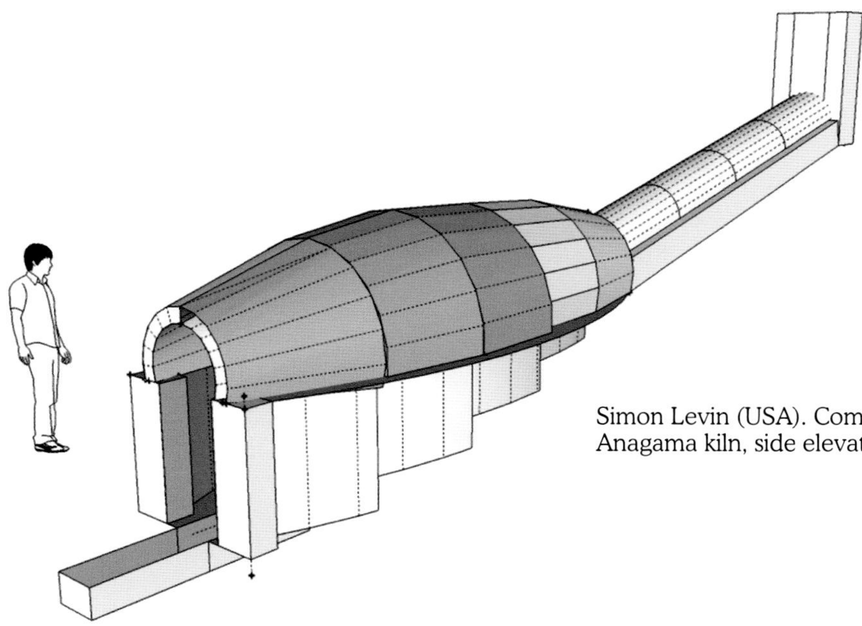

Simon Levin (USA). Computer rendering, Anagama kiln, side elevation

Simon Levin (USA). Computer rendering, anagama kiln, kiln pad

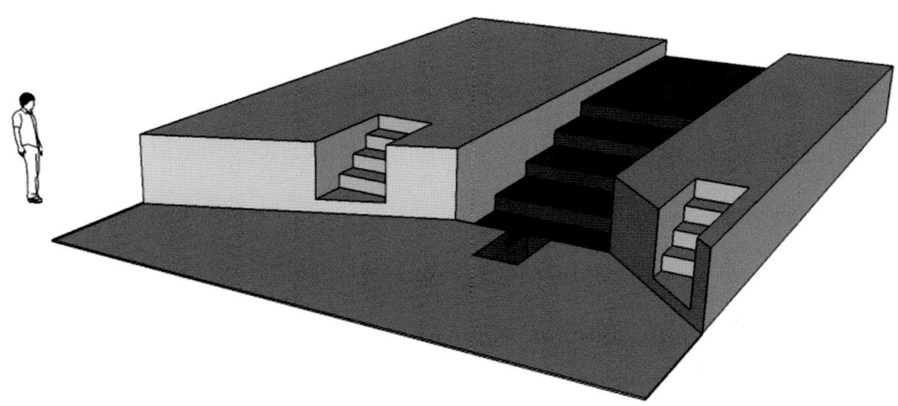

Jordan Taylor, Chapel Hill, North Carolina

Anagama Kiln

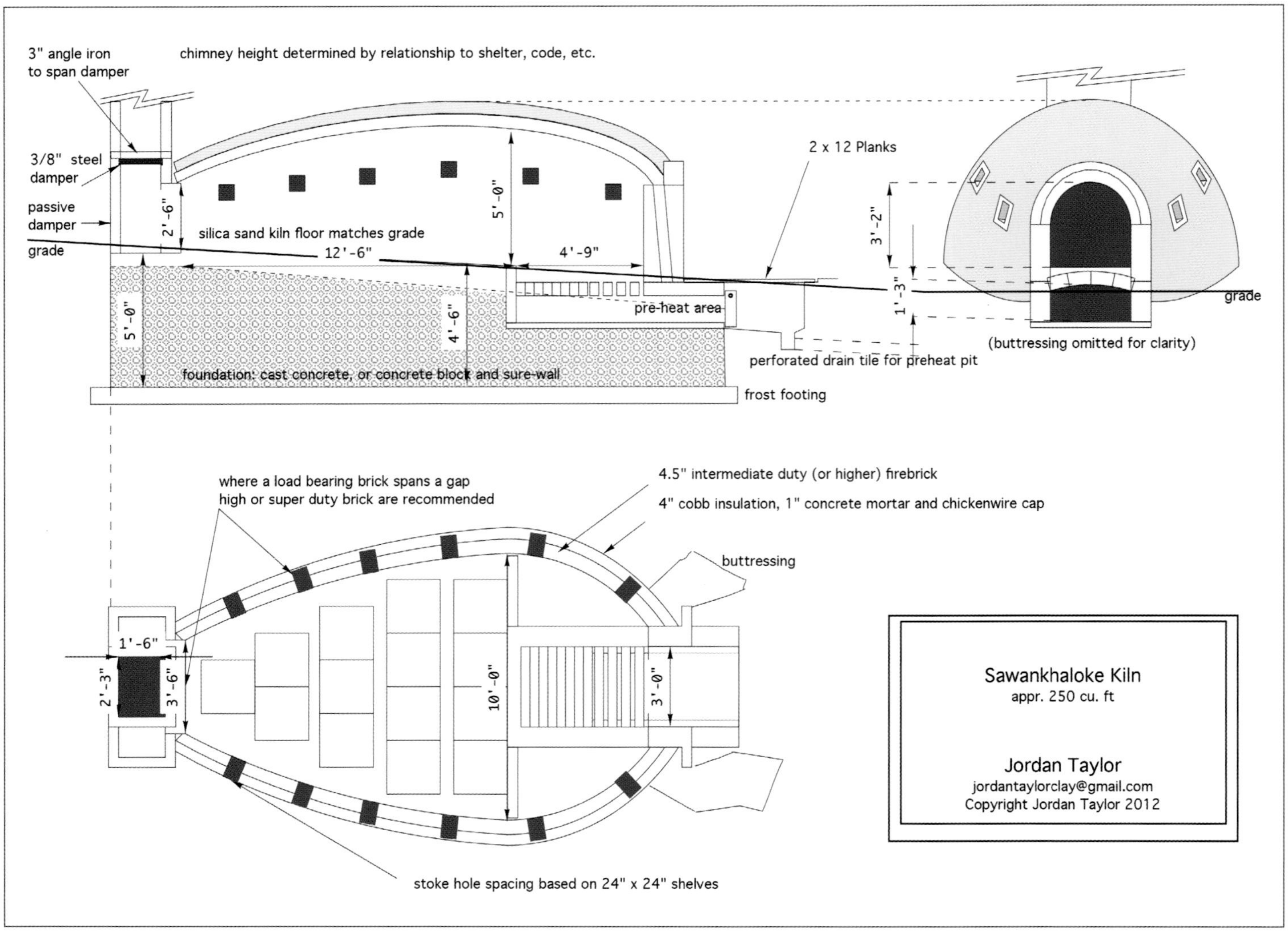

Jordan Taylor (USA). *Sawankhaloke Kiln,* anagama kiln, kiln plans

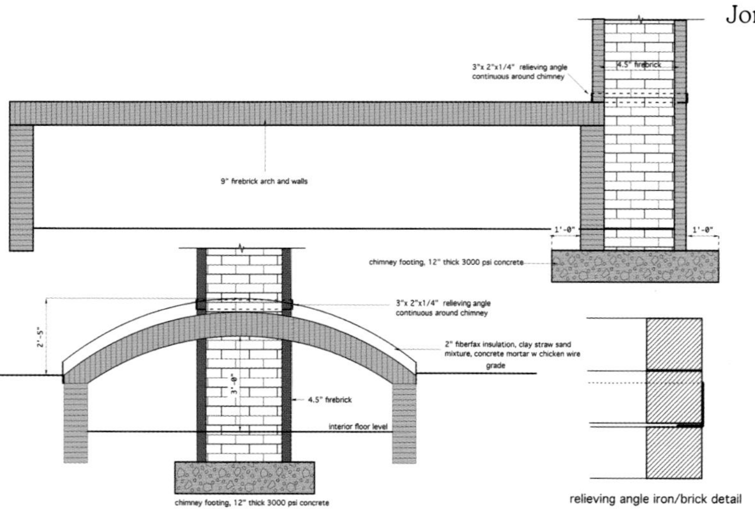

Kiln: Section

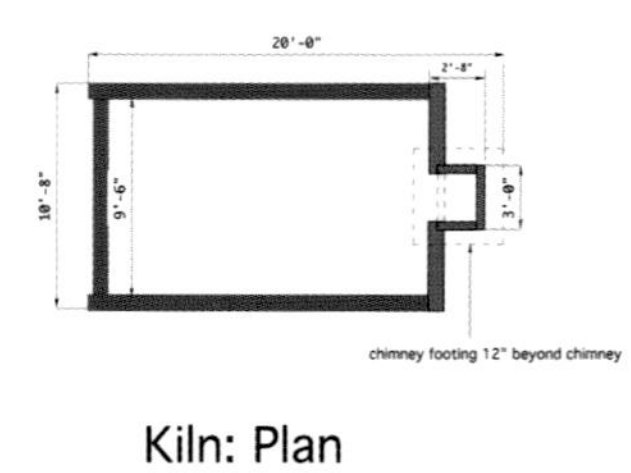

Kiln: Plan

Jordan Taylor (USA). *Sawankhaloke Kiln,* anagama kiln, kiln shed plans

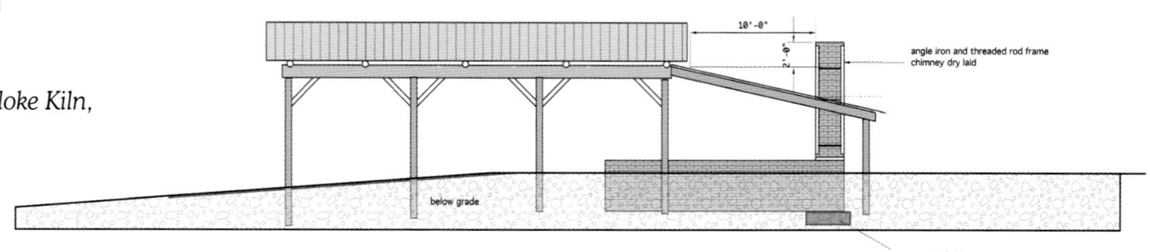

Kiln: South Elevation

David Wright, Thrussington, Leicester, England

Downdraft Kiln

David Wright (England). *Back Yard* Downdraft kiln

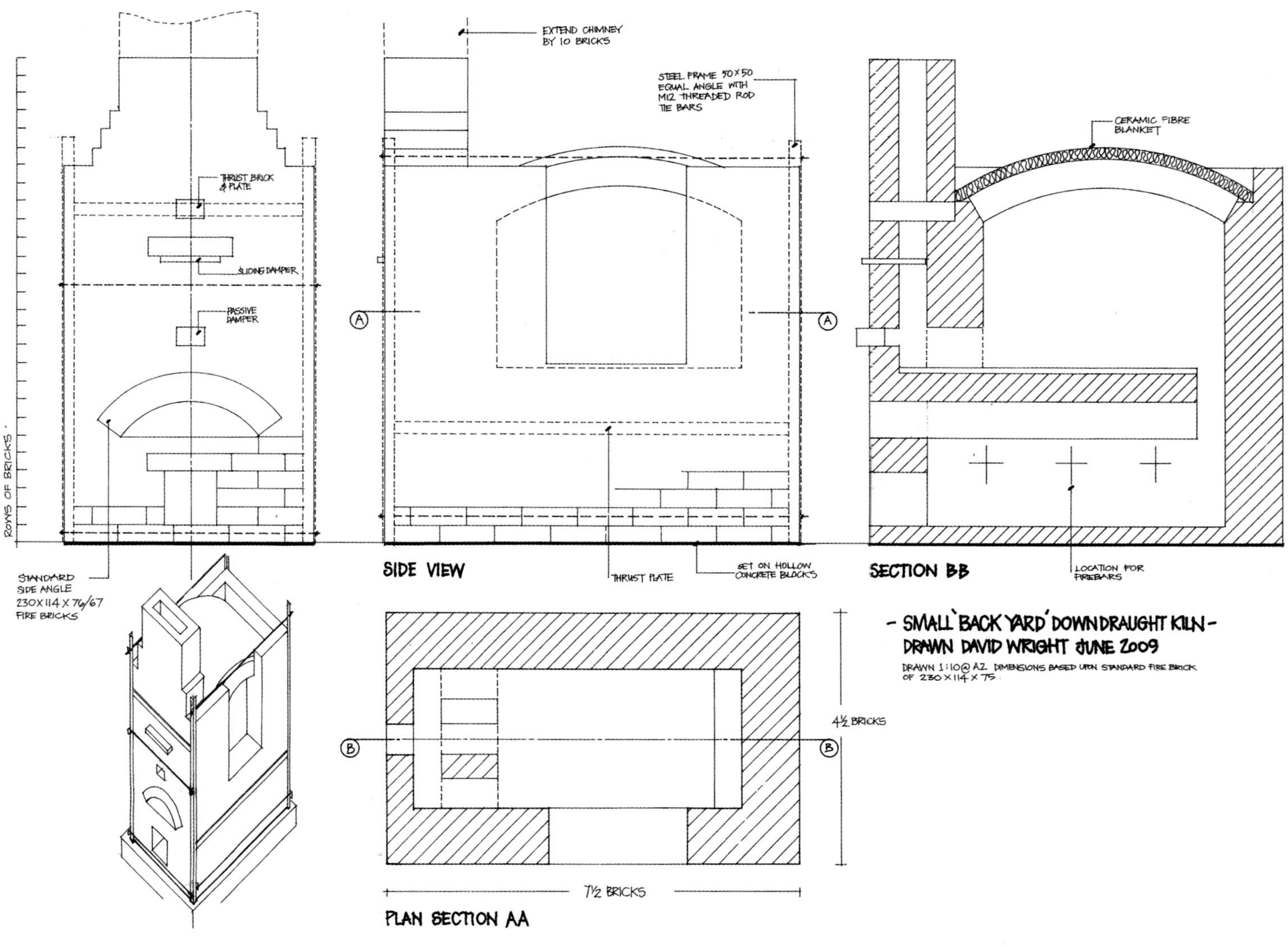

Appendix 2

Kiln Logs

Clay Cunningham, Council Bluffs, Iowa

Anagama Kiln

Kiln Log ,"*Iowagama*" July 6–9, 2012

Shift Time and Firers	Temp and Cones	Notes
Friday Noon–4 P.M. Clay, Tyler, Amanda	Door Bricked at 11:30 A.M. Cones 2, 4, 6, 8, 10 on upper right in Rear and upper left and lower right in Front. Small fire started outside at 12 Noon.	Fire started, kept on outside of kiln. Lit a fire in the chimney to get kiln drawing. Got draw, damper closed to ½" Of the 4 entry flues, only 2 middles ones are open. Outside 2 are closed up.
Friday 4 P.M.–8 P.M. Clay, Tyler	4 P.M. 200F Front Pyrometer	Keeping kiln cool to release moisture from winter. First firing of the year.
Friday 8 P.M.–Midnight Clay, Michael	8 P.M. 400F Front Pyrometer	Kiln appears dry and pots at front of kiln are safe. Moving the pyrometer to the rear of the kiln. 10 P.M.–Coals pushed a few inches into the firebox
Saturday 12 A.M.–4 A.M. Tyler, Amanda, Michael	Midnight 600F Rear Pyrometer	A few coals into firebox, keeping logs on hot coals outside of kiln. Waiting for C to OK stoking.
Saturday 4 A.M.–8 A.M. Clay, Tyler, Kim	4 A.M. 900F Rear Pyrometer	Hot coals pushed into kiln. One thin log (3" x 3") put into top stoke hole. Letting it burn to coals, then repeating. About every 30 min.
Saturday 8 A.M.–Noon Clay, Ling, Alli	8 A.M. 1200F Rear Pyrometer	Damper open to 1 inch. Stoking larger logs (5" x 5" approx.) Three at a time. Building up coals. No stoking pattern yet.
Saturday Noon–4 P.M. Ling, Tyler	Noon 1400F Rear Pyrometer Cone 2 Down in Front	Good coals, getting fresh air and very red. C says to keep stoking the bigger logs.
Saturday 4 P.M.–8 P.M. Tyler, Kelsey	4 P.M. 1550F Rear Pyrometer 6 P.M. 1525F Rear Pyrometer	Having trouble getting it to climb. Tried closing the damper in, made it worse. Drop in temp. Opened back to 1", got slight rise.
Saturday 8 P.M.–Midnight Clay, Tyler, Kelsey, Michael	8 P.M. 1550F Rear Pyrometer	Damper open to 1.5" Start stoking 5 smaller logs. 9 P.M. Stoking pattern established Stoking at 1. No sign of smoke or flame AND/OR 2. Stall of temp/ drop of temp AFTER rising.
Sunday 12 A.M.–4 A.M. Michael, Amanda	Midnight 1775F Rear Pyrometer Cone 4 down in Front	Keeping stoke pattern, temps rising slowly. Coals are pretty deep in firebox. Starting to rake and stir coals to keep fresh air getting to them.
Sunday 4 A.M.–8 A.M. Clay, Michael, Kim, Robert	4 A.M. 1925F Rear Pyrometer Cone 6 down in Front Cone 8 bending Cone 2 down in Rear	Stoking heavier now (7-9 THIN logs). Using palate wood burning in bottom stoke holes. All 4 entry flue inlets open. Raking after every other stoke to distribute ash.
Sunday 8 A.M.–Noon Clay, Ling, Alli, Tyler	8 A.M. 2050F Rear Pyrometer Cone 4 down in Rear Cone 10 nearly flat in front	Gave last big rake at 9 A.M. NO MORE RAKING! 10 A.M. Damper opened to 2 inches.
Sunday Noon–4 P.M. Clay, Ling, Tyler, Michael	Noon 2150F Rear Pyrometer Cone 6 down in Rear Cone 8 bending in Rear Cone 10 FLAT in Front.	Begin side stoking, alternating stokes with the front of the kiln. Damper closed back to 1.5 inches. Probably won't gain heat, but will maintain.
Sunday 4 P.M.–8 P.M. Clay, Tyler, Michael	4 P.M. 2200F Rear Pyrometer Cone 10 bending in rear	Continue side stoking sticks smaller than 1 inch in diameter. Mostly small limbs and cut offs from local saw mill. Also adding small 'burritos' of salt through side stoke holes.
Sunday 8 P.M.–Midnight Estimated Kiln Closing Shift Clay, Tyler, Michael, Kelsey, Bridget	8 P.M. 2225F Rear Pyrometer Cone 10 touching, but not flat. 10 P.M. 2250F Rear Pyrometer	8 P.M. Very heavy stoke up front to induce reduction cooling. Covered kiln front in slip covered newspaper, multiple layers. Continue side stoking. 10 P.M. Very heavy side stoke, added last of salt (total of 6 lbs). Closed up kiln.
Monday Morning 10 A.M. Tree Planting Ceremony	An annual tradition, we plant a new tree on the property to give back to the cycle that has provided us with beauty.	Kiln will be unloaded on Thursday evening, ensuring 72 hours of slow cooling.

Stephen Driver, Little Rock, Arkansas

Modified Fast Fire Kiln

August 5-6, 2012

Time	Temp	Inlet Flue Damper #1	Inlet Flue Damper #2	Active Damper #1	Active Damper #2	Passive Damper
7:30	75F	Open	Closed	Open	Open	Closed
Small fire started in front of firebox. Draft started. Plan to climb 100° per hour. Fair number of re-fires.						
8:30	150F	Open 4"	Closed	Open 2"	Closed	Open all
9:30	250F					
10:30	350F					
11:30	450F					
Slowly pull inlet damper to full open.						
12:30	500F					
1:00	600F					
2:00	788F					
3:00	900F					
Inlet flu #1 full open.						
3:30	1005F	Open full		Open 2"		
4:00	1128F					
6:00	1300F		Open 4"			
Start fire second firebox.						
7:00	1300F		Open full	Open 4"	Open 4"	Close all
Time for more draft.						
8:15	1455F					
9:15	1620F					
9:52	1807F			Close to 2"	Close to 2"	Top 3 out
Start body reduction. Fill firebox above and below grate and side stoke. Burn down and climb and repeat until cone 04.						
10:30	1910F			Open 4"	Open 4"	
Cone 08 down.						
11:00	1990F					
Cone 1 down.						
11:45	2110F					Top 3 back in
12:40	2207F					
Cone 9 bending.						
1:41	2296F					Top 3 out
Cone 9 down						
3:00	2306F					Pull bottom 3
Side stoke each stoke cycle, one thin piece. Cone 10 bending, cutting back to hold temperature.						
4:00	2245F					Bottom 3 back
Chimney cooled too much; need heat.						
4:45	2360F					Pull bottom 3
Cone 12 down. Hold temp.						
6:30	2360F					
Pyrometer is a little off from the last couple of firings, and kiln is hotter. Cone 13 over. Hold temp.						
7:30	2355F					
Kiln off, close off dampers.						

Note: Cones – 3 packs 012, 08, 04, 1 + 6.8, 9, 10 + 11, 12, 13

Gary Hootman, Swisher, Iowa

Anagama Kiln

Kiln Log, *"Hootmanama II"* Feb. 23–March 2.

People that helped are:	**Man hours**
Chris Irvin	60 total hours fired 172
Jim Kasper	30
Chuck Hindes	5
Samuel Johnson	5
Eric Gorder	12
Conifer Smith	9
Shumpei Yamaki	5
Andy Dalrumple	5
Gary Hootman	86
Alan Jung	8

Total man hours fired: 225

Took four days to load with the help of Chris Irvin each days 10 hours (man hours 40): 2.5 hours to put door up; ten cone packs made up, three for back wall placed top middle and bottom 7, 9 and 10.

Two set front 25" off floor back of firebox 09, 01, 2, 3, 5

Two sets of 7, 9, 10

Two sets of 11, 12, 13

Notes of interest: wadding and seashells were used to separate pieces. Wadding 70% silica sand, 20% mesh fire clay, seashells by size to fit pieces.

Remember, not in a hurry, kiln will get hot when ready.

Write down time of cone when down (not bent a little).

If there is a question as what to do, fire at a 25° per hour rate or close to it.

Whenever cone 7 goes down flat, take a reading on the pyrometer, will add 50° to that and hold somewhere within 40° of that for two day. This is an exception to the 25° per hour during soak time.

If you would write down hourly time and temp on pyrometer, I would like that and make notes as to what is happening (snow, rain, three pieces of wood per stoke, etc.) whatever you think would be of interest.

Started fire in lower firebox (both) small amount of wood keeping the pyrometer reading below 200° for first 24 hours of firing, then begin 25° per hour rate.

MONDAY FEB. 24 PM Temp 255°

I've been working on pre-heat, just driving off moisture, putting one big chunk of oak in lower fireboxes about every hour some times adding a smaller piece to help the burn.

Not in a hurry, there is a great deal of moisture to be driven off at the 36 hour point. I will start to bring up kiln up 25° per hour.

1:00 temp peak still at 255°

2:00 temp 273° Chris came on talked with him as to what I was doing, I'm finally going to get some sleep

2:30 Chuck called, asked what and how firing was going, and said he would call back tomorrow

2:40 temp 269° add two small pieces to the left side; add a bigger piece to right side with two small cones

3:20 temp 279° same thing as above, moved larger chunks of wood from wood pile to a place near the kiln to have close by when it gets dark

4:00 temp 287° adding just two small pieces this time to each box, getting cold and windy

4:35 temp 291° adding bigger pieces to both boxes, trying to keep same amount of coals and fire in each

5:10 added two small pieces of wood, need to add a little more wood more often

5:45 sun going down, will be going to dinner soon, going to add extra wood on and go eat

6:00 shook up coals, added two more bigger pieces then left temp 310°

6:45 temp 329° back to adding big wood for a while till coal bed gets back up

7:50 temp 328° Gary had said not to worry about getting 25° per hour, just keep trying to learn how the wood burns now; that it will be useful later in the firing so I've been just doing as I have been, adding two small and one large about every 30 to 40 minutes

8:25 temp 332° adding two small pieces

8:55 temp 329° put in two big pieces, I think temp didn't move because I couldn't get a big piece in the last stoke

9:25 temp 336° put two small pieces in each side of the kiln; still pushing moisture out of it (it's cold out, news said that it currently is 2°) and you can feel it

9:55 temp 351° put in big piece on each side

10:30 temp 359° added two small pieces to each side, the rhythm seems to be now at one big piece and two small pieces every half hour

11:05 temp 355° put one piece on both sides, seems to still have nice coal bed (Gary showed up and taking a look and talking with me about how things went)

11:35 temp 351° added one small piece to each side, lots of coal, just soaking

FEB 25 A.M.

12:00 temp 360° Gary is on, took off covers to lower firebox, been 48 hours time to get serious about the firing, more oxygen, put more wood on, checked at 12:20 temp is 374° so put lids back on, supported by a brick, laid sideways to allow more air, damper still at 4 inches, it's cold, the radio just said it is 10 below zero! 12:45 temp 379°

1:00 temp 384° added four pieces of oak, putting wood in about every 20 minutes

2:00 temp 422° put in big chunks, not a good idea for temp to rise need smaller, more often, to keep a rise—drop in temp to 409° at 2:15

3:00 temp 434° adding two at each side

4:00 temp 425° adding wood every 20 minutes, making sure to crisscross

5:00 temp 475°

6:00 temp 485° not putting in enough wood, increased to three pieces, each lower firebox

7:00 temp 513° started pushing burnt wood forward to make room for more wood

8:00 temp 547° Jim comes on now, explained what's going on before leaving him in charge

9:00 temp 562° adding wood in singles or multi pulls as needed

10:00 temp 587°

11:00 temp 615°

12:00 temp 643° dropped one piece of wood through main firebox just over the edge, still firing the lower firebox

12:30 note: need flame in lower firebox to burn main firebox, so keep the lower firebox going

1:00 temp 674°

2:00 temp 705° Chris is coming on now, talked with him as to what is going on

3:00 temp 741° moving pieces into lower firebox more, then dropping a piece through main box on each side every 10 to 15 minutes as needed, one piece each place

4:00 temp 767° ready to move flame into firebox, pushed coals forward, placed the shelves lower to aid in moving around kiln freely
5:00 temp 783° had trouble with getting rise with lids down flat so placed a brick 4.5 side up as spacer that helped, still putting a piece of wood in lower firebox (Gary was here helping)
5:15 changed damper to increase airflow, still no smoke from chimney
6:00 temp 827° stoking only through main firebox doors only one piece each (Gary took over for me while I went to eat)
7:00 temp 864° throwing two pieces then one piece in a stoke pattern for now, Gary talked with me as to what he was doing and how to place the wood through the doors
8:00 temp 891° same program
9:00 temp 912° things look good with stoke pattern, about every 10 to 12 minutes
10:00 temp 933° no change in stoking pattern, looks good
11:00 temp 964° still going good dropped a little in temp while getting an extension cord for Gary to hook up the radio on out here by the kiln. Gary is on now

FEB 26 A.M.

12:00 temp 1027° still the same stoke pattern—Eric Gorder stopped by for a chat about when he was going to fire, etc.
1:00 temp 1065° oak stoking one or two every other stoke
2:00 temp 1120° need to slow down a little
3:00 temp 1152° hard to hold the kiln back, it wants to go, need to let the temp back up a little before next stoke to get more control and a closer stoking pattern
4:00 temp 1185° better, still hard to get a good rhythm to control rise a little more
5:00 temp 1223°
6:00 temp 1243° much better (external temp is at 4°, much better than last night)
7:00 temp 1272° (external temp now 3°)
8:00 temp 1305° (now 8° outside) Jim is on, talked briefly on what I was doing
9:00 temp 1325° using one piece at a time, varying by weight of piece for stoking, varying about 10° with a nice rise
10:00 temp 1348° checked ash pit, still ok, no smoke from chimney
11:00 temp 1375° work is glowing behind Gerry Eskins' piece, still blowing off a lot of steam out of kiln
12:00 temp 1400°
12:30 noted coloring the top of arch and wood using has changed, ash pits with coal still
1:00 temp 1425° much smaller temp swings can hold at any given temp now easily
2:00 temp 1455° Chris has come on, explained to him what I was doing before leaving
3:00 temp 1483° throwing one piece in at a time, the temp is rising well and holds ok
4:00 temp 1507° stoking as needed to keep steady rise
5:00 temp 1543° change to throwing in a couple to three pieces per stoke, maybe a little too much (peaked at 1449°)
6:00 temp 1560° Gary came on so I could eat
7:00 temp 1593° one piece in at a time (getting hotter around the kiln, you can smell the earth getting hot and cooking)
8:00 temp 1619° color is moving back in the kiln nicely
9:00 temp 1646° cones have not moved, no smoke from chimney
10:00 temp 1666°
11:00 temp 1700° not moving up as easy, Gary here now opened the damper about 1.5 inches total, said need to put less wood in more often to get better rise

FEB 27 A.M.

12:00 temp 1770° going too fast, need to slow down; movement from cone 09 on the right side firebox as you look at the kiln, no movement yet on the left side
1:00 temp 1776° just held it here for last hour to make up for such a large rise in previous hour, no change in cone 09 on left side but continues to move on right side
2:00 temp 1820° both 09 cones on left and right sides are down flat
3:00 temp 1838° temp going up very easy
4:00 temp 1865° first signs of smoke out of kiln, as mouse grey very light smoke
5:00 temp 1900°
6:00 temp 1937°
7:00 temp 1970° I'm very tired loosing the ability to keep a rhythm
7:45 temp 2000° Jim came in to take over some drop in temp
8:00 temp 2033° stoking two pieces, letting smoke clear following rise then as peaks adding one or two pieces depending on the size of wood
8:10 cones are on the move 01 both places half over
8:30 temp 2015° (peak 2033°) cones 2 and 3 are half over; 01 cone is down
8:45 temp 2010° cones 2 and 3 half over still
9:00 temp 2025° cones 2 and 3 are down in both places, changing stoke pattern to alternating stokes from left to right firebox with a clearing between stokes, much better kiln response in temp rise; cone 5 still straight up
10:00 temp 2050° peak and 2030° low point, kiln is changing so change things to accommodate those changes, taking bigger stokes three pieces sometimes there is more smoke which needs to clear, pulled the two secondary air intakes below the primary stoke holes, also raised the lower firebox doors to 2.5 inches for more primary air intake
11:00 temp 2065° peak to 2050°, bottom swing still alternating stokes, trying to drop wood straight down in front of secondary air intakes, about three or four pieces of wood to each box depending on size of wood (clay over kiln starting to push moisture off even more than before with the ever increasing temp), cone 5 starting to move
11:30 holding right (kiln wanting to climb with the changes made 2070-2075° left and right, cone 5 moving about the same)
12:00 temp 2095° peak 2075-2080° low end during this period both cones about half over
12:35 both cones 5 tip down temp at 2100°
1:00 temp holding between 2100 and 2090° Gary is mudding up the door and arch, preparing the kiln for the end of the firing to come in just a few more days from now, cone 5 getting flatter during this soak time the last hour
2:00 temp after two hours of around 2100° has started to rise again, Chris has come on, cone 7 on the left side has started to move, nothing on the right, peak temp 2115°
3:00 temp 2137° cone 7 same as at 2:00
4:00 temp 2143° not much move in temp (Gary was here looking around as to how things were going, decided that there was a need to change the damper a little more at this stage of the firing)
5:00 temp 2165° both of cone 7 about 3:00 o'clock, throwing two or three pieces in each firebox door, alternating left to right
5:45 temp 2173 ° cones still bending just need a little longer (Gary is here so I can go eat)
6:00 2201 ° peak temp, Chris is back from eating
7:15 temp 2195° cone 7 down on both sides, cone 9 soft and bending slightly, will hold here till Gary comes on
7:45 cone 9 has still been moving both at 3:00 o'clock must still be too hot will back up to 2180° and hold
9:00 temp 2183° holding, cones look like they are holding, using big chunks of oak to hold temp
10:00 temp 2185° cone 9 has moved to flat position on right side and about 5:00 o'clock on the left, but 10 is still straight up, still trying to hold around 2180°
11:00 temp 2186° looks good, no movement in cones for an hour

FEB 28 A.M.

12:00 temp 2195° cone 9 both down, some bend in cone 10, Gary is on, talked to me about firing, agreed we should hold below

2180° for the next two days, that I did a good job. The damper move was a good one; it's easy to hold the kiln

1;00 temp been isolating between 2180-2210° its next to cone pack holder on the left – still maybe too hot to hold here?

2:00 temp 21080-2190° about the same as before

3:00 temp letting up, there be a bigger swing in temperature from 2169 to 2190°, not trying to go much hotter, but if I do, not for very long duration, still alternating the stokes from left to right firebox

4:00 same

5:00 same

6:00 same

7:00 same, putting in four or five pieces of wood, alternating left to right (some pieces are pine now)

8:00 same, Chuck, Sam and Andy came on for their shift, told them what I was doing, to have fun just don't drop cone 10, let it go down for a couple three hours then bring back up and hold for 30 minutes, then let go down again

10:20 two hours at 2160° then slowly up to 2217° for 30 minutes, peaked at 2230°, then repeat

1:00 Chris came on, noted that cone 10 was down, still in holding mold, alternating stokes as before, going up and down

2:00 temp 2181°

5:00 temp holding at 2180°, closed the damper supports on the lower firebox to see what would happen

6:00 Gary came on so I could eat

7:00 temp holding pattern 2180°, moved wood from under the other shed to have it closer to get at; cleaned out the lower firebox for the first time; Gary made a hoe to clean it out

8:00 still holding at 2180° after the clean out

10:00 held at 2160° for two hours going up to 2200° for a half hour

11:00 Gary has come on, everything is going good

MARCH 1 A.M.

12:00 been holding around 2190° and letting it drop back to 2170° then bringing it back up slowly, putting in three to four pieces of wood, depending on size, mixing oak and pine at this point in the firing, we have burnt close to 11 cords of wood

Started snowing again around 3:00 and stopped around 5:30

8:00 Jim and Eric came on, told them same old same old, they wrote down what cones were still standing, cone 11 had at one time or another gotten soft enough to move slightly, 12 and 13 stiff; stoking was to be alternated between left and right, three pieces of wood each time

10:00 Eric firing, Jim watching now; 9-9:30 brought up to 2238° peak 9:30 to 10:00 held 2210 to 2230°, looking closely at cones, 9 are fully flat and cones 10 are lying on top of 9s, letting come back down to 2180-2190° again. ERIC IS THE MAN!!

11:00 Jim is firing now – stoking two to three pieces and holding between 2170-2180°, cones are holding the same

Jim still on, temp 2185°, starting to climb back to 2230°

12:00 temp 2230-2220° held for one half hour, dropped cone 10 more this time, fully down now, going back to 2180°

2:00 Chris and Conifer are on kiln watch

4:20 brought up temp to 2230°, held for one half hour, then let drop back to 2270°

6:00 Gary came on so we could go up and eat

7:00 still holding; back from dinner, it was good as always

9:30 still holding; but will bring it up around 10:30

11:00 held at 2230° for 25 minutes and have went back down to 2180; 11:15 Gary showed up, Al, Chris and Conifer still here

11:45 Conifer left went home; Gary and Al are on

MARCH 2 A.M.

12:00 same thing, hold a couple of hours then top out 2230° then drop back to 2180°

3:00 holding at 2190-2230° up and down, but a little hotter then before on the swings

5:00 starting to go up for final push to finish, looking to go up to 2230° and hold

6:30 temp 2280° and holding

7:45 temp 2298° and holding, opened lower firebox by putting in a brick 4.5 opening to allow more air, starting to get touchy to get rise

8:00 Jim and Eric on, Jim at the helm; windy outside, dropped 20 degrees to 2280°; having problems finding air to fuel ratio 9:30 peak at 2309°, using more pine more often through each door to keep grates covered in wood. Lower firebox covers opened to 3 inches; watching for smoke out of stack then the pyrometer for a rise in temp and stoke at the peak

10:30 Gary came back down, temp 2270° is watching the stack, waiting for smoke to clear. Hit again with two pieces, quicker stokes, alternating doors, temp is steady at 2283°, opened damper5" 2" 5", using larger pine for now, still one piece in at a time

12:00 waiting for smoke to just clear again – medium size pine 4x4x18, kiln has been above 2300° for past hour and above 2330 ° for last half hour GARY IS FIRE KING!!

12:30 cone 11 flat, both sides, cone 12 is at 2:00 o'clock on the left, peak temp 2348° for last two hour, been above 2320°. Goal was to reach 2310° and hold above for two hours

1:00 Firing done!! Time to close up; started by sealing lower fireboxes with clay all around them and the bricks leading to the kiln. Then filling both fireboxes with large oak chunks, then sealing the doors. This was done by first placing six bricks in the door opening, then a piece of fiber and finally wet clay to seal; then the metal door was closed. The damper was then closed down to two inches where it would remain till final closing. Secondary air plugs are still open

1:30 about one gallon of water was introduced through secondary air plugs, pulling the steam into the kiln, this was done twice, about ten minutes apart, then closed damper and sealed secondary air ports. Sealing with clay wherever there was a leak, including the chimney which was also sealed

FINISHED

2:00 final look around, temp 2173°, came in for a drink

3:30 temp 1993° left for the night

MARCH 3 A.M.

8:00 temp 1540°

5:30 temp 1316°

Note: total wood consumed was 12 cords of wood using the formula of 4' x 4' x 8' as a cord. Ten cords was white oak and two cords was a mix of pine, cedar and elm.

***Special note, after this firing, I decided that five days was plenty long. I had to grind off too much ash off the feet of pots because of the amount of ash and the longer firing.

Conifer Smith, Cedar Rapids, Iowa

Kirkwood Community College–Anagama Kiln

Kiln Log April 5-8, 2011

TIME	TEMP.	COMMENTS
THURDAY, APRIL 5TH 7:00 P.M.		LIT KILN
9:15 P.M.	142 F	
10:10 P.M.	141 F	
11:10 P.M.	147 F	
FRIDAY, APRIL 6TH MIDNIGHT	171 F	
12:45 A.M.	221 F	
1:50 A.M.	270 F	
2:15 A.M.	300 F	
2:35 A.M.	321 F	
3:00 A.M.	340 F	
4:00 A.M.	363 F	
5:00 A.M.	377 F	
5:30 A.M.	400 F	
6:00 A.M.	405 F	
7:00 A.M.	410 F	
8:05 A.M.	581 F	MOVED FIRE INSIDE; STOKE MAIN STOKE HOLE
9:00 A.M.	645 F	MOVED FIRE INSIDE
10:00 A.M.	690 F	
11:00 A.M.	780 F	
NOON	900 F	
1:00 P.M.	1000 F	IN REDUCTION; LOTS OF WOOD
1:30 P.M.	1100 F	IN REDUCTION
2:10 P.M.	1117 F	IN REDUCTION, CONE 012 DOWN
2:40 P.M.	1216 F	OUT OF REDUTION
3:00 P.M.	1233 F	SOME COLOR IN FRONT SIDE SPY
3:31 P.M.	1287 F	
4:00 P.M.	1318 F	
4:30 P.M.	1335 F	
5:00 P.M.	1397 F	SOME COLOR IN MIDDLE OF KILN (SECOND SPY)
5:30 P.M.	1450 F	FAIRLY BRIGHT ORANGE IN SECOND SPY; OPENED DAMPER 1" (1/2" ON EACH SIDE)
6:00 P.M.	1490 F	
6:30 P.M.	1534 F	
7:00 P.M.	1597 F	
7:30 P.M.	1628 F	OPENED DAMPER A COUPLE INCHES EACH
8:00 P.M.	1696 F	
8:30 P.M.	1731 F	
9:00 P.M.	1810 F	CONE 5 DOWN
9:30 P.M.	1863 F	CONE 7 DROOPED; SPY HOLES YELLOW

TIME	TEMP.	COMMENTS
10:00 P.M.	1043 C	CONE 9 DROOPING; SPY HOLES YELLOW
10:30 P.M.	1065 C	CONE 9 DOWN
11:00 P.M.	1070 C	CONE 10 NOT MOVING YET
11:30 P.M.	1093 C	CONE 10 NOT MOVING YET
SATURDAY APRIL 7TH MIDNIGHT	1095 C	MOVED DAMPERS IN A BIT; PUT BRICKS INTO PRIMARY AIR TO SLOW THINGS DOWN A BIT; 888 C IN BACK
1 A.M.	1090 C	
1:30 A.M.	1090 C	BACK COOLED TO 780 C; PULL PASSIVES
2:00 A.M.	1105 C	BACK 905 C
2:30 A.M.	1118 C	
3:00 A.M.	1126 C	BACK 945 C; CONE 10 IS BENDING; STOKING PATTERN OF 3-4 SLATS & 1 MED/ BIG
3:30 A.M.	1134 C	BACK 945 C; LOW ON WOOD—KICK IT, BACK 1050 C
4:00 A.M.	1140 C	CONE 10 DOWN; CONE 11 SOFT
5:00 A.M.	1147 C	STOKING PATTERN OF A HANDFUL OF LATH & A BIGGIE; BACK 1068 C; BEAUTIFUL COAL BED
5:30 A.M.	1154 C	CONE 11 BENDING; WHITE HEAT!
6:00 A.M.	1161 C	
7:00 A.M.		PUT SMALL BRICKS IN PASSIVES IN CHIMNEY; PUT TWO ¼" BRICKS AND A ½" & A ¼" BRICK IN PRIMARY AIR TO CURB AIR INTAKE—TO SLOW THINGS DOWN; STOKING ONLY SMALL WOOD
9:00 A.M.	1158 C	BACK 1216 C; CONE 11 NOT QUITE HALF DOWN; PRIMARY AIR HAS ONLY ONE BRICK IN & NEED LOTS OF POKING TO KEEP IT UNCLOGGED
10:00 A.M.	1193 C	BACK 1170 C
10:35 A.M.	1200 C	
11:30 A.M.	1210 C	
NOON	1218 C	
12:30 A.M.	1176 C	
1:30 A.M.	1175 C	
2:00 P.M.	1182 C	CONE 11 DOWN; CONE 12 IS HALF DOWN
3:00 P.M.	1175 C	
5:15 P.M.		
5:30 P.M.		KILN SHUT DOWN; KAOLIN WOOL AND KILN SHELF PLACED AT TOP OF CHIMNEY
7:00 P.M.	950 C	3 STICKS FOR REDUCTION CCOLING
11 P.M.	1600 F	STOPPED REDUCTION COOLING AND WENT HOME

Notes: PYROMETER HALFWAY THROUGH THE KILN
STAY 200° UNTIL MIDNIGHT, THEN 50° PER HOUR
START WITH DAMPER IN, WITH 2 INCHES BETWEEN, FOR EVEN FLAME
START WITH PASSIVES CLOSED

Renata Wadsworth, Spencer, New York

Modified Fast Fire Kiln

Kiln Log, Firing #17 in *"Margaret"* fast-fire July 30–31, 2012.

(I use cones 5, 8, 9 and 10, 11, 12, as well as cones 012 and 010 for early reduction. My pyrometer doesn't seem to move beyond 2200 degrees, and is unreliable after 2000 degrees, at which point I rely on the cones to determine temperature, and the pyrometer gives me something to look at as the hours tick by.)

July 30th	preheat
8 P.M.	lit fire east side
9 P.M.	230 degrees, closed damper. Good night.

July 31st	
6:15 A.M.	122 degrees - lit fire west side
7 A.M.	216 degrees
8 A.M.	550 degrees
9 A.M.	800 degrees
10 A.M.	1100 degrees
11 A.M.	1365 degrees
11:30 A.M.	1485 degrees cone 012 going hot spots, start reduction, damper in ¾ of the way, fireboxes covered after stokes
12 P.M.	1520 degrees
1 P.M.	1600 degrees
2:30 P.M.	1790 degrees
3 P.M.	1830 degrees damper out an inch
3:30 P.M.	1916 degrees
3:45 P.M.	1930 degrees damper out another inch
4:15 P.M.	2000 degrees! Yay! Damper all the way out
5 P.M.	2070 degrees
5:45 P.M.	2100 degrees
6:40 P.M.	cones 8&9 bending hot spots cone 5 down cool spots start soda spray (approx 10 oz. soda ash dissolved in 1.5 gallon water, sprayed between 4 stokes, alternating sides.)
7:15 P.M.	done with soda spray
8:15 P.M.	2147 degrees cone 10 down hot spots, 9 bending cool spots
9:40 P.M.	cone 11 bending hot spots, cone 10 looking soft in cool spots
10:00 P.M.	cone 9 down cool spots, cone 11 bending hot spots, last stokes!
10:15 P.M.	all sealed up.

Appendix 3

Clays, Glazes, & Slips

Please note that the quantities of materials found in the following recipes are supplied by the artists, they include metric and avoirdupois weights, and are represented in both batch recipes and percentages.

Clays

James Tingey, Athens, Ohio

Clay Bodies	(cone 9-12)
Tingey Light "Gumbo"	
Tile	6
Kaolin	20
Helmar Kaolin	20
Tennessee	10
Ball Clay	15
XX Saggar	5
Flint	20
NC 4 Soda Feldpar	15
Nepheline Syenite	5

Tingey Light Stoneware

Helmer Kaolin	25
Hawthorne Fireclay	25
Tennessee 10 Ballclay	15
XX Sagger	10
NC 4 Soda Feldspar	15
Flint	10
Silica Sand (60 mesh)	5-10
Grog	5-10

Grollegy Porcelain

Grolleg	55
NC 4 Soda Feldspar	15
G-200Feldspar	7
Flint	23

Wollanstonite Stoneware

Tile 6	30
Tennesee 10	15
XX Saggar	20
Minspar	15
Wollastonite	20

Dick Lehman, Elkhart, Indiana

Lehman Recipe 13-D

English Grolleg	55
Nepheline Syenite	25
#401 Yellow Banks	15
(80 mesh) (or any higher-firing ball clay)	
Flint (325 mesh)	10
Bentonite	2
Water	33

This body is very fine in texture and tends to be quite light in color. Although I suppose the Yellow Banks clay is largely responsible for some of the caramel colors we see in the areas of transition between heavy ash deposits, and light ash deposits. There is often some "metallic oil spotting" in these transition areas. This body also tends to trap some carbon, making for areas of shiny black in the lighter ashed areas (backsides of pots), where the nepheline syenite has fluxed on the surface of the pot. It is very responsive to atmosphere and ash thickness ... one of my favorite recipes.

Justin Lambert, Jupiter, Florida

Lambert Shop Wood Body

Helmar Kaolin	40
Foundry Hill Cream	25
Goldart	13
Custer Spar	15
Silica	5
Redart	2

J. Daniel Murphy, Logan, Utah

Dan's 50/50 Freddy-iron Rich Clay

50 Fredericksburg Fireclay
50 Lizella

Murph's Iron Helmer Body

Helmer	100
Custer feldspar (pot spar)	50
Silica	15
Redart	16
Lizella	16

I strive to achieve iron red surfaces from the wood kiln. This clay body has the ability to enhance surface effects and promote red, maroons ... in the wood kiln when cooled properly. I generally fire dark clays in the wood kiln and cool this body in reduction to 1450° F (using my pyrometer). The firing cycle is about twenty-four hours to peak temperature. Cooling takes about six to eight hours!

Murph's Porcelaneous

Bentonite	2
Grolleg	54
Nepheline Syenite	24
Helmer	15
Silica	10

Shaner's Helmer Body

Helmer kaolin	100
Custer Feldspar (pot spar)	50
Roseville stoneware	10
Silica	15

Murph's Stoneware

Fireclay (a.p. green)	100
Yellowbanks 401 (stoneware)	50
Helmer	50
Custer feldspar (pot spar)	50
Silica	25

I have been using these porcelaneous stonewares for some time and have found them to be my favorites in this wood-firing process.

Kevin Lehman, Lancaster, Pennsylvania

Clay recipe for *"Monoliths: Stable Growth"*

Standard 2161	100
OM-4 Ball Clay	15
NC-4 Feldspar	5

Added

Silica Sand	35

Perry Haas,Logan, Utah

Shaner's Helmer Body

Helmer	100
Custer	50
Silica	25
Roseville	10

Murph's Porcelaneous (revised)

Grolleg	55
Custer	10
Helmer	15
Silica	15
Bentonite	2

Pioneer Day

Helmer	36
Goldart	28
Custer	16
Silica	12
China sand	8

Steven Driver, Little Rock, Arkansas

Porcelain

Kaolin (Helmer)	40
Clay (Old Hickory)	7
Soda Feldspar	24
Flint (200 mesh)	20
Pyrax HS	8
Molochite	8
Bentonite	1.6

This is working really well in reduction cooling firings

Fred Herbst, Painted Post, New York

Fred's Wood-fire Body

McNamee kaolin	50
Helmer kaolin	15
Custer feldspar	20
Nepheline syenite	5
Silica	10
Add:	
Red Art	2
Silica sand	3

Glazes

Dick Lehman, Elkhart, Indiana

Horseman Temoku

G-200 Feldspar 58
Whiting
12
Flint
12
Zinc Oxide
8
Red Iron Oxide
8
Rutile
2

Here is a glaze that I inherited from someone ... lovely in the back of kilns where it receives lots of work heat and just a little ash. Great color modulation including black, maroon, carrot, red-orange, gold, gray.

James Tingey, Athens, Ohio

Glazes (cone 10)
Sweetness Celadon
Blue Grey transparent celadon, holds up well to ash
deposits.

Flint	26
G-200	46
EPK Kaolin	10
Whiting	18
Bentonite	2
Red Iron Ox	.75
Bone Ash	3

Honey Plus
Highly variable glaze, earthy orange to grey purple, largely dependent on ash deposits and cooling cycle.

G-200	36
EPK Kaolin	22
Dolomite	18
Flint	14
Gerstley Borate	4
Whiting	4
Rutile	2
Yellow Ochre	3

J. Daniel Murphy, Logan, Utah

Murph's ambygonite shino glaze

Nepheline Syenite	44.1
Kona F-4	10.6
Amblygonite	
or LM spodumene	14.9
Yellowbanks 401	14.7
Helmer	9.8
Soda ash	3.9
Bentonite	2.0

Oestreich shino

F-4	10.8
Nepheline Syenite	45
Spodument	15.2
EPK Kaolin	10
OM-4	15
Soda ash	4
Zirco Pax	3

Annetes gold @ USU

Redart	55
Silica	5
Ball clay	5
Manganese dioxide	45
Copper oxide	5
Cobalt oxide	5

Note: this glaze can run with a passion, beware.

RJ Dry Shino #16

Nepheline Syenite	38.8
Spodumene	29.1
Ball Clay	29.1
Soda Ash	2.9

Yellow Salt
Nepheline Syenite
60
Dolomite
20.15
Zinc ox.(zirco)
152.8
OM 4
4.11
Bentonite
3.92
Also add RIO
.96
To 10,000 grams add 1/2 C. Muriatic Acid

Perry Haas,Logan, Utah

Pinhead Shino (Liner Glaze)

Nepheline Syenite	60
OM4 Ball Clay	25
Spodumene	15
Soda Ash	3%

RJ Dry Shino #16

Nepheline Syenite	39
Spodumene	29
Ball Clay	29
Soda Ash	3

Frazer Celadon
G-200 Feldspar
34

Silica	21
Grolleg	15
Wollastonite	27
Talc	3
Red Iron Oxide	0.5

Fred Herbst,Painted Post, New York

Currie Shino liner

Nepheline Syenite	70
Kaolin	30
Soda ash	3

Derek Larsen, Kitano, Kita-ku, Kyoto, Japan

Horsley Sinter Shino

Nepheline Syenite	30
Calcined Alumina	55
Spodumene	5

Tony Nankervis Blaze

Nepheline Syenite	29.2
Calcined Alumina	29.2
Spodumene	8.8

Karatsu

Soda Feldspar	33
Pine Ash	33
Rice Straw Ash	33

Iraho

Soda Spar	7.9
Kaolin	20.79
Silica	36.54
Magnasite	3.44
Strontium Carbonate	31.75
Iron Oxide	8
Bone Ash	8

JoAnne Dekuster, Taos, New Mexico

Gustn Shino

Nepheline Syenite	1800g
Kono Fspar	432g
Spogemine	608g
Ball clay	600g
EPK Kaolin	400g
Soda ash	160g

The only glaze we use is Gustn Shino. The great thing about the glaze is that we achieve a wide variety of colors with just one glaze. Depending on where it is in the kiln we get great reduction, great carbon trapping, and looks great with any amount of ash build up.

Rena Hamilton, Flagstaff, Arizona

Rob's Green *(exterior use only)*

Custer Feldspar	45.9
Whiting	21.2
Silica	17.6
EPK Kaolin	9.4
Gerstley Borate	5.9
Add:	
Copper Carbonate	5.9
Rutile	3.5
Bentonite	2.4

Val's Satin Matt (Blue-Pink)

Cornwall Stone	46
Whiting	34
EPK Kaolin	20
Add:	
Titantium Dioxide	6
Cobalt Carbonate	.25
Chrome Oxide	.125

Coleman's Teadust

Custer Feldspar	40
Silica	25
Whiting	16
OM4	12
Talc	7
Add:	
Red Iron Oxide	10

Mark's Tenmoku base
(Translucent Glossy Olive)

Custer Feldspar	45
Silica	27
Whiting	17
#6 Tile	11
Add:	
Red Iron Oxide	3

Semler's Turquoise Matt
(Opaque Blue Matt with Strontium)

Nepheline Syenite	53.1
Strontium Carbonate	31.9
Tenessee Ball #10	6.7
Silica	8.3
Add:	
Copper Carbonate	4
Bentonite	1

Pizzaz Black (Brush on Exterior Accents)

Red Art	55
Nepheline Syenite	20
Talc	15
Strontium Carbonate	10
Add:	
Manganese Dioxide	2
Cobalt Carbonate	2
Chrome Oxide	1

Coffman's Yellow
(Shino-like Slip; Apply Thin on Bisque)

Nepheline Syenite	59.4
EPK Kaolin	39.6
Soda Ash	1.0
Add:	
Titanium Dioxide	4.95

Peter Meyer, Bend, Oregon

LAU Luster Shino
Cone 9-10

Nepheline Syenite	53
EPK Kaolin	25
Silica	10
Whiting	2
Soda Ash	5
Lithium Carb	5
Tin Ox	2
Bentonite	2

COCC Shino

Nepheline Syenite	50
Kona F-4	14.6
Spodumene	12.5
Ball Clay	16.7
EPK Kaolin	2.9
Soda Ash	3.4

Good Clear

Custer Feldspar	40
Silica	30
Whiting	25
EPK Kaolin	10

Celadon

Custer Feldspar	40
Silica	30
Whiting	20
EPK Kaolin	10
Add:	
Iron Oxide	1–1.5

Rhodes 32
A dry matte glaze

Custer Feldspar	49
Dolomite	21.5
EPK Kaolin	25
Whiting	3.5

Slips

Bede Clark, Columbia, Missouri

Red Art Wash

Red Art Clay	54%
Whiting	23%
Custer Feldspar	16%
Ball Clay	7%
Add:	
Rutile	3%

Helmar/EPK Flashing Slip
(apply to bisque)

Helmar	23%
EPK Kaolin	23%
Nepheline Syenite	54%
Add:	
Soda Ash	2%
Titanium	.5%

J. Daniel Murphy, Logan, Utah

Laterite Slip: Reduction cool

Laterite	100

Apply very thin application to start, perhaps like non-fat milk.
Test 1st. Apply to Bisqueware

Crocus Martis: Reduction cool

Crocus	100

Test 1st. Apply to Bisqueware
Apply very thin application to start, perhaps like non-fat milk.

Stick to anything white slip

EPK Kaolin	500
Ball Clay	500
Flint	500
Cornish Stone	250
Nepheline Syenite	250
Opax	50
Frit: 3195	50

Test first, can be applied to greenware … please test first.

Newman Slip

Grolleg	50
Newman Red	10
EPK Kaolin	10
Nepheline Syenite	30

Peter Meyer, Bend, Oregon

White Slip Base

EPK Kaolin	45
Ball Clay	15
Custer Feldspar	20
Silica	10
Whiting	12
Zircopax	5
Gum (dry)	2.5

Hellmer Slip

Helmer Kaolin	30
EPK Kaolin	15
Ball Clay	15
Custer Feldspar	20
Silica	10

Whiting	12
Zircopax	5
Gum (dry)	2.5

Flashes Orange

Crackle Bisque Slip

Calcined EPK Kaolin	20
EPK Kaolin	15
OM 4 Ball Clay	15
Custer Feldspar	20
Silica	20
Zircopax	5
Borax or	
Gerstley Borate	5

Use on Dry Greenware or Bisque

Crackle Slip: Helmer Variation

Calcined Hellmer	20
Hellmer Kaolin	15
OM 4 Ball Clay	15
Custer Feldspar	20
Silica	20
Zircopax	5
Gerstley Borate	
Or Borax	5

Michael Poness, Tacoma Park, Maryland

Bauer Flashing Slip

EPK Kaolin	41.9
Ball Clay	41.9
Borax	5.7
Zircopax	10.5

Rust/Orange/Tan Wash

Red Iron Oxide	33.3
Rutile	33.3
Gerstley Borate	33.3

Medium Green Wash

Cobalt Carbonate	25
Rutile	53
Chrome Oxide	2
Gerstley Borate	20

Orange/Tan/Straw Wash

Rutile	50
Titanium Dioxide	17
Gerstley Borate	33

Conifer Smith, Cedar Rapids, Iowa

Titanium Yellow

Nepheline Syenite	32
OM4 Ball Clay	63
Flint	5
Super Pax	5
Titanium	10

Red Slip

Nepheline Syenite	15
Sapphire Kaolin	20
Helmer Kaolin	65

Alumina Slip

Helmer Kaolin	25
EPK Kaolin	25
Nepheline Syenite	25
Alumina	25

Fake Avery

Nepheline Syenite	48
EPK Kaolin	74
Georgia Kaolin	74
Newman Red	9
(use thin)	

Geo. Lowe's White Crackle Slip

Borax	5
Zinc	5
Custer Feldspar	20
OM4 Ball Clay	15
EPK Kaolin	15
Calcined EPK Kaolin	15
Flint	20

Hindes Shino Slip

Kona F4 Feldspar	11
Spodumene	15
Soda Ash	4
Nepheline Syenite	45
Ball Clay	15
EPK Kaolin	30

Bauer Orange

Borax	5.7
Zircopax	10.5
EPK Kaolin	41.9

Waterman's Yellow Slip

EPK Kaolin	50
Nepheline Syenite	20
Whiting	20
Ball Clay	20
Bentonite	2

Mix this 50/50 with Wollastonite

Hakeme Ship for Dampware

Custer Feldspar	25
Flint	25
Ball Clay	25
EPK Kaolin	25

Mark Pharis Dry Slip for Greenware 1

OM4 Ball Clay	33.3
EPK Kaolin	33.3
Alumina Hydrate	33.3

Mark Pharis Dry Slip for Greenware 2

EPK Kaolin	36.84
OM4 Ball Clay	31.58
Custer Feldspar	21.05
Flint	10.53

J's Salt Slip, damp or dry

Helmer Kaolin	25
EPK Kaolin	10
Ball Clay	25
Soda Feldspar	15
Whiting	5
Flint	15
Borax	5

Helmer Slip

Helmer Kaolin	60
Soda Ash	15
Nepheline Syenite	25

Crackle White Slip for Bisqueware

Borax	5
Zircopax	5
Custer Feldspar	20
OM4 Ball Clay	45
Calcined EPK Kaolin	15
EPK Kaolin	15
Flint	20

Ruggles and Rankin Slip

Tile 6 Ball Clay	14
Grolleg	3
Flint	1
Nepheline Syenite	2
Bentonite	3 oz.

White Slip

Grolleg	50
Pot Spar	25
Flint	25
Bentonite	2

Grolleg Slip

Nepheline Syenite	50
Grolleg	50
Titanium Dioxide	3

Sam Chung's Mustard Slip

Apply to Bisque

OM4 ½ Calcined	90
Nepheline Syenite	10
Titanium Dioxide	10

Glick's White Slip

Borax	3.5
Potash Feldspar	25
Kaolin	25
Ball clay	25
Flint	25

Glick's Blue Slip

Albany clay	90
Cobalt Carb.	10

Glick's Rust

Red Iron Oxide	10
Rutile	5

SP Albany Slip

Albany	180
Cornwallstone	50
Nepheline Syenite	25
Barium	15
Red Iron Oxide	13
Rutile	2

Appendix 4

Wood-firing Resources

Due to space limitations I have included a sampling of the many wood-firing resources that are available. This section is intended as a starting point for further investigation.

Art Centers with Wood-firing Kilns

Anderson Ranch Arts Center
5263 Owl Creek Rd, Snowmass Village, CO 81615
970- 923-3181, www.andersonranch.org

Archie Bray Foundation for the Ceramic Arts
2915 Country Club Avenue, Helena, MT 59602
406-443-3502, www.archiebray.org

Arrowmont School of Arts and Crafts
556 Parkway Gatlinburg, TN 37738
865-436-5860, www.arrowmont.org

Auckland Studio Potters
96 Captain Springs Rd, Onehunga,
P.O. Box 13 195, Auckland, NZ
Ph.096343622, www.ceramics.co.nz

Baltimore Clay Works
5707 Smith Avenue, Baltimore, MD 21209
(410) 578-1919, www.baltimoreclayworks.org

Canton Clayworks
150 Cherry Brook Rd (Rte 179N) Canton, CT 06019
860- 693-1000 , www.canonclayworks.com

Chester Springs Studios
1671 Art School Rd, Chester Springs, PA 19425
610- 827-7277

The Clay Studio of Missoula
1106 Hawthorne Street, Missoula, MT 59802
406- 543-0509, www.theclaystudioofmissoula.org

Cub Creek Foundation
4411 Wheeler Spring Road, Appomattox, VA 24522
434-248-5074, www.cubcreek.org

Energy X Change
66 Energy X change Drive, Burnsville, NC 28714
828-675-5541, info@energyxchange.org

The Hambidge Center
105 Hambidge Court, Rabun Gap, GA 30568
(706) 746-5718, www.hambidge.org

Mendocino Art Center
45200 Little Lake Street, Mendocino, CA 95460
707-937-5818, www.mendocinoartcenter.org

Ox-Bow School of the Arts
3435 Rupprect Way, Saugatuck, MI 49453
269- 857-5811, www.ox-bow.org

Penland School of Crafts
67 Doras Trail, Penland, NC 28765
828-765-2359, www.penland.org

Perkins Center for the Arts
395 Kings Highway (Main St.), Moorestown, NJ 08057
856.235.6488, www.perkinscenter.org

Peters Valley Craft Center
19 Kuhn Road, Layton, NJ 07851
973-948-5200, www.petersvalley.org

Pottery West
5026 North Pioneer Way, Las Vegas, NV 89149
702- 685-7573, www.potterywest.com

St. Petersburg Clay Company
420 22nd Street South, St. Petersburg, FL 33712
727-896-2529, www.stpeteclay.com

Watershed Center for the Ceramic Arts
19 Brick Hill Rd., Newcastle, ME 04553
207.882.6075

Ceramics Publications, Websites, and Other Sources

Anagama-West.com
Information and resources on anagama firing. Includes a gallery of work.
www.anagama-west.com

Australian Ceramics
www.australianceramics.com

Ceramic Review
www.ceramicreview.com

Clay Art
www.ceramics.org/clayart

Clay Art Web Guide
www.clayartwebguide.com
Clay Art Web Guide is a leading online ceramic portal serving the clay art community. It is a great resource for all types of ceramic arts.

Ceramics Monthly
www.ceramicartsdaily.org/ceramics-monthy

Ceramic Arts Daily
www.ceramicartsdaily.org

Ceramics Art and Perception
http://www.ceramicart.com.au

Ceramics Today
www.ceramicstoday.com

The Log Book
www.thelogbook.net

Edited and published by Coll Minogue and Robert Sanderson, *The Log Book* is the quarterly international publication for wood-firers and those interested in wood-firing and wood-fired ceramics—professional potters/artists, educators, students, and enthusiasts.

Now in its thirteenth year of publication, *The Log Book* is recognized as a significant resource documenting the aesthetics, materials, and processes of international wood-fired ceramics.

Articles are published on historical, traditional, and contemporary wood-firing, together with reports on specialist wood-fire related events worldwide. Reviews of books, videos, and DVDs of relevance to wood-firing are also included.

Supported only by subscription, *The Log Book* is a non-profit publication with no commercial advertising.

~Coll Minogue and Robert Sanderson

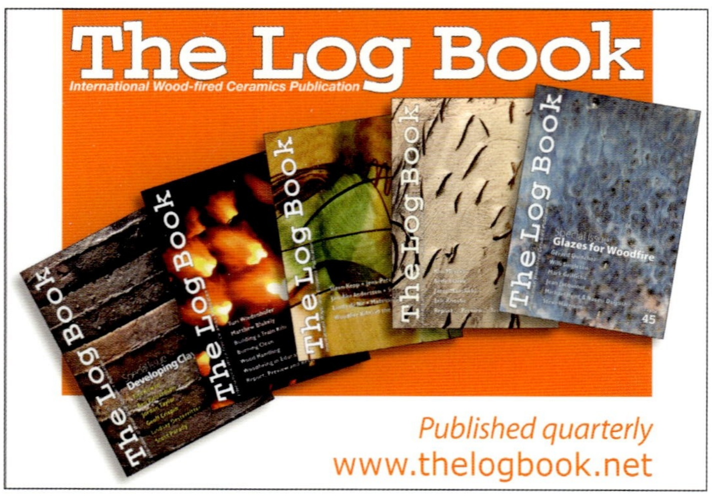

The Log Book, edited and published by Coll Minogue and Robert Sanderson.

NCECA, National Council on Education for the Ceramic Arts
www.nceca.net

New Ceramics, The European Ceramics Magazine
www.neue-keramik.de

New Zealand Potters Website
www.nzpotters.com

Pottery Making Illustrated
www.ceramicartsdaily.org/pottery-making-illustrated

Potters Council
www.ceramicartsdaily.org/potters-council

Revue de la Céramique et du Verre
www.revue-ceramique-verre.com

SideStoke
www.sidestoke.com
An online magazine for wood-firers.

Studio potter
www.studiopotter.org

UK Potters
www.ukpotters.com

Wood-fired Kilns: World Map
Created by Simon Levin, USA

For Simon Levin's map, see: www.simonlevin.com/worldmap

Years ago at an international wood-fire conference held in Iowa, participants marked a world map with the location of their kilns. The variety and proliferation of kilns was informative, but not available outside the conference. In 2009, using Google mapping technology, I created a similar map that I hoped could serve as an accessible community resource. The interactive quality of this map is some of what the Internet does best. The pins are placed accurately on the roof of the kiln shed, the information includes type of kiln, contact information, an image and often a link to the potter's website. With a click you can get directions to the kiln.

Although I don't think wood-fire tourism will ever be a major force, this map does allow one to see the wood-fire potters in and around an area. I use it when I travel and potters who are looking for a place to fire use it. It has become a good way to promote individuals as well as the wood-firing community as a whole. Since the map's inception it has had over 90,000 views and has been published in *Ceramics Monthly*.

Because of privacy issues, kilns are only posted by request of the owner. To be included on the map or embed it in your own site, email Simon@SimonLevin.com. Listings of wood kilns in Central and South America, Africa, as well as most of Asia, are distinctly lacking. It is my hope over the next few years that word will spread and this map will continue to become a vital community resource.

~Simon Levin

Schools, Colleges, and Universities With Wood-Firing Kilns

Alfred University
1 Saxon Drive, Alfred, NY 14802
607- 871-2111, www.alfred.edu

Central Oregon Community College
2600 Northwest College Way, Bend, OR 97701
541-383-7700, www.cocc.edu

College for Creative Studies
201 East Kirby Street, Detroit, MI 48202
313- 664-7425, www.collegeforcreativestudies.edu

College of the Redwoods
7351 Tompkins Hill Rd. Eureka, CA 95501
1-800-641-0400, www.redwoods.edu

Eastern Kentucky University
521 Lancaster Avenue, Richmond, KY 40475
859- 622-1000, www.art.eku.edu

Florida Atlantic University
777 Glades Rd, Boca Raton, FL 33431
561-297-3000, www.fau.edu

Goshen College
1700 South Main Street, Goshen, IN 46526
574- 535-7000, www.goshen.edu

Guldagergaard International Ceramic Research Center
Heilmannsvej 31 A DK-4230 Skælskør Denmark
+45 5819 0016 M, ceramic@ceramic.dk

Jingdezhen Ceramic Institute
Jingdezhen City, JiangXi Province, China 333001
+86-798-8499600, www.jci.edu.cn/english

Kent State University
161 Schwartz Center, Kent, Ohio 44242
330-672-2444, www.kent.edu

Lewis Central High School
3504 Harry Langdon Blvd., Council Bluffs, IA 51503
712-366-8222, www.lchs.lewiscentral.org

Linfield College
900 SE Baker St, McMinnville, OR 97128
503-883-2261, www.linfield.edu

Millbrook School
131 Millbrook School Road , Millbrook, NY 12545
845-677-8261, www.millbrook.org

Napa Valley College
2277 Napa Vallejo Highway, Napa, CA 94558
707-253-3020, www.napavalley.edu

Noble and Greenough School
10 Campus Drive, Dedham, MA 02026
781.326.3700, www.nobles.edu

Nicholls State University
906 East 1st Street, Thibodaux, LA 70301
985-446-8111, www.nicholls.edu

Northern Arizona University
South San Francisco Street, Flagstaff, AZ 86011
928-523-9011,www.nau.edu

Sacramento State University
6000 J Street, Sacramento, CA 95819
916-278-6011, www.csus.edu

San Diego State University, CA
5500 Campanile Drive, San Diego, CA 92182
619-594-5200, www.sdsu.edu

Shigaraki Ceramic Cultural Park
2188-7 Shigarakicho-Chokushi Koka City Shiga Pref. Japan 529-1804
+81-748-83-0909, www.sccp.jp/e

Solano College
4000 Suisun Valley Road, Fairfield, CA 94534
707-864-7000, www.solano.edu

University of Arkansas at Little Rock
2801 South University Avenue, Little Rock, AR 72204
501-569-3000, www.ualr.edu

University of Kentucky
410 Administration Dr, Lexington, KY 40506
859-257-9000, www.uky.edu

University of Louisville
550 South Jackson Street, Louisville, KY 40292
502- 852-5555, www.louisville.edu

University of Missouri
A125 Fine Arts Center Columbia, MO 65211
573-882-3555, www.missouri.edu

University of North Carolina
One University Heights, Asheville, NC 28804
828.251.6600, www.unca.edu

University of North Dakota
221 Centennial Drive, Grand Forks, ND 58202
701-777-4314, www.und.edu

University of Notre Dame in Notre Dame
306 Riley Hall of Art & Design, Notre Dame, IN 46556
574-631-7602, www.nd.edu

University of Montana
32 Campus Drive, Missoula, MT 59812
406-243-0211, www.umt.edu

West Virginia University
1601 University Avenue, Morgantown, WV 26506
304-293-0111, www.wvu.edu

Bibliography

Berensohn, Paulus. *Finding Ones Way with Clay*. New York, New York: Simon and Schuster, 1972.

Ceramic Art of Japan, One Hundred Masterpieces from Japanese Collections. Kodansha International, LTD., 1972.

Ceramics Arts Daily. *Glossary* (www.ceramicartdaily.org/daily/glossary/).

Finch, Joe. *Kiln Construction, A Brick by Brick Approach*. London, England: A&C Black, 2006, Philadelphia, Pennsylvania: University of Pennsylvania press, 2006.

Gregory, Ian. *Kiln Building*. London, England: Pitman Publishing Limited,1977.

Gregory, Ian. *Kiln Building*. London, England: A&C Black, 1995; Oviedo, Florida: Gentle Breeze Publishing, 1995.

Minogue, Coll and Robert Sanderson. *Wood-fired Ceramics, Contemporary Practices*. London, England: A&C Black, 2000; Philadelphia, Pennsylvania: University of Pennsylvania press, 2000.

Minogue, Coll and Robert Sanderson. *The Log Book (www.thelogbook.net)* Ennis, Ireland: The Print Den.

Olsen, Frederick L. *The Kiln Book*. Randor, Pennsylvania: Chilton Book Company, 1982.

Rhodes, Daniel. *Clay and Glazes for the Potter*. Randor, Pennsylvania: Chilton Book Company, 1973.

Rhodes, Daniel. *Kilns, Design, Construction, and Operation*. Radnor, Pennsylvania: Chilton Book Company, 1968.

Snyder, Jeffrey B. *Ceramics Today*. Atglen, Pennsylvania: Schiffer Publishing LTD., 2010.

Speight, Charlotte F. *Hands in Clay, An Introduction to Ceramics*. Palo Alto, California: Mayfield Publishing Company, 1983.

Troy, Jack. *Wood-Fired Stoneware and Porcelain*. Randor, Pennsylvania: Chilton Book Company, 1995.

Wood-firing Journeys and Techniques, A Collection of Articles from Ceramics Monthly. Westerville, Ohio: The American Ceramic Society, 2001.